Guidebook to the
Cytoskeletal and Motor Proteins

A computer system will be available from October 1993 to accompany
Guidebook to the cytoskeletal and motor proteins
and
Guidebook to the extracellular matrix and adhesion proteins

Due to the rapid pace of biological research, the editors and publishers of this book believe it is important that its readers are kept informed of recent developments on these proteins. For this purpose, we have established a computer database that can be accessed throughout the world-wide computer Internet system. This computer database will not include the full entries shown in this book and its companion volume; instead the authors have been asked to add, periodically, any new information on their protein that has been obtained or published since they wrote their original entry. Authors will be asked to deposit updates in this database from October 1993.

The protein update system can be accessed through the Internet by the information server program called Gopher, developed by the University of Minnesota Computer and Information Services Department. If you do not have the Gopher client program installed on your computer, contact a computer system specialist at your institution for more information. You can also receive Gopher software for a variety of computer platforms via anonymous FTP from **boombox.micro.umn.edu** in the **/pub/gopher** directory. Questions or comments about Gopher can also be sent by e-mail to **gopher@boombox.micro.umn.edu**.

Instructions for accessing the protein update system:

1. Using the gopher program, contact the Gopher server account located at **itsa.ucsf.edu** (port 70). On some systems, type **gopher.itsa.ucsf.edu**.
2. Choose **Reseacher Tools**.
3. Choose **Protein Data Base** (for the *Guidebook* series).
4. Follow the instructions.

This protein update directory will also be made available by anonymous FTP from **itsa.ucsf.edu** (use your own e-mail address as the password). The file will be located in the directory **/pub/protein**.

Guidebook to the
Cytoskeletal and Motor Proteins

Edited by
Thomas Kreis

University of Geneva,
Geneva, Switzerland

and

Ronald Vale

University of California,
San Francisco, USA

OXFORD   NEW YORK   TOKYO
OXFORD UNIVERSITY PRESS
1993

Oxford University Press, Walton Street, Oxford OX2 6DP

Oxford New York Toronto
Delhi Bombay Calcutta Madras Karachi
Kuala Lumpur Singapore Hong Kong Tokyo
Nairobi Dar es Salaam Cape Town
Melbourne Auckland Madrid

and associated companies in
Berlin Ibadan

Oxford is a trade mark of Oxford University Press

Published in the United States
by Oxford University Press Inc., New York

A catalogue record for this book is available from the British Library

Library of Congress Cataloging-in-Publication Data
Guidebook to the cytoskeletal and motor proteins/ edited by Thomas Kreis and Ronald Vale.
1. Cytoskeletal proteins.   2. Tubulins.   I. Kreis, Thomas.
II. Vale, Ronald.
[DNLM: 1. Cytoskeletal Proteins.   2. Muscle Proteins.   QU 55 G946 1993]
QP552.C96G85   1993   612'.01575—dc20   93–9654
ISBN 0-19-859932-3
ISBN 0-19-859931-5 (Pbk)

Typeset by
Selwood Systems, Midsomer Norton
Printed in Great Britain by
Butler & Tanner Ltd Frome and London

Contents

PART 1 ACTIN AND ASSOCIATED PROTEINS

INTRODUCTION
Actin and Actin Binding Proteins *T. D. Pollard*   3

Actins *J. S. Vanderkerckhove*   13
Actin Binding Protein-50 (ABP-50) *F. Yang and J. Condeelis*   15
Actin Binding Protein-120 (ABP-120) *A. R. Bresnick and J. Condeelis*   16
Actin Binding Protein-280 (ABP-280; Nonmuscle Filamin)
   J. Gorlin and J. Hartwig   18
Actin Depolymerizing Factor (ADF) *J. R. Bamburg*   20
α-Actinins *D. R. Critchley*   22
Actobindin *M. R. Bubb and E. D. Korn*   23
Actolinkin *I. Mabuchi*   25
Annexins *J. R. Dedman and M. A. Kaetzel*   26
Caldesmons *J. Bryan and C.-L. A. Wang*   29
Calponin *M. Gimona, A. Draeger, M. P. Sparrow and J. V. Small*   32
Capping Proteins *J. A. Cooper, J. F. Amatruda, C. Hug and D. A. Schafer*   34
Cofilin *K. Moriyama and E. Nishida*   35
Coronin *E. L. de Hostos and G. Gerisch*   37
C-proteins *D. A. Fishman*   39
Dematins *A. Husian-Chishti and D. Branton*   42
Depactin *I. Mabuchi*   43
Dystrophin *T. J. Byers and L. M. Kunkel*   45
Ezrin *A. Bretscher*   47
Fascin *J. Bryan, J. Otto and R. E. Kane*   49
Fimbrin *A. Bretscher*   50
gCap39 (Macrophage Capping Protein, MCP) *F.-X. Yu and H. L. Yin*   51
Gelsolins *H. L. Yin*   52
Hisactophilin *M. Schleicher*   54
Insertin *A. Gaertner and A. Wegner*   55
MARCKS *P. A. Janmey and J. H. Hartwig*   56
Myomesin and M-Protein *J.-C. Perriard*   58
Nebulin *K. Wang*   59
Nuclear Actin Binding Protein (NAB) *D. L. Rimm*   60
Paramyosin *C. Cohen*   62
Ponticulin *E. J. Luna*   64
Profilins *L. M. Machesky and T. D. Pollard*   66
Proteins 4.1 *V. T. Marchesi*   68
Radixin *S. Tsukita and S. Tsukita*   70
Sarcomeric M-Creatine Kinase *J.-C. Perriard*   71
Severin *A. A. Noegel*   73
Small Actin Crosslinking Proteins *M. Fechheimer*   74
Spectrins (Fodrin) *J. S. Morrow*   76
Tenuin *S. Tsukita and S. Tsukita*   78

Thymosin β4 (Tβ4) *D. Safert and K Nochmias*   79
Titin *K. Wang*   81
Tropomodulin *V. Fowler*   83
Tropomyosins *L. B. Smillie*   85
Troponins *J. Gergeley, Z. Garbarek and T. Tao*   87
Villin *E. Friedrich and D. Louvard*   89
Vitamin D Binding/Gc Protein (DBP/Gc) *P. J. Goldschmidt*   92
25kDa Inhibitor of Actin Polymerization (25 kDa IAP) *B. Geiger*   94
43kDa Protein *H.-O. Ngheim and J.-P. Changeux*   95

PART 2 TUBULIN AND ASSOCIATED PROTEINS

INTRODUCTION

Tubulin and Associated Proteins *D. W. Cleveland*   101

Chartins *F. Solomon*   107
MAP1A *G. S. Bloom*   108
MAP1B/MAP5 *N. J. Cowan and A. Matus*   110
MAP2 *P. D. Walden, S. A. Lewis and N. J. Cowan*   111
MAP3 *G. Huber and A. Matus*   113
MAP4 (MAP-U) *J. B. Olmsted and H. Murofushi*   115
MARPs *M. Affolter, A. Hemphill and T. Seebeck*   117
Pericentrin *S. J. Doxsey and M. Kirschner*   118
Radial Spoke Proteins *A. M. Curry and J. L. Rosenbaum*   119
Sea Urchin MAPs and Microtubule Motors *J. M. Scholey and R. J. Leslie*   120
STOPs *D. Job and R. L. Margolis*   122
Syncolin *G. Wiche*   124
Tau *D. N. Drechsel and M. Kirschner*   125
α/β-Tubulin *E.-M. Mandelkow and E. Mandelkow*   127
γ-Tubulin *B. R. Oakley*   130
Tubulin Tyrosine Ligase (TTL) and Tubulin
   Carboxypeptidase (TCP) *J. Wehland*   131
X-MAP (Xenopus) *D. L. Gard*   132
205K MAP (Drosophila) *A. Pereira and L. S. B. Goldstein*   133

PART 3 THE INTERMEDIATE FILAMENTS

INTRODUCTION

The Intermediate Filaments and Associated Proteins *W. W. Franke*   137

Cytokeratins *J. Kartenbeck and W. W. Franke*   145
Desmin *J. V. Small*   148
Epinemin *D. Lawson*   151
Filaggrins *S.-Q. Gan and P. M. Steinert*   152
Filensin *S. Georgatos*   153
GFAP *R. A. Quinlan*   155
α-Internexin *R. K. H. Liem*   157
Lamins *L. Gerace*   158
Nestin *L. B. Zimmerman and D. G. McKay*   160

Neurofilament Triplet Proteins (NF-L, NF-M, NF-H) *K. Weber*   161
Paranemin *B. L. Granger*   164
Peripherin *M.-M. Portier and F. Landon*   165
Plectin *G. Wiche*   166
Synemin *B. L. Granger*   167
Vimentin *M. Osborn*   169

PART 4 MOTOR PROTEINS

INTRODUCTION
Motor Proteins *R. D. Vale*   175

Axonemal Dyneins *I. Gibbons*   185
Brush Border Myosin I *M. S. Mooseker and J. S. Wolenski*   187
Caltractin *V. D. Lee and B. Huang*   188
Cytoplasmic Dynein (MAP1C) *R. B. Vallee*   191
Cytoplasmic Myosin II *H. M. Warrick and J. A. Spudich*   193
Dynactin *T. A. Schroer*   195
Dynamin *R. B. Vallee*   197
Kinesin *R. D. Vale*   199
Kinesin Related Proteins *M. D. Rose*   201
MYO2 Myosin *R. A. Singer and G. C. Johnson*   204
Protozoan Myosin I *J. A. Hammer III*   206
Sarcomeric Myosins *R. Cooke*   207
Scallop Myosin *A. G. Szent-Györgyi*   209
Smooth Muscle Myosin *K. M. Trybus and S. Lowey*   211

PART 5 CYTOSKELETAL ANCHOR PROTEINS

INTRODUCTION
Cytoskeletal Anchor Proteins *B. Geiger*   215

Adducin *D. M. Gilligan, R. Joshi, and V. Bennett*   219
Ankyrins *V. Bennett*   221
Band 6 Polypeptide *J. Kartenbeck and W. W. Franke*   223
Catenins *R. Kemler*   224
Cingulin *S. Citi*   225
Desmocalmin *S. Tsukita and S. Tsukita*   227
Desmoplakins *J. Kartenbeck and W. W. Franke*   228
Micro-Calpain and Milli Calpain (EC 3.4.22.17) *D. E. Croall*   230
Paxillin *C. E. Turner*   231
Pemphigoid Antigens *K. Owaribe*   233
Plakoglobin *P. Cowin*   235
pp60[c-src] *T. Hunter and S. Simon*   236
Talin *K. Burridge*   239
Tensin *S. Lin*   240
Vinculin *B. Geiger*   243
ZO-1 *D. A. Goodenough*   245
Zyxin *A. Crawford and M. C. Beckerle*   247

PART 6 ORGANELLE MEMBRANE ASSOCIATED STRUCTURAL PROTEINS

INTRODUCTION
Organelle Membrane Associated Structural Proteins *T. E. Kreis*   251

Coat Proteins

AP180 *E. Ungewickell*   255
Auxilin *E. Ungewickell*   256
Clathrin *B. M. F. Pearse*   257
Clathrin Adaptor Proteins *E. Ungewickell*   259
β-COP *R. Duden and T. E. Kreis*   261

Cytoplasmic linker proteins

CLIP-170 *J. E. Rickard and T. E. Kreis*   263
58K *G. S. Bloom*   264
Synapsin *M. Bähler and P. Greengard*   266

PART 7 OTHER PROTEINS

Major Sperm Proteins *T. M. Roberts*   271
Tektins *R. W. Linck*   272

Contributors

Alan Aderem, Laboratory of Cellular Physics and Immunology, Rockefeller University, USA

Marianne Affolter, Institute for General Microbiology, 3012 Bern, Switzerland

James F. Amatruda, Department of Cell Biology and Physiology, Washington University Medical School, St Louis, MO, USA

Martin Bähler, Laboratory of Molecular and Cellular Neuroscience, The Rockefeller University, New York, USA

James R. Bamburg, Department of Biochemistry, Colorado State University, Fort Collins, CO 80523, USA

Mary C. Beckerle, Department of Biology, University of Utah, Salt Lake City, UT 84112, USA

Vann Bennett, Departments of Medicine and Biochemistry, Howard Hughes Medical Institute, Duke University Medical Center, Durham NC 27710, USA

George S. Bloom, Department of Cell Biology and Neuroscience, University of Texas, Southwestern Medical Center, 5323 Happy Hines Blvd., Dallas, TX 75235, USA

Daniel Branton, The Biological Laboratories, Harvard University, Cambridge, MA02138, USA

Anne R. Bresnick, Department of Anatomy and Structural Biology, Albert Einstein College of Medicine, Bronx, New York, USA

Anthony Bretscher, Section of Biochemistry, Molecular and Cell Biology, Cornell University, Ithaca, NY, USA

Joseph Bryan, Baylor College of Medicine, Department of Cell Biology, Houston, TX 77030, USA

Michael R. Bubb, Laboratory of Cell Biology, National Heart Lung and Blood Institute, National Institutes of Health, Bethesda, MD, USA

Keith Burridge, Department of Cell Biology and Anatomy, University of North Carolina at Chapel Hill, Chapel Hill, NC27599, USA

Timothy J. Byers, Howard Hughes Medical Institute, Children's Hospital, Medical Center and Harvard Medical School, Boston, MA, USA

John Candeelis, Department of Anatomy and Structural Biology, Albert Einstein College of Medicine, Bronx, New York, USA

Jean-Pierre Changeux, Institut Pasteur, Neurobiologie Moléculaire, 25 rue du Docteur Roux, Paris 75015, France

Sandra Citi, Department of Cell Biology and Anatomy, Cornell University Medical College, New York, USA

Don W. Cleveland, Department of Biological Chemistry, Johns Hopkins Medical School, Baltimore, MD 21205, USA

Carolyn Cohen, Rosenstiel Basic Medical Sciences Research Center, Brandeis University, Waltham, MA, USA

Roger Cooke, Department of Biochemistry and Biophysics and the CVRI, University of California, San Francisco, USA

John A. Cooper, Department of Cell Biology and Physiology, Washington University Medical School, St Louis, MO, USA

Nicholas J. Cowan, Biochemistry Department, New York University Medical School, New York, USA

Pamela Cowin, Departments of Cell Biology and Dermatology, NYU Medical Center, New York, USA

Aaron W. Crawford, Department of Biology, University of Utah, Salt Lake City, UT 84112, USA

David R. Critchley, Department of Biochemistry, University of Leicester, Leicester LE1 7RH, UK

Dorothy E. Croall, Department of Biochemistry, Microbiology and Molecular Biology, University of Maine, Orono, Maine, USA

Alice M. Curry, Department of Biology, Yale University, USA

John R. Dedman, Department of Physiology and Cell Biology, University of Texas Medical School, Houston, Texas 77225, USA

Stephen J. Doxsey, Department of Biochemistry and Biophysics, University of California, San Francisco, USA

Annette Draeger, Institute of Molecular Biology, Austrian Academy of Sciences, Salzburg, Austria

David N. Drechsel, Department of Biochemistry and Biophysics, UC San Francisco, USA

Rainer Duden, EMBL, Meyerhofstrasse, 6900 Heidelberg, Germany

Marcus Fechheimer, Department of Zoology, University of Georgia, USA

Donald A. Fischman, Department of Cell Biology and Anatomy, Cornell University Medical College, New York, NY, USA

Velia Fowler, Department of Molecular Biology, Research Institute of Scripps Clinic, La Jolla, CA, USA

Werner W. Franke, Institute of Cell and Tumour Biology, German Cancer Research Centre, 6900 Heidelberg, Germany

Evelyne Friederich, Institut Pasteur, Department of Molecular Biology, 25 rue du Docteur Roux, Paris 75015, France

Andrea Gaertner, Institute of Physiological Chemistry, Ruhr-University, Bochum, Germany

Song-Qing Gan, Laboratory of Skin Biology, National Institute of Arthritis and Musculoskeletal and Skin Diseases, NIH, Bethesda, MD 20892, USA

David L. Gard, Department of Biology, University of Utah, Salt Lake City, Utah, USA

Benjamin Geiger, Department of Chemical Immunology, The Weizmann Institute of Science, Rehovot, Israel

Spyros Georgatos, EMBL, Heidelberg, Germany

Larry Gerace, The Scripps Research Institute, La Jolla, CA 92037, USA

John Gergely, Boston Biomedical Research Institute, Department of Muscle Research, Boston, MA 02114, USA

Günter Gerisch, Abteilung Zellbiologie, Max Planck-Institut für Biochemie , 8033 Martinsried, Germany

Ian Gibbons, Pacific Biomedical Research Center, University of Hawaii, Honolulu, HI 96822, USA

Diana M. Gilligan, Departments of Medicine and Biochemistry, Howard Hughes Medical Institute, Duke University Medical Center, Durham NC 27710, USA

M. Gimona, Institute of Molecular Biology, Austrian Academy of Sciences, Salzburg, Austria

Pascal J. Goldschmidt-Clermont, Division of Cardiology, Department of Medicine, Johns Hopkins Medical School, Baltimore, MD 21205, USA

Lawrence S. B. Goldstein, Department of Cellular and Developmental Biology, Harvard University, Cambridge, MA, USA

Daniel A. Goodenough, Department of Anatomy and Cellular Biology, Harvard Medical School, 220 Longwood Avenue, Boston, MA 02115, USA

Jed Gorlin, Experimental Medicine Division, Brigham and Women's Hospital, Boston, MA, USA

Zenon Grabarek, Boston Biomedical Research Institute, Department of Muscle Research, Boston, MA 02114, USA

Bruce L. Granger, Veterinary Molecular Biology, Montana State University, Bozeman, MT 58717, USA

Paul Greengard, Laboratory of Molecular and Cellular Neuroscience, The Rockefeller University, New York, USA

John A. Hammer III, Laboratory of Cell Biology, National Heart, Lung, and Blood Institute , National Institutes of Health, Bethesda, MD, USA

John Hartwig, Experimental Medicine Division, Brigham and Women's Hospital, Boston, MA, USA

John A. Hartwig, Department of Medicine , Harvard Medical School, USA

Andrew Hemphill, Institute for General Microbiology, 3012 Bern, Switzerland

Eugenio L. de Hostos, Abteilung Zellbiologie, Max Planck-Institut für Biochemie , 8033 Martinsried, Germany

Bessie Huang, Department of Cell Biology, The Scripps Research Institute, La Jolla, USA

Gerda Huber, Pharmaceutical Research, Hoffmann-La Roche, Basel, Switzerland

Christopher Hug, Department of Cell Biology and Physiology, Washington University Medical School, St Louis, MO, USA

Tony Hunter, The Salk Institute; P.O. Box 85800, San Diego, USA

Athar Husain-Chishti, The Biological Laboratories, Harvard University, Cambridge, MA02138, USA

Paul A. Janmey, Department of Medicine , Harvard Medical School, Boston, MA, USA

Didier Job, Département de Biologie Moléculaire et Structurale, Centre d'Etudes Nucléaires, Grenobles, France

Gerald C. Johnston, Department of Biochemistry and Department of Microbiology, Dalhousie University, Halifax, N.S. Canada

Rashmi Joshi, Departments of Medicine and Biochemistry, Howard Hughes Medical Institute, Duke University Medical Center, Durham, Nc 27710, USA

Marcia A. Kaetzel, Department of Physiology and Cell Biology, University of Texas Medical School, Houston, Texas 77225, USA

Robert E. Kane, University of Hawaii, Pacific Biomedical Res. Center, Honolulu, HI 96822, USA

Juergen Kartenbeck, Institute of Cell and Tumour Biology, German Cancer Research Centre, 6900 Heidelberg, Germany

Rolf Kemler, Max-Planck Institut für Immunbiologie, 7800 Freiburg, Germany

Marc Kirschner, Department of Biochemistry and Biophysics, University of California, San Francisco, USA

Edward D. Korn, Laboratory of Cell Biology, National Heart Lung and Blood Institute, National Institutes of Health, Bethesda, MD, USA

Thomas E. Kreis, Department of Cell Biology, Sciences III, University, CH-1211 Geneva, Switzerland

Louis M. Kunkel, Howard Hughes Medical Institute, Children's Hospital, Medical Center and Harvard Medical School, Boston, MA, USA

Francoise Landon, Collège de France, Biochimie Cellulaire, Paris, France

Durward Lawson, Biology Department, Medawand Building, University College London, Gower Street, London WCE 6BT, UK

Vincent D. Lee, Department of Cell Biology, The Scripps Research Institute, La Jolla, USA

Roger J. Leslie, Department of Zoology, University of California, Davis, CA 95616-8755, USA

Sally A. Lewis, Department of Biochemistry, NYU Medical Center, New York, USA

Ronald K. H. Liem, Departments of Pathology and Anatomy and Cell Biology, Columbia University, New York, NY, USA

Shin Lin, Department of Biophysics, Johns Hopkins University, Baltimore, MD 21218, USA

Richard W. Linck, Department of Cell Biology and Neuroanatomy, University of Minnesota, Minneapolis, MN, USA

Daniel Louvard, Institut Pasteur, Department of Molecular Biology, 25 rue du Docteur Roux, Paris 75015, France

Susan Lowey, Rosenstiel Research Center, Brandeis University, Waltham, MA, USA

Elizabeth J. Luna, Worcester Foundation for Experimental Biology, Shrewsbury, MA, USA

Issei Mabuchi, Department of Biology, College of Arts and Sciences, University of Tokyo, Tokyo, Japan

Laura M. Machesky, Department of Cell Biology, Johns Hopkins Medical School, Baltimore, MD, USA

Eva-Maria Mandelkow, Max-Planck-Unit for Structural Molecular Biology, Hamburg, Germany

Eckhard Mandelkow, Max-Planck-Unit for Structural Molecular Biology, Hamburg, Germany

Vincent. T. Marchesi, Boyer Center for Molecular Medicine, Yale University, New Haven, CT06510, USA

Robert L. Margolis, The Fred Hutchinson Cancer Research Center, Seattle, USA

Andrew Matus, Friederich Miescher-Institut, Basel, Switzerland

Ronald D. G. McKay, Department of Biology, Massachusetts Institute of Technology, Boston, MA, USA

Mark. S. Mooseker, Department of Biology, Yale University, New Haven, CT, USA

Kenji Moriyama, Department of Biophysics and Biochemistry, Faculty of Science, University of Tokyo, Tokyo, Japan

Jon S. Morrow, Department of Pathology, Yale Medical School, New Haven, USA

Hiromu Murofushi, Department of Biophysics and Biochemistry, Faculty of Science, University of Tokyo, Tokyo, Japan

Vivianne T. Nachmias, Department of Anatomy, School of Medicine, University of Pennsylvania, Philadelphia, PA 19104-6058, USA

Hoàng-Oanh Nghiêm, Institut Pasteur, Neurobiologie Moléculaire, 25 rue du Docteur Roux, Paris 75015, France

Eisuke Nishida, Department of Biophysics and Biochemistry, Faculty of Science, University of Tokyo, Tokyo, Japan

Angelika A. Noegel, Max-Planck Institute for Biochemistry, 8033 Martinsried, Germany

Berl R. Oakley, Department of Molecular Genetics, Ohio State Universty, Columbus, Ohio, USA

Joanna. B. Olmsted, Department of Biology, University of Rochester, Rochester, NY 14627, USA

Mary Osborn, Max-Planck Institute for Biophysical Chemistry, 3400 Göttingen, Germany

J. Otto, Purdue University, Department of Biological Sciences, W. Lafayette, IN 47907, USA

Katsushi Owaribe, Department of Molecular Biology, School of Science, Nagoya University, Nagoya 464-1, Japan

Barbara M. F. Pearse, Medical Research Council, Laboratory of Molecular Biology, Hills Road, , Cambridge CB2 2QH, UK

Andrea Pereira, Department of Cellular and Developmental Biology, Harvard University, Cambridge, MA, USA

Jean-Claude Perriard, Institute for Cell Biology, Swiss Federal Institute of Technology, 8093 Zurich, Switzerland

Thomas D. Pollard, Department of Cell Biology, Johns Hopkins Medical School, Baltimore, MD, USA

Marie-Madeleine Portier, Collège de France, Biochimie Cellulaire, Paris, France

Roy A. Quinlan, Department of Biochemistry, The University, Dundee DD1 4HN, UK

Janet E. Rickard, Department of Cell Biology, Sciences III, University, CH-1211 Geneva, Switzerland

David L. Rimm, Department of Pathology, Yale Medical School, New Haven, CT, USA

Thomas M. Roberts, Department of Biological Science, Florida State University, USA

Mark D. Rose, Department of Molecular Biology, Princeton University, Princeton, New Jersey 08544-1014, USA

Joel L. Rosenbaum, Department of Biology, Yale University, USA

Daniel Safer, Department of Anatomy, School of Medicine, University of Pennsylvania, Philadelphia, PA 19104-6058, USA

Dorothy A. Schafer, Department of Cell Biology and Physiology, Washington University Medical School, St Louis, MO, USA

Jonathan M. Scholey, Department of Zoology, University of California, Davis, CA 95616-8755, USA

Trina A. Schroer, Department of Biology, The Johns Hopkins University, Baltimore, MD 21218, USA

Thomas Seebeck, Institute for General Microbiology, 3012 Bern, Switzerland

Suzanne Simon, The Salk Institute, P.O. Box 85800, San Diego, USA

Richard A. Singer, Department of Biochemistry and Department of Microbiology, Dalhousie University, Halifax, Nova Scotia, Canada

Victor J. Small, Institute of Molecular Biology, Austrian Academy of Sciences, Salzburg, Austria

Lawrence B. Smillie, MRC Group in Protein Structure and Function, Department of Biochemistry, University of Alberta, Edmonton, Canada T6G 2H7

Frank Solomon, Department of Biology, M.I.T., Cambridge, USA

M. P. Sparrow, Institute of Molecular Biology, Austrian Academy of Sciences, Salzburg, Austria

James A. Spudich, Departments of Cell Biology and Developmental Biology, Stanford University School of Medicine, Stanford, CA, USA

Peter M. Steinert, Laboratory of Skin Biology, National Institute of Arthritis and Musculoskeletal and Skin Diseases, NIH, Bethesda, MD 20892, USA

Andrew G. Szent-Györgyi, Department of Biology, Brandeis University, Waltham, Massachusetts, USA

Terence Tao, Boston Biomedical Research Institute, Department of Muscle Research, Boston, MA 02114, USA

Kathleen M. Trybus, Rosenstiel Research Center, Brandeis University, Waltham, MA, USA

Sachiko Tsukita, Department of Information Physiology, National Institute for Physiological Sciences, Okazaki, Japan

Shoichiro Tsukita, Department of Information Physiology, National Institute for Physiological Sciences, Okazaki, Japan

Christopher E. Turner, Department of Anatomy and Cell Biology, Suny Health Science Center at Syracuse, Syracuse, NY 13210, USA

Ernst Ungewickell, Max-Planck Institut für Biochemie, 8033 Martinsried, Germany

Ronald D. Vale, Department of Pathology, Washington University, School of Medicine, 660 South Euclid Avenue, St Louis, MO, USA

Richard B. Vallee, Cell Biology Group, Worcester Foundation for Experimental Biology, Shrewsbury, MA, USA

Joël S. Vandekerckhove, Laboratory of Physiological Chemistry, State University of Ghent, 9000 Ghent, Belgium

Paul D. Walden, Department of Biochemistry, NYU Medical Center, New York, USA

C. L. Albert Wang, Boston Biomedical Research Institute, Department of Muscle Research, Boston, MA 02114, USA

Kuan Wang, Department of Chemistry and Biochemistry, University of Texas, Austin, TX, USA

Hans M. Warrick, Departments of Cell Biology and Developmental Biology, Stanford University School of Medicine, Stanford, CA, USA

Klaus Weber, Department of Biochemistry, Max-Planck Institute for Biophysical Chemistry, 3400 Göttingen, Germany

Albrecht Wegner, Institute of Physiological Chemistry, Ruhr-University, Bochum, Germany

Juergen Wehland, National Research Centre for Biotechnology, 3300 Braunschweig, Germany

Gerhard Wiche, Institute of Biochemistry, University of Vienna, 1090 Vienna, Austria

J. S. Wolenski, Department of Biology, Yale University, USA

Fan Yang, Department of Anatomy and Structural Biology, Albert Einstein College of Medicine, Bronx, New York, USA

Helen L. Yin, Department of Physiology, University of Texas, Southwestern Medical Center, Dallas, Texas, USA

Fu-Xin Yu, Department of Physiology, University of Texas, Southwestern Medical Center, Dallas, Texas, USA

Lyle B. Zimmerman, Department of Biology, Massachusetts Institute of Technology, USA

Preface

The biology of the 1980s and 1990s may well come to be remembered as the era of discovery of new proteins. Recent advances in molecular biology, genetics, and protein purification have conspired to accelerate the rate at which cellular proteins and their amino acid sequences are being identified. Such efforts, combined with information from genome sequencing projects, will ultimately lead to the identification of the entire repertoire of proteins that govern the workings of the cell.

Now that the floodgate of discovery of new proteins has been opened wide, the amount of new information on cellular proteins is exceeding the capacity of assimilation of most scientists. At the same time, it has become imperative for research workers to expand their knowledge base, since interactions between previously unconnected sets of proteins are being uncovered at a rapid pace.

These considerations motivated us to compile the 'Guidebook to the Cytoskeletal and Motor Proteins' and the Guidebook to the Extracellular Matrix and Adhesion Proteins' which should serve both seasoned scientists and students alike. Each class of proteins is prefaced by a general introduction that describes their overall functions and some of the interesting questions that challenge workers in the field. The biological and structural attributes of about 200 individual proteins, or groups of closely related proteins, are concisely described by in-

vestigators who participated in their discovery or characterization. Information about purification methods, assays of activity, and reagents available to study the proteins (such as antibodies and cDNA clones) are provided, together with a list of key review and research articles. We have included only the well characterized 'structural' proteins in these books; primarily those which have been purified, sequenced, and characterized with specific antibodies. Regulatory proteins that modulate the cytoskeleton or cell adhesion have, by and large, not been included in this edition. We also acknowledge that our coverage of structural proteins is regrettably incomplete, and we apologize to investigators whose protein of interest is not found in these volumes. We are happy to receive suggestions for new entries or other improvements that can be incorporated into the next edition.

We are indebted to the more than 240 authors of the entries without whose collaboration and contributions this project would not have been possible. We would also like to thank Drs D. Cleveland, B. Geiger, D. Louvard, M. Osborn, T. Pollard, L. Reichardt, J.P. Thiery, J. Tooze and K. Weber for their advice and helpful assistance. We are very grateful to L. Hymowitz and C. Kjaer for excellent secretarial services.

November, 1992   Thomas E. Kreis (Geneva)
                              Ronald D. Vale (San Francisco)

Actin and Associated Proteins

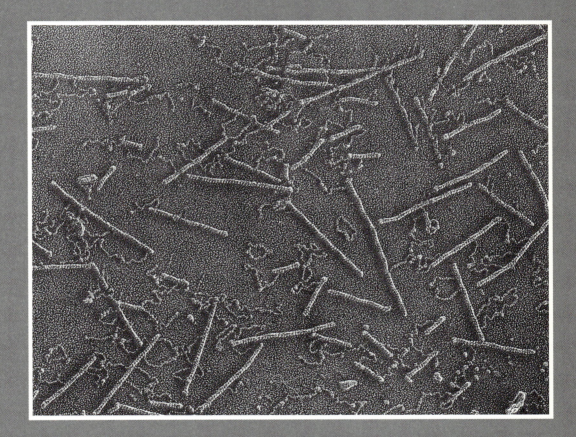

Short actin filaments (typical of the lengths found in dense actin filament meshworks such as in the cell cortex) mixed in vitro with filamin and then spread out on mica, in preparation for freeze-drying and platinum replication. Filamin appears as the irregular strands that interconnect adjacent filaments or loops out from single filaments.

(Courtesy of Dr John Heuser, Washington University, St Louis.)

1

Actin and Actin Binding Proteins

The actin based cytoskeletal and motility system can now be viewed as one of the hallmarks of eukaryotic organisms. The major building block of the system is the actin molecule, a protein of 385 residues arranged into four domains surrounding a deep cleft containing ATP or ADP together with a tightly bound divalent cation (Figure 1)[1]. Actin polymerizes to form filaments (Figure 2)[2] that serve two functions. First, the interaction of actin filaments with the mechanochemical enzyme, myosin, produces the forces for many types of cellular movements including muscle contraction, cytokinesis, cytoplasmic streaming and amoeboid motion. These actin based movements complement those produced by motor proteins operating along microtubule tracks. Second, a network of crosslinked actin filaments accounts for many of the viscoelastic properties of cytoplasm. This actin cytoskeleton complements (and may physically interact with) cytoskeletal structures composed of microtubules or intermediate filaments.

The **actin** system is not only very old in an evolutionary sense but is also highly abundant. All studied eukaryotic species have actin. Both actin and **myosin** are thought to rank among the five most abundant proteins on earth and in the best studies cases, actin and all of its associated proteins constitute more than 25% of protein in nonmuscle cells; in muscle the figure is over 60%.

The actin system is also complex. Table I lists 48 different classes of actin binding proteins. Since the mid-1970's two or more new classes of actin binding proteins have been discovered every year, and we see no sign that the inventory is complete. In addition, new variants of each class appear annually as investigators broaden the search for these proteins across the biological kingdoms.

This field of investigation is now so large and complex that no one can master every aspect. Therefore, this introduction will not summarize the field, but will seek answers to a few questions about the general properties of the actin system. For more detailed reviews of the subject as it develops, I recommend the annual issue of the journal "Current Opinion in Cell Biology" dealing with Cytoplasm and Cell Motility.

◼ WHAT IS THE MOST PARSIMONIOUS CLASSIFICATION OF THE PROTEINS IN THE ACTIN SYSTEM ?

To provide a framework for the questions that follow, I have attempted to classify all of the known actin binding proteins in Table I. In many cases we know from the primary structures that all of the proteins in a given group are homologs. In other cases the groupings seem valid based on molecular size and functional characteristics, but are without sequence confirmation. Some of this classification may need revision in the future, but for now this appears to be the minimum number of classes. Remarkably, even this minimal classification yields 48 distinct classes of actin binding proteins!

Fortunately, these numerous classes can be grouped into a more manageable number of families based on functional characteristics. For example, six classes fall into the family of actin monomer binding proteins and another ten classes are all actin filament crosslinking proteins.

Alternatively, one could classify these proteins based on primary (and in the future three dimensional) structure. For example, at least six proteins share the actin binding domain originally identified in α-**actinin**, the two classes of myosin use similar motor domains and at least four classes of barbed end capping proteins are constructed of homologous domains about the size of profilin. At the present time a classification system based on sequence homologies appears to match closely the groupings based on function.

◼ HOW MANY DIFFERENT TYPES OF ACTIN BINDING UNITS HAVE EVOLVED?

We now recognize four distinct protein structures that have been used in more than one class of actin binding protein. They are the myosin head; the α-actinin head; the **profilin**, **gelsolin** domain; and the **depactin**, **cofilin** motif. These four structural units provide the actin binding sites for 16 of the 48 classes. Representatives of most of the other 30 classes have been sequenced and do not appear homologous with each other or any of the four multiply used actin binding units. Many of these proteins, such as **tropomyosin**, represent unique structures that can bind to actin. However, I expect that some simplification will emerge from further study, because the evolution of more than 30 different actin binding sites seems unlikely to me.

The best characterized (and largest) actin binding unit is the myosin head. A typical head consists of a heavy chain of about 850 residues associated with one or more light chains[3]. The **myosin-I** class has one of these heads attached to a highly variable tail domain that interacts with other ligands, including membranes[4]. The **myosin-II** class has two heavy chains which bind together in an extended α-helical coiled-coil with the two heads at one end[5]. The primordial myosin may have had one head like myosin-I; the tail of myosin-II may have been obtained during a fusion of the heavy chain gene with a gene for an α-helical coiled-coil protein similar to tropomyosin or **paramyosin**. The nucleotide binding site on myosin heads is related in sequence to other ATP binding pro-

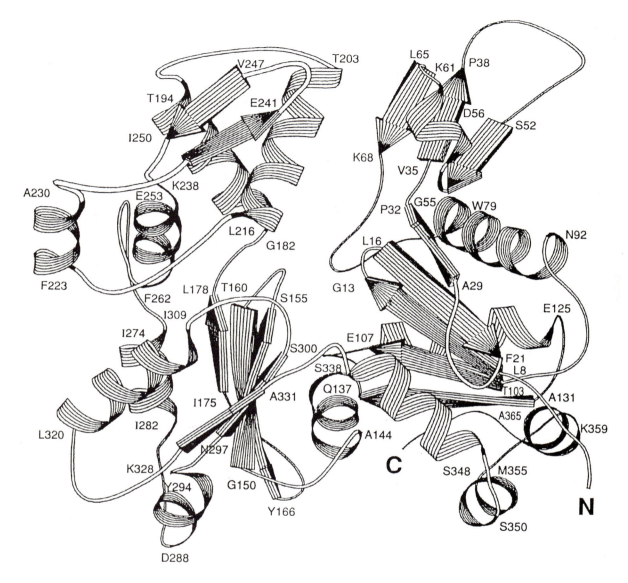

Figure 1. A ribbon diagram of the 3-D structure of actin viewed roughly perpendicular to the flat face of the molecule from the work of Kabsch et al. (1990)[6]. The molecule is thought to have a similar orientation in a filament with its long axis vertical. The right side of the molecule is thought to be exposed on the outer surface of the filament. The left side is thought to be near the axis of the filament. The polarity of the filament defined by the arrowhead shape of its complex with myosin would have the pointed end at the top and the barbed end at the bottom.

teins. In the absence of ATP myosin heads bind to actin filaments with high affinity (K_d = 1 nM). The binding site is on the exposed lateral surface of the actin filament and myosin can be crosslinked to sites at the N- and C-termini of the actin molecule. In the presence of ATP, myosin heads bind to actin filaments weakly in a rapid equilibrium. When the products of ATP hydrolysis dissociate from the myosin, a transition to a strongly bound form is cou-

pled with the production of force for movement of the head along the actin filament. The chemistry of these reactions is well characterized, but the mechanical events remain obscure[6].

The α-actinin head represents a second primordial actin binding unit. This structure consisting of about 245 amino acids is used in a remarkable number of different ways to crosslink actin filaments[7,8]. Placed on the end of an α-heli-

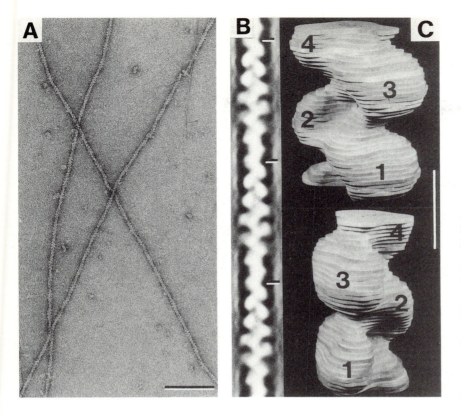

Figure 2. (A) An electron micrograph of negatively stained actin filaments. Bar 100 nm. (B) A filtered image of a micrograph of a negatively stained actin filament. (C) Two views of 3-D reconstructions of four subunits from an actin filament. The barbed end is at the top. Bar 5.5 nm. [From the work of U. Aebi and colleagues, modified from T.D. Pollard and J.A. Cooper (1986) Annu. Rev. Biochem. 55, 987-1035].

cal repeat of variable length, it forms α-actinin, **spectrin** and **dystrophin**. Alternatively, it can appear at the N-terminal end of a repeated β-sheet structure of variable length to form **ABP-120** and **ABP-280**. All of these polypeptides form antiparallel dimers capable of crosslinking two actin filaments. In contrast, **fimbrin** has two of these actin binding domains in tandem on one polypeptide chain.

Proteins in the gelsolin family are composed of three to six domains which share limited sequence homology and vary in actin binding capacity. Each of these domains is about the size of the invertebrate profilins and some of the domains that bind actin monomers share with *Acanthamoeba* profilin about 20 residues near the C-terminus that can be chemically crosslinked to residue 364 near the C-terminus of actin[9]. All of these proteins bind to the barbed end of actin filaments at a site that is buried, at least in part, in the actin filament. I expect that the fragmin-60 and scinderin/adseverin classes will be added to this family when their sequences are determined. One attractive possibility is that all of these proteins evolved from an ancient precursor the size of profilin. Through successive gene duplications, we now have progeny with one (profilins), three (**fragmin**, **gCAP39**/MCP) or six (gelsolin, **villin**) tandem domains. The sequence homologies among these domains are weak, so the family of proteins and the individual domains must have diverged long ago.

For example, while it is possible to align the putative actin binding sites of *Acanthamoeba* profilin and vertebrate gelsolin and the corresponding sequences of *Acanthamoeba* profilin and vertebrate profilin, the vertebrate profilin and gelsolin sequences have little or no homology.

The homology between the depactin class and cofilin is extensive, but the proteins have strikingly different properties. They must have had a common ancestor. There is weak homology with **ABP-50**, but the extent of this structural family will not be known until some three dimensional structures are available. Like myosin, these proteins bind near the N- and C-termini of actin but in contrast this binding site is not exposed in the actin filament.

■ HOW OLD IS THE ACTIN SYSTEM?

Actin is one of the most conserved and ubiquitous proteins in nature. It is present in all eukaryotes and no prokaryotes. Some have argued that the development of an actin filament gel in the cortex of cells provided the mechanical support of the plasma membrane that allowed the earliest eukaryotes to dispense with their cell walls.

The actin binding proteins must also be very ancient. The ancestral genes for actin binding proteins that were postulated above and the duplications, fusions and

Table I. Actin binding proteins

Protein	Synonyms	Distribution*	Subunits (N × kD)	Ligands with dissociations constants			Sequence	Sequence Homologies	Diseases or Mutations
				Actin Monomers	Actin Filaments	Other Ligands			
Monomer binding proteins									
Thymosin-beta-4	FX	V	1×5	0.7 μM			H, R	Actob	—
Actobindin		Ac	1×9.7	3.3 μM, 20 μM			Ac	Prof, actob	—
Profilin		V, Ec, Ac, Dd, Ph, Y, Vi	1×12-15	1-10 μM		PIP2 (1 μM); PIP; polyproline (30 μM); DNA; Ca	V, Ec, Ac, Dd, Ph, Y	Gel, Frag	Yst
DNaseI		V	1×29	1 nM	P-end				—
ABP-50	Elongation factor 1a	V, Dd	1×50	0.2 μM	2 μM	GTP (0.5 μM); aminoacyl tRNA; ribosome	V, Dd	Depact, A, actob	—
Vitamin D-binding Gc-protein	DBP/Gc	V	1×55	1 nM		Vitamin D (50 nM); complement component C5a	H, R		—
Small severing proteins									
Depactin	ADP, actophorin, destrin	V, Ac, Ec	1×18	0.1 μM	weak	PIP2; PIP	Ch, Sw, Ec	EF-1a, Cof, TM	—
Cofilin									
Cofilin		V	1×21	0.1-0.2 μM	0.1-1.0 μM	Polyphosphoinositides	H, Sw, M, Ch	Depact, prof, gel	—
Barbed end capping proteins									
Capping protein	Cap Z, capactin	V, Ac, Dd, Y, Xn	1×32-36; 1×28-32	weak	B-end 1nM	PIP2; PIP	Ch, Xn, Nem Dd, Y		Yst
Radixin		V	1×82		B-end		—		—
Insertin	?Tensin								—
Barbed end capping/severing proteins									
gCAP39/MCP	Mbbl	V	1×39	3 μM	B-end 1 nM	Ca (? μM); PIP2; PIP	M	Gel, Frag, Vil	
Fragmin	Severin	Ph, Dd, Ee	1×40		B-end 1 nM	Ca (? μM)	Dd, Ph	Gel, Prof	Dd
Fragmin 60		Ph	1×60		B-end	Ca (? μM)			
Scinderin	Adseverin	V	1×74		B-end	Ca (0.7 μM); PIP2; PI; PS			
Gelsolin		V, Ec	1×80 or 1×83		B-end 1 pM	Ca (1 μM); PIP2; PIP	H, M, Sw	Vil, Frag, Prof	H amyloidosis
Villin		V	1×90		B-end ? nM	Ca (? μM); PIP2; PIP	H, Ch	Gel, Frag, Prof	—

Table I. Actin binding proteins (*Continued*)

Protein	Synonyms	Distribution*	Subunits (N × kD)	Ligands with dissociations constants			Sequence	Sequence Homologies	Diseases or Mutations
				Actin Monomers	Actin Filaments	Other Ligands			
Lateral binding proteins									
Hisactophilin	—	Dd	1×13	0.2 µM	? µM	Palmitic acid (covalent)	Dd	aA, annexins	—
Calponin	—	V, Ch	1×34			TM; CM	Ch	EF, MLC	—
Troponin	—	V	1×18 (TNC); 1×21 (TNI); 1×31 (TNT)		Side ? µM; Side ? µM; Side ? µM	TNI; TNT; TNT; Tropomyosin; TNT; TNC	V; V; V		Dros
Tropomyosin	—	V, Ec, Ph, Y	2×32 (ms); 2×28 (nonms)		Side ? µM	Tropomyosin; TNT; TNC; Tropomyosin; troponin; caldesmon; calponin	H, R, Ch, Dr, Y		Dros, Yst
Tropomodulin	—	V	2×41			Tropomyosin (? µM)			—
Caldesmon	—	V	1×87 (ms); 1×60 (nonms)		Side ? µM	Tropomyosin; CM; myosin; S100	Ch		—
Adducin	PK1, PK2, 115/110, CamBP	V	1×100+1×105		Side 0.3 µM	Spectrin (cooperative)			
Crosslinking Proteins									
ABP-30	Gelactins (?)	Dd, Ac, V, Dr	1×34		Side 0.2 µM	Ca (0.1-1.0 µM)	Dd	EF, aA, ABP120, ez, caldes	—
Dematin	Band 4.9	V	3×48-52		Side 0.2 µM	Red blood cell membranes			—
Fascin		Ec	?×58		Side ? µM				—
Fimbrin	Plastin, SAC6(Y)	V, Y	1×71		Side ? µM		H, Ch, Y	aA, Spec Dys, Fil, EF	—
MARCKS		V	?×88		Side ? µM		M, Bv, Ch		—
ABP-120	Gelation factor	Dd	2×92		Side <1 µM	CM (? µM)	Dd	aA, Spec, Dys, Fil	Dd
Alpha-actinin		V, Ec, Ac, Dd	2×100		Side 1-25 µM	Vinculin (? µM); Ca (variable); phospholipids	H, Ch, Dd	Fil, Dys, Spec, EF, APB120	Dros, Dd
Spectrin	Fodrin, TW260/240 Calspectin	V, Ec, Ac, Dr	2×280+2×246		Side 1-25 µM	Ankyrin (0.01-0.1 µM); spectrin (0.1 µM); band 4.1 (10 µM); adducin (0.1 µM)	H, R, Ch, Dr	aA, Dys, Fil,	H bered sphero-cytosis & eliptocytosis
ABP-280	Filamin	V, Ec	2×280		Side ? µM	GPIb/Ix	H	aA, Spec Dys, APB120	—
Dystrophin		V	?×427		Side ? µM	Muscle sarcolemmal glycoproteins	H, M, Ch	aA, Dys, Fil, Spec	HDMD, BMD

(Table continued)

Table I. Actin binding proteins (Continued)

Protein	Synonyms	Distribution*	Subunits (N × kD)	Ligands with dissociations constants			Sequence	Sequence Homologies	Diseases or Mutations
				Actin Monomers	Actin Filaments	Other Ligands			
Membrane associated actin binding proteins									
Ponticulin		Dd, V	1×17	>10 μM	0.3 μM	Plasma membrane	Dd		—
Actolinkin		Ec	1×20		B-end ? μM	Plasma membrane			—
Synapsins		V	Ia = 84; Ib = 80; IIa = 74; IIb = 55		Side 2 μM	Synaptic vesicles (10 nM); MT (5 μM); acidic phospholipids (14 μM); NF	H, R, Bv		—
Miscellaneous									
Nuclear ABP		Ac	2×34		0.25 μM	DNA			
Yeast ABP-1		Y	I = 1×39				Y		
Annexins	Synexin, calpactin calcimedin lipocortin	V	II = 2×38 + 2×10		0.2 μM	Ca (1-40 μM); acidic Phospholipids (10 nM)	H, M, Ch, Bv, R, Dr, Dd		—
		V, Dr	III, IV, V = 1×35; VI = 1×67; VII = 1×53						
Coronin		Dd	1×55				Dd	G-proteins	
Ezrin	Cytovillin, p81	V	1×66	?	?		H	Band 4.1, talin	—
Band 4.1		V, Ch, Dr	1×80; isoforms 30 to 175			Spectrin (10 μM); CM (1 μM); glycophorin; PIP2; PKA; PKC	H, Ch		H bered ellipto-cytosis

Table I. Actin binding proteins (Continued)

Protein	Synonyms	Distribution*	Subunits (N × kD)	Ligands with dissociations constants			Sequence	Sequence Homologies	Diseases or Mutations
				Actin Monomers	Actin Filaments	Other Ligands			
Microtubule binding proteins									
Tau		V			Side ? µM	Microtubules	Ch		
MAP-2		V			Side ? µM	Microtubules; PKA; intermediate filaments	M		
Myosins									
Myosin-I	110 K/CM	V, Ac, Dd, Dr, Y	1×110-130 + LC		Side variable	ATP (µM); membrane lipids (PS)	Bv, Ch, Ac, Dd, Y	Myosin-II, src (SH-3)	Mouse, Dros, Dd, Yst
Myosin-II		All eukaryotes	2×~200 + LC		Side variable	ATP (µM); myosin; paramyosin; C-protein; titin	V, Dr, Ac, Dd, Y	Myosin-I	H cardio-myopathy, Dros, Dd, Yst

Table I. Classification of actin binding proteins, including a summary of the properties of each class. Details are provided in the individual entries in this section. I thank many of the authors of the following entries on the individual proteins for providing data for this table. Abbreviations: V=Vertebrate; Ac=*Acanthamoeba*; Dd=*Dictyostelium discoideum*; Dr=*Drosophila*; Y=Yeast; Ec=Echinoderm; Ph=*Physarum*; Vi=Virus; Xn=*Xenopus*; LC=Light Chains; ms=muscle; nonms=nonmuscle; B-end=Barbed end; P-end=Pointed end; PS=Phosphatidylserine; PIP$_2$=Phosphatidylinositol bisphosphate; PIP=Phosphatidylinositol phosphate; PI=Phosphatidylinositol; CM=Calmodulin; PKA=Protein Kinase A; PKC=Protein Kinase C; Bv=Bovine; Ch=Chicken; H=Human; R=Rabbit; Sw=Swine; M=Mouse; Nem=Nematode; Actob=Actobindin; Prof=Profilin; Gel=Gelsolin; Frag=Fragmin; A=Actin; Depact=Depactin; Cof=Cofilin; TM=Tropomyosin; Vil=Villin; aA=α-Actinin; EF=EF hand Calcium binding site; MLC=Myosin Light Chain; Ez=Ezrin; Caldes=Caldesmon; Spec=Spectrin; Dys=Dystrophin; Fil=Filamin/ABP-280.

subsequent divergence of these genes must have taken place before the radiation of the eukaryotic kingdoms, since many of the contemporary descendants of these ancient proteins are found across the phylogenetic tree. Myosin-I, myosin-II, α-actinin, **fimbrin, spectrin,** profilin and **fragmin/gCAP39/MCP** are found in protozoa, yeast, myxomycetes and vertebrates, so they must have been present in primitive eukaryotes at the base of the phylogenetic tree.

■ ARE ALL CLASSES OF ACTIN BINDING PROTEINS UNIVERSAL?

Ten years ago, many investigators would have guessed that some classes of actin binding proteins are unique to certain phyla or eukaryotic kingdoms, maybe even unique to a particular species. Now it seems more likely that unique actin binding proteins will be the exception rather than the rule. Negative evidence for uniqueness is no longer persuasive. There are now many examples including profilins, myosins and the depactin/actophorin classes where antibodies fail to crossreact between homologous actin binding proteins.

■ WHY ARE ACTIN BINDING PROTEINS SO NUMEROUS AND WHY DO THEIR FUNCTIONS APPEAR REDUNDANT?

The diversity of actin binding proteins is probably the most striking feature of this system. Given 48 classes of actin binding proteins and more expected just over the horizon, no wonder actin has a reputation as a "sticky" protein. In fact, it is not very "sticky". Many of the actin binding proteins have a relatively low affinity for actin monomers or filaments.

One view of this question is that the "actin binding proteins" function primarily as actin binding proteins. Assuming this position, one wonders why the 48 different classes of actin binding proteins should appear to have such a limited repertoire of functions. For example, why should a cell need more than one actin filament crosslinking protein? Each protein could have a different activity, but knock out mutations in *Dictyostelium* suggest that the system is redundant[10]. For example, *Dictyostelium* tolerates the loss of either α-actinin or **ABP-120** remarkably well. The null mutants have only subtle defects in behaviour and growth rate, but these defects might give wild type cells a competitive advantage outside the lab. This evidence suggests that the actin system is highly redundant with multiple proteins sharing overlapping functions. This could make the system fail-safe. Alternatively, many of the proteins that appear similar *in vitro* may actually differ substantially in their functions *in vivo*. Their superficial similarity may reflect the limitations of the available *in vitro* assays for their activities.

A second view supported by a growing body of evidence is that the actin binding function of many of these proteins is incidental to other more important functions. The glycolytic enzymes provide one example of proteins engaged primarily in intermediary metabolism that may

incidently perch on actin filaments. Profilins appear to regulate both actin and phospholipid metabolism. **ABP-50** is elongation factor-1α. Most other actin binding proteins have not been tested for other activities, so the number of multifunctional proteins remains to be determined. From this point of view, the list of actin binding proteins is so large because many proteins use actin as a secondary binding site.

■ WHY DO MANY PROTEINS BIND TO ACTIN WITH LOW AFFINITY?

Two thirds of the actin binding proteins have relatively low affinities for actin monomers or filaments, with $K_d >$ 0.1 μM. K_d's in this range usually indicate high dissociation rate constants on the order of $1s^{-1}$ or greater. This means that at steady state, the bonds between these actin binding proteins and actin are broken and reformed on a second or even subsecond time scale.

These weak bonds may allow the actin cytoskeleton to remodel and change shape on a fast time scale that is compatible with cellular locomotion and the rapid changes in shape that accompany activation of platelets. For example, the rapid (50-$100s^{-1}$) equilibrium binding of intermediates in the actomyosin cycle prevents heads that are not producing force from interfering with the sliding of an actin filament past a myosin filament at rates >1 μm/sec[6]. Similarly, gels of actin filaments and α-actinin are much more rigid when deformed rapidly than slowly[11], presumably because the crosslinks can rearrange if given sufficient time.

■ WHAT ARE THE RESEARCH CHALLENGES FOR THE IMMEDIATE FUTURE?

Although the field has grown rapidly during its first 15 years, there are still really many more questions than answers. The areas in most serious need include completion of the molecular inventory, determination of key molecular structures, assignment of physiological functions, elucidation of physiological regulatory mechanisms, and exploration of interfaces with other cellular systems.

Several times over the last decade I thought that we already had a large enough inventory of actin binding proteins - that any new proteins would be minor variants of well characterized proteins. Each time I have been proven wrong. Last year the **thymosin-β4** story[12] emerged completely by surprise. Not only had this actin monomer binding protein been missed, but now thymosin-β4 appears to have supplanted profilin as the major monomer sequestering protein. This has caused a reevaluation of the role that profilin and other monomer binding proteins play in cells. The inescapable conclusion is that one cannot play without a full deck of cards. I expect that the missing components can be found with available biochemical assays, but hope that molecular and classical genetics[13] will speed the process along.

Given that the actin cytoskeleton is a complex three

dimensional molecular jigsaw puzzle, knowledge of the molecular structures of the components will not only be essential for understanding the mechanisms involved, but also sharpen our experimental design. The recent determination of the structure of actin[1] has already had a major impact on our thinking. Both thymosin-β4[14] and **actobindin**[15] have been studied by NMR and found to be largely unfolded in solution. Presumably they fold upon binding to actin. We can hope that atomic resolution structures of some of the other major building blocks will be available before long. Crystals of the myosin head[16], the actin-profilin complex[17], profilin[18] and **actophorin**[19] are under investigation.

Assignment of physiological functions is best determined by genetic knockouts in suitable organisms. Already yeast, *Dictyostelium* and *Drosophila* have been cooperative[10]. Profilin, myosin-I and myosin-II null mutants in yeast have interesting phenotypes. The myosin-II null mutant of *Dictyostelium* fails in cytokinesis but moves remarkably well. α-actinin, ABP-120 and **severin** mutants have subtle phenotypes. Myosin-II light chain mutants in *Drosophila* fail during embryogenesis, perhaps due to defective cytokinesis. A whole range of sarcomeric contractile protein mutations in *Drosophila* produce defects in flight muscle assembly, stability and contractility. In some cases, expression of antisense RNA or antibody injections may prove valuable, but genetic knockouts will generally provide the most definitive results.

Perhaps the least well developed part of this field concerns the mechanisms that regulate the system in live cells. There are exceptions such as our detailed knowledge regarding excitation-contraction coupling in striated muscles, but it is sobering to consider that in spite of years of work by many laboratories that we lack a complete understanding of the mechanisms regulating contraction of smooth muscle[20]. To be sure, we now appreciate that myosin light chain phosphorylation initiates contraction, but we do not understand how tension is maintained long after light chain phosphorylation returns to resting levels, or the roles of either **caldesmon** or **calponin**. I am confident that the spatial and temporal regulation of motility in nonmuscle cells will be more complex than in smooth muscle. Fortunately, great strides are being made in understanding signalling mechanisms at the molecular and cellular levels. The actin system will get caught up in this excitement before long.

A recent major trend has been the realization that some of the components of the actin system also have roles in other cellular processes. For example, profilin, gelsolin and a growing number of other actin binding proteins bind to polyphosphoinositides (Table I). Originally these interactions were thought to regulate the actin system[21], but profilin may actually be part of the mechanism coupling growth factor receptors to the productions of the second messengers inositol trisphosphate and diacylglycerol[22]. Even more striking, the actin bundling protein ABP-50 from *Dictyostelium* turns out to be elongation factor 1α, a component of the ribosomal protein synthetic machinery[23]. Actin may regulate protein synthesis! These are but two examples of how the actin system will be found to be integrated into the cellular function above and beyond its roles in cytoplasmic structure and motility.

Suggestions for additional reading on actin and actin binding proteins.

Cooper, J.A. (1991) Ann. Rev. Physiol. 53, 585-606. The role of actin polymerization in cell motility.

Janmey, P.A. (1991) Curr. Opin. Cell Biol. 3, 4-11. Mechanical properties of cytoskeletal proteins.

Carlier, M.-F. (1991) Curr. Opin. Cell Biol. 3, 12-17. Nucleotide hydrolysis in cytoskeletal assembly.

Wang, Y.-L. (1991) Curr. Opin. Cell Biol. 3, 27-32. Dynamics of the cytoskeleton in live cells.

Luna, E.J. (1991) Curr. Opin. Cell Biol. 3, 120-126. Molecular links between the cytoskeleton and membranes.

■ REFERENCES

1. Kabasch, W., Mannherz, H.G., Suck, D., Pai, E.F. and Holmes, K.C. (1990) Nature 347, 37-44.
2. Carlier, M.F. (1991) J. Biol. Chem. 266, 1-4.
3. Warrick, H.M. and Spudich, J.A. (1987) Annu. Rev. Cell Biol. 3,
4. Pollard, T.D., Doberstein, S.K. and Zot, H.G. (1991) Annu. Rev. Physiol. 53, 653-681.
5. Korn, E.D. and Hammer, J.A. III. (1988) Annu. Rev. Biophys.Biophys. Chem. 17, 23-45.
6. Hibberd, M.G. and Trentham, D.R. (1986) Annu. Rev. Biophys. Biophys. Chem. 15, 119-61.
7. Hartwig, J.H. and Kwiatkowski, D.J. (1991) Curr. Opin. Cell Biol. 3, 87-97.
8. Matsudaira, P. (1991) Trends Biochem. Sci. 16, 87-92.
9. Vandekerckhove, J. (1990) Curr. Opin. Cell Biol. 2, 41-50.
10. Noegel, A.A. and Schleicher, M. (1991) Curr. Opin. Cell Biol. 3, 18-26.
11. Sato, M., Schwarz, W.H. and Pollard, T.D. (1987) Nature 325, 828-830.
12. Safer, D., Golla, R. and Nachmias, V.T. (1990) Proc. Natl. Acad. Sci. 87, 2536-2540.
13. Titus, M.A., Warrick, H.M. and Spudich, J.A. (1990) Curr. Opin. Cell Biol. 2, 116-120.
14. Zarbock, J., Oschkinat, H., Hannappel, E., Kalbacher, H., Voelta, W. and Holak, T.A. (1990) Biochem. 29, 7814-7821.
15. Bubb, M.R., Knutson, J.R., Bax, A. and Korn, E.D. (1990) Personal communication.
16. Winkelman, D.A., Mekeel, H. and Rayment, I. (1985) J. Mol. Biol. 181, 487-501.
17. Schutt, C.E., Lindberg, U., Myslik, J. and Strauss, N. (1989) J. Mol. Biol. 209, 735-746.
18. Magnus, K.A., Lattman, E.E., Sato, M. and Pollard, T.D. (1986) J. Biol. Chem. 261, 13360-13361.
19. Magnus, K.A., Maciver, S.K. and Pollard, T.D. (1988) J. Biol. Chem. 263, 8143-8144.
20. Trybus, K. (1991) Cell Motil. Cytoskel. 18, 81-85.
21. Stossel, T.P. (1989) J. Biol. Chem. 264, 8261-8264.
22. Goldschmidt-Clermont, P.J., Kim, J.W., Machesky, L.M., Rhee, S.G. and Pollard, T.D. (1991) Science 251, 1231-1233.
23. Yang, F., Demma, M., Warren, V., Dharmawardhane, S. and Condeelis, J. (1990) Nature 347, 494-496.

■ *Thomas D. Pollard:*
Department of Cell Biology and Anatomy,
Johns Hopkins Medical School,
Baltimore, MD 21205, USA

Actins

Actin was originally purified from skeletal muscle tissue[1] and further recognized as the major protein of the microfilament system of eukaryotic cells. It interacts with a large number of actin associated proteins and this results in a variety of stable or transiently regulated different supramolecular organizations. Actin is also one of the key proteins in various cell motility processes which are either based on actin gel-sol transitions or ATP-dependent actomyosin interactions.

Actin is found in all eukaryotic cells examined so far. It contains 374/375 amino acids[2] and has a sequence which is highly conserved throughout evolution[3], suggesting its central pivotal role in important cellular processes. Warm blooded vertebrates (mammals and birds) express six distinct isoproteins. They are expressed solely in a tissue specific manner, independent of the species; two striated muscle actins (skeletal muscle and cardiac actin), two smooth muscle actins (vascular and visceral actin) and two coexpressed nonmuscle actins[4]. Isoelectric focusing separates these isoforms into three isoelectric variants, referred to as α-, β- and γ- actin[5]. They form a typical abundant protein triplet in a 2-D gel system[6] (pI ≈ 5.1, M_r = 42,000). No differences were found in their polymerization properties[7] and their segregation into different cellular structures is not clearly established[8,9]. Isoactins could interact differently with enzymes such as the ADP-ribosyl-transferase[10] or actin binding proteins such as **myosin**[11].

Actins are all N-terminally acetylated and this modification follows two different pathways, depending on the actin type[12]. Most actins also contain a 3-methyl-histidine residue located at a very conserved position (His-73)[13]. Other types of covalent modifications are ADP-ribosylation[10] phosphorylation[14,15], methylation[16] and ubiquitination[17]. Some of these modifications affect F-actin polymerization[18,19].

The 3-D structure of actin in the actin-DNase 1 complex has been obtained at 2.8Å resolution[20]. The molecule consists of two nearly equal halves each further subdivided into two subdomains. This structure is extremely similar to that of the N-terminal ATPase fragment of the 70 kDa heat shock cognate protein[21]. Models for F-actin have been constructed in accordance with fibre diffraction patterns and image analyses[22,23]. Most of the effects of chemical modification, crosslinking data and effects of site-directed mutagenesis previously obtained could be explained by these models[24,25]. Another model of actin organization has been deduced from the X-ray diffraction patterns of the **profilin**-actin complex[26].

Actin gene expression is regulated in a developmental and tissue specific manner both in lower[27-29] and in higher eukaryotes[30,31]. Tissue specific regulatory sequences have been recognized[32,33].

Actin is an essential protein as illustrated by the lethal effect following disruption of the single yeast actin gene[34]. Mutations of actin cause defects in muscle[35,36] or cytoskeletal organization[37] and, in some cases are accom-

panied by increased tumorigenicity[38]. An unusual actin mutant is the v-*fgr* oncogene product which is a hybrid protein containing the first 129 amino acids of γ-nonmuscle actin linked to a tyrosine specific kinase[39].

■ PURIFICATION

Actin is most easily purified from skeletal muscle sources in a procedure involving repeated cycles of polymerization-depolymerization[40]. Nonmuscle actin purifications are more complex because of the higher ratio of actin binding proteins[41]. They are either purified by multiple column separations[7] or by affinity chromatography on immobilized DNase 1[42]. Some of the actin isoforms can be separated on hydroxylapatite[43].

■ ACTIVITIES

In low salt buffers, actin is a monomeric protein (G-actin) but it associates under physiological conditions into a double helical 10 nm thick filament structure (F-actin)[44]. Actin polymerization is accompanied by hydrolysis of the actin bound ATP, although nucleotide hydrolysis is not obligatory for addition of monomers to polymers[45]. F-actin filaments are formed by adding monomers at both ends with different kinetics. There is a fast and a slow growing end[46], and this polarity can be visualized by decoration with heavy meromyosin into arrowhead-like structures pointing towards the slow growing end[47]. Several drugs bind to actin and exhibit a profound effect on the polymer structure. Phalloidin binds strongly to F-actin in a ratio of one molecule of phalloidin per protomer[48]. Another class of fungal metabolites (the cytochalasins) bind to the barbed end of the filament[24]. Information on the overall actin polymerization process is generally obtained by viscometry[49], absorbance at 232 nm[50], fluorescence enhancement of derivatized actin[51,52] and by the DNase 1 inhibition assay[53].

■ ANTIBODIES

Rabbit polyclonal antisera were raised against denatured SDS-PAGE purified actin[54]. These antibodies showed a broad specificity for the actins from species as diverse as plants[55], and mammals. More specific antibodies could be obtained by using the specific variable actin N-terminus as antigen[56-58] or by differential immunoabsorption[9,59].

■ GENES

The first actin cDNA sequence was reported for *Dictyostelium discoideum*[60]. During recent years an overwhelming amount of cDNA and genomic sequence information has become available of which we can only list a few examplary results in the context of this short review.

These studies covered species as diverse as yeast[61], plants[62], slime molds[63], sea urchin[64], *Tetrahymena*[65], *C. elegans*[66], *Drosophila*[67], mammals[68] and birds[69]. Actins are generally encoded by large gene families (e.g.: *Dictyostelium*[63], *Drosophila*[27], mouse[70]). In mammals only b- and γ-actin genes are multicopy[71,72]. Actin genes are scattered over different chromosomes[73] in the human genome and partially linked in the sea urchin genome[74].

The updated GenBank contains at least 165 actin sequences. Early literature starts from the late 70's with *Dictyostelium* and yeast actin DNA sequences. Actin cDNA and genomic sequences can be searched for, by using conserved stretches of actin sequences and allowing a 50% similarity. A muscle specific actin sequence can be found under the GenBank number CHKACASK.

Genes encoding divergent forms of actin (50-60% identity) have been identified in *S. cerevisiae* (ACT 2)[75], *S. pombe*[76], and vertebrates (centractin/actin-RPV)[77,78]. Centractin/actin-RPV is a component of the centrosome and a dynein regulatory complex.

■ REFERENCES

1 Straub, F.B. (1942) Studies, University of Szeged II, 3-15.
2. Elzinga, M., Collins, J.H., Kuehl, W.M. and Adelstein, R.S. (1973) Proc. Natl. Acad. Sci. (USA) 70, 2687-2691.
3. Vandekerckhove, J. and Weber, K. (1984) J. Mol. Biol. 179, 391-413.
4. Vandekerckhove, J. and Weber, K. (1979) Differentiation 14, 123-133.
5. Whalen, R.G., Butler-Browne, G.S. and Gros, F. (1976) Proc. Natl. Acad. Sci. (USA) 75, 588-599.
6. Garrels, J.I. and Gibson, W. (1976) Cell 9, 793-805.
7. Korn, E.D. (1978) Proc. Natl. Acad. Sci. (USA) 75, 588-599.
8. Otey, C.A., Kalnoski, M.H., Lessard, J.L. and Bulinski, J.C. (1986) J.Cell Biol. 102, 1726-1737.
9. Pardo, J.V., Pittenger, M.F. and Craig, S.W. (1983) Cell 32, 1093-1103.
10. Aktories, K., Bärmann, M., Ohishi, I., Tsuyama, S., Jakobs, K.H. and Habermann, E. (1986) Nature 322, 390-392.
11. Rubenstein, P.A. (1981) Arch. Biochem. Biophys. 210, 598-608.
12. Solomon, L.R. and Rubenstein, P.A. (1985) J. Biol. Chem. 260, 7659-7664.
13. Elzinga, M. (1971) Biochemistry 10, 224-229.
14. Sonobe, S., Takahashi, S., Hatano, S. and Kuroda, K. (1986) J. Biol. Chem. 261, 14837-14843.
15. Ampe, C. and Vandekerckhove, J. (1987) EMBO J. 6, 4149-4157.
16. Vandekerckhove, J., Lal, A.A. and Korn, E.D. (1984) J. Mol. Biol. 172, 141-147.
17. Ball, E., Karlik, C.C., Beall, C.J., Saville, D.L., Sparrow, J.C., Bullard, B. and Fyrberg, E.A. (1987) Cell 51, 221-228.
18. Aktories, K. and Wegner, A. (1989) J. Cell Biol. 109, 1385-1387.
19. Maruta, H., Knoerzer, W., Hinssen, H. and Isenberg, G. (1984) Nature 312, 424-427.
20. Kabsch, W., Mannherz, H.G., Suck, D., Pai, E.F. and Holmes, K.C. (1990) Nature 347, 37-44.
21. Flaherty, K.M., De Luca-Flaherty, C. and McKay, D. (1990) Nature 346, 623-628.
22. Holmes, K.C., Popp, D., Gehhard, W. and Kabsch, W. (1990) Nature 347, 44-49.
23. Milligan, R.A., Whittaker, M. and Safer, D. (1990) Nature 348, 217-221.
24. MacLean-Fletcher, S. and Pollard, T. (1980) Cell 20, 329-341.
25. MacLean-Fletcher, S. and Pollard, T.D. (1980) J. Cell Biol. 85, 414-428.
26. Schutt, C.E., Lindberg, U., Myslik, J. and Strauss, N. (1989) J. Mol. Biol. 209, 735-746.
27. Fyrberg, E.A., Mahaffey, J.W., Bond, B.J. and Davidson, N. (1983) Cell 33, 115-123. (1983) Cell 33, 115-123.
28. Garcia, R., Paz-Aliaga, B., Ernst, S.G. and Crain Jr., W.R. (1984) Mol. Cell. Biol. 4, 840-845.
29. Cox, K.H., Angerer, L.M., Lee, J.J., Davidson, E.H. and Angerer, R.C. (1986) J. Mol. Biol. 188, 159-172.
30. Minty, A.J., Alonso, S., Caravatti, M. and Buckingham, M.E.. (1982) Cell 30, 185-192.
31. Rizucka, D.L. and Schwartz, R.J. (1988) J. Cell Biol. 107, 2575-2586.
32. Minty, A.J. and Kedes, L. (1986) Mol. Cell. Biol. 6, 2125-2136.
33. Petropoulos, C.J., Rosenberg, M.P., Jenkins, N.A., Copeland, N.G. and Hughes, S.H. (1989) Mol. Cell. Biol. 9, 3785-3792.
34. Shortle, D., Haber, J. and Botstein, D. (1982) Science 217, 371-373.
35. Karlik, C.C., Coutu, M.D. and Fyrberg, E.A. (1984) Cell 38, 711-719.
36. Drummond, D.R., Peckham, M., Sparrow, J.C. and White, D.C.S. (1990) Nature 348, 440-442.
37. Leavitt, J., Bushar, G., Kakunaga, T., Hamada, H., Hirakawa, T., Goldman, D. and Merril, C.. (1982) Cell 28, 259-268.
38. Lin, C.-S., NG, S.-Y., Gunning, P., Kedes, L. and Leavitt, J. (1985) Proc. Natl. Acad. Sci. (USA) 82, 6995-6999.
39. Naharro, G., Robbins, K.C. and Reddy, E.P. (1984) Science 223, 63-66.
40. Pardee, J.D. and Spudich, J.A. (1982) Meth. Cell Biol. 24, 271-302.
41. MacLean-Fletcher, S. and Pollard, T.D. (1980) Biochem. Biophys. Res. Commun. 96, 18-27.
42. Zechel, K. (1981) Eur. J. Biochem. 119, 209-213.
43. Segura, M. and Lindberg, U. (1984) J. Biol. Chem. 259, 3949-3954.
44. Pollard, T.D. (1990) Curr. Opinion Cell Biol. 2, 33-40.
45. Carlier, M.-F. (1989) Int. Rev. Cytol. 115, 139-170.
46. Wegner, A. (1976) J. Mol. Biol. 108, 139-150.
47. Woodrum, D.T., Rich, S.A. and Pollard, T.D. (1975) J. Cell Biol. 67, 231-237.
48. Wieland, T. and Faulstich, H. (1978) C.R.C. Crit. Rev. Biochem. 5, 185-260.
49. Grumet, M. and Lin, S. (1980) Biochem. Biophys. Res. Commun. 92, 1324-1334.
50. Spudich, J.A. and Cooke, R. (1975) J. Biol. Chem. 250, 7485-7491.
51. Cooper, J., Walker, S. and Polland, T. (1983) J. Muscle Res. Cell Motil. 4, 253-262.
52. Tellam, R. and Frieden, C. (1982) Biochemistry 21, 3207-3214.
53. Carlsson, L., Markey, F., Blikstad, I., Persson, T. and Lindberg, U. (1979) Proc. Natl. Acad. Sci. (USA) 76, 6376-6380.
54. Lazarides, E. and Weber, K. (1974) Proc. Natl. Acad. Sci. (USA) 71, 2268-2272.
55. Metcalf, T.N., Szabo, L.J., Schubert, K.R. and Wang, J.L. (1980) Nature 285, 171-172.
56. Bulinski, J.C., Kumar, S., Titani, K. and Hauschka, S.D. (1983) Proc. Natl. Acad. Sci. (USA) 80, 1506-1510.
57. Skalli, O., Ropraz, P., Trzeciak, A., Benzonana, G., Gillessen, D. and Gabbiani, G. (1986) J. Cell Biol. 103, 2787-2796.

58. Roustan, C., Benyamin, Y., Boyer, M. and Cavadore, J.C. (1986) Biochem. J. 233, 193-197.
59. Lubit, B.W. and Schwartz, J.H. (1980) J. Cell Biol. 86, 891-897.
60. McKeown, M., Taylor, W.C., Kindle, K.L., Firtel, R.A., Bender, W. and Davidson, N. (1978) Cell 15, 789-800.
61. Gallwitz, D. and Sures, I. (1980) Proc. Natl. Acad. Sci. (USA) 77, 2546-2550.
62. Shah, D.M. (1982) Proc. Natl. Acad. Sci. (USA) 79, 1022-1026.
63. Romans, P. and Firtel, R.A. (1985) J. Mol. Biol. 186, 321-335.
64. Akhurst, R.J., Calzone, F.J., Lee, J.J., Britten, R.J. and Davidson, E.H. (1987) J. Mol. Biol. 194, 193-203.
65. Cupples, C.G. and Pearlman, R.F. (1986) Proc. Natl. Acad. Sci. (USA) 83, 5160-5164.
66. Files, J.G., Carr, S. and Hirsh, D. (1983) J. Mol. Biol. 164, 355-375.
67. Sanchez, F., Tobin, S.L., Rdest, U., Zulauf, E. and McCarthy, B.J. (1983) J. Mol. Biol. 163, 533-551.
68. Zakut, R., Shani, M., Givol, D., Neuman, S., Yaffe, D. and Nudel, U. (1982) Nature 298, 857-859.
69. Fornwald, J.A., Kuncio, G., Peng, I. and Ordahl, C.P. (1982) Nucl. Acid. Res. 10, 3861-3875.
70. Minty, A.J., Alonso, S., Guenet, J.-L. and Buckingham, M.E. (1983) J. Mol. Biol. 167, 77-101.
71. Ponte, P., Gunning, P., Blau, H. and Kedes, L. (1983) Mol. Cell. Biol. 3, 1783-1791.
72. Engel, J., Gunning, P. and Kedes, L. (1982) Mol. Cell. Biol. 2, 674-684.
73. Czosnek, H., Nudel, U., Mayer, Y., Barker, P.E., Pravtcheva, D.D., Ruddle, F.H. and Yaffe, D. (1983) EMBO J. 2, 1977-1979.
74. Scheller, R.H., McAllister, L.B., Crain Jr., W.R., Durica, D.S., Posakony, J.W., Thomas, T.L., Birtten, R.J. and Davidson, E.H. (1981) Mol. Cell. Biol. 1, 609-628.
75. Schwob, E. and Martin, R.P. (1992) Nature 355, 179-182.
76. Lees-Miller, J.P., Henry, G. and Helfman, D. M. (1992) Proc. Natl. Acad. Sci (USA) 89, 80-83.
77. Lees-Miller, J.P., Helfman, D.M. and Schroer, T.A. (1992) Nature 359, 244-246.
78. Clark, S.W. and Meyer, D.I. (1992) Nature 359, 246–250.

■ Joël S. Vandekerckhove:
Laboratory of Physiological Chemistry,
State University of Ghent,
9000 Ghent, Belgium

Actin Binding Protein-50 (ABP-50)

Actin binding protein-50 (ABP-50) was isolated originally from Dictyostelium discoideum as an actin binding protein of 50 kDa[1]. It was identified later as elongation factor 1 α (EF-1α)[2], a protein involved in eukaryotic protein synthesis. The dual function of ABP-50 may act to link protein synthesis activity with the actin cytoskeleton. This may provide a mechanism for the spatial and temporal regulation of protein synthesis in eukaryotic cells.

ABP-50 has a Stokes radius of 3.1 nm and an apparent mass of 50 kDa on SDS-PAGE suggesting a globular monomeric protein in solution[1]. It is a basic protein based on its behaviour on ion exchange columns[1]. ABP-50 constitutes approximately 1% of total cell protein[3]. In resting amoebae of *Dictyostelium discoideum*, approximately 90% of ABP-50 is present in the cytosol and the remaining 10% is associated with the cytoskeleton. During chemotactic stimulation, the amount of cytoskeletal ABP-50 doubles at 90 sec after stimulation when filopodia are formed and ABP-50 becomes localized in filopodia[3].

ABP-50 cosediments with **actin** filaments (F-actin) as bundles[1] (Figure). It is also found in the filopodia and other cortical regions of the vegetative amoebae that contain F-actin bundles[1,3]. It binds to monomeric actin (G-actin) *in vivo* and *in vitro* as well[3]. The filament packing in the ABP-50 bundles can be liquid, hexagonal or square. ABP-50 crosslinks neighbouring actin filaments that are rotated by 90° relative to each other[4]. This bonding rule is different from that of other actin bundling proteins in which neighbouring actin filaments are aligned unrotated[4].

The cDNA sequence of ABP-50 reveals its identity as EF-1α of *Dictyostelium discoideum*.[2]. EF-1α complexes with GTP and amino acyl-tRNA to catalyze the codon-dependent placement of the amino acyl-tRNA in the "A site" of the ribosome followed by hydrolysis of GTP to GDP. A region of ABP-50 adjacent to its GTP-binding pocket is

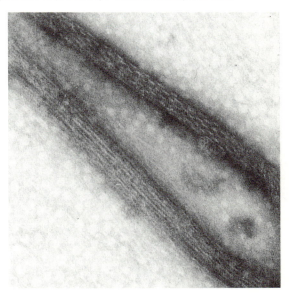

Figure. Electron micrograph (150,000x) of negative stained bundles formed by mixing equalmolar amounts of ABP-50 and actin together in the presence of 50 mM KCl and 2 mM $MgCl_2$. The two types of bundles shown here are square packed and interconvertible through image reconstruction.

homologous to the actin binding motif of **depactin** from starfish oocytes[5]. A potential G-actin binding activity of EF-1α from a broad range of species has been inferred from the depactin-like region which shows 25-35% identity with depactin and the highest homology around the most critical residues for actin binding[2]. Consistent with this is the preliminary observation that EF-1α from rabbit reticulocytes has actin binding activity. The binding of ABP-50 to G-actin may inhibit GTP exchange. The binding of G-actin to ABP-50 can be inhibited by 1 mM GTP but not GDP[3]. The K_d for binding of ABP-50 to G-actin is similar to the K_d's for the interaction of EF-1α and GTP/GDP[6,7].

The reversible association of ABP-50 with the actin cytoskeleton may provide a mechanism for regulation of protein synthesis in eukaryotic cells.

■ PURIFICATION

ABP-50 is an abundant protein and can be purified by conventional biochemical techniques[1,8]. ABP-50 is collected in the flow through of a DEAE-anion exchange column and is eluted from a Fast-S Sepharose-cation exchange column at about 180 mM NaCl. Further purification with hydroxylapatite column generates 95% pure ABP-50.

■ ACTIVITIES

ABP-50 cosediments with F-actin. The K_d for binding of ABP-50 to F-actin is 2×10^{-6} M based on the analysis of cosedimentation assays by the method of Scatchard[1]. The binding of ABP-50 to G-actin was measured in vitro with purified G-actin coupled to CNBr-sepharose, which results in a K_d around $1-2 \times 10^{-7}$ M[3]. ABP-50 bundles F-actin within seconds of mixing the two proteins in vitro[1]. Cytosolic ABP-50 is complexed with monomeric actin in a 1:1 molar ratio based on immunoprecipitation studies[3]. The protein synthetic activity of ABP-50 has been characterized with a poly-U mediated in vitro translation assay[2]. In preliminary experiments ABP-50 catalyzes polyphenylalanine synthesis in a concentration dependent fashion in a manner similar to that of EF-1α from rabbit reticulocytes[2].

■ ANTIBODIES

Affinity purified polyclonal antibodies against ABP-50 have been prepared in this laboratory. These antibodies do not crossreact with bacterial EF-Tu or the 30 kDa **small actin-crosslinking protein** from Dictyostelium discoideum under the stringency used[1]. They do crossreact with EF-1α from rabbit reticulocytes.

■ GENES

Two distinct cDNA clones coding for ABP-50 have been cloned and sequenced[2]. They are different in the third position of 34 codons but still code for an identical polypeptide of the ABP-50, which indicates the presence of two genes, D. discoideum EF1-I (GenBank X55973) and EF1-II (GenBank X55972) respectively.

■ REFERENCES

1. Demma, M., Warren, V., Hock, R., Dharmawardhane, S. and Condeelis, J. (1990) J. Biol. Chem. 265, 2286-2291.
2. Yang, F., Demma, M., Warren, V., Dharmawardhane, S. and Condeelis, J. (1990) Nature 374, 494-496.
3. Dharmawardhane, S., Demma, M., Yang, F. and Condeelis, J. (1991) Cell Motil. Cytoskel. 20, 279-288.
4. Owen, C., DeRosier, D. and Condeelis, J. (1992) submitted.
5. Sutoh, K. and Mabuchi, I. (1989) Biochemistry 28, 102-106.
6. Nagata, S., Iwasaki, K. and Kaziro, Y. (1977) J. Biochem. 82, 1633-1646.
7. Caravalno, M., Caravalno, J. and Merrick, W. (1984) Arch. Biochem. & Biophy. 234, 603-611.
8. Bresnick, A. and Condeelis, J. (1991) Method. Enzym. 196, 70-83.

■ *Fan Yang and John Condeelis:*
Department of Anatomy and Structural Biology,
Albert Einstein College of Medicine,
Bronx, New York, USA

Actin Binding Protein-120 (ABP-120)

Actin binding protein-120 (ABP-120) is an actin binding protein from Dictyostelium discoideum that crosslinks actin filaments into orthogonal arrays in vitro[1]. These networks closely resemble actin networks found in situ in ABP-120 rich pseudopods[2]. Localization of ABP-120 in the cell cortex, and more specifically in pseudopods[3,4], suggests that this protein may function in the formation of filament networks during pseudopod extension. Network expansion is proposed to constitute part of the driving force for pseudopod extension[5].

Hydrodynamic studies[6] indicate that ABP-120 is a homodimer of 240 kDa. Although the cDNA sequence predicts a molecular weight of 92.2 kDa for each monomer[7], the protein migrates at 120 kDa on SDS-PAGE. The monomers are arranged in an antiparallel orientation[8]. Rotary shadowing[6] of ABP-120 reveals a flexible rod 35 nm in length that contains globular domains. In addition, Garnier analysis of the cDNA sequence[7] predicts six β-sheet repeats in the C-terminal two-thirds of the molecule (Figure 1).

Limited tryptic digestion followed by **actin** cosedimentation has demonstrated that residues 89-115 are essential for actin binding activity in ABP-120[9]. The 27 amino acids

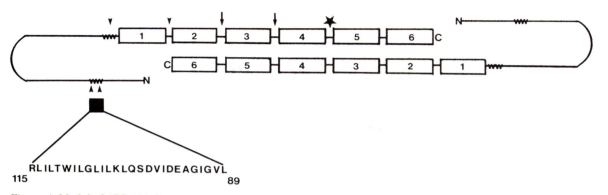

Figure 1. Model of ABP-120. Each monomer is defined as having a head and tail based on susceptibility to tryptic digestion. The head consists of the N-terminal 10 kDa and a 31 kDa tryptic peptide which begins at residue 89 and contains the first β-sheet repeat. The tail consists of the five remaining structural repeats. *Arrows*, predicted sites of tryptic cleavage from the cDNA sequence which have been confirmed by analysis of tryptic digests; *star*, predicted site of tryptic cleavage from the cDNA sequence not seen in tryptic digests; *arrowheads*, predicted sites of cleavage which have been confirmed by analysis of tryptic digests and protein sequencing.

essential for actin binding show high sequence identity (>60%) with regions near the N-termini of **ABP-280/filamin**, β-**spectrin**, α-**actinin** and **dystrophin**[9] (Figure 2). More recently, similar 27-mer sequences have been localized in **fimbrin** and plastin[10]. A synthetic peptide of the 27-mer sequence binds to F-actin, synthetic 27-mer peptides inhibit the actin binding and crosslinking of ABP-120, and Fab-fragments from an antibody to the 27-mer sequence inhibit the binding of ABP-120 to F-actin half-maximally at seven Fab-fragments per ABP-120 monomer[11]. These results indicate that the 27-mer is the actin binding site of ABP-120.

Immunofluorescence reveals that ABP-120 is preferentially localized to newly formed pseudopods following cAMP stimulation[3]. Deletion of ABP-120 by chemical mutagenesis[8] or homologous recombination[12] does not inhibit morphogenesis. However, ABP-120-minus cells locomote and chemotax poorly and their pseudopod extension activity is significantly depressed after cAMP stimulation[12]. This phenotype is consistent with the proposed function of ABP-120 as the actin crosslinking protein which assembles orthogonal networks of filaments in growing pseudopods.

■ PURIFICATION

ABP-120 is routinely purified from *Dictyostelium* amoebae by ammonium sulphate precipitation and conventional column chromatography[13]. ABP-120 binds to DEAE-cellulose, Mono-Q and hydroxylapatite resins.

■ ACTIVITIES

ABP-120 binds to F-actin and crosslinks it into orthogonal arrays. Under identical conditions, the specific crosslinking activity of ABP-120 is greater than that of α–actinin[6]. The actin binding activity of ABP-120 can be measured using high speed actin cosedimentation assays[1] and falling ball assays[6]. ABP-120 also inhibits the actin-activated Mg^{2+}-ATPase of **myosin II**[6].

```
                                                                    % IDENTITY
ABP-120               (89)  LVGIGAEDIVDSQLKLILGLIWTLILR (115)          100
D.d. α-ACTININ       (100)  LVGIGAEELVDKNLKMTLGMIWTIILR (126)           70
CHICK SK. α-ACTININ  (117)  LVSIGAEEIVDGNVKMTLGMIWTIILR (143)           66
HUMAN ABP/FILAMIN     (94)  LVSIDSKAIVDGNLKLILGLIWTLILH (120)           70
HUMAN DYSTROPHIN      (91)  LVNIGSTDIVDGNHKLTLGLIWNIILH (117)           63
CHICK DYSTROPHIN      (95)  LVNIGSSDIVDGNHKLTLGLIWNIILR (121)           63
FLY ß-SPECTRIN       (126)  LENIGSHDIVDGNASLNLGLIWTIILR (152)           63
Consensus                   Lv.Igs.diVDgn.kltLGlIWtiILr
                            *    *   *    *   *    *  *
```

Figure 2. Alignment of the 27-mer sequence in ABP-120 with 27-mer sequences in other actin binding proteins. The positions of the first and last residues of the aligned sequences for each protein are in parentheses. *Consensus* indicates the most frequently used amino acid at that position. *Upper case letters*, absolutely conserved residues; *period*, no clear consensus; *star*, conservative replacements. Alignment of the 27-mer sequences reveals that most of the different amino acids are conservative replacements.

■ ANTIBODIES

Rabbit polyclonal antisera against ABP-120 have been described[14]. A rabbit polyclonal antibody to the 27-mer sequence (residues 89-115) is available[11]. A number of monoclonal antibodies to ABP-120 have also been described[7,8].

■ GENES:

The full length cDNA sequence for ABP-120 (GenBank X15430) is published[7].

■ REFERENCES

1. Condeelis, J., Geosits, S. and Vahey, M. (1982) Cell Motil. 2, 273-285.
2. Wolosewick, J.J. and Condeelis, J. (1986) J. Cell. Biochem. 30, 227-243.
3. Condeelis, J., Hall, A., Bresnick, A., Warren, V., Hock, R., Bennett, H. and Ogihara, S. (1988) Cell Motil. Cytoskel. 10, 77-90.
4. Ogihara, S., Carboni, J. and Condeelis, J. (1988) Dev. Genet. 9, 505-520.
5. Condeelis, J., Bresnick, A., Demma, M., Dharmawardhane, S., Eddy, R., Hall, A., Sauterer, R. and Warren, V. (1990) Dev. Genet. 11, 333-340.
6. Condeelis, J., Vahey, M., Carboni, J., Demey, J. and Ogihara, S. (1984) J. Cell Biol. 99, 119S-126S.
7. Noegel, A., Rapp, S., Lottspeich, F., Schleicher, M. and Stewart, M. (1989) J. Cell Biol. 109, 607-618.
8. Brink, M., Gerisch, G., Isenberg, G., Noegel, A., Segall, J., Walraff, E. and Schleicher, M. (1990) J. Cell Biol. 111, 1477-1489.
9. Bresnick, A., Warren, V. and Condeelis, J. (1990) J. Biol. Chem. 265, 9236-9240.
10. de Arruda, M., Watson, S., Lin, C., Leavitt, J. and Matsudaira, P. (1990) J. Cell Biol. 111, 1069-1079.
11. Bresnick, A., Janmey, P. and Condeelis, J. (1991) J. Biol. Chem. 266, 12989-12993.
12. Cox, D., Condeelis, J., Wessels, D., Soll, D., Kern, H. and Knecht, D. A. (1992) J. Cell Biol. 116, 943-955.
13. Bresnick, A. and Condeelis, J. (1991) Meth. Enzym. 196, 70-83.
14. Carboni, J.M. and Condeelis, J. (1985) J. Cell Biol. 100, 1884-1893.

■ *Anne R. Bresnick and John Condeelis:*
Department of Anatomy and Structural Biology,
Albert Einstein College of Medicine,
Bronx, New York, USA

Actin Binding Protein-280 (ABP-280; Nonmuscle Filamin)

Actin binding protein-280 (ABP-280) is a ubiquitous dimeric actin crosslinking phosphoprotein of peripheral cytoplasm where it promotes orthogonal branching of actin filaments and links actin filaments to membrane glycoproteins.

Cloning of human endothelial cell ABP cDNA predicts a polypeptide subunit chain of 2647 amino acids, corresponding to 280 kDa, the mass derived from physical measurements of the native protein[1]. ABP-280 molecules are homodimers that appear as thin strands having contour lengths of ~160 nm in electron micrographs (Figure). The important functional domains of each subunit include an F-**actin** binding region, a self-association domain, and a membrane glycoprotein binding region. An N-terminal actin binding domain comprises 274 amino acids and is similar to the N-termini of α-**actinin**, β-**spectrin**, **dystrophin**, and **ABP-120**. The remaining 90% of the sequence comprises 24 repeats with cross-β-structure. Each repeat motif in ABP-280 is ~96 residues long, and is predicted to have six to eight runs of six to nine residues with antiparallel β-sheet secondary structure alternating with three to four amino acids with high turn potential. These runs of β-structure form a planar repeat that has hydrophobic and hydrophilic faces because residues forming the β-runs alternate polar and nonpolar side chains.

The first 23 repeats of each subunit interact by overlapping hydrophobic faces with its neighbours in a staggered fashion to yield a backbone with the proper dimensions. ABP molecules dimerize at the extreme C-terminus of each chain as documented by genetic deletion experiments where removing 65 residues from the C-terminus of ABP-280 inhibits dimer formation. An insertion just proximal to the final repeat may contain a membrane glycoprotein binding site.

Evidence for tissue specific isoforms includes changing 2-D peptide maps during myocyte differentiation[2] and discrepancies between amino acid sequence derived from endothelial cell cDNA and direct sequencing of the human uterine (smooth muscle) protein analog[3].

■ PURIFICATION

ABP-280 has been purified from macrophages[4], neutrophils, platelets, smooth muscle, hog thyroid, and toad oocytes[5] by gel filtration through Bio-Gel A15, 200-400

Mesh (Bio-Rad) followed by ion exchange chromatography on DEAE-Sepharose (Pharmacia). All buffers should include EGTA at millimolar levels to inhibit **calpain**.

■ ACTIVITIES

ABP-280 crosslinks actin filaments into orthogonal networks in the cortical cytoplasm of cells[6]. In platelets it links the glycoprotein Ib/IX complex to actin filaments[7-9] and in myeloid cells, it binds directly to the high affinity Fc receptor (FcgRI, CD64)[10]. The importance of ABP-280 for cell motility has been established by genetic rescue[11].

■ ANTIBODIES

Four monoclonal antibodies against human endothelial cell ABP have been prepared and characterized[1].

■ GENES

The complete cDNA sequence is available for human endothelial cell actin binding protein (GenBank 53416)[1]. The gene has been localized to the X-chromosome Xq2.4-Xqter.

■ REFERENCES

1. Gorlin, J., Yamin, R., Egan, S., Stewart, M., Stossel, T., Kwiatkowski, D. and Hartwig, J. (1990) J. Cell Biol. 111, 1089-1105.
2. Gomer, R.H. and Lazerides, E. (1983) J. Cell Biol. 96, 321-329.
3. Hock, R., Davis, G. and Speicher, D. (1990) Biochem. 29, 9441-9451.
4. Hartwig, J. and Stossel, T. (1981) J. Mol. Biol. 145, 563-581.
5. Corwin, H.L. and Hartwig, J.H. (1983) Devel. Biol. 99, 61-74.
6. Hartwig, J. and Shevlin, P. (1986) J. Cell Biol. 103, 1007-1020.

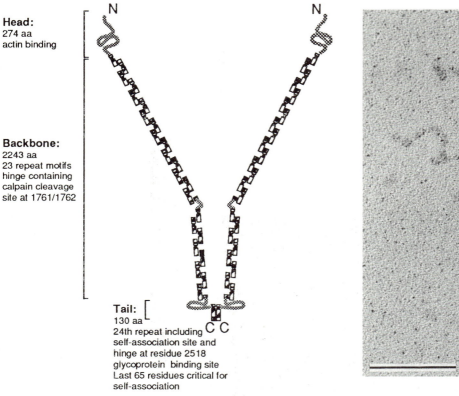

Head:
274 aa
actin binding

Backbone:
2243 aa
23 repeat motifs
hinge containing
calpain cleavage
site at 1761/1762

Tail:
130 aa
24th repeat including
self-association site and
hinge at residue 2518
glycoprotein binding site
Last 65 residues critical for
self-association

Figure. Schematic model of the predicted dimeric structure of the ABP molecule. Each monomer chain contains three domains: head, backbone, and tail. Each head contains an actin filament binding site within 274 residues of the N-terminus. The backbone contains repeat blocks of 96 amino acids composed of the last half of each sequence repeat and the beginning of the following sequence repeat. Repeats associate through hydrophobic intrachain bonding until repeat 16 which is preceded by a 25 residue insertion that contains a calpain cleavage site. This region may act as a flexible hinge allowing movement of ABP's heads to accommodate actin binding. Intrarepeat bonding continues in motifs 16 to 23 and completes the backbone of the subunit. The tail region (motif 24) is offset by a large (~35 amino acid) insertion and the repeat abruptly terminates before the end of a functional repeat, thereby disrupting further intrachain bonding. The unpaired hydrophobic surfaces of motif 24 from two ABP subunits therefore are free to self-associate. (Right) Human uterus ABP molecules after glycerol spraying, vacuum drying, and metal coating with tantalum-tungsten at 5°. Note the minimal region of monomer overlap. Bar 0.1 μm.

7. Ezzell, R., Kenney, D., Egan, S., Stossel, T. and Hartwig, J. (1988) J. Biol. Chem. 263, 13303-13309.
8. Fox, J. (1985) J. Biol. Chem. 260, 11970-11977.
9. Okita, L., Pidard, D., Newman, P., Montogomery, R. and Kunicki, T. (1985) J. Cell Biol. 100, 317-321.
10. Ohta, Y., Stossel, T.P. and Hartwig, J.H. (1991) Cell 67, 275-282.
11. Cunningham, C., Gorlin, J., Kwiatkowski, D., Hartwig, J., Janmey, P., Byers, R. and Stossel, T. (1992) Science 255, 325-327.

■ Jed Gorlin and John Hartwig:
Experimental Medicine Division,
Brigham and Women's Hospital
Boston, MA, USA

Actin Depolymerizing Factor (ADF)

Actin depolymerizing factor (ADF), a 19 kDa protein, was originally identified in embryonic chick brain based upon its ability to depolymerize rapidly filamentous actin and sequester actin monomers in a 1:1 complex[1]. ADF expression is developmentally regulated in muscle where ADF levels decline as muscle specific genes are activated[2]. Found distributed throughout the cytoplasm of cells[2,3], ADF is presumed to play a role in the delivery of actin to its site of assembly (Figure).

Multiple charge isoforms of ADF have been detected by 2-D gel analysis of extracts of tissues and cultured cells[4]. Purified chick brain ADF gives rise to two charged species on 2-D gels differing in apparent pI by 0.3 to 0.5 pH units. The major isoform has a pI of about 7.9. The second isoform is probably an artifact of the electrophoretic method, since purified bacterially expressed ADF[5] also gives rise to a second spot with similar charge characteristics. An isoform of immunoreactive ADF with a pI of about 6.5 has been isolated from embryonic brain and cultured cells. This form of ADF is inactive in depolymerizing **actin** and contains phosphate[6]; therefore, it represents a post-translationally inactivated form of ADF. Two mRNAs of about 900 and 2100 bases encoding ADF have been identified in chick tissues[5,7]. Sequence analysis of their respective cDNAs suggest that the smaller mRNA differs from the larger one only in that an alternative polyadenylation site is used. Southern blot analysis of chicken genomic DNA suggests ADF is encoded by a single gene[5,7].

ADF contains seven cysteine residues, all in the reduced

A   B

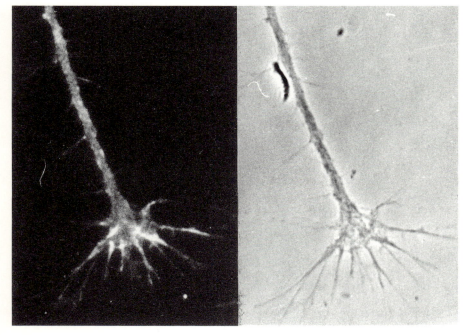

Figure: Immunofluorescent localization of ADF in growth cone and neurite fiber of embryonic chick dorsal root ganglion neuron cultured for 24 h on glass. (A) Indirect immunofluorescence using Texas-Red streptavidin. (B) Phase contrast micrograph of same growth cone.

form. Modification of more than one cysteine with sulfhydryl reactive reagents completely inhibits the ability of ADF to depolymerize F-actin[4]. Within the ADF-actin complex, only a single cysteine is available for modification. This cysteine residue can be crosslinked to cys-374 in actin by N,N'-paraphenylene-di-maleimide, but not by the orthoisomer or diamide, indicating the two cysteine residues are separated by 0.6 to 1.4 nm[4,8].

ADF is developmentally regulated in muscle, the levels declining from 0.2% of the total soluble protein in ten day embryonic chick muscle to undetectable amounts by two weeks posthatching[2], the period over which F-actin increases from 60% to over 98% of the total actin[9]. Cultured myocytes are unable to down-regulate ADF expression but instead they accumulate large amounts of the inactive, posttranslationally modified form of ADF[10]. Thus, myofibril assembly both *in vivo* and *in vitro* is temporally correlated with a decline in the molar ratio of active ADF to actin. ADF is a component of slow axonal transport in chicken sciatic nerve[11]. Quantitative studies on the radiolabelled proteins in slow component b showed labelled ADF present in close to a 1:1 molar ratio with labelled actin[11], suggesting that actin and ADF may be cotransported as a complex.

Proteins with similar actin depolymerizing activity to ADF include **depactin**[12,13] from echinoderms and actophorin[14] from *Acanthamoeba*. A protein (**cofilin**)[15-17] with over 70% sequence homology to ADF, but somewhat different activity, also occurs in avian and mammalian tissues. A region of sequence homology exists between all these proteins in the presumptive actin binding domain which contains the DAIKKK sequence that is also found in **tropomyosin**.

■ PURIFICATION

ADF can be purified from embryonic chick brain by subsequent ion exchange, gel filtration and dye matrix (Green A) chromatography[4,6]. Alternatively, ADF can be obtained by affinity chromatography of tissue extracts on DNase 1 agarose followed by conventional chromatographic fractionation of the eluted actin binding proteins[3,18].

■ ACTIVITIES

ADF induces rapid disassembly of F-actin through a weak Ca^{2+}-independent F-actin severing activity most readily detected in Mg^{2+}-containing low ionic strength buffers[6], but it does not cap or cosediment with F-actin[3,18,19]. ADF binds to ATP-actin monomers with a K_d of 0.1 µM and inhibits the exchange of the actin bound nucleotide[19]. Tropomyosin from both muscle[20] and brain[21] blocks the ability of ADF to depolymerize F-actin. A protein with similar activity and high sequence homology to ADF was subsequently isolated from mammalian tissues and named destrin[18,22].

■ ANTIBODIES

Rabbit polyclonal antisera against chick ADF[2,3] recognizes ADF in many vertebrates, especially mouse and rat[2], but do not crossreact with cofilin[6], a protein which shares over 70% sequence homology with ADF[7,23].

■ GENES

Complete cDNA sequences for chicken ADF (GenBank J02912, J02915)[5,7] and porcine destrin (GenBank J05290)[24] have been published. The chicken ADF cDNA coding region was used to screen a human cDNA library and the isolated cDNA clone had a coding sequence identical to porcine destrin[25]. The chicken ADF genomic sequence containing two large introns is currently being completed.

■ REFERENCES

1. Bamburg, J.R., Harris, H.E. and Weeds, A.G. (1980) FEBS Lett. 121, 178-182.
2. Bamburg, J.R. and Bray, D. (1987) J. Cell Biol. 105, 2817-2825.
3. Abe, H. and Obinata, T. (1989) J. Biochem. (Tokyo) 106, 172-180.
4. Giuliano, K.A., Khatib, F.A., Hayden, S.M., Daoud, E.W.R., Adams, M.E., Amorese, D.A., Bernstein, B.W. and Bamburg, J.R. (1988) Biochemistry 27, 8931-8938.
5. Adams, M.E., Minamide, L.S., Duester, G. and Bamburg, J.R. (1990) Biochemistry 29, 7414-7420.
6. Bamburg, J.R., Minamide, L.S., Morgan, T.E., Hayden, S.M., Giuliano, K.A. and Koffer, A. (1991) Methods Enzymol. 196, 125-140.
7. Abe, H., Endo, T., Yamamoto, K. and Obinata, T. (1990) Biochemistry 29, 7420-7425.
8. Daoud, E.W., Hayden, S.M. and Bamburg, J.R. (1988) Biochem. Biophys. Res. Commun. 155, 890-894.
9. Shimizu, N. and Obinata, T. (1986) J. Biochem. (Tokyo) 99, 751-759.
10. Morgan, T.E. (1990) Ph.D. Thesis, Colorado State University, Fort Collins, CO.
11. Bray, J.J., Fernyhough, P., Bamburg, J.R. and Bray, D. (1992) J. Neurochem. 58, 2081-2089.
12. Mabuchi, I. (1981) J. Biochem. (Tokyo) 89, 1341-1344.
13. Takagi, T., Konishi, K. and Mabuchi, I. (1988) J. Biol. Chem. 263, 3097-3102.
14. Cooper, J.A., Blum, J.D., Williams Jr., R.C. and Pollard, T.D. (1986) J. Biol. Chem. 261, 477-485.
15. Nishida, E., Maekawa, S. and Sakai, H. (1984) Biochemistry 23, 5307-5313.
16. Yonezawa, N., Nishida, E. and Sakai, H. (1985) J. Biol. Chem. 260, 14410-14412.
17. Abe, H., Ohshima, S. and Obinata, T. (1989) J. Biochem. (Tokyo) 106, 696-702.
18. Nishida, E., Maekawa, S., Muneyuki, E. and Sakai, H. (1984) J. Biochem. (Tokyo) 95, 387-398.
19. Hayden, S.M. (1988) Ph.D. Thesis, Colorado State University, Fort Collins, CO.
20. Bernstein, B.W. and Bamburg, J.R. (1982) Cell Motil. Cytoskel. 2, 1-8.
21. Bamburg, J.R. and Bernstein, B.W. (1990) In: "The Neuronal Cytoskeleton", R.D. Burgoyne, ed., Wiley-Liss, Inc. NY. pp. 121-159.
22. Nishida, E., Muneyuki, E., Maekawa, S., Ohta, Y. and Sakai, H. (1985) Biochemistry 24, 6624-6630.

23. Matsuzaki, F., Matsumoto, S., Yahara, I., Yonezawa, N., Nishida, E. and Sakai, H.(1988) J. Biol. Chem. 263, 11564-11568.
24. Moriyama, K., Nishida, E., Yonezawa, N., Sakai, H., Matsumoto, S., Iida, K. and Yahara, I. (1990) J. Biol. Chem. 265, 5768-5773.
25. Hawkins, M. and Weeds, A.G. (1992) submitted.

■ James R. Bamburg:
Department of Biochemistry,
Colorado State University,
Fort Collins, CO 80523, USA

α-Actinins

α-Actinin is an F-actin binding and crosslinking protein originally isolated from skeletal muscle, where it is a major component of the Z-disc. Distinct isoforms of the protein are found in smooth muscle and nonmuscle cells. α-Actinin may function as an actin bundling protein and/or a linking protein attaching actin filaments to a variety of intracellular structures.

α-Actinin is a homodimer with a subunit molecular weight of 94-103 kDa in which the subunits are antiparallel in orientation. It is visualised in the electron microscope as a rod-shaped molecule with dimensions of 3-4 nm by 30-40 nm. It has a sedimentation coefficient of 6 S, and CD data suggest a high α-helical content (62-74%) with very little β-structure[1]. The molecule (Figure) can be divided into three domains, an N-terminal **actin** binding domain (approximately residues 1-245), four internal 120 residue repeats, and a C-terminal region containing two EF-hand Ca^{2+}-binding motifs[1,2]. The actin binding domain is homologous to that found in β-**spectrin**[3], **dystrophin**[1], **filamin**[4], **fimbrin**[5] and **ABP-120** (*Dictyostelium discoideum* gelation factor)[1], and the repeats are homologous to those found in spectrin and dystrophin[1]. Apart from actin, α-actinin has been reported to bind to the cytoskeletal proteins **vinculin**[6], **nebulin**[7], **clathrin**[8], and to the cytoplasmic domain of the β1-family of *integrins*, receptors for extracellular matrix proteins[9].

Chick smooth and nonmuscle isoforms of α-actinin are identical except for a region of 27 amino acids covering the C-terminal region of the first EF-hand, and arise by alternative splicing of the primary transcript of a single gene[10]. This difference may account for the fact that binding of the nonmuscle isoform of α-actinin to actin is Ca^{2+}-sensitive whereas binding of the smooth muscle isoform is Ca^{2+}-insensitive[11]. Chick pectoralis skeletal muscle α-actinin shows about an 80% sequence similarity to the smooth and nonmuscle isoforms, and is encoded by a separate gene[10]. It is likely that there are several skeletal muscle isoforms. The structure of α-actinin expressed in cardiac muscle has not been determined.

In smooth muscle, α-actinin is localized in membrane associated dense plaques and cytoplasmic dense bodies, whilst in cardiac tissue, it is found in the fascia adherens of the intercalated discs as well as in Z-discs. In nonmuscle cells, α-actinin is localized in adherens-type cell-cell and cell-extracellular matrix junctions, and is distributed along the actin filaments found in cultured fibroblasts. Here, the distribution is antiperiodic with respect to **tropomyosin**[1].

In Nemeline myopathy, α-actinin is found as a component of Nemeline rods which are deposited in the skeletal muscle cell[12]. However, there is no evidence that the defect resides in the α-actinin gene.

■ PURIFICATION

α-Actinin can be readily purified from smooth and skeletal muscle by standard procedures[13], 100 g of tissue yielding 10-20 mg of protein. Yields of the protein from nonmuscle tissues are much lower (~1 mg/100 g).

■ ACTIVITIES

The actin binding and crosslinking properties of α-actinin have been demonstrated by a variety of methods including sedimentation, viscometry[11] and electron microscopy[1].

■ ANTIBODIES

Rabbit antibodies to chick skeletal and smooth muscle α-actinin have been produced[14], and antibodies to the latter are commercially available (Sigma Chem. Co.). The antibodies are crossreactive but can be rendered isoform specific by adsorption. They also crossreact with mammalian α-actinins as detected by indirect immunofluorescence. Monoclonal antibodies to *D. discoideum* α-actinin[15] and chicken smooth muscle α-actinin[16] have been reported, and Serotec (UK) market a monoclonal anti-α-actinin antibody (BM 75.2).

■ GENES

cDNAs encoding skeletal (chick[10], *Drosophila*[17], EMBL X51753), smooth muscle (chick[2], GenBank J03486) and nonmuscle (chick[10], human[18] EMBL X15804, *D. discoideum*[19] GenBank Y00689) α-actinin have been sequenced. In humans, the gene encoding the nonmuscle isoform is on chromosome 14 close to the gene for β-spectrin[20], GenBank M31300.

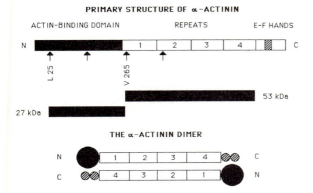

PRIMARY STRUCTURE OF α-ACTININ

Figure: Structure of α-actinin. The diagram shows the relative positions of the actin binding domain (1-245), the spectrin-like repeats (246-712), and the two EF-hand sequences (750-814) within the primary sequence of chick smooth muscle α-actinin (887 amino acids)[2]. The positions of the two predominant thermolysin cleavage sites which have been defined are indicated, along with two additional less susceptible sites (arrows). Thermolysin digestion of α-actinin gives rise to a monomeric 27 kDa fragment capable of binding actin, and a 53 kDa fragment containing the spectrin-like repeats, which is dimeric, and accounts for the rod shape of α-actinin. The arrangement of the α-actinin subunits in an antiparallel orientation is also illustrated.

■ REFERENCES

1. Blanchard, A., Ohanion, V.O. and Critchley, D.R. (1989) J. Muscle Res. and Cell Motil. 10, 280-289.
2. Baron, M.D., Davison, M.D., Jones, P. and Critchley, D.R. (1987) J. Biol. Chem. 262, 17623-17629.
3. Byers, T.J., Husain-Christi, A., Dubreuil, R.R., Branton, D. and Goldstein, L.S.B. (1989) J. Cell Biol. 109, 1633-1641.
4. Gorlin, J.B., Yamin, R., Egan, S., Stewart, M., Stossel, T.P., Kwiatkowski, D.J. and Hartwig, J.H. (1990) J. Cell Biol. 1089-1105.
5. deArruda, M.V., Watson, S., Lin, C.-S., Leavitt, J. and Matsudaira, P. (1990) J. Cell Biol. 111, 1069-1079.
6. Wacchstock, D.H., Wilkins, J.A. and Lin, S. (1987) Biochem. Biophys. Res. Commun. 146, 554-560.
7. Nave, R., Furst, D.O. and Weber, K. (1990) Febs Lett. 269, 163-166.
8. Merisko, E.M., Welch, J.K., Chen, T.-Y. and Chen, M. (1988) J. Biol. Chem. 263, 15705-15712.
9. Otey, C.A., Pavalko, F.M. and Burridge, K. (1990) J. Cell Biol. 111, 721-729.
10. Arimura, C., Suzuki, T., Yanagisawa, M., Imamura, M., Hamada, Y. and Masaki, T. (1988) Eur. J. Biochem. 177, 649-655.
11. Duhaiman, A.S. and Bamburg, J.R. (1984) Biochemistry 23, 1600-1608.
12. Hashimoto, K., Shimizu, T., Nonaka, I. and Mannen, T. (1989) J. of the Neurological Sciences 93, 199-209.
13. O'Halloran, T., Molony, L. and Burridge, K. (1986) Methods Enzymol. 134, 69-77.
14. Endo, T. and Masaki, T. (1984) J. Cell Biol. 99, 2322-2332.
15. Schleicher, M., Noegel, A., Schwarz, T., Wallraff, E., Brink, M., Faix, J., Gerisch, G. and Isenberg, G. (1988) J. Cell Sci. 90, 59-71.
16. Imamura, M., Endo, T., Kuroda, M., Tanaka, T. and Masaki, T. (1988) J. Biol. Chem. 263, 7800-7805.
17. Fyrberg, E., Kelly, M., Ball, E., Fyrberg, C. and Reedy, M.C. (1990) J. Cell Biol. 110, 1999-2011.
18. Millake, D.B., Blanchard, A.D., Patel, B. and Critchley, D.R. (1989) Nucleic Acids Rs. 17, 6725.
19. Noegel, A., Witke, W. and Schleicher, M. (1987) FEBS Lett. 221, 391-396.
20. Youssoufian, H., McAfee, M. and Kwiatkowski, D.J. (1990) Am. J. Hum. Genet. 47, 62-72.

■ Dr. David R. Critchley,
Department of Biochemistry,
University of Leicester,
Leicester LEI 7RH, UK

Actobindin

Actobindin from Acanthamoeba castellanii is a potent inhibitor of the polymerization of actin under certain conditions. Although the mechanism of action remains experimentally unproven, the marked delay in actin polymerization seen in the presence of actobindin can best be explained by assuming that, in addition to binding to actin monomers, actobindin can bind to actin oligomers, effectively preventing the oligomers from serving as nuclei for filament elongation.

Actobindin is a 9.7 kDa protein purified from the protozoan, *Acanthamoeba castellanii*, originally by the technique described by Reichstein and Korn for the purification of **profilin**[1-3]. The entire 88 amino acid sequence of actobindin has been determined and is most notable for a nearly identically repeated segment of approximately 33 amino acids (Figure 1)[4]. Although the sequence predicts the presence of extensive α-helical portions, the CD spectrum of actobindin in aqueous solution shows little α-helix or β-sheet[4]. As isolated, actobindin appears to have a loosely folded structure as indicted by its considerable flexibility in 2-D NMR and fluorescence lifetime experiments, and its sensitivity to proteolysis[5]. The 33-residue repeated sequences of actobindin contain regions of similarity to an 11 amino acid peptide near the N-terminus of *Acanthamoeba* profilin, an 11 amino acid

```
          1                  10                  20                  30            m      40
Ac-M N P E L Q S A I G Q G A A L K H A E T V D K S A P Q I E N V T V K K V D R S S F L E E
                  50                  60                  70       m            80                  88
V A K P H E L K H A E T V D K S G P A I P E D V H V K K V D R G A F L S E I E K A A K Q
```

Figure 1. The amino acid sequence of actobindin. Residues 15-47 and 51-84 in italics are almost identical except a proline is inserted at position 65. The underlined hexapeptides are common to several other actin binding proteins.

peptide surrounding a trimethyllysine residue in elongation factor 1a (**actin bonding protein 50**), and a hexapeptide (residues 15-20 and 51-56) common to several other **actin** binding proteins, including **myosin** heavy chain, α-**actinin**, **tropomyosin**, and **paramyosin**[4].

■ PURIFICATION

The purification of actobindin has been recently reviewed in detail[2]. Conventional chromatographic techniques result in a highly purified protein that is stable in storage after lyophilization or rapid freezing. When very pure, actobindin has little absorbance at 280 nm, and thus the fine structure of phenylalanine (the only aromatic residues in actobindin are two phenylalanines) is readily apparent (Figure 2). The molar extinction coefficient of actobindin is ~6.5 x 10^2 M^{-1} cm^{-1} at 258 nm[2].

■ ACTIVITIES

Quantification of the decrement in F-actin concentration caused by actobindin at steady-state, as determined by either light scattering or fluorescence techniques, suggests that actobindin sequesters monomeric actin with an apparent K_d of 3-8 μM, assuming a 1:1 molar complex[1]. Even though the rate of formation of elongating actin oligomers depends critically on the concentration of available G-actin (to the third or fourth power), the affinity of actobindin for G-actin is too low to explain the ability of actobindin to inhibit the rate of actin polymerization[6]. Either actobindin must bind more actin than the steady-state experiments suggest or actobindin must interfere with an early stage of polymerization.

Covalent crosslinking experiments suggest that actobindin might have two binding sites for actin. 1-ethyl-3-(3-dimethylaminopropyl)-carbodiimide crosslinks Lys-16 and Lys-52 of actobindin to both the N-terminus of actin and the spatially proximate Glu-100[7,8]. These lysine residues are located at corresponding positions in the repeated segments of actobindin within the hexapeptide that is similar to sequences in other actin binding proteins, as discussed above[7]. *In vitro*, actobindin binds two actin monomers with a slight degree of negative cooperativity[3]. The first actin monomer binds to actobindin with a K_d of about 3.3 μM, and the binding of the second actin monomer is five to eight times weaker[3].

The presence of two binding sites for actin could make it possible for actobindin to bind actin oligomers very tightly. If the two binding sites of actobindin could simultaneously interact with a single actin oligomer, the affinity of actobindin for actin oligomers would be expected to be many times greater (probably more than 1000-fold) than for actin monomers[9]. We believe that by tightly binding to actin oligomers, possibly at the stage of an actin dimer, actobindin is able to sequester actin filament nuclei and thus delay filament elongation.

■ ANTIBODIES

Polyclonal antibodies prepared in this laboratory are currently being used in an attempt to immunolocalize actobindin.

■ GENES

No clones are available.

■ REFERENCES

1. Lambooy, P.K. and Korn, E.D. (1986) J. Biol. Chem. 261, 17150-17155.
2. Bubb, M. R. and Korn, E. D. (1990) Meth. Enzymol. 196, 119-125.
3. Bubb, M.R., Lewis, M.S. and Korn, E.D. (1991) J. Biol. Chem. 266, 3820-3826.
4. Vandekerckhove, J., Van Damme, J., Vancompernolle, K., Bubb, M.R., Lambooy, P.K. and Korn, E.D. (1990) J. Biol. Chem. 265, 12801-12805.
5. Bubb, M.R., Knutson, J.R., Bax, A. and Korn, E.D. (1990) unpublished observations.
6. Lambooy, P.K. and Korn, E.D. (1988) J. Biol. Chem. 263, 12836-12843.
7. Vancompernolle, K. Vanderkerckhove, J., Bubb, M.R. and Korn, E.D. (1991) J. Biol. Chem. 266, 15427-15431.
8. Kabsch, W., Mannherz, H.G., Suck, D., Pai, E.F. and Holmes, K.C. (1990) Nature 347, 37-44.
9. Chateliar, R.C. (1987) Biophys. Chem. 28, 121-129.

■ *Michael R. Bubb and Edward D. Korn:*
Laboratory of Cell Biology, National Heart Lung and Blood Institute,
National Institutes of Health,
Bethesda, MD, USA

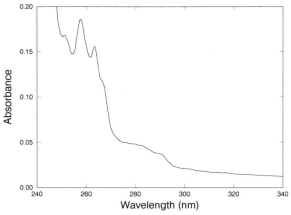

Figure 2. The UV absorption spectrum of 280 μM actobindin in 5.0 mM Tris, pH 8.0.

Actolinkin

Actolinkin[1] is found in echinoderm eggs and seems to anchor actin filaments at their barbed ends to the inner surface of the plasma membrane .

Actolinkin from sea urchin eggs and starfish oocytes is a monomeric protein that migrates with an apparent molecular weight of 20,000 on SDS-PAGE and has a Stokes radius of 2.4 nm. Its pI is 5.0. The majority of actolinkin in

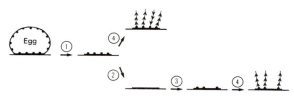

Figure. An experiment which shows the possible role of actolinkin. Step 1, Application of a jet stream of an appropriate buffer solution to remove cytoplasm of an unfertilized sea urchin egg which is attached to a protamine coated glass surface. Step 2, Extraction with a 0.6 M KCl solution. Step 3, Addition of actolinkin-actin complex. Step 4, Addition of G-actin under actin polymerizing conditions, followed by addition of **myosin** subfragment 1 to visualize the polarity of the actin filaments.

the egg is considered to exist in a 1:1 complex with **actin**. This actin-actolinkin complex is very stable and is not dissociated by 4 M urea, 1 M KCl, or 0.6 M KI, but can be dissociated by 7 M urea. Actolinkin binds to an N-terminal segment of actin. Both the actolinkin-actin complex and free actolinkin bind to the barbed end of the actin filament and block actin monomer addition at that end.

Actolinkin localizes exclusively in the cortical fraction of the sea urchin egg[1]. After actolinkin is extracted with 0.6 M KCl, the cortex loses its ability to induce actin polymerization. Upon addition of actolinkin-actin complex to the extracted cortex, the ability to polymerize actin is restored[1] (Figure). All actin filaments are attached to such a reconstituted cortex by their barbed ends. Thus, actolinkin is believed to play a role in anchoring actin filaments to the inner surface of the plasma membrane.

■ PURIFICATION

The actolinkin-actin complex is extracted from an insoluble fraction obtained from sea urchin eggs using a 0.6 M KCl containing physiological buffer[2]. The complex is then purified by ammonium sulphate precipitation, Sephacryl S-300 gel filtration, DNaseI affinity and hydroxylapatite column chromatography. Actolinkin was separated from

actin by gel filtration in the presence of 7 M urea. The activity of actolinkin-actin complex during the course of purification can be assayed by low shear viscometry[3].

■ ACTIVITIES

Since the actolinkin-actin complex or free actolinkin is an actin barbed end capping protein, the most appropriate method to detect its activity is the elongation-inhibition assay using acrosomal actin bundles as seeds for polymerization of actin[4-6]. The activity can also be measured by high-shear viscometry[2], light scattering[2], absorbance at 234 nm[2], or fluorescence measurement of pyrene-labelled actin[2]. The actolinkin-actin complex accelerates the rate of polymerization of actin, increases the critical concentration of actin for polymerization and inhibits reannealing of fragmented F-actin at substoichiometric amounts and independent of the Ca^{2+} concentration. Free actolinkin does not accelerate the polymerization of actin.

■ ANTIBODIES

Polyclonal antibodies against *Hemicentrotus pulcherrimus* egg actolinkin were raised in a mouse and used for localization of actolinkin in the egg cortex[1]. The antibodies detected actolinkin in eggs of other species of sea urchins and oocytes of a starfish[2].

■ GENES

Actolinkin has not yet been cloned and sequenced.

■ REFERENCES

1. Ishidate, S. and Mabuchi, I. (1988) Eur. J. Biochem. 46, 275-281.
2. Ishidate, S. and Mabuchi, I. (1988) J. Biochem. 104, 72-80.
3. MacLean-Fletcher, S.D. and Pollard, T.D. (1980) J. Cell Biol. 85, 414-421.
4. Bonder, E.M. and Mooseker, M.S. (1983) J. Cell Biol. 96, 1097-1107.
5. Mabuchi, I. (1983) J. Biochem. 94, 1349-1352.
6. Hosoya, H. and Mabuchi, I. (1984) J. Cell Biol. 99, 994-1001.

■ *Issei Mabuchi:*
Department of Biology,
College of Arts and Sciences,
University of Tokyo,
Tokyo, Japan

Annexins

The annexins are a family of Ca^{2+}-dependent phospholipid binding proteins. They are differentially expressed and demonstrate unique subcellular localizations. These proteins may be mediators of the intracellular calcium signal.

Several groups have characterized proteins of ~35 kDa which bind acidic phospholipids in a Ca^{2+}-dependent manner. The proteins were identified by Ca^{2+}-dependent binding to chromaffin granule membranes[1,2], to synaptosomes prepared from *Torpedo* electric organ[3], and to detergent-insoluble membrane fractions[4,5]. The p35 substrate for the EGF receptor kinase was also found to associate with membranes in a Ca^{2+}-dependent manner[6]. By the use of a different approach, several proteins in addition to calmodulin were isolated by Ca^{2+}-dependent binding to phenothiazine resins[7]. More recently, several laboratories have identified protein fractions that inhibited phospholipase A2 and blood coagulation[8]. Sequence data demonstrated identities between these protein groups[9]. Independent identifications led to different names for these proteins, including synexin, calelectrins, chromobindins, calcimedins, calpactins, and lipocortins. However, the term "annexin," suggested by Geisow[10], is widely accepted[11]. Sequence data indicate that there are eight unique proteins in the mammalian annexin family (see Table I). The proteins are also present in plants[12] and invertebrates[13].

The sequence organization of the family is highly conserved. Annexins I-V and VII are comprised of four repeated domains, whereas annexin VI is comprised of eight domains; each domain is 70 amino acids in length. The sequence conservation for each domain ranges between 40% and 60% when individual family members are compared. The N-terminus of each protein is unique, suggesting that this region may confer functional differences to the proteins (Figure 1).

The annexins are differentially expressed in tissues (Figure 2). The unique subcellular distribution of individual annexins further suggests that they regulate specified cellular functions (Figure 3). Burns et al.[14] have shown that annexin VII, *in vitro*, forms voltage dependent Ca^{2+} channels; Ross et al.[15] have reported annexin III to hydrolyze inositol 1,2-cyclic phosphate; Ali and Burgoyne[16] have demonstrated a stimulation of catecholamine secretion with annexin II; and Diaz-Munoz et al.[17] have demonstrated a marked Ca^{2+}-dependent alteration in the gating properties of the sarcoplasmic reticulum Ca^{2+}-release channel by annexin VI.

The distinguishing features of the annexin family are the ability to bind phospholipids and the ability to bind Ca^{2+} without using the classical EF-hand motif found in the calmodulin/**troponin C** family. Both Ca^{2+} and phospholipids may modulate the function of these proteins and provide an independent mechanism for the mediation of intracellular calcium[18].

■ PURIFICATION

The purification of the annexins is based on their Ca^{2+}-

TABLE 1

Calcium/phospholipid-binding proteins			
Annexin	I	II	III	IV
Previous terminology	Lipocortin I p35 Calpactin II Chromobindin 9 GIF	Calpactin I Lipocortin II p36 Chromobindin 8 Protein I PAP-IV	Lipocortin III PAP-III 35-α Calcimedin	Endonexin I Protein II 32.5 Calelectrin Lipocortin IV Chromobindin 4 PAP-II PP4-X 35-β Calcimedin
Annexin	V	VI	VII	VIII
Previous terminology	PAP-I IBC Lipocortin V 35K Calelectrin Endonexin II PP4 VAC-α 35-γ Calcimedin Calphobindin I Anchorin CII	p68,p70,73K 67K Calelectrin Lipocortin VI Protein III Chromobindin 20 67K Calcimedin Calphobindin II	Synexin	VAC-β

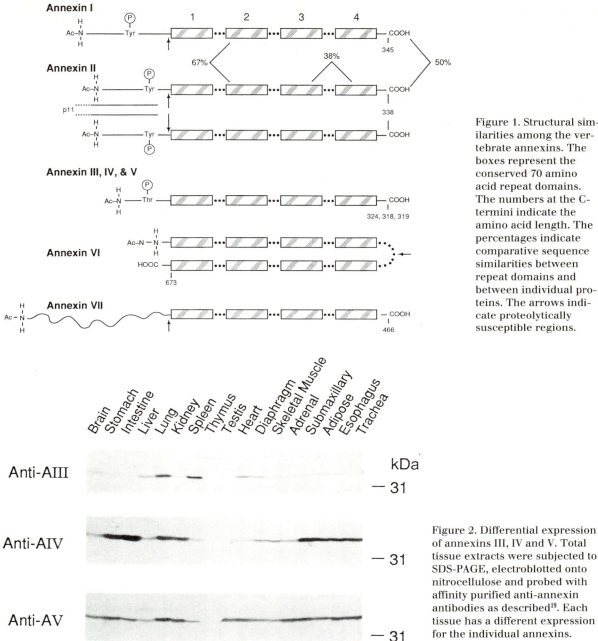

Annexin I

Annexin II

p11

Annexin III, IV, & V

324, 318, 319

Annexin VI

673

Annexin VII

466

67%   38%   50%

Figure 1. Structural similarities among the vertebrate annexins. The boxes represent the conserved 70 amino acid repeat domains. The numbers at the C-termini indicate the amino acid length. The percentages indicate comparative sequence similarities between repeat domains and between individual proteins. The arrows indicate proteolytically susceptible regions.

Brain, Stomach, Intestine, Liver, Lung, Kidney, Spleen, Thymus, Testis, Heart, Diaphragm, Skeletal Muscle, Adrenal, Submaxillary, Adipose, Esophagus, Trachea

Anti-AIII

Anti-AIV

Anti-AV

kDa
— 31

— 31

— 31

Figure 2. Differential expression of annexins III, IV and V. Total tissue extracts were subjected to SDS-PAGE, electroblotted onto nitrocellulose and probed with affinity purified anti-annexin antibodies as described[19]. Each tissue has a different expression for the individual annexins.

dependent association with phospholipids and their individual charge properties. Annexins in tissue extracts can be separated by phenyl-Sepharose and Mono-Q FPLC chromatography[19,20].

ties include inhibition of phospholipase A_2 and blood coagulation[8], hydrolysis of inositol 1,2-cyclic phosphate[15] and gating of Ca^{2+}-release channel[17].

■ ACTIVITIES

The primary biochemical identity of the annexins is Ca^{2+}-dependent binding to acidic phospholipids. Other activi-

■ ANTIBODIES

Most of the laboratories listed in the references have produced poly- or monoclonal antibodies. Kaetzel and Dedman (Univ. Texas Medical School, Houston, FAX 713-

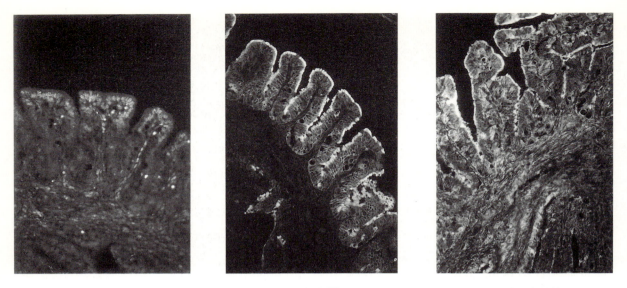

| Anti-AIII | Anti-AIV | Anti-AV |

Figure 3. Differential localization of annexins III, IV and V in rat fallopian tube. Tissue was fixed in 10% formalin, paraffin embedded and sectioned at 4 µm. The deparaffinated sections were incubated with affinity purified anti-annexin antibodies[19]. Each annexin has a unique localization pattern.

794-1349) have developed antibodies to annexins I-VI and have a site-directed consensus antibody which recognizes all annexins[21].

■ GENES

The cDNA sequences of all of the annexins from several species have been submitted to GenBank. Many are indexed under previous terminologies (see Table I).

■ REFERENCES

1. Creutz, C.E. (1981) Biochem. Biophys. Res. Comm. 103, 1395-1400.
2. Geisow, M.J. and Burgoyne, R.D. (1982) J. Neurochem. 38 1735-1741.
3. Walker, J.H. (1982) J. Neurochem. 39, 815-823.
4. Davies, A.A. Wigglesworth, N.M., Allan, D. and Crumpton, M.J. (1981) Biochem. Soc. Trans. 9, 565-566.
5. Gerke, V. and Weber, K. (1984) EMBO J. 3, 277-233.
6. Fava, R.A. and Cohen, S. (1984) J. Biol. Chem. 259, 2636-2645.
7. Moore, P.B. and Dedman, J.R. (1982) J. Biol. Chem. 257, 9663-9667.
8. Hauptman, R., Maurer-Fogy, I., Krystek, E., Bodo, G., Andree, H. and Reutelingsperger, C.P.M. (1989) Eur. J. Biochem. 185, 63-71.
9. Crompton, M.R., Moss, S.E. and Crumpton, M.J. (1988) Cell 55, 1-3.
10. Geisow, M.J. (1986) FEBS Lett 203, 99-103.
11. Crumpton, M.J. and Dedman, J.R. (1990) Nature 345, 212.
12. Boustead, C.M., Smallwood, M., Small, H., Bowles, D.J. and Walker, J.H. (1989) FEBS Lett. 244 456-460.
13. Johnson, P.A., Perin, M.S., Reynolds, G.A., Wasserman, S.A. and Sudhof, T.C. (1990) J. Biol. Chem. 265 11382-11388.
14. Burns, A.L., Magendzo, K., Shirvan, A., Arivastava, M., Rojas, E., Alijani, M.R. and Pollard, H.B. (1989) Proc. Natl. Acad. Sci. (USA) 86, 3798-3802.
15. Ross, T.S., Tait, J.F. and Majerus, P.W. (1990) Science 248, 605-607.
16. Ali, S.M. and Burgoyne, R.D. (1990) Cell Signalling 2, 265-272.
17. Diaz-Munoz, M., Hamilton, S.L., Kaetzel, M.A., Hazarika, P. and Dedman, J.R. (1990) J. Biol. Chem. 265, 15894-15899.
18. Smith, V.L., Kaetzel, M.A. and Dedman, J.R. (1990) Cell Regulation 1, 165-172.
19. Kaetzel, M.A., Hazarika, P. and Dedman, J.R. (1989) J. Biol. Chem. 264, 14463-14470
.20. Seaton, B.A., Head, J.F., Kaetzel, M.A. and Dedman, J.R. (1990) J. Biol. Chem 265, 4567-4569.
21. Kaetzel, M.A. and Dedman, J.R. (1989) Biochem. Biophys. Res. Commun. 3, 1233-1237.

■ *John R. Dedman and Marcia A. Kaetzel:*
Department of Physiology and Cell Biology,
Univ. Texas Medical School,
Houston, Texas 77225, USA

Caldesmon

Caldesmon[1,2] is an actin binding protein found in smooth muscle and many nonmuscle cells. Interactions with calmodulin[1,2], myosin[3-5], tropomyosin[6-8], troponin C[9], and S100 proteins[10] have been described. The most remarkable property of caldesmon is an inhibitory effect on the ATPase activity of actomyosin which is reversible by calmodulin in the presence of Ca^{2+}. Because of this property caldesmon is thought to be involved in the regulation of smooth muscle contraction.

Smooth muscle caldesmon isolated from chicken gizzard migrates as a tight doublet with an apparent molecular weight of 140-142,000 on SDS gels; cDNA sequencing gives molecular weights of 86,974[11] and 88,743[12] for two gizzard isoforms. Hydrodynamic measurements, ~93 kDa, are in agreement with the lower values[13]. The protein sequence shows a long predicted central α-helical region containing a repeated 13 (or 15) amino acid motif; the two isoforms differ by one of these repeats. The number of repeats is 8-12 or more, depending on the degree of degeneracy allowed. The 60,174 daltons nonmuscle caldesmon is smaller than the muscle form[14]. Antibodies specific for N- and C-terminal peptides show both ends are conserved and CNBr peptide mapping[15] suggests a loss of the central region of the molecule; cDNA sequencing[14] of a nonmuscle caldesmon indicates deletion of 232 amino acids beginning near residue 200 and including the region of repeats. Initial EM observations[16] indicated smooth muscle caldesmon was highly asymmetric and quite flexible; recent rotary shadowed images show a more rigid structure in the central portion of the molecule consistent with a helical middle[17] (Figure 1). Hydrodynamic studies[13] suggest caldesmon is an extended, 75 nm long molecule in solution. Preliminary physical studies on the central region are consistent with a helical structure. The overall picture is of a rather extended, dumbbell shaped molecule.

The emerging overall picture from domain mapping is that the N-terminal head of the dumbell contains a **myosin** binding domain[18] and a weak calmodulin binding site[19], while the C-terminal head contains **actin**, calmodulin and **tropomyosin** binding domains[20-22]. The extended central region acts as a spacer between the heads and is dispensed with in the nonmuscle form. Figure 2 illustrates the overall model. Effort is being directed toward defining the precise sequence of the C-terminal binding domains. The result is an increasingly complicated picture with one calmodulin binding site flanked by two actin binding domains. The available evidence is consistent with the extreme C-terminal actin binding site, plus the calmodulin binding domain, being necessary for inhibition of the ATPase of actomyosin[23,24]. The precise boundaries of the binding domains and their organization is disputed[23-25]. We[24] have suggested that the C-terminal calmodulin binding domain lies between Val-629 and Ser-666 and the two actin binding segments, from Leu-597 to Val-629, and from Arg-711 to Pro-756. Ca^{2+}-calmodulin binding to the C-terminal head affects its binding properties, but it is not clear whether Ca^{2+}-calmodulin affects myosin binding to the N-terminal head. A putative tropomyosin binding site has been assigned tentatively, based on sequence similarity with **troponin T**, but recent experimental evidence brings this assignment into question[23,26,27].

How caldesmon inhibits myosin ATPase activity has been disputed. Marston[28] proposed that caldesmon inhibits the V_{max} of ATP hydrolysis without weakening the binding of the myosin to actin, while Hemric and Chalovich[4] have argued caldesmon can competitively displace the myosin-ATP complex from actin. Recent experiments[29] support the latter mechanism and demonstrate clearly that the myosin binding fragment is not required for inhibition. Smith and Marston[2] have reviewed the earlier functional studies and Marston[30] has reviewed the evidence for a role for caldesmon in "latch". Lehman et al.[31] have presented a detailed model for the organization of caldesmon and tropomyosin in native thin filaments. Nonmuscle caldesmon has been shown to be phosphorylated during mitosis, presumably by cdc2 kinase, resulting in its release from actin filaments[32].

■ PURIFICATION

Caldesmon can be purified by boiling tissue homogenates, followed by DEAE chromatography of a 30-50% ammonium sulphate fraction[33]. Affinity chromatography on calmodulin-Sepharose can be used as a final purification step.

■ ACTIVITIES

Caldesmon has been assayed principally through its binding properties: (1) interaction with calmodulin-Sepharose, (2) cosedimentation with F-actin, or F-actin-tropomyosin, and inhibition of this interaction by calmodulin, and (3) inhibition of the actin-activated ATPase activity of phosphorylated smooth muscle myosin and skeletal muscle myosin.

■ ANTIBODIES

Rabbit polyclonal antibodies have been used to localize nonmuscle caldesmon on stress fibres[34-36] and the large form in smooth muscle[37]. Murine monoclonal antibodies have been prepared against, and used to localize human platelet caldesmon[36] and are also available for several regions of smooth muscle caldesmon[38].

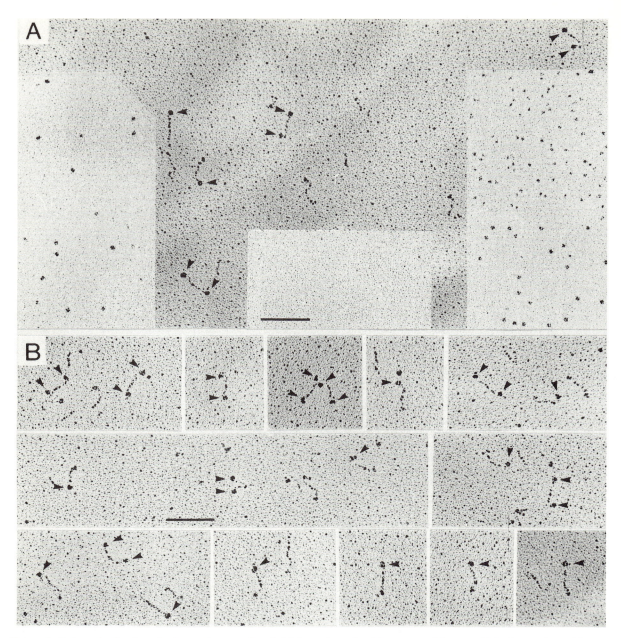

Figure 1. Rotary shadowing electron micrographs of chicken gizzard caldesmon and its crosslinked complex with calmodulin.

■ GENES

Two cDNA sequences have been reported for chicken gizzard caldesmon[11,12]. The GenBank name is CHKSMCA. The translated protein sequences differ by 15 amino acids, which includes one of the 13 amino acid repeat units, suggesting the two clones are two different molecular weight isoforms. The mass difference is about 1800 daltons, in good agreement with an apparent M_r difference of about 2000 on SDS gels. In addition, one nucleotide sequence[11] has about 1.5 kB of 3'-untranslated sequence absent from the other. The cDNA sequence[14] from a chicken nonmuscle caldesmon indicates identity except for a 232 amino acid deletion beginning at Val-200. We have sequenced the human smooth muscle and nonmuscle forms of caldesmon[39]. The N- and C-terminal binding domains are very similar to those in the chicken, the repeat sequence is largely preserved, but the number of

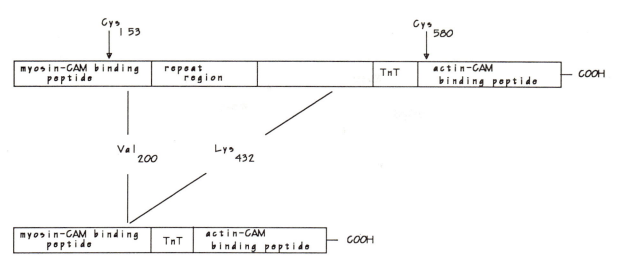

Figure 2. Binding domains and relationship of smooth muscle (upper) and nonmuscle (lower) caldesmons.

repeats is only 3-4. The i, i+4 motif in the central region is preserved predicting a helical character.

■ REFERENCES

1. Sobue, K., Muramoto, Y., Fujita, M. and Kakiuchi, S. (1981) Proc. Natl. Acad. Sci. (USA) 78, 5652-5655.
2. Marston, S.B. and Smith, C.W.J. (1985) J. Muscle Res. 6, 669-708.
3. Lash, J.A., Sellers, J.R. and Hathaway, D.R. (1986) J. Biol. Chem. 261, 16155-16160.
4. Hemric, M.E. and Chalovich, J.M. (1988) J. Biol. Chem. 263, 1878-1885.
5. Ikebe, M. and Reardon, S. (1988) J. Biol. Chem. 263, 3055-3058.
6. Graceffa, P. (1987) FEBS Lett. 218, 139-142.
7. Fujii, T., Ozawa, J., Ogama, Y. and Kondo, Y. (1988) J. Biochem. (Tokyo) 104, 734-737.
8. Horiuchi, K.-Y. and Chacko, S. (1988) Biochem. 27, 8388-8393.
9. Skripnokova, E.V. and Gusev, N.B. (1989) FEBS Lett. 257, 380-382.
10. Fujii, T., Machino, K., Andoh, H., Sato, H.T. and Kondo, Y. (1990) J. Biochem. (Tokyo) 107,133-137.
11. Bryan, J., Imai, M., Lee, R., Moore, P., Cook, R.G. and Lin, W.-G. (1989) J. Biol. Chem. 264, 13873-13879.
12. Hayashi, K., Kanda, K., Kimizuka, F., Kato, I. and Sobue, K. (1989) Biochem. Biophys. Res. Comm. 164, 503-511.
13. Graceffa, P., Wang, C.-L.A. and Stafford, W.F. (1988) J. Biol. Chem. 263, 14196-14202.
14. Bryan, J. Saavedra-Alanis, V., Wang, C.-L.A., Wang, L.-W. and Lu, R.C. (1990) J. Musc. Res. Cell Motil. 11, 434.
15. Ball, E.H. and Kovala, T. (1988) Biochem. 27, 6093-6098.
16. Furst, D.O., Cross, R.A., Mey, J.D. and Small, J.V. (1986) EMBO J. 5, 251-257.
17. Mabuchi, K. and Wang, C.-L.A. (1990) J. Musc. Res. Cell Motil.12, 145-151.
18. Riseman, F.M., Lynch, W.P., Nefsky, B. and Bretscher, A. (1989) J. Biol. Chem. 264, 2869-2875.
19. Wang, C.-L.A. (1988) Biochem. Biophys. Res. Commun. 156, 1033-1038.
20. Szpacenko, A. and Dabrowska, R. (1986) FEBS Lett. 202, 182-186.
21. Fujii, T., Imai, M., Rosenfeld, G.C. and Bryan, J. (1987) J. Biol. Chem. 262, 2757-2763.
22. Yazawa, M., Yagi, K. and Sobue, K. (1987) J. Biochem. (Tokyo) 102, 1065-1073.
23. Bartegi, A., Fattoum, A., Derancourt, J. and Kassab, R. (1990) J. Biol. Chem. 265, 15231-15238.
24. Wang, C.-L.A., Wang, L.-W.C., Xu, S., Lu, R.C., Saavedra-Alanis, V. and Bryan, J. (1990) J. Biol. Chem. 266, 9166-9172.
25. Takagi, T., Yazawa, M., Ueno, T., Suzuki, S. and Yagi, K. (1989) J. Biochem. (Tokyo) 106, 778-783.
26. Watson, M.H., Kuhn, A.E. and Mak, A.S. (1990) Biochim. Biophys. Acta 1054, 103-113.
27. Watson, M.H., Kuhn, A.E., Novy, R.E., Lin, J.J.-C. and Mak, A.S. (1990) J. Biol. Chem. 265, 18860-18866.
28. Marston, S. (1988) FEBS Lett. 238, 147-150.
29. Velaz, L., Ingraham, R.H. and Chalovich, J.M. (1990) J. Biol. Chem. 265, 2929-2934.
30. Marston, S. (1989) J. Musc. Res. Cell Motil. 10, 97-100.
31. Lehman, W., Craig, R., Lui, J. and Moody, C. (1989) J. Musc. Res. Cell Motil. 10, 101-112.
32. Yanashiro, S., Yamakita, Y., Ishikawa, R. and Matsumura, F. (1990) Nature 344, 675-678.
33. Lynch, W. and Bretscher, A. (1986) Methods Enzymol. 134, 37-42.
34. Owada, M.K., Hakura, A., Lida, K., Yahara, I., Sobue, K. and Kakuichi, S. (1984) Proc. Natl. Acad. Sci. 81, 3133-3137.
35. Bretscher, A. and Lynch, W. (1985) J. Cell Biol. 100, 1656-1663.
36. Dingus, J., Hwo, S. and Bryan, J. (1986) J. Cell Biol. 102, 1748-1757.
37. Kossmann, T., Furst, D. and Small, J.V. (1987) J. Musc. Res. Cell Motil. 8, 135-144.
38. Lin, J.J.-C., Lin, J.L.-C., Davis-Nanthakumar, E.J. and Lourim, D. (1988) Hybridoma 7, 273-288.
39. Humphrey, M.B., Herrera-Sosa, H., Gonzalez, G., Lee, R. and Bryan, J. (1992) Gene 112, 197-204.

■ Joseph Bryan:
Baylor College of Medicine,
Department of Cell Biology,
Houston, TX 77030, USA
■ C.-L. Albert Wang:
Boston Biomedical Research Institute,
Department of Muscle Research,
Boston, MA 02114, USA

Calponin

Calponin is a basic, low molecular weight (34 kDa) actin and calmodulin binding protein first isolated from avian gizzard[1]. It is smooth muscle specific[2,3] being expressed coordinately with metavinculin and the heavy isoform of caldesmon during differentiation[3]. It has been demonstrated in various vertebrate smooth muscles, as well as in Aplysian gut[2-4]. In vitro data suggest that calponin acts as a thin filament linked regulator of smooth muscle contraction, but the mechanism of regulation is unclear.

Calponin is abundant in smooth muscle, occurring in amounts comparable to **tropomyosin**[1]. Smooth muscle thin filaments contain estimated stoichiometries of **actin**:tropomyosin:**caldesmon**:calponin of 7:0.9:0.6:0.7[5]. Calponin binds to affinity columns carrying calmodulin or tropomyosin[1,3,6,7] and cosediments with F-actin alone or F-actin-tropomyosin[1]; only the binding to calmodulin is Ca^{2+}-dependent. The binding stoichiometry of calponin to actin *in vitro* is 1:3-4 in 100 mM KCl[1].

Calponin occurs as a monomer in solution with a Stokes radius of 27.1-27.8 Å and exhibits an $S^{\circ}_{20,w}$ value of 3.16[6]. The periodic association of calponin along tropomyosin paracrystals indicates a binding position 17 nm from the N-terminus of tropomyosin, as found for **troponin T** of skeletal muscle. Antibodies to troponin T show a weak crossreaction with calponin[6].

Several isoforms of calponin have been identified in the pI range of 8.4-9.1[6] (Gimona et al., unpublished; Figure 1); the number increases from one in early stages of smooth muscle differentiation to three or four in adult tissue (Gimona et al., unpublished). In addition, a smaller variant of calponin (~28 kDa) with corresponding isoforms is expressed in considerable amounts in smooth muscles of the human urogenital tract (Draeger et al., unpublished).

Antibodies against calponin localize the molecule on the stress fibre bundles of smooth muscle cells in primary culture[3] and on the actomyosin fibrils of isolated, differentiated smooth muscle cells (Draeger et al., unpublished; Figure 2).

■ PURIFICATION

Calponin can be readily purified on the basis of its heat stability. In the original procedure[1] smooth muscle tissue from chicken gizzard or bovine aorta was boiled, the protein extracted with 300 mM KCl and precipitated with 30% ammonium sulphate. Further purification was performed using (SP-Sephadex C-50) and gel filtration (Ultrogel AcA 44) in the presence of 6 M urea. An alternative method[8] employs microwave treatment of minced smooth muscle (chicken gizzard or porcine stomach) followed by extraction with 300 mM KCl at neutral pH and dialysis of the 30% ammonium sulphate precipitate in MES buffer at pH 5.4 to remove most of the other heat stable components (caldesmon, calmodulin, tropomyosin). Final purification is then performed on S-

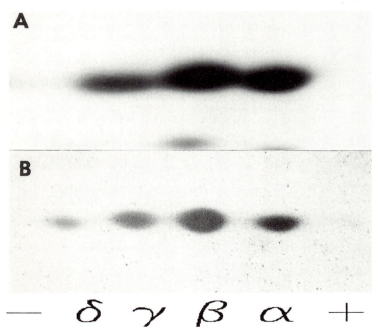

Figure 1. 2-D gel electrophoresis (NEPHGE) patterns of purified calponin from turkey gizzard (A) and porcine stomach (B) showing 3 and 4 isoforms respectively.

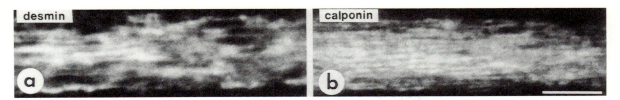

Figure 2. Calponin is localized in contractile fibrils (b) and not in the **desmin** rich cytoskeleton. as seen by double immunofluorescence confocal microscopy (Draeger et al., unpublished). Bar 5 μm.

Sepharose followed by gel filtration on FPLC Superose[8] in the MES buffer used for dialysis. In this buffer the protein is stable for about a week. Calponin may be stored freeze-dried from this buffer at -70°C.

ACTIVITIES

In a reconstituted system containing phosphorylated **smooth muscle myosin**, actin and tropomyosin, the actin activated Mg-ATPase is inhibited 40%[9] - 80%[10] by calponin in a Ca^{2+}-independent fashion. This inhibition is enhanced by caldesmon and reversed by Ca^{2+}-calmodulin[9]. More recent data (Makuch et al., personal communication) suggest that calponin and caldesmon may compete for occupancy on F-actin. Winder and Walsh[10] have shown that Ca^{2+}-dependent phosphorylation of calponin in vitro regulates these proteins' inhibitory activity. Studies on living muscle indicate, however, that calponin is not phosphorylated in vivo[11] (Gimona et al., unpublished). Abe et al.[9] have likened calponin functionally to troponin I of skeletal muscle. More data is required to clarify the conformation of the regulatory complex(es) on the smooth muscle thin filament(s) and the precise role of calponin.

ANTIBODIES

A monoclonal antibody[3] specific for avian calponin is commercially available from Sigma (CP-93).

GENES

cDNA clones encoding chicken gizzard calponin suggest the existence of two variants of the molecule of M_r 32.333 and 28.127 Da (GenBank/EMBL Data Bank numbers are M63559 and M63560). Amino acid sequences derived from the cDNA clones and direct sequencing [8,12] reveal high homology to SM-22 a[13], a smooth muscle specific pro-

tein of unknown function. Other sequence elements show homologies to the actin binding domain of α-**actinin**, to **annexin** (p36) and the muscle specific products of the Unc 87 gene of C. elegans and the m20 gene of Drosophila[8].

REFERENCES

1. Takahashi, K., Hiwada, K. and Kokubu, T. (1986) Biochem. Biophys. Res. Comm. 141, 20-26.
2. Takahashi, K., Hiwada, K. and Kokubu, T. (1987) Life Sciences 41, 291-296.
3. Gimona, M., Herzog, M., Vandekerckhove, J. and Small, J.V. (1990) FEBS Lett. 274, 159-162.
4. Winder, S.J., Sutherland, C. and Walsh, M.P. (1991) In Regulation of smooth muscle:"Progress in Solving the Puzzle" (Moreland, R.S. ed.) Plenum Publishing Corp. New York, in press.
5. Nishida, W., Abe, M., Takahashi, K. and Hiwada, K. (1990) FEBS Lett. 268, 165-168.
6. Takahashi, K., Hiwada, K. and Kokubu, T. (1988) Hypertension 11, 620-626.
7. Takahashi, K., Abe, M., Hiwada, K. and Kokubu, T. (1988) J. Hypertension 6, 40-43.
8. Vancompernolle, K., Gimona, M., Herzog, M., Van Damme, J., Vandekerckhove, J. and Small, J.V.(1990) FEBS Lett. 274, 146-150.
9. Abe, M., Takahashi, K. and Hiwada, K. (1990) J. Biochem. 108, 835-838.
10. Winder, S.J. and Walsh, M.P. (1990) J. Biol. Chem. 265, 10148-10155.
11. Bárány, M., Rokolya, A. and Bárány, K. (1991) FEBS Lett. 279, 65-68.
12. Takahashi, K. and Nadal-Ginard, B. (1991) J. Biol. Chem. 266, 13284-13288.
13. Pearlstone, J.R., Weber, M., Lees Miller, J.P., Carpenter, M.R. and Smillie, L.B. (1987) J. Biol. Chem. 262, 5985-5991.

■ M. Gimona, A. Draeger, M.P. Sparrow and J.V. Small: Institute of Molecular Biology, Austrian Academy of Sciences, Salzburg, Austria

Capping Proteins

Capping proteins[1] are a family of actin binding proteins that cap the barbed ends of actin filaments, nucleate polymerization of actin monomers, do not sever filaments and do not require Ca^{2+}. Capping protein from muscle is called CapZ because it is found at the Z line[2], where it probably mediates the attachment of the barbed ends of thin filaments to the Z line. In nonmuscle cells, capping protein may regulate the assembly and organization of the actin cytoskeleton.

Capping proteins are heterodimers with subunits of M_r 32-36 kDa (α) and 28-32 kDa (β) and are found in all eukaryotes examined, including yeast[3], protozoa[4,5], birds[6], and mammals[7]. In chickens, the α-subunit is encoded by two expressed genes, and the β-subunit by one[8-10]. Additional isoforms are found in purified protein[9]. Proteins purified from different sources have similar structural and functional properties, and their classification as a family was confirmed by protein sequence derived from cDNA cloning for chicken[8-10] and *Dictyostelium*[11]. These sequences have been used to identify the gene for the β-subunit of capping protein in budding yeast, deletion of which leads to an altered **actin** cytoskeleton[3].

The physiologic role of CapZ, the name for capping protein of muscle, is probably the attachment of actin filaments to Z lines via their barbed ends, since CapZ binds to the barbed ends of actin filaments *in vitro*[2] (Figure) and is located at the Z line[2]. The physiologic role of capping protein in nonmuscle cells is less clear. The altered actin cytoskeleton of the yeast capping protein mutant indicates that capping protein may regulate actin assembly and/or organization. The chicken genes are expressed in all muscle and nonmuscle tissues[9,10], so it is probable that nonmuscle cells contain the same capping protein as muscle cells and that it therefore subserves a similar physiologic role. In *Acanthamoeba*, capping protein is found in a distribution similar to that of actin[5]. Nuclear staining is a prominent feature of localization in work with *Xenopus* cells[12]. Barbed ends of actin filaments might be attached to membranes or other structures via capping protein. Nonmuscle cells contain a large number of relatively short actin filaments organized into an isotropic network [13]. Capping protein may cap the barbed ends of these filaments. This network is most likely the major element controlling the mechanical properties[14] and porosity[15] of cytoplasm. Since capping proteins can nucleate actin polymerization[16,17], they may help control the number and length of actin filaments, key parameters in the properties of the actin network.

■ PURIFICATION

Capping protein is easier to purify from muscle than from other sources because chaotropes are required to solubilize it from the Z line. Two similar protocols for the purification of CapZ from chicken muscle, which employ ion exchange chromatography, gel filtration chromatography, ammonium sulphate precipitation and sucrose gradient sedimentation, have been described[18]. Similar purification schemes have been developed for nonmuscle sources[4,5,7].

■ ACTIVITIES

Capping protein caps the barbed end of actin filaments (Figure), as shown by inhibition of both polymerization and depolymerization of actin by fluorescence assays and by electron microscopy[17]. Capping protein also nucleates the polymerization of actin filament from monomers, as detected by using similar fluorescence assays[16,17]. These assays have also been used to document capping protein's lack of ability to sever actin filaments and lack of sensitivity to Ca^{2+}, which distinguishes the capping proteins from members of the **gelsolin** family. Capping protein is inhibited by anionic phospholipids, including PIP_2[19].

■ ANTIBODIES

Polyclonal antisera against chicken CapZ[2], *Dictyostelium* cap 34/32[11], *Xenopus* capping protein a[12] and *Acanthamoeba* capping protein[5] have been described and used for blotting and localization. A few mouse monoclonal antibodies have also been prepared (Cooper and Pollard, 1982, unpublished; Hug, Miller and Cooper, 1990, unpublished) .

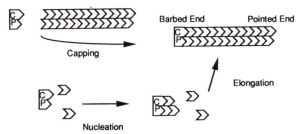

Figure. A diagram of the effects of capping protein on actin polymerization *in vitro*. Capping protein binds to the barbed ends of actin filaments ("caps"), which prevents the addition and loss of actin subunits there. In addition, capping protein binds to actin monomers to form new filaments, which nucleate actin polymerization.

■ GENES

cDNA's for chicken (GenBank OV:Chkcapza, M36882, and Un:J04959)[8-10], *Dictyostelium* (GenBank IN:Ddicap34a & IN:Ddicap32a)[11], *Xenopus*[12] and genes for *Saccharomyces cerevisiae*[3] (GenBank M31720 and EMBL X61398) are published.

■ REFERENCES

1. Pollard, T.D. and Cooper, J.A. (1986) Ann. Rev. Biochem. 55, 987-1035.
2. Casella, J.F., Craig, S.W., Maack, D.J. and Brown, A.E. (1987) J. Cell Biol. 105, 371-379.
3. Amatruda, J.F., Cannon, J.F., Tatchell, K., Hug, C. and Cooper, J.A. (1990) Nature 344, 352-354.
4. Schleicher, M., Gerisch, G. and Isenberg, G. (1984) EMBO J. 3, 2095-2100.
5. Cooper, J.A., Blum, J.D. and Pollard, T.D. (1984) J. Cell Biol. 99, 217-225.
6. Casella, J.F., Maack, D.J. and Lin, S. (1986) J. Biol. Chem. 261, 10915-10921.
7. Kilimann, M.W. and Isenberg, G. (1982) EMBO J. 1, 889-894.
8. Casella, J.F., Casella, S.J., Hollands, J.A., Caldwell, J.E. and Cooper, J.A. (1989) Proc. Natl. Acad. Sci. (USA) 86, 5800-5804.
9. Caldwell, J.E., Waddle, J.A., Cooper, J.A., Hollands, J.A., Casella, S.J. and Casella, J.F. (1989) J. Biol. Chem. 264, 12648-12652.
10. Cooper, J.A., Caldwell, J.E., Gattermeir, D.J., Torres, M.A., Amatruda, J.F. and Casella, J.F. (1991) Cell Motil. Cytoskeleton 18, 204-214..
11. Hartmann, H., Noegel, A.A., Eckerskorn, C., Rapp, S. and Schleicher, M. (1989) J. Biol. Chem. 264, 12639-12647.
12. Ankenbauer, T., Kleinschmidt, J.A., Walsh, M.J., Weiner, O.H. and Franke, W.E. (1989) Nature 342, 822-824.
13. Hartwig, J.H. and Shevlin, P. (1986) J. Cell Biol. 103, 1007-1020.
14. Elson, E.L. (1988) Annu. Rev. Biophys. Biophys. Chem. 17, 397-430.
15. Luby-Phelps, K. and Taylor, D.L. (1988) Cell Motil Cytoskeleton 10, 28-37.
16. Cooper, J.A. and Pollard, T.D. (1985) Biochemistry 24, 793-799.
17. Caldwell, J.E., Heiss, S.G., Mermall, V. and Cooper, J.A. (1989) Biochemistry 28, 8506-8514.
18. Casella, J.F. and Cooper, J.A. (1991) Meth. Enzymol. 196, 140-154.
19. Heiss, S.G. and Cooper, J.A. (1991) Biochemistry, 30, 8753-8758.

■ *John A. Cooper, James F. Amatruda, Christopher Hug and Dorothy A. Schafer:*
Dept. of Cell Biology and Physiology,
Washington University Medical School,
St. Louis, MO, USA

Cofilin

Cofilin is a phosphoinositide-sensitive actin-modulating protein which can reversibly regulate actin polymerization and depolymerization in a pH dependent manner. It forms intranuclear and/or cytoplasmic actin-cofilin rods in cultured cells exposed to various stress conditions. Cofilin may play a regulatory role in reorganization of the actin cytoskeleton in response to a variety of extracellular signals.

Cofolin was originally identified as a 21 kDa protein that binds to G- and F-**actin** stoichiometrically[1,2]. At near neutral pH, cofolin binds to F-actin in a 1:1 molar ratio of cofolin to actin protomer in the filament and induces partial depolymerization of F-svtin[1,3], while at higher pH (pH>7.3) cofolin depolymerizes F-actin completely[3]. Cofilin inhibits the ability of **tropomyosin** or **myosin** to interact with F-actin[1]. Cofilin is able to bind to G-actin in a 1:1 molar ratio with a dissociation constant of 0.1–0.2μM at pH 7.1[2]. The interactions of cofilin with actin are inhibited specifically by various phosphoinsotides PIP_2, PIP, and PI[4].

Cofilin consists of 166 amino acid residues[5-7, 8] and contains a sequence homologous to a nuclear localization signal sequence of SV40 large T antigen and a hexapeptide identical to the N-terminal sequence (residues 2-7) of tropomyosin. The hexapeptide sequence may be involved in the binding of cofilin to F-actin[9]. Cofilin can be chemically crosslinked to an N-terminal acidic portion of actin[10]. *Lys* 112 or *Lys* 114 of cofilin is responsible for this chemical crosslinking[11]. A synthetic dodecapeptide patterned on the sequence around this crosslinking site of cofilin (*Trp* 104 ~Met *105*) interferes with the binding of cofilin to actin[11]. Moreover, this peptide acts as a potent inhibitor of actin polymerization[11,12], and it binds to PIP_2 molecules and inhibits PIP_2 hydrolysis by phospholipase C[12]. A pH independent actin depolymerizing protein, whose amino acid sequence is closely related (>70% identical) to that of cofilin, exists in mammalian[13,14] and avian[7,15] tissues. The activity of this protein, called **ADF** (avian protein) or **destrin (mammalian protein)**, is also sensitive to phosphoinositides[4].

Cofilin exists in a wide variety of mammalian and avian cells and tissues[14,16.] Incubation of cultured fibroblastic cells under specific conditions results in disorganization of stress fibres and at the same time induction of intracellular actin paracrystal-like structures called actin rods, with which cofilin is specifically associated[17]. For example, incubation in an isotonic NaCl buffer solution induces cytoplasmic actin-cofilin rods whereas heat shock or dimethylsulfoxide treatment induces intranuclear rods[17] (Figure). Intriguingly, under these conditions nuclear accumulation of cofilin, which is supposed to be the cause of nuclear transport of actin, is well correlated with-

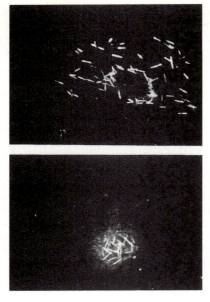

Figure. Immunofluorescence staining of cytoplasmic (top) and intranuclear (bottom) actin-cofilin rods. Mouse C3H-2K cells were exposed to an isotonic NaCl buffer solution (top) or heat shock (bottom), and stained with an anti-cofilin antibody.

dephosphorylation of cofilin[18.] In chicken skeletal muscle, cofilin is suggested to play some transient roles in the regulation of actin assembly during development[19].

■ PURIFICATION

Cofolin is purified by ammonium sulphate precipitation and subsequent sequential chromatography on Toyo Pearl and butyl-Toyo pearl hydrophobic columns, hydroxylapatite, phosphocellulose and gel filtration on Sephadex G-75[17]. DNase 1 affinity column chromatography can be used for rapid small scale purification[1,20]. Recombinant cofilin can be expressed in *E. coli* and purified in an active form.

■ ACTIVITIES

Binding of cofilin to F-actin was assayed by cosedimentation with F-actin[1]. Cofilin induced depolymerization of F-actin or inhibition of actin polymerization can be assayed by a DNase 1 inhibition assay[1], absorbance change at 237 nm or viscometry[20]. Binding of cofilin to G-actin[1] or phosphoinositide[4] was assayed by gel filtration on Sephadex G-100.

■ ANTIBODIES

Rabbit polyclonal antisera against porcine brain cofilin[17] and a monoclonal antibody (MAB22) to chick muscle cofilin[19] have been described.

■ GENES

Cofilin cDNAs of pig (GenBank J03917)[5], mouse (D00472)[6], chick (J02915)[7] and human (D00682)[8] have been published.

■ REFERENCES

1. Nishida, E., Maekawa, S. and Sakai, H. (1984) Biochemistry 23, 5307-5313.
2. Nishida, E. (1985) Biochemistry 24, 1160-1164.
3. Yonezawa, N., Nishida, E. and Sakai, H. (1985) J. Biol. Chem. 260, 14410-14412.
4. Yonezawa, N., Nishida, E., Iida, K., Yahara, I. and Sakai, H. (1990) J. Biol. Chem. 265, 8382-8386.
5. Matsuzaki, F., Matsumoto, S., Yahara, I., Yonezawa, N., Nishida, E. and Sakai, H. (1988) J. Biol. Chem. 263, 11564-11568.
6. Moriyama, K., Matsumoto, S., Nishida, E., Sakai, H. and Yahara, I. (1990) Nucl. Acid. Res. 18, 3053.
7. Abe, H., Endo, T., Yamamoto, K. and Obinata, T. (1990) Biochemistry 29, 7420-7425.
8. Ogawa, K., Tashima, M., Yumoto, Y., Okuda, T., Sawada, H., Okuma, M. and Maruyama, Y. (1990) Nucl. Acids Res. 18, 7169.
9. Yonezawa, N., Nishida, E., Ohba, M., Seki, M., Kumagai, H. and Sakai, H. (1989) Eur. J. Biochem. 183, 235-238.
10. Muneyuki, E., Nishida, E., Sutoh, K. and Sakai, H. (1985) J. Biochem. (Tokyo) 97, 563-568.
11. Yonezawa, N., Nishida, E., Iida, K., Kumagai, H., Yahara, I. and Sakai, H. (1991) J. Biol. Chem. 266, 10485-10489.
12. Yonezawa, N., Homma, Y., Yahara, I., Sakai, H. and Nishida, E. (1991) J. Biol. Chem. 266, in 17218-17221.
13. Nishida, E., Muneyuki, E., Maekawa, S., Ohta, Y. and Sakai, H. (1985) Biochemistry 24, 6624-6630.
14. Moriyama, K., Nishida, E., Yonezawa, N., Sakai, H., Matsumoto, S., Iida, K. and Yahara, I. (1990) J. Biol. Chem. 265, 5768-5773.
15. Adams, M.E., Minamide, L.S., Duester, G. and Bamburg, J.R. (1990) Biochemistry 29, 7414-7420.
16. Yonezawa, N., Nishida, E., Koyasu, S., Maekawa, S., Ohta, Y., Yahara, I. and Sakai, H. (1987) Cell Struct. Funct. 12, 443-452.
17. Nishida, E., Iida, K., Yonezawa, N., Koyasu, S., Yahara, I. and Sakai, H. (1987) Proc. Natl. Acad. Sci. (USA) 84, 5262-5266.
18. Ohta, Y., Nishida, E., Sakai, H. and Miyamoto, E. (1989) J. Biol. Chem. 264, 16143-16148.
19. Abe, H., Ohshima, S. and Obinata, T. (1989) J. Biochem. (Tokyo) 106, 696-702.
20. Maekawa, S., Nishida, E., Ohta, Y. and Sakai, H. (1984) J. Biochem. (Tokyo) 95, 377385.

■ *Kenji Moriyama and Eisuke Nishida:*
Department of Biophysics and Biochemistry,
Faculty of Science, University of Tokyo,
Tokyo, Japan

Coronin

Coronin is an actin binding protein from Dictyostelium discoideum which has sequence similarities to the β-subunits of trimeric G-proteins. Dictyostelium mutants lacking coronin are defective in cytokinesis and cell motility.

Coronin is a 55 kDa **actin** binding protein from *Dictyostelium discoideum* which may represent a new class of cytoskeletal proteins[1]. Immunofluorescence microscopy using monoclonal antibody labelling of growth-phase cells, shows that coronin is concentrated in crown-shaped projections of the actin-rich cell cortex on the dorsal surface of cells (Figure 1 a,c). These "crowns" are clearly visible by scanning electron microscopy (Figure 2) and the protein was, therefore, named "coronin". In developing cells which aggregate by responding chemotactically to the cAMP pulses of neighbouring cells[2], labelling is pronounced at the moving fronts where pseudopods are extended.

The protein sequence, derived from a cDNA clone encoding coronin, lacks homology to known actin binding proteins but shows similarity to the β-subunits of trimeric G-proteins[1]. This similarity suggests that coronin could be part of a G-protein messenger directly coupled to the cytoskeleton.

One of the most fundamental questions in the study of

chemotaxis is the link between cell surface receptors, which bind the chemoattractants, and the cytoskeleton, which mediates the changes in cell shape and locomotion. Second messengers such as Ca^{2+} and inositol phosphates are involved in signal transduction to the cytoskeleton by regulating the activity of proteins which affect the structure of actin filaments or the polymerization of G-actin[3]. Mutants of *Dictyostelium* defective in **severin**[4], **ABP-120**[5], and α-**actinin**[6] show essentially normal chemotaxis suggesting that key proteins remain to be identified which link chemotactic stimulation and the cytoskeleton. Dictyostelium mutants lacking coronin are defective in cytokinesis and cell motility but have only a moderate reduction in their ability to orient chemotactically[8].

■ PURIFICATION

The purification of coronin involves the preparation of an actomyosin complex *in vitro* from a concentrated cyto-

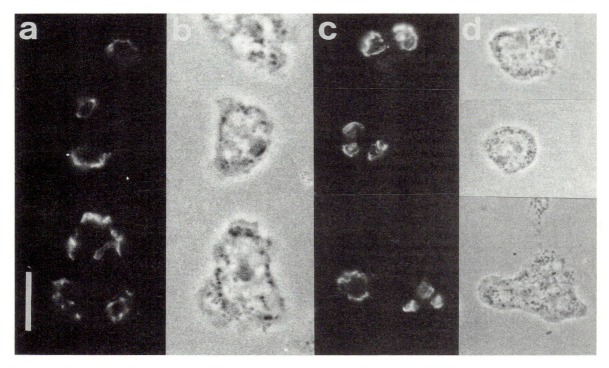

Figure 1. Localization of coronin by immunofluorescence labelling (a,c) and phase-contrast microscopy (b,d) of growth-phase cells with monoclonal antibody 176-2-5. Bar 10 μm.

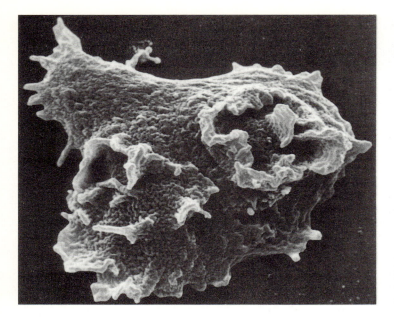

Figure 2. Scanning electron micrograph of the "crowns" in a *Dictyostelium* ameba. (Micrograph courtesy of Dr. R. Guggenheim, Universität Basel).

plasmic supernatant fraction. The complex consists mainly of actin, **myosin II**, a 30 kDa **small actin crosslinking protein**[7] and two proteins of 55 and 17 kDa, respectively. The complex is similar to the "contracted pellet" which has been used as the starting material for the purification of the 30 kDa bundling protein[7]. After the dissociation of the complex in a low-salt buffer, the components can be resolved by chromatography over DEAE-cellulose.

ACTIVITIES

In cosedimentation experiments coronin binds to actin filaments in a Ca^{2+}-independent manner, but does not significantly affect the viscosity of actin solutions as assayed by low-shear viscometry. It is not known if coronin affects the kinetics of actin polymerization, as do other actin binding proteins.

ANTIBODIES

Monoclonal antibodies against coronin have been prepared. The two antibodies that gave the best results (mAb 176-2-5 and mAb 176-3-6) on immunoblots of total cell protein were used in immunofluorescence experiments[1]; both antibodies show the same labelling pattern. It is not known if both recognize the same epitope. These antibodies have not been tested for crossreactivity with proteins from other organisms.

GENES

Using a probe generated by PCR, we have obtained several cDNA clones from a λ-gt11 library (courtesy of R. Kessin, Columbia University) and have sequenced one which contains the entire coding region of a coronin transcript[1]. The sequence has been deposited in the EMBL Data Library with the accession number X61480.

REFERENCES

1. de Hostos, E.L., Bradtke, B., Lottspeich, F., Guggenheim, R. and Gerisch, G. (1991) EMBO J. 10, 4097-4104.
2. McRobbie, S.M. (1986) CRC Crit. Rev. Microbiol. 14, 335-375.
3. Newell, P.C., Europe-Finner, G.N. and Small, N.V. (1987) Microbiological Sciences 4, 5-9.
4. André, E., Brink, M., Gerisch, G., Isenberg, G., Noegel, A., Schleicher, M., Segall, J.E. and Wallraff, E. (1989) J. Cell Biol. 108, 985-995.
5. Brink, M., Gerisch, G., Isenberg, G., Noegel, A.A., Segall, J., Wallraff, E. and Schleicher, M. (1990) J. Cell Biol. 111, 1477-1489.
6. Wallraff, E., Schleicher, M., Modersitzki, D., Rieger, D., Isenberg, G. and Gerisch, G. (1986) EMBO J. 5, 61-67.
7. Fechheimer, M. and Taylor, L.D. (1984) J. Biol. Chem. 259, 4514-4520.
8. de Hostos, E.L. Rehfuess, C., Bradtke, B., Wattel, D.R., Albrecht, R., Murphy, J. and Gerisch, G. (1993) T. Cell Biol. 120, in press.

Eugenio L. de Hostos and Günther Gerisch,
Abteilung Zellbiologie, Max-Planck-Institut für Biochemie,
8033 Martinsried,
Federal Republic of Germany

C-Proteins

C-proteins[1-3] define an isoform family[4-10] of thick filament associated proteins located within the A-bands of vertebrate cross-striated muscles. The proteins confer a series of 43 nm transverse stripes on the C-zone of the A-band[11,12]. Although the cardiac isoform is phosphorylated by a catecholamine dependent kinase[13-15], phosphorylation of the skeletal isoforms is uncertain. Binding to both the rod region of myosin[16] and to actin[17,18] have been demonstrated. Primary sequence analyses reveal (Figure 2) that C-protein is structurally related to a group of myosin-binding proteins (twitchin[19], titin[20,21], smooth muscle light chain kinase[22], skelemin[23], 86 kDa protein[24], projectin[25] and M-protein[26] which all share internally repetitive C2-immunoglobulin and type III fibronectin domains[27-29]. Physiological analyses with myofibre bundles depleted of C-protein indicate that C-protein may alter myosin crossbridge movements at low levels of Ca^{2+}-activation[30].

C-protein was first identified in and purified from crude **myosin** preparations from rabbit skeletal muscle[2,31]. It has a single subunit of 140-150 kDa[14], is virtually devoid of α-helix[14], has a high content of proline[2,32] and hydrodynamic studies indicate an asymmetric molecule with an axial ratio of about 10. Rotary shadowed electron micrographs of C-protein from cardiac[33] and skeletal muscle[34] indicate a rod-shaped molecule, 3x35-40 nm, with a flexible hinge midway along the shaft. Fast, slow and cardiac isoforms have been identified in adult striated muscle[4,7-9] and additional isoforms have been noted during embryogenesis[10]. It is not known if the different isoforms are products of one or more genes. The most striking feature of C-protein is its distribution in the A-band. The C-zone within the crossbridge bearing region of the A-band contains 11 transverse stripes, 43 nm apart. Depending upon the muscle fibre type, seven to nine of these stripes decorate with monoclonal or polyclonal antibodies to C-protein[12,35] (Figure 1) and in some myofibrils, identical stripes contain more than one C-protein isoform. Additional stripes in the C-zone can be labelled with antibodies to the related myosin-binding proteins, 86 kDa-protein[36] and H-protein[35]. In the pectoralis muscle of developing chickens, different isoforms are expressed sequentially: a cardiac-type isoform is expressed first, followed by a slow form and this is eventually replaced by a fast-type isoform[9]. During myofibrillogenesis within cultured cardiac myocytes, C-protein is first detectable by immunofluorescence at stages when thick filaments laterally align in definitive sarcomeres[37].

■ PURIFICATION

C-protein can be purified from the void volume of crude myosin extracts after chromatography on a DEAE-Sephadex column[2]. Separation of C-protein isoforms from each other and from other myosin associated proteins is achieved by hydroxylapatite chromatography[8]. A detailed protocol is available[38].

■ ACTIVITIES

The biological functions of C-protein remain uncertain.

Binding to both the myosin rod and **actin** have been demonstrated *in vitro*[16,17] and the binding to native thin filaments is Ca^{2+}-sensitive[18]. The protein activates or depresses the actin activated, Mg^{2+}-ATPase of myosin, depending on the ionic strength of the assay[39]. Recent studies with permeabilized muscle fibre bundles have shown that 70% of C-protein can be selectively and reversibly extracted from rabbit psoas myofibrils. Since C-protein depleted fibres exhibit an enhanced rate of shortening and greater maximal tension at low concentrations of Ca^{2+}, it has been proposed that the protein tethers myosin crossbridges to the thick filament shaft, possibly looping around the S2 region of the molecule[30]. There is no high resolution information bearing on the precise placement of C-protein within thick myofilaments. C-protein modifies the polymerization properties of myosin, favouring the formation of longer synthetic filaments and lowering the myosin critical concentration[40].

■ ANTIBODIES

Polyclonal antibodies have been reported against rabbit skeletal C-protein[2,3], H-protein and X-protein[35] and rabbit cardiac C-protein[7]. Monoclonal antibodies have been generated against chicken fast, slow and cardiac C-proteins[8,41]. A polyclonal antibody to chicken 86 kDa-protein has also been described[36].

■ GENES

A partial CDNA sequence has been reported for fast-type chicken skeletal C-protein (GenBank M31209)[32].

■ REFERENCES

1. Offer, G. (1972) Cold Spring Harbor Symp. Quant. Biol. 37, 87-93.
2. Offer, G., Moos, C. and Starr, R. (1973) J. Mol. Biol. 74, 653-676.
3. Pepe, F.and Drucker, B. (1975) J. Mol. Biol. 99, 609-617.
4. Callaway, J.E.and Bechtel, P.J. (1981) Biochem. J. 195, 463-469.
5. Jeacocke, S. and England, P. (1980) FEBS Lett. 122, 129-132.
6. Starr, R. and Offer, G. (1983) J. Mol. Biol. 170, 675-698.
7. Yamamoto, K. and Moos, C. (1983) J. Biol. Chem. 258, 8395-8401.

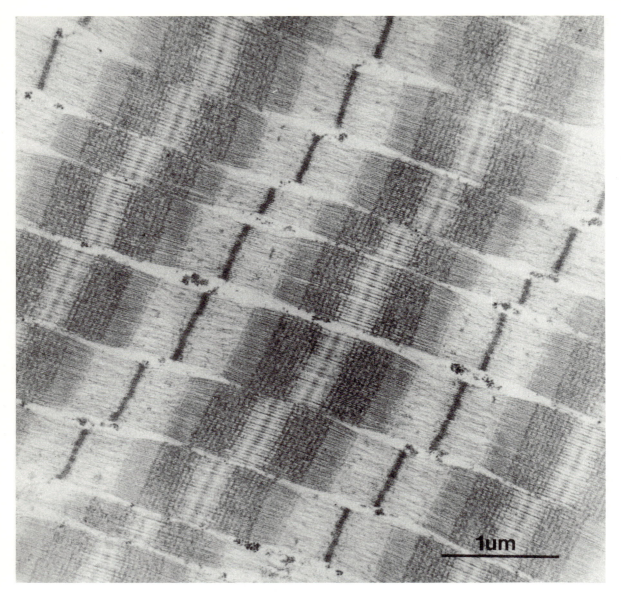

Figure 1. Electron micrograph of chicken posterior latissimus dorsi (PLD) muscle labelled with monoclonal antibody ALD-66 (specific for slow-type C-protein). Each half A-band is decorated with 9 transverse stripes 43 nm apart. Bar 1 μm. (Micrograph provided by J.E. Dennis and D.A. Fischman)

8. Reinach, F.C., Masaki, T., Shafiq, S., Obinata, T. and Fischman, D.A. (1982) J. Cell Biol. 95, 78-84.

9. Obinata, T., Reinach, F.C., Bader, D.M., Masaki, T., Kitani, S. and Fischman, D.A. (1984) Dev. Biol. 101, 116-124.

10. Takano-Ohmuro, H., Goldfine, S.M., Kojima, T., Obinata, T. and Fischman, D.A. (1989) J. Musc. Res. Cell Motil. 10, 369-378.

11. Craig, R. and Offer, G. (1976) Proc. R. Soc. Lond. B. 192, 451-461.

12. Dennis, J.E., Shimizu, T., Reinach, F.C. and Fischman, D.A. (1984) J. Cell Biol. 98, 1514-1522.

13. Hartzell, H. and Titus, L. (1982) J. Biol. Chem. 257, 2111-2120.

14. Hartzell, H. and Glass, D. (1984) J. Biol. Chem. 259, 15587-15596.

15. Hartzell, H. (1985) J. Mol Biol. 186, 185-195.

16. Moos, C., Offer, G., Starr, R. and Bennett, P. (1975) J. Mol. Biol. 97, 1-9.

17. Moos, C., Mason, C., Besterman, J., Feng, I. and Dubin, J. (1978) J. Mol. Biol. 124, 571-586.

18. Moos, C. (1981) J. Cell Biol. 90, 25-31.

19. Benian, G.M., Kiff, J.E., Neckelmann, N., Moerman, D.G. and Waterston, R.H. (1989) Nature 342, 45-50.

20. Wang, K., McClure, J. and Tu, A. (1979) Proc. Natl. Acad. Sci. (USA) 76, 3698-3702.

A

Myosin-associated Ig C-2 family

```
C-2:    **LTC****P***L*W************K/R**********L*L******D*G*Y*C*A*-N
         |    |           |                          |   |       |   |  |  |
C-pro:  **L*C****P*P*VTW*K*G************K/R*********SL*I*****D*G*Y*C**V*N
         |    |           |                          |   |       |   |  |  |
86-kD:  **L*C****P***I*W*************-K-*********SL*I*****D*G*Y*C*AV-N
         |    |           |                          |   |       |   |  |  |
smLCK:  ****C****P*P*V*W*K***************--****----SL*I*****D***Y*C*AV-N
         |    |           |                          |   |       |   |  |  |
Titin:  **L*V****P*P*V*W*K*G****--*****--********--L*******D*G*Y**TL*-N
         |    |           |                          |   |       |   |  |  |
Twthn:  ***V****P*P*V*W*K*G***********K/R*********L*I*****D*G*Y***A--N
         #  #      ^ # # ^ # #              #          #^  #     ^ # ^ # ## ^
```

B

Myosin-associated FN III family

```
C-pro:  PPQ***V*EV******L*W*PP*DDGNA*I*GYTVQK*D****EW**V****R*-TR**V**L**G*EY*FRV*S*N**G*S*E***
         |         |          |  |    |  |      |            |  |  |  |||  |    |
86-kD:  PPQ***L*DV******L*W*PP*DNGNS*I*GYTVQK*D****KW**V****T*-TS**I**L**G*TY*FRV*S*N**G*S*T***
         |         |          |  |    |  |      |            |  |  |  |||  |    |
smLCK:  PAG***A*DI******L*W*GS*YDGGS*V*SYTVEI*N****KW**L-***R*-TS**V*DL**D*EY*FRV*A*N**G*S*P***
         |         |          |  |    |  |      |            |  |  |  |||  |    |
Twthn:  P**-**V*DV******L*W*PP*DDGGA*I**YVVEK*D*****W**V******T***V**L**G*EY*FRV*A*N**G*S*P***
         |         |          |  |    |  |      |            |  |  |  |||  |    |
Titin:  P**-**V**V******L*W**P**DGGS*I*GYIVEK*D*****W******---****V**L**G*EY*FRV*A*N**G*G*P***
         ^#     #  ##       ^  ^    # ##^## #   ^#^^# #      ^    #     ## #   ^   # #^ ^^^  #  ^   ^ # #
```

Figure 2. Amino acid consensus alignments of the myosin associated proteins that belong to the immunoglobulin (Ig) C-2[27,28] and *fibronectin* (FN) type III families[29]. (A) The consensus sequence for the C-2 family is aligned with the predicted amino acid sequences for C-protein (C-pro), 86-kDa protein, smooth muscle light chain kinase (smLCK), **titin** and twitchin (Twthn). (B) The consensus sequences of the type III FN repeats in these same myosin-associated proteins are also aligned. For each figure, those amino acids that are identical in every member of the family are linked by a vertical bar and underscored with a caret. Those amino acids present in at least three members of each class have also been marked(#). Asterisks have been inserted in place of nonidentical amino acids, and dashes have been inserted where the sequences have been shifted to facilitate vertical matches between the protein repeats.

21. Labeit, S., Barlow, D.P., Gautel, M., Gibson, T., Holt, J., Hsieh, C.-L., Francke, U., Leonard, K., Wardale, J., Whiting, A. and Trinick, J. (1990) Nature 345, 273-276.
22. Olson, N., Pearson, R.B, Needleman, D.S., Hurwitz, J.Y., Kemp, B.E. and Means, A.R. (1990) Proc. Natl. Acad. Sci. (USA) 87, 2284-2288.
23. Price, M.G. (1987) J. Cell Biol. 104, 1325-1336.
24. Bahler, M., Eppenberger, H.M. and Wallimann, T. (1985) J. Mol. Biol. 186, 381-391.
25. Saide, J.D., Chin-Bow, S., Hogan-Sheldon, J., Busquets-Turner, L., Vigoreaux, J.O., Valgeirsdottir, K. and Pardue, M. (1989) J. Cell Biol. 109, 2157-2167.
26. Masaki, T. and Takaiti, O. (1973) J. Biochem (Tokyo) 75, 367-380.
27. Cunningham, B.A., Hemperly, J.J., Murray, B.A., Prediger, E.A., Brackenbury, R. and Edelman, G.M. (1987) Science 236, 799-806.
28. Williams, A. F. and Barclay, A .N. (1988) Ann. Rev. Immunol. 6, 381-405.
29. Petersen, T.E., Thogersen, H.C., Skortengard, K., Vibe-Pedersen, K., Sahl, P., Sottrup-Jensen, L. and Magnusson, S. (1983) Proc. Natl. Acad. Sci. (USA) 80, 137141.
30. Hofmann, P.A., Greaser, M.L. and Moss, M.L. (1991) J. Physiol. 439, 701-715.
31. Starr, R. and Offer, G. (1971) FEBS Lett. 15, 40-44.
32. Einheber, S. and Fischman, D.A. (1990) Proc. Natl. Acad. Sci. (USA) 87, 2157-2161.

33. Hartzell, H.C. and Sale, W.S. (1985) J. Cell Biol. 100, 208-215.
34. Swan, R.C. and Fischman, D.A. (1986) J. Musc. Res. Cell Motil. 7, 160-166.
35. Bennett, P., Craig, R., Starr, R. and Offer, G. (1986) J. Musc. Res. Cell Motil. 7, 550567.
36. Bahler, M., Eppenberger, H.M. and Wallimann, T. (1985) J. Mol. Biol. 186, 393-401.
37. Schultheiss, T., Lin, Z., Lu, M.-H., Murray, J., Fischman, D.A., Weber, K., Masaki, T., Imamura, M. and Holtzer, H. (1990) J. Cell Biol. 110, 1159-1172.
38. Starr, R. and Offer, G. (1982) Methods Enzymol. 85, 130-138.
39. Moos, C. and Feng, I.M. (1980) Biochim. Bophys. Acta 632, 141-149.
40. Davis, J.S. (1988) J. Musc. Res. Cell Motility 9, 174-183.
41. Obinata, T., Kawashima, M., Kitani, S., Saitoh, Q., Masaki, T., Bader, D.M. and Fischman, D.A. (1986) *In*: Molecular Biology of Muscle Development, edited by Fischman, D.A., Emerson, C., Nadal-Ginard, B. and Siddiqui, M.A.Q. New York: Alan R. Liss, Inc. 293-307.

■ Donald A. Fischman:
Department of Cell Biology and Anatomy,
Cornell University Medical College,
New York, NY, USA

Dematins

Dematin (erythrocyte protein 4.9) is a stable component of the mature erythrocyte membrane skeleton that was originally identified by its ability to bundle actin filaments in vitro[1]. Dematin's role in vivo is unknown but it may participate in cytoskeletal reorganization during erythroblast maturation, and it may stabilize and help link the membrane skeleton to the plasma membrane[1,2].

While immunoblotting and immunofluorescence studies show that dematin subunits are present in many vertebrate tissues[3], only the protein from human erythrocyte membranes has been purified and biochemically characterized[4]. Human erythrocyte dematin is a trimer composed of subunits that migrate as 48 and 52 kDa polypeptides on SDS-PAGE[1]. The two polypeptides are sequence related phosphoproteins, and both participate to form a molecule with an apparent molecular weight of 145,000 that has a Stokes radius of 59 Å and a three-lobed structure that has been visualized by electron microscopy[1] (Figure 1). Avian erythrocytes contain immunoreactive subunit polypeptides of dematin with apparent molecular masses of 44, 47, 49, 50 and 52 kDa. These variants are differentially synthesized and accumulated during embryonic development. The stable form of avian dematin, consisting of the 50 and 52 kDa variants, is synthesized late during erythropoiesis, and the timing of its expression suggests it plays a role in stabilizing the terminally differentiated erythrocyte membrane skeleton.

Purified human erythroid dematin bundles **actin** filaments *in vitro (Figure 2)*. The 48 and 52 kDa subunits, which may be individually purified and renatured from SDS gels, are capable of reassociating into trimers that bundle actin filaments[5]. Quantitative immunoassays show that human erythrocytes contain 43,000 trimers of dematin per cell, which is equivalent to one trimer for each of the 43,000 short (37 nm long[6]) actin oligomers per red blood cell[5]. *In vivo* as well as *in vitro*, dematin is a substrate of many protein kinases[7-11]. Bundling activity is reversibly abolished when dematin is phosphorylated by protein kinase A (cyclic AMP-dependent protein kinase); phosphorylation by protein kinase C is without effect[5]. Upon phosphorylation with protein kinase A, the 48 and 52 kDa polypeptides of dematin incorporated 3.0 and 1.5 moles of phosphate per mole of protein, respectively, whereas values of 0.6 and 0.3 mole/mole were obtained after phosphorylation by protein kinase C[4].

■ PURIFICATION

Dematin can be solubilized by incubating Triton X-100 extracted erythrocyte ghost skeletons at 37°C in low ionic strength buffer. Dematin is then purified on a DEAE-Sephacel anion exchange matrix[1,4], followed by FPLC (Fast Protein Liquid Chromatograph) on a Mono-Q column to remove residual contaminants including a 55 kDa polypeptide[4]. The resultant dematin has a basal phosphate content of 0.4 mole P/mole of protein[5]. Dematin can also be purified from the high salt extract of inside-out

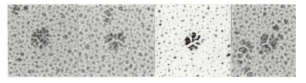

Figure 1. Low angle, rotary shadowed images of purified dematin. X 550,000.

erythrocyte membranes by chromatography on DEAE-Sephacel and hydroxylapatite columns, but this procedure gives a low yield of protein that is often contaminated with kinases[4].

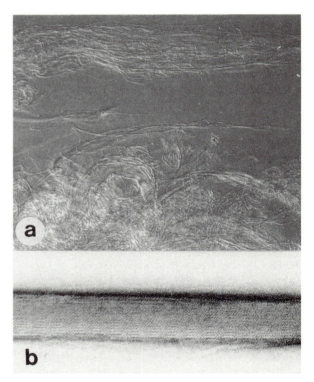

Figure 2. Bundling of F-actin by dematin. (a) Differential interference contrast micrograph of rabbit muscle actin incubated with dephosphorylated dematin. No such bundles were seen when actin filaments were incubated with phosphorylated dematin. X 1,000. (b) Electron micrograph of rabbit muscle actin bundled by dephosphorylated dematin. X 100,000.

ACTIVITIES

Dematin does not bind monomeric actin, does not alter critical actin concentration, but does increase the length of the lag phase and decrease the rate of elongation during actin polymerization[1]. As implied by its name ("δεμα" = a bundle), dematin is primarily recognized as a bundling protein. Bundling activity has been detected and measured by phase-contrast light microscopy, electron microscopy, falling ball viscometry and low speed centrifugation[1]. Because bundling activity is reversibly abolished by phosphorylation[5], care must be taken to relate measurements of bundling activity to dematin's state of phosphorylation. Dephosphorylated dematin also binds to a protein component on erythrocyte membrane vesicles that have been stripped of all peripheral membrane proteins[12]. The identity of this protein component is not known.

ANTIBODIES

Polyclonal antibodies raised against human erythrocyte dematin recognize both the 48 and 52 kDa dematin polypeptides of human erythrocytes as well as the dematin subunits from other species[2,3]. When purified by affinity to either the 48 or 52 kDa subunits, the purified antibodies continue to recognize both of these subunits on Western blots[4]. A series of monoclonal antibodies has been developed which recognize two nonoverlapping epitopes on the 48 kDa subunit[5] but cannot distinguish between the 48 kDa subunit and the 52 kDa subunit by Western blotting.

GENES

Cloning of the cDNA for dematin is currently under way.

REFERENCES

1. Siegel, D.L. and Branton, D. (1985) J. Cell Biol. 100, 775-785.
2. Faquin, W.C., Husain-Chishti, A. and Branton, D. (1990) Eur. J. Cell Biol. 53, 48-58.
3. Faquin, W.C., Husain, A., Hung, J. and Branton, D. (1988) Eur. J. Cell Biol. 46, 168-175.
4. Husain-Chishti, A., Faquin, W., Wu, C.-C. and Branton, D. (1989) J. Biol. Chem. 264, 8985-8991.
5. Husain-Chishti, A., Levin, A. and Branton, D. (1988) Nature 334, 718-721.
6. Byers, T.J. and Branton, D. (1985) Proc. Natl. Acad. Sci. (USA) 82, 6153-6157.
7. Palfrey, H.C. and Waseem, A. (1985) J. Biol. Chem. 260, 16021-16029.
8. Cohen, C.M. and Foley, S.F. (1986) J. Biol. Chem. 261, 7701-7709.
9. Horne, W.C., Leto, T.L. and Marchesi, V.T. (1985) J. Biol. Chem. 260, 9073-9076.
10. Horne, W.C., Miettinen, H. and Marchesi, V.T. (1988) Biochem. Biophys. Acta 944, 135-143.
11. Faquin, W.C., Chahwala, S.B., Cantley, L.C. and Branton, D. (1986) Biochem. Biophys. Acta 887, 142-149.
12. Husain-Chishti, A. and Branton, D. (1989) J. Cellular Biochem. CD 010, pp. 210.

Athar Husain-Chishti and Daniel Branton:
The Biological Laboratories,
Harvard University,
Cambridge, MA 02138, USA

Depactin

Depactin[1,2] is an actin depolymerizing protein originally isolated from starfish oocytes. Similar proteins have been isolated from sea urchin eggs[3] and Acanthamoeba castellanii[4]. The latter protein is called actophorin.

In echinoderm eggs, G-**actin** comprises about 50% of the total actin[5]. This form is stabilized as a 1:1 depactin-actin complex[2] and 1:1 **profilin**-actin complex[6]. When the actin filaments are required by the cell, these complexes are thought to dissociate and release free actin. The factors that dissociate these complexes in the cell have not, however, been identified. *In vitro* experiments have shown that **myosin** can dissociate the actin-depactin complex in the absence of ATP[7]. **Tropomyosin** reduces the rate of depolymerization of actin induced by depactin, but cannot block it[7]. A polyamine, spermine, is also able to polymerize actin from the depactin-actin complex[8]. Depactin binds to both the N- and C-terminal segments of actin, which are also the segments that bind myosin[9,10]. The actin binding site of depactin is in its N-terminal domain[11]. The amino acid sequence of depactin is not similar to other actin binding proteins known so far[12]. However, a segment in depactin has homology to mammalian **actin depolymerizing factor** (ADF)[13]. The molecular weight of starfish oocyte depactin deduced from its sequence is 17,590[12] and its pI is ~6[2]. The *Acanthamoeba* form of depactin (actophorin) has been crystallized[14].

PURIFICATION

Depactin is purified from starfish oocyte extracts by ammonium sulphate precipitation, DEAE-cellulose, hydroxylapatite and gel filtration column chromatography[1,2].

ACTIVITIES

The effects of depactin on actin polymerization are assayed either by high-shear viscometry[1], light scattering, absorbance at 238 nm[2], fluorescence monitoring of

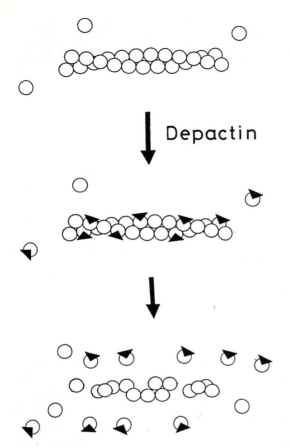

Figure 1. Mode of interaction of depactin with actin. Triangles represent depactin molecules.

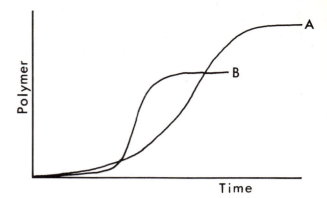

Figure 2. Actin polymerization in the absence (A) or presence (B) of depactin.

pyrene-labelled actin[8], birefringence measurement[2], DNase 1 inhibition assay[7], or electron microscopy using a negative staining technique[2]. Depactin depolymerizes F-actin rapidly by binding at a 1:1 ratio to actin monomers in F-actin and destabilizing the filament[2] (Figure 1). Depactin also binds to free G-actin[2]. When depactin is present during KCl-induced polymerization of actin, the rate of actin polymerization is initially slower than the control, but then abruptly accelerates and rapidly reaches completion[2] (Figure 2). Binding of depactin to free G-actin accounts for the initial slow rate; the subsequent abrupt polymerization is thought to be due to depactin induced cutting of long F-actin at the early stage of polymerization, which generates seeds for polymerization. The binding constant of depactin to actin monomer is $2-5 \times 10^6$ M^{-1} [2,7]. The action of depactin is not regulated by Ca^{2+} or pH[2].

■ ANTIBODIES

Rabbit polyclonal antisera against starfish oocyte depactin and *Acanthamoeba catellanii* actophorin have been produced. Anti-egg depactin antibodies do not crossreact with actophorin and anti-amoeba actophorin antibodies do not crossreact with depactin (Mabuchi, I. and Pollard, T.D., unpublished). Antibodies against an N-terminal segment (15 residues) and a C-terminal segment (12 residues) have also been produced and used as probes in determination of actin binding sites of depactin[11].

■ GENES

cDNA clones are not yet available.

■ REFERENCES

1. Mabuchi, I. (1981) J. Biochem. 89, 1341-1344.
2. Mabuchi, I. (1983) J. Cell Biol. 97, 1612-1621.
3. Hosoya, H., Mabuchi, I. and Sakai, H. (1982) J. Biochem. 92, 1853-1862.
4. Cooper, J.A., Blum, J.D., Williams Jr., R.C. and Pollard, T.D. (1986) J. Biol. Chem. 261, 477-485.
5. Mabuchi, I. and Spudich, J.A. (1980) J. Biochem. 87, 785-802.
6. Mabuchi, I. and Hosoya, H. (1982) Biomedical Res. 3, 465-476.
7. Mabuchi, I. (1982) J. Biochem. 92, 1439-1447.
8. Mabuchi, I. (1986) Int. Rev. Cytol. 101, 175-213.
9. Sutoh, K. and Mabuchi, I. (1984) Biochemistry 23, 6757-6761.
10. Sutoh, K. and Mabuchi, I. (1986) Biochemistry 25, 6186-6192.
11. Sutoh, K. and Mabuchi, I. (1989) Biochemistry 28, 102-106.
12. Takagi, T., Konishi, K. and Mabuchi, I. (1988) J. Biol. Chem. 263, 3097-3102.
13. Abe, H., Endo, T., Yamamoto, K. and Obinata, T. (1990) Biochemistry 29, 7420-7425.
14. Magnus, K.A., Maciver, S.K. and Pollard, T.D. (1988) J. Biol. Chem. 263, 18143-18144.

■ *Issei Mabuchi:*
Department of Biology, College of Arts and Sciences, University of Tokyo, Tokyo, Japan

Dystrophin

Dystrophin is the protein that, when defective, gives rise to the Duchenne/Becker muscular dystrophies (DMD/BMD). It is a high molecular weight member of the spectrin superfamily of cytoskeletal proteins and associates with the plasma membrane of muscle and neuronal tissues.

Dystrophin is a very large (427 kDa) protein that consists of four major structural domains[1] (Figure 1). Domain I bears strong sequence similarity to the **actin** binding regions of α-**actinin** and β-**spectrin**[2]. Domain II is a series of repeats that share a consensus profile with the spectrin repeat motif but are much more variable[3]. This domain is predicted to form a series of short α-helical barrels and turns that nest in succession to form an elongated, flexible structure[4]. Breaks in the repeat motif may represent

Figure 1. Schematic model of dystrophin with domains indicated above.

hinge segments (H's in Figure 1) that confer additional flexibility[3]. Domain III is rich in cysteine residues and contains sequence similarity to the EF-hand region of α-actinin, but the EF-hands in dystrophin are degenerate and are not predicted to bind Calcium[1]. The C-terminal domain IV is unique to dystrophin and to a recently described protein called DRP. DRP is similar in size to dystrophin, but is encoded by an autosomal gene and is expressed in a broader range of tissues than dystrophin[5,6]. In dystrophin, alternate promoters in muscle and brain give rise to different N-termini[7] and differential splicing in the C-terminal region predicts several different protein isoforms[8].

At ~2.5 million base pairs and at least 70 exons, the X-linked dystrophin gene is the largest described to date. The unusually high mutation rate (30% of DMD patients

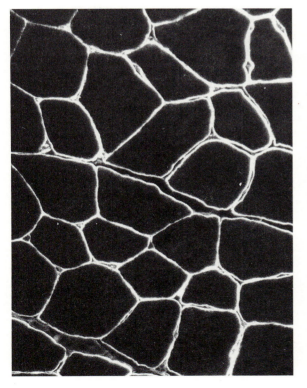
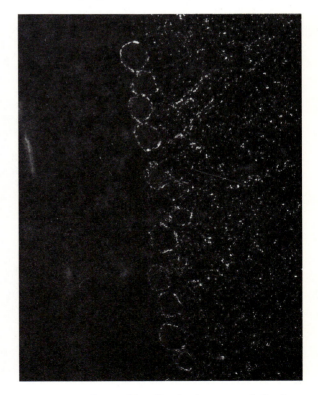

Figure 2. Immunofluorescence using anti-dystrophin antibodies shows sarcolemmal localization in mouse skeletal muscle (left), and to membranes of Purkinje cell bodies and dendrites in mouse cerebellum (right). No staining is seen in the dystrophin deficient *mdx* mouse (not shown). (Micrographs provided by Dr. S.C. Watkins and Dr. H.G.W. Lidov)

have new mutations) may be explained by the fact that the gene is such a large target for deletion mutation (65% of cases). Nonsense or frameshifting deletion mutations generally result in the absence of dystrophin and the severe Duchenne phenotype. In-frame deletions or duplications give rise to a molecule that is altered in size and/or abundance and usually result in the milder Becker phenotype. The mildest phenotypes result from in frame deletions in domain II while loss of the region that includes domain III is always severe[9]. A dystrophin deficient mouse strain exists (called *mdx*), but these mice display few of the physical consequences that result in human DMD.

■ PURIFICATION

After subcellular fractionation of mammalian skeletal muscle, dystrophin is found primarily associated with the plasma membrane[10]. It has been partially purified by virtue of its association with other membrane proteins in a WGA- lectin binding complex[11], but does not itself bind to WGA. Rabbit skeletal muscle dystrophin appears as linear aggregates of dumbbell-shaped molecules in rotary shadow EM[12].

■ ACTIVITIES

Because of its membrane localization in all muscle types[13-16] (Figure 2), the progressive pattern of damage to the muscle membrane in DMD[17] and by analogy to spectrin, it is thought that dystrophin plays a structural role in providing resilience to the muscle membrane through cycles of contraction and relaxation[18]. It may also assist in linking actin filaments to the membrane at myotendinous junctions of skeletal muscle[19], but such a role is unlikely in the intercalated disks of cardiac muscle and in dense plaques of smooth muscle because of the absence of dystrophin in these locals[16]. In the brain, the highly restricted localization of dystrophin[20] argues for a more limited function in synaptic organization rather than a general structural role in strengthening the plasma membrane.

■ ANTIBODIES

Polyclonal antisera have been produced against bacterially expressed polypeptides of mouse[21] and polypeptides and synthetic peptides of human[3,14,15,20,22,23] dystrophin. Polyclonal antisera have also been produced against chicken dystrophin[24] and *Torpedo* dystrophin[25]. Dystrophin antibodies sometimes react with a protein of similar mobility in a variety of vertebrate species[26] and some dystrophin antibodies have also been found to crossreact with known related proteins[27].

■ GENES

Full cDNA sequence is available for human dystrophin (GenEMBL M18533), chicken dystrophin (X13369) and mouse dystrophin (submitted to GenBank, J. Chamberlain,

personal communication). Partial sequence is also available for *Torpedo* dystrophin (M37645). cDNA probes representing the entire dystrophin coding sequence are available from ATCC.

■ REFERENCES

1. Koenig, M., Monaco, A.P. and Kunkel, L.M. (1988) Cell 53, 219-226.
2. Karinch, A.M., Zimmer, W.E. and Goodman, S.R. (1990) J. Biol. Chem. 265, 11833-11840.
3. Koenig, M. and Kunkel, L.M. (1990) J. Biol. Chem. 265, 4560-4566.
4. Cross, R.A., Stewart, M. and Kendrick -Jones, J. (1990) FEBS Lett. 262, 87-92.
5. Love, D.R., Hill, D.F., Dickson, G., Spurr, N.K., Byth, B.C., Marsden, R.F., Walsh, F.S., Edwards, Y.H. and Davies, K.E. (1989) Nature 339, 55-58.
6. Khurana, T.S., Hoffman, E.P. and Kunkel, L.M. (1990) J. Biol. Chem. 265, 16717-16720.
7. Nudel, U., Zuk, D., Einat, P., Zeelon, E., Levy, Z., Neuman, S. and Yaffe, D. (1989) Nature 337, 76-78.
8. Feener, C.A., Koenig, M. and Kunkel, L.M. (1989) Nature 338, 509-511.
9. Koenig, M., Beggs, A.H., Moyer, M., Scherpf, S., Heindrich, K., Bettecken, T., Meng, G., Muller, C.R., Lindlof, M., Kaariainen, H., de la Chapelle, A., Kiuru, A., Savontaus, M.-L., Gilgenkrantz, H., Recan, D., Chelly, J., Kaplan, J.-C., Covone, A.E., Archidiacono, N., Romeo, G., Liechti-Gallati, S., Schneider, V., Braga, S., Moser, H., Darras, B.T., Murphy, P., Francke, U., Chen, J.D., Morgan, G., Denton, M., Greenberg, C.R., van Ommem, G.J.B. and Kunkel, L.M. (1989) Am. J. Hum. Genet. 45, 498-506.
10. Ohlendieck, K., Ervasti, J.M., Snook, J.B. and Campbell, K.P. (1991) J. Cell Biol. 112, 135-148.
11. Campbell, K.P. and Kahl, S.D. (1989) Nature 338, 259-262.
12. Murayama, T., Sato, O., Kimura, S., Shimizu, T., Sawada, H. and Maruyama, K. (1990) Proc. Japan Acad. 66B, 96-99.
13. Watkins, S.C., Hoffman, E.P., Slayter, H.S. and Kunkel, L.M. (1988) Nature 333, 863-866.
14. Zubrzycka-Gaarn, E.E., Bulman, D.E., Karpati, G., Burghes, A.H., Belfall, B., Klamut, H.J., Talbot, J., Hodges, R.S., Ray, P.N. and Worton, R.G. (1988) Nature 333, 466-469.
15. Arahata, K., Hoffman, E.P., Kunkel, L.M., Ishiura, S., Tsukahara, T., Ishihara, T., Sunohara, N., Nonaka, I., Ozawa, E. and Sugita, H. (1989) Proc. Natl. Acad. Sci. (USA) 86, 7154-7158.
16. Byers, T.J., Kunkel, L.M. and Watkins, S.C. (1991) J. Cell Biol. 115, 411-421.
17. Rowland, L.P. (1980) Muscle and Nerve 3, 3-20.
18. Hoffman, E.P. and Kunkel, L.M. (1989) Neuron 2, 1019-1029.
19. Tidball, J.G. and Law, D.J. (1991) Am. J. Pathol. 138, 17-21.
20. Lidov, H.G.W., Byers, T.J., Watkins, S.C. and Kunkel, L.M. (1990) Nature 348, 725-728.
21. Hoffman, E.P., Brown Jr., R.H. and Kunkel, L.M. (1987) Cell 51, 919-928.
22. Kao, L., Krstenansky, J., Mendell, J., Rammohan, K.W. and Gruenstein, E. (1988) Proc. Natl. Acad. Sci. (USA) 85, 4491-4495.
23. Nicholson, L.V., Davison, K., Johnson, M.A., Slater, C.R., Young, C., Bhattacharya, S., Gardner Medwin, D. and Harris, J.B. (1989) J. Neurol. Sci. 94, 137-146.
24. Jasmin, B.J., Cartaud, A., Ludosky, M.A., Changeux, J.P. and Cartaud, J. (1990) Proc. Natl. Acad. Sci. (USA) 87, 3938-3941.
25. Chang, H.W., Bock, E. and Bonilla, E. (1989) J. Biol. Chem. 264, 20831-20834.

26. Hoffman, E.P., Beggs, A.H., Koenig, M., Kunkel, L.M. and Angelini, C. (1989) Lancet 2, 1211-1212.
27. Hoffman, E.P., Watkins, S.C., Slayter, H.S. and Kunkel, L.M. (1989) J. Cell Biol. 108, 503-510.

■ Timothy J. Byers and Louis M. Kunkel:
Howard Hughes Medical Institute, Children's Hospital Medical Center and Harvard Medical School, Boston, MA, USA

Ezrin

Ezrin is a membrane associated protein that is enriched in some surface structures that contain an F-actin cytoskeleton, such as microvilli and membrane ruffles. Although its function is currently unknown, the cDNA derived sequence of the N-terminal domain of ezrin shows homology to the equivalent domain of erythrocyte protein 4.1, suggesting that it is involved in linking actin filaments to the plasma membrane. Ezrin is a substrate for the tyrosine-kinase activity of the EGF receptor and other kinases.

Ezrin was originally identified (at Ezra Cornell University) as a minor M_r ~80,000 component of the isolated micro-filament bundle of intestinal microvilli and purified from this structure[1]. Antibodies to chicken intestinal ezrin revealed immunologically related species in a wide variety of cells, and localization studies showed that these are preferentially localized to surface structures that contain an F-**actin** cytoskeleton[1]. It is present in microvilli,

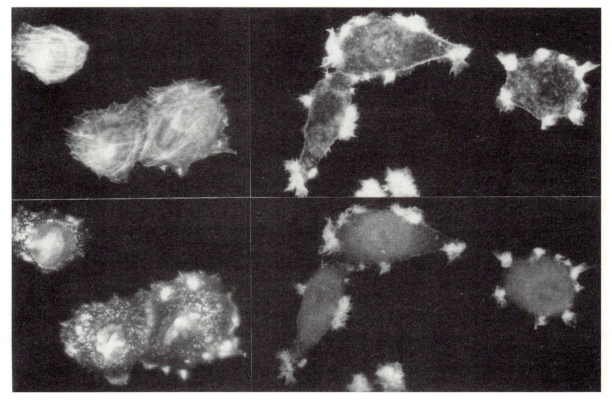

A
B
C
D

Figure. Localization of actin (A and C) and ezrin (B and D) in CHO cells (A and B) and in human carcinoma KB cells stimulated with EGF (C and D). Note in (A) and (B) that ezrin does not colocalize with the most abundant actin containing structures, but is seen enriched in cell surface structures (microvilli and ruffles), as seen in (B) where the plane of focus is on the top of the cells. KB cells ruffle dramatically in response to epidermal growth factor: under these conditions actin and ezrin are both found highly enriched in these structures (C and D). During this rapid process, ezrin is phosphorylated.

microspikes, and membrane ruffles of cultured cells[1-5], (Figure) in placental microvilli[6], in growth cones of cultured neurons[7], and in the marginal band of avian erythrocytes[8]. It is also present in the microvillus surface of acid secreting parietal cells of the gastric glands[9-11] where it is present as a family of isoelectric variants[9]. Immunoelectron microscopy has shown that ezrin is enriched just inside the plasma membrane of microvilli, but not on more planar aspects of the membrane, in human choriocarcinoma cells[2].

Ezrin was independently identified using an antibody to a synthetic peptide deduced from a cloned human endogenous retroviral *gag*-related DNA sequence, called *erv-1* [12]. Antibodies to this peptide identified a 75 kDa component present in microvilli of human choriocarcinoma and other cells[2-4]. This protein was termed "cytovillin"[3]; antibody crossreactivity[13] and sequence analysis revealed its identity to ezrin[14,15].

Ezrin is identical to a previously identified protein (called p81)[16] that becomes phosphorylated on tyrosine residues in human carcinoma A431 cells in response to the addition of EGF[17]. In these cells there is a correlation between the phosphorylation of ezrin and the appearance of cell surface structures (microvilli and membrane ruffles) that contain ezrin[5]. It is also a substrate for a kinase in rabbit parietal cells, probably the cAMP-dependent protein kinase[9].

The cDNA for the human protein predicts a polypeptide of 585 residues[14,15]. The N-terminal ~260 residues show 34% sequence identity[15] to the corresponding region of human **protein 4.1**[18], which links the actin-**spectrin** network to glycophorin in the red blood cell membrane[19]. Although the function of ezrin remains to be elucidated, its localization as a peripheral membrane protein and its homology to protein 4.1 strongly suggest that it plays a cytoskeletal-membrane linking role in cell surface structures.

■ PURIFICATION

Ezrin has been purified from chicken intestinal brush borders[1,20], human placenta[5], human choriocarcinoma cells[21] and rabbit parietal cells[9]. In all cases it has been isolated as a freely soluble protein in the absence of detergents; two of the protocols[1,5] make use of hydroxylapatite chromatography, a resin to which it binds extraordinarily tightly.

■ ACTIVITIES

None are known, except that it is a substrate for certain tyrosine protein kinases[17] and the catalytic subunit of the cAMP-dependent protein kinase (unpublished).

■ ANTIBODIES

Polyclonal antibodies to chicken intestinal ezrin[1], human placental ezrin[5] and human choriocarcinoma cell ezrin[3]

(cytovillin) have been described. These antibodies are suitable for immunoblotting, immunofluorescence microscopy, immunoelectron microscopy and immunoprecipitations. Monoclonal antibodies to chicken brain ezrin[8], chick retina ezrin[22] and rabbit parietal cell ezrin[10] have been described and used in immunoblotting and immunofluorescence studies.

■ GENES

Full length cDNAs for human ezrin have been isolated and sequenced ("cytovillin": GenBank J05021[14]; ezrin: GenBank X51521[15]). There appears to be a single ezrin gene in the human genome[14,15] located on the long arm of chromosome 6[14].

■ REFERENCES

1. Bretscher, A. (1983) J. Cell Biol. 97, 425-432.
2. Pakkanen, R., Hedman, K., Turunen, O., Wahlstrom, T. and Vaheri, A. (1987) J. Histochem. Cytochem. 35, 809-816.
3. Pakkanen, R. (1988) Europ. J. Cell Biol. 46, 435-443.
4. Pakkanen, R. (1988) J. Cell Biochem. 38, 65-75.
5. Bretscher, A. (1989) J. Cell Biol. 108, 921-930.
6. Edwards, H.C. and Booth A. G. (1987) J. Cell Biol. 105, 303-311.
7. Goslin, K., Birgbauer, E., Banker, G. and Solomon, F. (1989) J. Cell Biol. 109, 1621-1631.
8. Birgbauer, E. and Solomon F. (1989) J. Cell Biol. 109, 1609-1620.
9. Urushidani, T., Hanzel, D.K. and Forte, J.G. (1989) Am. J. Physiol. 256, G1070-1081.
10. Hanzel, D.K., Urushidani, T., Usinger, W.R., Smolka, A. and Forte, J.G. (1989) Am. J. Physiol. 256, G1082-1089.
11. Hanzel, D., Reggio, H., Bretscher, A., Forte, J.G. and Mangeat, P. (1991) EMBO J. 10, 2363-2373.
12. Suni , J., Narvanen, A., Wahlstrom, T., Aho, M., Pakkanen, R.,Vaheri, A., Copeland, T., Cohen, M. and Oroszian, S. (1984) Proc. Natl. Acad. Sci. (USA) 81, 6197-6201.
13. Pakkanen, R. and Vaheri, A. (1989) J. Cell Biochem. 41, 1-12.
14. Turunen, O., Winqvist, R., Pakkanen, R., Grzeschik, K-H., Wahlstrom, T. and Vaheri, A. (1989) J. Biol. Chem. 264, 16727-16732.
15. Gould, K.L., Bretscher, A., Esch, F.S. and Hunter, T. (1989) EMBO J. 8, 4133-4142.
16. Hunter, T. and Cooper, J.A. (1981) Cell 24, 741-752.
17. Gould, K.L., Cooper, J.A., Bretscher, A. and Hunter, T. (1986) J.Cell Biol. 102, 660-669.
18. Conboy, J.G., Kan, Y.W., Shohet, S.B. and Mohandaas, N. (1986) Proc. Natl. Acad. Sci. (USA) 83, 9512-9516.
19. Anderson, R.A. and Marchesi, V.T. (1985) Nature 307, 655-658.
20. Bretscher, A. (1986) Methods Enzymol. 134, 24-37.
21. Narvanen, A. (1985) Biochem. J. 231, 53-62.
22. Everett, A.W. and Nichol, K.A. (1989) J. Histochem. Cytochem. 38, 1137-1144.

■ Anthony Bretscher:
Section of Biochemistry,
Molecular and Cell Biology,
Cornell University,
Ithaca, NY, USA

Fascin

Fascin is an actin filament crosslinking protein originally isolated from structured actin gels induced in sea urchin egg extracts by elevated temperature[1,2]. Fascin has been identified in several echinoderm cells including eggs and coelomocytes[3-8] where it is associated with actin filaments in microvilli and filopodia and presumably is responsible for their organization[9].

Purified fascin consists of a single species with an apparent molecular weight of 58,000 on SDS-PAGE. Reconstitution with F-**actin** results in the formation of highly crosslinked filament bundles with a 33-35 nm crossbanding pattern (Figure). Similar bundles with an identical banding pattern have been demonstrated in the structured actin gels which form in sea urchin egg extracts upon warming[1,2,10] and have also been demonstrated in the actin filament cores of coelomocyte filopodia[4,5] and egg microvilli[3]. Immunolocalization studies have demonstrated the protein in sea urchin egg microvilli[4,5], in coelomocyte filopodia[4-7] and in the actin containing "spikes" that form in starfish oocytes following 1-methyladenine treatment[8]. There is a report of fascin in starfish sperm[11].

Optical diffraction and image reconstruction studies of the bundles suggest an approximately parallel arrangement of filaments arranged in a hexagonal lattice having an 8.3 nm spacing. Two characteristic views of negatively stained bundles are apparent in electron micrographs; in one view the filaments and 33-35 nm strips are clearly visible (Figure), in the other view filaments are more difficult to make out and an 11-13 nm transverse striping pattern is prominent[12]. A model with nine "links" per 41 actin monomers has been proposed to account for the observed geometry[12]. The predicted stoichiometry of monomers to links is 41 to 9, in agreement with the chemical determination[13] if each link is a fascin monomer. An example of an actin-fascin bundle is given in the Figure.

■ PURIFICATION

Fascin can be purified by dissolving structured egg actin gel with KI, followed by chromatography on A-5m agarose to remove high molecular weight actin binding proteins. Fractions containing fascin and actin are collected and can be resolved on DEAE-Sephacel[9,14]. There is a report of an actin bundling protein with the same molecular weight as fascin, isolated from porcine brain using a similar purification strategy[15].

■ ACTIVITIES

Beyond the actin bundling activities described above, no other activity has been ascribed to fascin.

■ ANTIBODIES

Rabbit polyclonal antisera against fascin[3] from *Tripneustes gratilla* recognize a 58 kDa protein in *Strongylocentrotus purpuratus*, *Dendraster excentricus* and *Pisaster ochraceus* and have been used to localize fascin in these species as outlined above.

■ GENES

Fascin cDNA's have yet to be cloned.

■ REFERENCES

1. Kane, R.E. (1975) J. Cell Biol. 66, 305-315.
2. Kane, R.E. (1976) J. Cell Biol. 71, 704-714.
3. Otto, J.J., Kane, R.E. and Bryan, J. (1979) Cell 17, 285-293.
4. Otto, J.J., Kane, R.E. and Bryan, J. (1980) Cell Motil. 1, 31-40.
5. Otto, J.J. and Bryan, J. (1981) Cell Motil. 1, 179-192.
6. Edds, K.T. (1977) J. Cell Biol. 73, 479-491.
7. Edds, K.T. (1979) J. Cell Biol. 83, 109-115.
8. Otto, J.J. and Schroeder, T.E. (1984) Dev. Biol. 101, 263-273.
9. Bryan, J. and Kane, R.E. (1978) J. Mol. Biol. 125, 207-224.
10. Mabuchi, I. and Nomomura, Y. (1981) Biomed. Res. 2, 143-153.
11. Maekawa, S., Endo, S. and Sakai, H. (1982) J. Biochem. (Japan) 92, 1959-1972.
12. DeRosier, D., Mandelkow, E., Silliman, A., Tilney, L. and Kane, R.E. (1977) J. Mol. Biol. 113, 679-695.
13. Bryan, J. and Kane, R.E. (1982) Meth. Cell Biol. 25, 176-199.
14. Bryan, J. (1986) Meth. Enzymol. 134, 13-23.

Figure. Negatively stained actin-fascin needle prepared by reconstitution of purified sea urchin egg actin with the "S" solution described[1,2]. The upper half of the needle shows the 33- to 35-nm repeat quite clearly. In this view the projected spacing of the individual filaments, which are well resolved, is d3/2 where d is the center-to-center spacing between hexagonally packed filaments. About midway down the needle the bundle becomes slightly twisted and the projected spacing of the poorly resolved individual filaments is now $d/2$. In this view, the 11- to 13 nm repeat is enhanced and becomes clearly visible. The needle was stained with 1% uranyl acetate. 210,000X.

15.Maekawa, S., Endo, S. and Sakai, H. (1983) J. Biochem. (Japan) 94, 1329-1337.

■ Joseph Bryan:
Baylor College of Medicine,
Dept. Cell Biology,
Houston, TX 77030, USA

■ J.J. Otto:
Purdue University,
Dept. of Biological Sci.
W. Layfayette, IN 47907, USA

■ Robert E. Kane:
University of Hawaii,
Pacific Biomedical Res. Center,
Honolulu, HI 96822, USA

Fimbrin

Fimbrin is a monomeric F-actin bundling protein that is found enriched in microvilli, microspikes, stereocilia, membrane ruffles and adhesion sites of nonmuscle cells. The cDNA derived amino acid sequence of the chicken intestinal protein shows high homology to human plastins.

Fimbrin was originally identified as a major 68 kDa component of the microfilament core bundle of microvilli found on the apical surface of intestinal epithelial cells[1]. Purified fimbrin is an F-**actin** bundling protein *in vitro*[2-4]. Bundling of actin filaments by **villin** and fimbrin results in structures in which all the filaments have the same polarity[3-5]. Interfilament spacing in fimbrin-F-actin bundles is about 9-10 nm[2,6], whereas in fimbrin-villin-F-actin bundles and in native microvillus cores it is about 12-13 nm[6].

Isolation and sequence analysis of the cDNA for chicken intestinal epithelial fimbrin predicts a protein of 630 residues with a molecular weight of 70,894[7]. The sequence

predicts an N-terminal domain of 115 residues that has homology to two of calmodulin's Ca^{2+}-binding sites, and a C-terminal domain of about 500 residues. This domain has two repeats that appear to make up two F-actin binding sites that are homologous to the actin binding domains of α-**actinin** and the N-terminal domains of **dystrophin**, actin binding protein and β-**spectrin**. Fimbrin is probably the chicken homologue of human T- and L-plastins, as fimbrin shows about 71% sequence identity with the plastins and antibodies to fimbrin crossreact with human T- and L-plastins[7].

Fimbrin has been found in essentially all nonmuscle

(A)   (B)

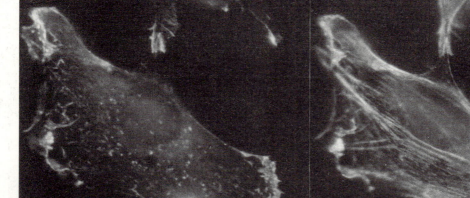

Figure. Localization of fimbrin (A) and actin (B) in a chicken embryo fibroblast. Note that fimbrin is enriched in cell surface structures (microspikes, ruffling membranes and surface microvilli) but is not seen associated with the stress fibre system.

cells; it is either absent or expressed at a very low level in striated and smooth muscles[1]. It is present in cultured non-muscle cells in surface structures that contain an actin cytoskeleton, such as microvilli, microspikes and membrane ruffles[1] (Figure) and stereocilia of auditory sensory cells (hair cells)[8,9]. It is also present in focal contacts of fibroblasts[1] and F-actin rich adhesion structures of other cells[10]. Its expression during development and intestinal epithelial cell differentiation are consistent with it playing a role in bundling actin filaments[11,12].

PURIFICATION

Fimbrin has so far only been purified from brush borders of avian[1-3,13] and vertebrate[14] intestinal epithelial cells. It is readily extracted from these structures and a combination of ammonium sulphate precipitation, gel filtration and chromatography on DEAE and/or Affigel-Blue yields a highly purified protein.

ACTIVITIES

No quantitative assays have been developed to measure the bundling activity of fimbrin. Currently the only assayable property is it's ability to crosslink F-actin *in vitro* into bundles that can be sedimented by low speed centrifugation[4].

ANTIBODIES

A number of polyclonal antibodies have been generated to purified chicken intestinal brush border fimbrin[1,11,12] which work well for immunoblotting, immunofluorescence and immunoprecipitation. At least some of them[1] are species specific. An antibody to porcine brush border fimbrin has also been described[14].

GENES

Full length cDNAs for the chicken intestinal protein have been isolated and sequenced (GenBank X5262)[7].

REFERENCES

1. Bretscher, A. and Weber, K. (1980) J. Cell Biol. 86, 335-340.
2. Bretscher, A. (1981) Proc. Natl. Acad. Sci. (USA) 78, 6849.
3. Bretscher, A. (1982) Cold Spring Harb. Symp. Quant. Biol. 46, 871-879.
4. Glenney, J. R., Kaulfus, P., Matsudaira, P.T. and Weber, K. (1981) J. Biol. Chem. 256, 9283-9288.
5. Coluccio, L.M and Bretscher, A. (1989) J. Cell Biol. 108, 495-502.
6. Matsudaira, P.T., Mandelkow, E., Renner, W., Hesterberg, L.K. and Weber, K. (1983) Nature 301, 209-214.
7. de Arruda, M.V., Watson, S., Lin, C.S., Leavitt, J. and Matsudaira, P. (1990) J. Cell Biol. 111, 1069-1079.
8. Flock, A., Bretscher, A. and Weber, K. (1982) Hear. Res. 6, 75-89.
9. Tilney, M.S., Tilney, L.G., Stephens, R.E., Merte, C., Drenckhahn, D., Cotanche, D.A. and Bretscher, A. (1989) J. Cell Biol. 109, 1711-1723
10. Carley, W., Bretscher, A. and Webb, W.W. (1986) Europ. J. Cell Biol. 39, 313-320.
11. Ezzell, R.M., Chafel, M.M. and Matsudaira, P.T. (1989) Development 106, 407-419.
12. Shibayama, T., Carboni, J.M. and Mooseker, M.S. (1987) J. Cell Biol. 105, 335-344.
13. Bretscher, A. (1986) Methods Enzymol. 134, 24-37.
14. Gerke, V. and Weber, K. (1983) Europ. J. Cell Biol. 31, 249-255.

Anthony Bretscher:
Section of Biochemistry, Molecular and Cell Biology,
Cornell University,
Ithaca, NY, USA

gCap39 (Macrophage Capping Protein, MCP)

gCap39 is a Ca^{2+} and polyphosphoinositide regulated actin filament capping protein. It is probably involved in regulating the actin polymerization/depolymerization cycles of cells during a variety of motile events.

gCap39 is a 39 kDa **actin** filament (plus) end capping protein which is particularly abundant in macrophages[1,2]. gCap39 has 48% sequence identity to **gelsolin**[3], but is distinct from gelsolin and other members of this family in that it does not sever actin filaments. Therefore, gCap39's primary function is capping and uncapping of actin filaments. gCap39 is activated by μM Ca^{2+} and inhibited by polyphosphoinositides (PPI), and so it is likely to be involved in regulating actin polymerization/depolymerization cycles in cells after agonist stimulation. The regulation of gCap39 uncapping differs from that of other

family members: gCap39 dissociates from the filament end either when Ca^{2+} is decreased to submicromolar concentrations, or PPI is increased without necessarily lowering Ca^{2+}-concentration[4]. Since gelsolin uncapping requires both an increase in PPI and a decrease in Ca^{2+}-concentration, gCap39 can generate actin nuclei by dissociating from filament ends under conditions in which gelsolin and other capping proteins remain associated. gCap39 is identical to a protein called MCP, macrophage capping protein[5-7], and mbh1, which has a nuclear as well as cytoplasmic localization[8].

PURIFICATION

gCap39 is most abundant in macrophages. It has been purified by conventional chromatography[5]. A simpler procedure is Ca^{2+}-dependent phenyl-Sepharose chromatography[2,9].

ACTIVITIES

gCap39 capping of actin filaments can be monitored fluorimetrically by its affects on the depolymerization of pyreneiodoacetamide conjugated actin. Other methods include solid state binding to actin-Sepharose.

ANTIBODIES

Goat anti-rabbit gCap39 antibodies[7] and rabbit anti-mouse gCap39 antibodies[2] have been generated.

GENES

Full length cDNA sequence for mouse gCap39 has been published (GenBank M38463). The chromosome localization has not been determined.

REFERENCES

1. Yu, F.-X., Johnston, P.A., Sudhof, T.C. and Yin, H.L. (1990) Science 250, 1413-1415.
2. Johnston, P.A., Yu, F.X., Reynolds, G.A., Yin, H.L, Moomaw, C.R., Slaughter, C.A. and Sudhof, T.C. (1990) J. Biol. Chem. 265, 17946-17952.
3. Kwiatkowski, D.P., Stossel, T.P., Orkin, S.H., Mole, J.E., Coltens, H.R. and Yin, H.L. (1986) Nature 323, 455-458.
4. Janmey, P.A., Iida, K., Yin, H.L. and Stollse, T.P. (1987) J. Biol. Chem. 262, 12228-12236.
5. Southwick, F.S. and DiNubile, M.J. (1986) J. Biol. Chem. 261, 14191-14195.
6. Young, C.L., Southwick, F.S. and Weber, A. (1990) Biochemistry 29, 2232-2240.
7. Young, C.L., Dabiri, G.A. and Southwick, F.S. (1990) J. Cell Biol. 111, 161a.
8. Prendergast, G.C. and Ziff, E.B. (1991) EMBO 10, 757-766.
9. Sudhof, T.C., Ebbecke, M., Walker, J.H., Fritsche, U. and Boustead, C. (1984) Biochemistry 23, 1103-1109.

Fu-Xin Yu and Helen L. Yin:
Department of Physiology,
University of Texas
Southwestern Medical Center,
Dallas, USA

Gelsolin

Gelsolin[1] is a Ca^{2+} and polyphosphoinositide regulated vertebrate actin filament severing protein. It is most likely involved in the restructuring of the actin cytoskeleton in a variety of motile events.

Gelsolin, an 80 kDa protein, is representative of a class of **actin** modulating proteins occurring widely, from lower eukaryotes to mammals, which has powerful effects on actin filament lengths[2]. These include **villin**[3], a microvilli-specific isoform of **gelsolin**, and nonvertebrate **severin**[4] and fragmin[5] which are half as large as gelsolin. Gelsolin has three main effects on actin: (1) it severs actin filaments; (2) it caps the ends of actin filaments and (3) it nucleates actin filament assembly. The combined effect of these interactions is to promote the formation of a large number of short actin filaments which are capped at their barbed ends. The ability to sever is unique to the gelsolin family of proteins, and permits gelsolin to rapidly change actin filament length and number distributions. Gelsolin-actin interactions are promoted by Ca^{2+} and inhibited by polyphosphoinositides[6]. Agonist stimulations of platelets and white blood cells modulate gelsolin: actin interactions[7] and induce gelsolin translocation towards the plasma membrane[8]. The function of gelsolin in cells has not yet been determined conclusively. Microinjection studies gave variable results[9,10] and *Dictyostelium* mutants lacking severin have apparently normal motility[11].

Nonetheless, gelsolin expression is modulated during differentiation[12] and transformation[13], and overexpression of gelsolin by cDNA-mediated transfection enhances stimulated migration of cultured fibroblasts[14].

A slightly larger form of gelsolin (83 kDa in human) is found in plasma at 0.2 mg/ml, and contains a 25 amino acid extension at its N-terminus compared with human cytoplasmic gelsolin. Plasma gelsolin is secreted primarily by muscles[15], and may be involved in the clearance of actin filaments released by damaged tissues. The two forms of gelsolin are derived by alternative transcriptional initiation and message processing from a single gene[16]. A mutation in the gelsolin gene is found in patients with familial amyloidosis, Finnish type, and the 12 kDa amyloid protein is a gelsolin fragment[17].

Recently, several other gelsolin-like proteins have been discovered. Scinderin or adseverin is a 74-79 kDa Ca^{2+}-regulated actin filament severing protein found in adrenal chromaffin cells and other neuroendocrine tissues[18,19]. It is immunologically distinct from gelsolin and has a different amino acid composition but may also have sequence homology to gelsolin. It is inhibited by polyphosphoinositides, but unlike gelsolin, it is also inhibited by phosphatidylserine[20]. **gCap39**, a 39 kDa protein, has extensive sequence homology to gelsolin but is unique in that it caps actin filament but does not sever them[21,22]. The coexistence of several Ca^{2+} and phospholipid modulated proteins suggest that they may act coordinately to regulate actin filament structure in cells.

■ PURIFICATION

Gelsolin is most easily purified from plasma or serum by affinity chromatography[23], Ca^{2+}-dependent elution from ion exchange columns[10], or ATP-elution from Affigel-Blue columns[24]. Recombinant gelsolin has also been expressed in *E. coli*[25] and COS cells[26], and purified in an active form.

■ ACTIVITIES

Gelsolin interactions with actin can be monitored by its effects on the depolymerization and polymerization of pyrene iodoacetamide conjugated actin[27]. Other methods include viscosity, flow birefringence and cosedimentation with actin, or solid state binding to actin-Sepharose.

■ ANTIBODIES

A variety of polyclonal and monoclonal antibodies against gelsolin have been described. A monoclonal antibody to human plasma gelsolin[7,23], which crossreacts with a number of species, is available from Sigma.

■ GENES

Full length cDNAs sequences for human[28], mouse[29], swine[30] gelsolin are published. The human gelsolin gene is localized to chromosome 9 q32-q34[31] and partial genomic sequence is published[16].

■ REFERENCES

1 Yin, H.L. (1987) BioEssays 7(4), 176-178.
2. Matsudaira, P. and Janmey, P.A. (1988) Cell 54, 139-140.
3. Pringault, E., Arpin, M., Garcia, A., Finidori, J. and Louvard, D. (1986) EMBO J. 5, 3119.
4. Andre, E., Lottspeich, F., Schleicher, M. and Noegel, A. (1988) J. Biol. Chem. 263, 722-727.
5. Ampe, C. and Vandekerckhove, J. (1987) EMBO J. 6, 4149-4157.
6. Janmey, P.A. and Stossel, T.P. (1987) Nature 325, 362-364.
7. Chaponnier, C., Yin, H.L. and Stossel, T.P. (1987) J. Exp. Med. 165, 97-106.
8. Hartwig, J.H., Chambers, K.A. and Stossel, T.P. (1989) J. Cell Biol. 108, 467-479.
9. Cooper, J.A., Bryan, J., Schwab III, B. and Loftus, D.J. (1987) J. Cell Biol. 104, 491-501.
10. Huckriede, A., Fuchtbauer, A., Hinssen, H., Chaponnier, C., Weeds, A. and Jockusch, B.M. (1990) Cell Motil. Cytoskel. 16, 229-238.
11. Andre, E., Brink, M., Gerisch, G., Isenberg, G., Noegel, A., Schleicher, J., Segall, E. and Wallraff, E. (1989) J. Cell Biol. 108, 985-995.
12. Kwiatkowski, D.P. (1988) J. Biol. Chem. 263, 13851-13862.
13. Vandekerckhove, J., Bauw, G., Vancompernolle, K., Honore, B. and Celis, J. (1990) J. Cell Biol. 111, 95-102.
14. Cunningham, C.C., Stossel, T.P. and Kwiatkowski, D.J. (1990) Science 251, 1233-1236.
15. Kwiatkowski, D.P., Mehl, R.A., Izumo, S., Nadal-Ginard, B. and Yin, H.L. (1988) J. Biol. Chem. 263, 8239-8243.
16. Kwiatkowski, D.P., Mehl, R.A. and Yin, H.L. (1988) J. Cell Biol. 106, 375-384.
17. Haltia, M., Prelli, F., Ghiso, J., Kiuru, S., Somer, H., Palo, J. and Frangione, B. (1990) Biochem. Biophys. Res. Commun. 167, 927-932.
18. Sakurai, T., Ohm, K., Kurodawa, H. and Nonomura, Y. (1990) Neuroscience 38, 743-756.
19. Rodriguez Del Castillo, A., Lemaire, S., Tchakarov, L., Jeyapragasan, M., Doucet, J.-P., Vitale, M.L. and Trifaro, J.-M. (1990) EMBO J. 9, 43-52.
20. Maekawa, S. and Sakai, H. (1990) J. Biol. Chem. 265, 10940-10942.
21. Yu, F.-X., Johnston, P.A., Sudhof, T.C. and Yin, H.L. (1990) Science, 250, 1413-1415.
22. Johnston, P.A., Yu, F.-X., Reynolds, G.A., Yin, H.L., Moomaw, C.R., Slaughter, C.A. and Sudhof, T.C. (1990) J. Biol. Chem. 265, 17946-17952.
23. Chaponnier, C., Janmey, P.A. and Yin, H.L. (1986) J. Cell Biol. 103, 1473-1481.
24. Yamamoto, H., Terabayashi, M., Egawa, T., Hayashi, E., Nakamura, H. and Kishimoto, S. (1989) J. Biochem. 105, 799-802.
25. Way, M., Gooch, J., Pope, B. and Weeds, A.G. (1989) J. Cell Biol. 109, 593-605.
26. Kwiatkowski, D.P., Janmey, P.A. and Yin, H.L. (1989) J. Cell Biol. 108, 1717-1726.
27. Janmey, P.A., Chaponnier, C., Lind, S.E., Zaner, K.S., Stossel, T.P. and Yin, H.L. (1985) Biochemistry 24, 3714-3723.
28. Kwiatkowski, D.P., Stossel, T.P., Orkin, S.H., Mole, J.E., Coltens, H.R. and Yin, H.L. (1986) Nature 323, 455-458.
29. Dieffenbach, C.W., SenGupta, D.N., Krause, D., Sawzak, D. and Silverman, R.H. (1989) J. Biol. Chem. 264, 13281-13288.
30. Way, M. and Weeds, A. (1988) J. Mol. Biol. 203, 1127-1133.
31. Kwiatkowski, D.J., Westbrook, C.A., Bruns, G.A.P. and Morton, C.C. (1988) Am. J. Hum. Genet. 42, 565-572.

■ *Helen L. Yin:*
Department of Physiology,
University of Texas
Southwestern Medical Center,
Dallas, USA

Hisactophilin

Hisactophilin is a histidine-rich actin binding protein from Dictyostelium discoideum. It binds to F-actin in a pH-dependent and saturable manner, induces actin polymerization in the absence of Mg^{2+} and K^+ and is enriched in the submembranous region of the amoeboid cells. Hisactophilin might act as an intracellular pH-sensor that links chemotactic signals to responses in the microfilament system.

Hisactophilin[1] is encoded by a single gene, the mRNA as well as the protein are present throughout growth and all developmental stages of *Dictyostelium*. It is detected in both soluble and particulate fractions of the cells. Immunofluorescence of cryosections shows an enrichment of hisactophilin close to the plasma membrane (Figure 1). The protein undergoes post-translational modification as indicated by *in vivo* acylation with palmitic acid[2] and by phosphorylation on serine residues[3].

The most characteristic feature of the protein is its content of 31 histidine residues out of 118 amino acids. This unusual amino acid composition seems to be the basis of the pH-dependent **actin** binding activity of hisactophilin. High yield expression of the protein in *Escherichia coli* and NMR spectroscopy[4] allow pH-titration studies that show different conformations at pH 6.0 and pH 7.0. Secondary structure experiments indicate extended β-strand conformations; the occurrence of α-helices in hisactophilin can be excluded. The three-dimensional structure of hisactophilin has been recently solved[6].

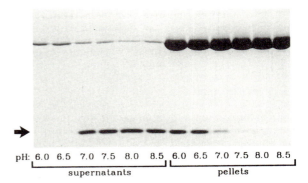

Figure 2. Binding of hisactophilin to actin[1]. G-actin (10 µM) was mixed with hisactophilin (4 µM) and polymerized at the pH-values indicated. The samples were sedimented in an airfuge and the supernatants and pellets analyzed by SDS-PAGE. The position of hisactophilin is indicated by an arrow.

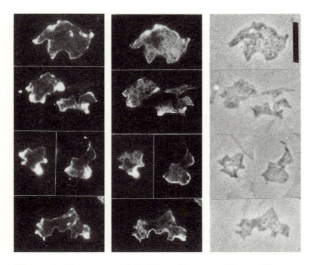

Figure 1. Fluorescent labelling of cryosections with anti-hisactophilin antibodies and phalloidin[1]. The same sections from aggregation competent cells as shown in phase-contrast optics (right) were photographed for FITC fluorescence representing indirect anti-hisactophilin antibody labelling (middle) and for rhodamine fluorescence representing direct phalloidin labelling of F-actin (left). Bar 10 µm.

■ PURIFICATION

The usual purification procedure of hisactophilin exploits its tendency to interact with various gel filtration resins[1]. Soluble fractions of *Dictyostelium* homogenates or bacterial extracts can be loaded directly onto a Sephacryl S-300 column, and essentially pure hisactophilin is eluted slightly behind the included volume. Ammonium sulphate precipitation and anion exchange chromatography can be used for final purification.

■ ACTIVITIES

Hisactophilin cosediments with F-actin in a strongly pH-dependent manner[1] within a critical range of pH 6.5 - 7.5 (Figure 2). At pH 6.5, binding of hisactophilin is essentially complete, at pH 7.5, almost all of the hisactophilin is recovered in the supernatant. The stoichiometry of actin-hisactophilin association suggests a molar ratio of 1:1.

In the presence of hisactophilin, actin polymerization as measured by high shear viscometry, occurs under nonpolymerizing salt conditions at pH 6.0 or 6.5. At pH 7.0, the hisactophilin induced polymerization is slowed down and remains incomplete[1]. These properties indicate that only the cationic form of hisactophilin binds to actin.

ANTIBODIES

Monoclonal antibodies against hisactophilin have been described[1,5]; they do not crossreact with proteins from other species.

GENES

A full length cDNA (GenBank/EMBL Data Bank accession number J04472) has been published[1].

REFERENCES

1. Scheel, J., Ziegelbauer, K., Kupke, T., Humbel, B.M., Noegel, A.A, Gerisch, G. and Schleicher, M. (1989) J. Biol. Chem. 264, 2832-2839.
2. Urban, M. and Gerisch, G. unpublished observations.
3. Müller-Taubenberger, A. and Gerisch, G. unpublished observations.
4. Holak, T. and Schleicher, M. in preparation.
5. Schleicher, M., Gerisch, G. and Isenberg, G. (1984) EMBO J. 3, 2095-2100.
6. Habazetti, J., Gondal, D., Wiltscheck, R., Otlewski, T., Schleicher, M. and Holak, T.A. (1992) Nature 359, 855-858.

■ *Michael Schleicher:*
Max-Planck-Institute for Biochemistry,
8033 Martinsried,
Germany

Insertin

Insertin is a 30 kDa protein isolated from chicken gizzard smooth muscle that binds to the barbed ends of actin filaments[1]. Insertin permits insertion of actin monomers between barbed ends and barbed end bound insertin[2].

Insertin isolated from chicken gizzard smooth muscle has a molecular mass of about 30 kDa according to SDS gel electrophoresis. The protein tends to aggregate. Substoichiometric amounts of insertin retard **actin** polymerization five-fold (at 100 mM KCl, 2 mM MgCl$_2$, pH 7.5, 37°C). Insertin binds strongly to the barbed ends of actin filaments. 10 nM insertin are sufficient to bring about maximal retardation of 1 μM polymeric actin. In contrast to capping proteins, insertin does not block polymerization and depolymerization at the barbed ends. According to a quantitative kinetic analysis, two insertin molecules bind cooperatively to the barbed end of an actin filament and remain bound to the terminal filament subunit during polymerization. Any mechanism can be excluded in which insertin hops off to allow attachment of a new terminal actin subunit and then rebinds. A model which is consistent with all experimental observations is depicted in the Figure. Insertin is suggested to allow polymerization and depolymerization from the plasma membrane bound filament end in the cell.

PURIFICATION

Insertin is purified from chicken gizzard smooth muscle by ammonium sulphate precipitation, DEAE-cellulose, gel filtration (Sephacryl S-300) and hydroxylapatite chromatography. The protein copurifies with **vinculin** and is separated from vinculin by the final hydroxylapatite chromatography step. On SDS-PAGE, insertin can be rendered visible by silver stain and appears as several distinct bands in the range of 29 to 35 kDa.

ACTIVITIES

The occurrence of insertin is assayed by retardation of

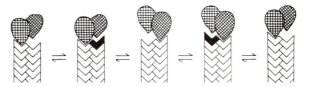

Figure. Mechanism of the insertion of actin monomers between a barbed end and barbed end-bound insertin. Newly incorporated subunits are drawn in black.

actin polymerization. Polymerization of G-actin onto barbed filament ends occupied by insertin is followed by the change of the fluorescence intensity of NBD-labelled actin. Both isolated insertin and insertin-vinculin complex have the same action on actin polymerization. Insertin separated from vinculin looses its retardation activity within a few hours or days.

GENES

No data on the genes of insertin are available.

REFERENCES

1. Ruhnau, K., Gaertner, A. and Wegner, A. (1989) J. Mol. Biol. 210, 141-148.
2. Gaertner, A. and Wegner, A. (1991) J. Muscle Res. Cell Motil. 12, 27-36.

■ *Andrea Gaertner and Albrecht Wegner:*
Institute of Physiological Chemistry,
Ruhr-University,
Bochum, Germany

MARCKS

The myristoylated alanine rich C-kinase substrate (MARCKS)[1] is a prominent substrate of protein kinase C (PCK) in neurons[2] and leukocytes[3], and is also phosphorylated during growth factor dependent mitogenesis in fibroblasts[4,5]. Following phosphorylation by PKC, MARCKS redistributes from the plasma membrane to the cell interior[6,7], and upon dephosphorylation, it reassociates with the membrane. MARCKS is a calmodulin binding protein[8], and it also binds and crosslinks actin filaments, an effect that is inhibited by both phosphorylation and calcium-calmodulin[9]. MARCKS probably has a role in linking cell activation via PKC to cytoskeletal changes.

MARCKS is a major acidic phosphoprotein of apparent M_r = 68,000-87,000 (depending on the cell type) that binds calmodulin and is specifically phosphorylated by PKC during neurosecretion[2,10], growth factor dependent mitogenesis[4,5], and leukocyte activation[3,11,12]. The migration of this protein on SDS-PAGE is anomalous since the polypeptide size predicted from cDNA clones of bovine[1], chicken[13], and murine[14] MARCKS is 27-32 kDa but transfection of each of the cloned cDNAs in MARCKS-negative TL-K cells produced a protein that was structurally and functionally indistinguishable from native MARCKS. The high degree of extended helix predicted from the primary structure may perturb its migration during electrophoresis[1,13-15]. Overall homology among bovine[1], chicken[13], and murine[14] MARCKS is approximately 70%, with two very highly conserved domains: an N-terminal myristoylation site which comprises the membrane binding domain[16] and a highly basic α-helical domain of about 25 residues in the middle of the polypeptide chain. This helical region contains all four of the C-kinase phosphorylation sites[1,14,17], a calmodulin binding site[8], and an **actin** binding site[9]. MARCKS synthesis, myristoylation and association with the plasma membrane in leukocytes are stimulated by bacterial lipopolysaccharides and tumour necrosis factor[3,11,12]. Cellular transformation by viral oncogenes is accompanied by down regulation of MARCKS[18,19].

MARCKS is encoded by a single gene[1,13,14]. Both immunoreactive protein and mRNA concentrations are high in brain, spinal cord, spleen and lung[1]. In brain, immunoreactivity is greatest in axons, axon terminals, glial cells, and small dendritic branches, but is not observed in large dendrites, somata, or nuclei[20]. In macrophages MARCKS colocalizes with **vinculin**, **talin**, and PKC in punctate structures on the substrate adherant surface of filopodia, and upon activation of PKC, MARCKS is released from these sites[7]. Immuno-EM studies detect MARCKS on the sides of actin filaments at points of contact with the plasma membrane[9]. In neuronal processes it also colocalizes with microtubules[20]. When PKC is activated by phorbol esters or other agonists, MARCKS is phosphorylated on two to four serines[17], redistributed from the plasma membrane to the cell interior[3,6,7], and upon its dephosphorylation, the protein reassociates with the plasma membrane[22].

■ PURIFICATION

The 87-90 kDa bovine brain protein has been purified to homogeneity by a sequence involving chromatography on DE-52, ammonium sulphate precipitation, and chromatography using DEAE-Sephacel, Bio-gel HTP, and Sephacryl S-400, followed by FPLC ProRPC chromatography in 0.1% CF_3COOH[15]. The purified protein is renatured by dialysis in 10 mM Tris-HCl, pH 7.6 with 14 mM 2-mercaptoethanol. MARCKS is partially degraded by freeze/thaw cycles but is stable at 4°C. Membrane bound MARCKS can be released by 1% Triton X-100 and behaves similarly on DE-52 and Biogel HTP. An alternate protocol utilizing the unusual properties of acid and heat stability has also been described[10].

■ ACTIVITIES

The physiologic function of MARCKS is unknown. Its association with actin, calmodulin, and perhaps microtubules, and the capacity of MARCKS to cycle to and from the plasma membrane depending on its phosphorylation state, suggests that it may reversibly link actin to the plasma membrane. Phosphorylation of MARCKS is frequently used as a marker for PKC activation in nerve cells, leukocytes, fibroblasts and other cell types.

■ ANTIBODIES

Polyclonal antibodies raised in rabbits against bovine and murine brain MARCKS[3,11,21] are effective in immunoprecipitating the protein in all species tested (human, mouse, bovine, chicken, and rat). However, the antibodies are useful in immunoblotting and immunofluorescence studies only of the species to which the antibody was raised.

■ GENES

Bovine[1] (GenBank M23738) chicken[13], and murine[14] (GenBank M60474) brain MARCKS have been cloned and sequenced.

■ REFERENCES

1. Stumpo, D.J., Graff, J.M., Albert, K.A., Greengard, P. and Blackshear, P.J. (1989) Proc. Natl. Acad. Sci. (USA) 86, 4012-4016.
2. Wu, W., Walaas, S., Naim, A. and Greengard, P. (1982) Proc. Natl. Acad. Sci. (USA) 79, 5249-5253.
3. Aderem, A.A., Albert, K.A., Keum, M.M., Wang, J.K., Greengard, P. and Cohn, Z.A. (1988) Nature 332, 362-364.
4. Rosengurt, E., Rodriguez-Pena, M. and Smith, K.A. (1983) Proc. Natl. Acad. Sci. (USA) 80, 7244-7248.

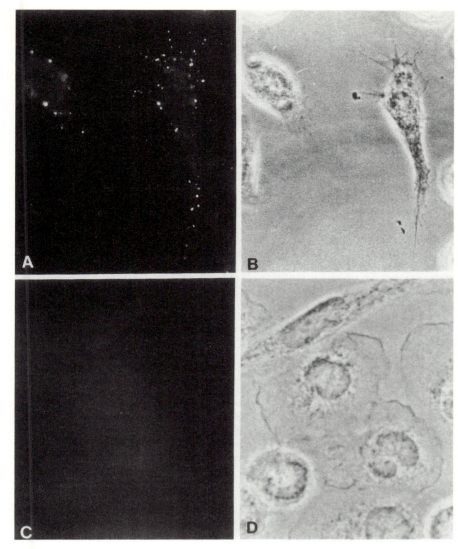

Figure. Phase micrograph (right panel) and indirect immunofluorescent staining (left panel) of murine macrophages with affinity purified rabbit antibodies to murine MARCKS visualized with FITC labelled goat anti-rabbit Fab'$_2$. (A,B) shows control, unstimulated macrophages. (C,D) shows PMA treated cells[7].

5. Blackshear, P., Wen, L., Glynn, B. and Witter, L. (1986) J. Biol. Chem. 261, 1459- 1469.

6. Wang, J.K., Walaas, S.I., Sihra, T.S., Aderem, A. and Greengard, P. (1989) Proc. Natl. Acad. Sci. (USA) 86, 2253-2256.

7. Rosen, A., Keenan, K., Thelen, M., AC, N. and Aderem, A. (1990) J. Exp. Med. 172, 1211-1215.

8. Graff, J.M., Young, T.N., Johnson, J.D. and Blackshear, P.J. (1989) J. Biol. Chem. 264, 21818-21823.

9. Hartwig, J.H., Janmey, P.A., Rosen, A., Thelen, M., Naim, A.C. and Aderem, A.(1990) J. Cell Biol. 111, 8a.

10. Patel, J. and Kligman, D. (1987) J. Biol. Chem. 262, 16686-16691.

11. Rosen, A., Naim, A.C., Greengard, P., Cohn, Z.A. and Aderem, A. (1989) J. Biol. Chem. 264, 9118-9121.

12. Thelen, M., Rosen, A., Naim, A.C. and Aderem, A. (1990) Proc. Natl. Acad. Sci. (USA) 87, 5603-5607.

13. Graff, J.M., Stumpo, D.J. and Blackshear, P.J. (1989) Mol. Endocrinol. 3, 1903-1906.

14. Seykora, J., Ravetch, J. and Aderem, A. (1991) Proc. Natl. Acad. Sci. (USA) 88, 2505-2509.

15. Albert, K.A., Naim, A.C. and Greengard, P. (1987) Proc. Natl. Acad. Sci. (USA) 84, 7046-7050.

16. Graff, J.M., Gordon, J.I. and Blackshear, P.J. (1989) Science, 246, 503-506.

17. Graff, J.M., Stumpo, D.J. and Blackshear, P.J. (1989) J. Biol. Chem. 264, 11912-11919.

18. Simek, S., Kligman, D., Patel, J. and Colbum, N. (1989) Proc. Natl. Acad. Sci. (USA) 86, 7410-7414.

19. Wolfman, A., Wingrove, T., Blackshear, P. and Macara, I. (1987) J. Biol. Chem. 262, 16546-16552.

20. Ouimet, C.C., Wang, J.K., Walaas, S.I., Albert, K.A. and Greengard, P. (1990) J. Neurosci. 10, 1683-1698.

21. Thelen, M., Rosen, A., Naim, A. and Aderem, A. (1991) Nature 351, 320-322.

22. Albert, K.A., Walaas, S.I., Wang, J.K. and Greengard, P.(1986) Proc. Natl. Acad. Sci. (USA) 83, 2822-2826.

■ Paul A. Janmey and John H. Hartwig:
Dept. of Medicine, Harvard Medical School,
Boston, Massachusetts, USA

■ Alan Aderem,
Lab. of Cell. Phys. and Immunol.,
Rockefeller Univ., New York, USA

Myomesin and M-Protein

Myomesin and M-protein are a family of structural sarcomeric proteins with molecular weights of 165-195,000 which typically localize in the M-band region of skeletal and heart myofibrils. The biological functions of these proteins are not yet elucidated but they may be involved in the stabilization of the thick filament bundles[1,2] and/or the linking of the M-band to titin, which then serves to connect the myofibrils to the Z-disc[3].

At least three protein components have been described to be localized in the M-band region of the sarcomere: myomesin, M-protein and M-type creatine kinase (MM-CK). The high molecular weight components were originally described as M-protein[4] until the heterogeneity of the fraction was revealed by monoclonal antibodies. The M-protein fraction consists of the 165 kDa M-protein and the 185 kDa protein myomesin[5,6]. Both make up ~0.3% of the high ionic strength extractable myofibrillar protein. Electron microscopic measurements of a rotary shadowed M-protein preparation from skeletal fast muscle, which very likely contained both M-protein and myomesin, yielded elongated molecular dimensions of 4 x 36 nm[7]. Although chicken heart tissue is devoid of an electron dense M-band[8] the presence of the high molecular weight M-band proteins was shown by immunological techniques[6,9] and the application of myomesin specific monoclonal antibodies demonstrated the existence of heart specific isoproteins with 190 and 195 kDa[10].

During development of striated muscle tissue in chicken, myomesin accumulation was observed in six to seven day old embryos, while M-protein expression lagged behind[10]. Independent regulation of the genes for M-protein and myomesin was also observed during development of chicken pectoralis[11]. Myomesin was found to be expressed in all types of chicken muscle fibres, but the M-protein was absent from the slow fibres type I, and type III[12]. For developing rat heart, myomesin is present in all myocytes, while M-protein and MM-CK only accumulate in animals older than about four weeks. In rat

skeletal muscle, however, all three types of M-band proteins had accumulated by birth and a possible different sequence for myofibrillogenesis in the skeletal muscle fibres was postulated. Myomesin and M-protein were always organized in striations in cultured muscle cells (Figure), indicating their usefulness as markers for muscle differentiation and myofibrillogenesis.

So far no direct assays of the function of these proteins are available and the study of their localization has been the only approach to functional analysis. Myomesin and M-protein are localized in the M-band region of sarcomeres from skeletal and cardiac myofibrils where these proteins are bound rather tightly to the rod segments packed in an antiparallel fashion. This localization persists even in isolated A-segments[1] or in purified bare zone assemblages[2]. Immunoelectron microscopy showed that myomesin and M-protein were localized along thick filaments between the substriations visible in the EM of the M-band named 6 and 6' where they possibly contribute to these transverse M-bridges crosslinking thick filaments in the M-band[13]. A potential role for myomesin and possibly also for M-protein as linkers of thick filaments to the longitudinal **titin** filament system has also been suggested[3].

■ PURIFICATION

The purification of myomesin and M-protein was achieved by classical column chromatography of extracted myofibrillar material[14]. Contaminating phosphorylase b can be removed by chromatography on 5'-AMP Sepharose[15]. M-

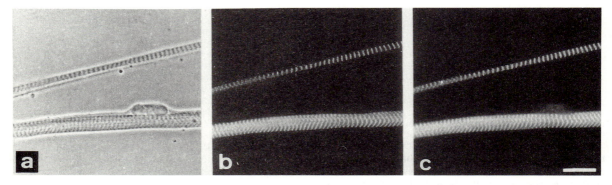

Figure. Accumulation of myomesin and M-protein in cultures of differentiated myotubes. Well differentiated myogenic cultures were fixed with 3% paraformaldehyde and visualized (a) by phase-contrast and double immunofluorescence (b,c). In (b) the primary antibody was the mouse monoclonal antibody B4 (anti-myomesin) and in (c) the staining was carried out with affinity purified rabbit antibodies against chicken M-protein. Bar 20 μm.

protein and myomesin can be partly separated by chromatography on DEAE-cellulose[5].

◼ ANTIBODIES

Polyclonal antibodies against the protein fraction isolated from skeletal muscle with an average molecular weight of 165,000 react with myomesin and M-protein[2,16] (this fraction probably also contained a 165 kDa degradation product of myomesin[5]). Monoclonal antibodies were generated against the 165 kDa fraction and they can be divided into antibodies that react with either M-protein or myomesin. Antibody mMaC myomesin B4 reacts specifically with myomesin, while mMaC M-protein A6 reacts with M-protein. Both these antibodies have been used for immunoblotting, immunohistochemistry and immunoprecipitation.

◼ GENES

cDNAs have been obtained recently and are now being sequenced (Perriard et al., unpublished).

◼ REFERENCES

1. Doetschman, T.C. and Eppenberger, H.M. (1984) Eur. J. Cell Biology 33, 265-274.
2. Bähler, M., Wallimann, T. and Eppenberger, H.M. (1985) J. Muscle Res. Cell Motil. 6, 783-800.
3. Nave, R., Fürst, D.O. and Weber, K. (1989) J. Cell Biol. 109, 2177-2187.
4. Masaki, T. and Takaiti, O. (1974) J. Biochemistry 75, 367-380.
5. Grove, B.K., Kurer, V., Lehner, C., Doetschman, T.C., Perriard, J.-C., Eppenberger, H.M. (1984) J. Cell Biol. 98, 518-524.
6. Eppenberger, H.M., Perriard, J.-C., Rosenberg, U.B. and Strehler, E.E. (1981) J. Cell Biol. 89, 185-193.
7. Woodhead, J.I. and Lowey, S. (1982) J. Mol. Biol. 157, 149-154.
8. Wallimann, T., Kuhn, H.J., Pelloni, G., Turner, D.C. and Eppenberger, H.M. (1977) J. Cell Biol. 75, 318-325.
9. Strehler, E.E., Pelloni, G., Heizmann, C. and Eppenberger, H.M. (1980) J. Cell Biol. 86, 775-783.
10. Grove, B.K., Cerny, L., Perriard, J.-C. and Eppenberger, H.M. (1985) J. Cell Biol. 101, 1413-1421.
11. Grove, B.K., Holmbom, B. and Thornell, L.-E. (1987) Differentiation 34, 106-114.
12. Grove, B.K., Cerny, L., Perriard, J.-C., Eppenberger, H.M. and Thornell, L.-E. (1989) J. Histochem. Cytochem. 37, 447-454.
13. Strehler, E.E., Carlsson, E., Eppenberger, H.M. and Thornell, L.-E. (1983) J. Mol. Biol. 166, 141-158.
14. Trinick, J. and Lowey, S. (1977) J. Mol. Biol. 113, 343-368.
15. Eppenberger, H.M. and Strehler, E.E. (1982) Meth. Enzymology 85, 139-149.
16. Strehler, E.E., Pelloni, G., Heizmann, C.W. and Eppenberger, H.M. (1979) Expte. Cell Res. 124, 39-45.

◼ Jean-Claude Perriard:
Institute for Cell Biology,
Swiss Federal Institute of Technology,
CH-8093 Zurich Switzerland

Nebulin

Nebulin is a giant actin binding protein that serves as a length regulating template for the thin filament. This protein ruler constitutes a fourth filament of the skeletal muscle sarcomere.

Nebulin is a family of giant muscle proteins with isoforms ranging from 650 to 850 kDa in mass in most vertebrate skeletal muscles[1-5]. Found exclusively in the skeletal muscle sarcomere and absent in the cardiac muscle, nebulin accounts for 2-3% of myofibrillar protein[3-5]. Sequencing studies of human nebulin cDNA indicate that the nebulin polypeptide is composed nearly entirely of a recurring motif of a 35-residue module and a higher order seven-module 240-residue repeat[6,7]. It is thought that the long chain of more than 200 modules forms a structural scaffold or template which matches the three dimensional contour and charge profiles of the **actin** filament[4-8].

In the sarcomere, a single nebulin polypeptide spans ~ 1 µm and attaches its C-terminus to the Z-line[4-6]. It constitutes a set of inextensible filaments that are associated with actins to form composite thin filaments[4,5,8-10]. Nebulin appears to serve as a length determining protein ruler in many skeletal muscles, since the characteristic length of thin filaments is proportional to the size of expressed nebulin isoform[5,7]. In developing muscles, nebulin may participate in the formation of I-Z-I bundles[11].

Nebulin is degraded somewhat in muscle tissues of Duchenne muscular dystrophy patients[12].

◼ PURIFICATION

Nebulin polypeptides are readily purified from SDS-solubilized myofibrils or muscle tissues by a novel NaCl fractionation step followed by gel filtration and ion exchange chromatography[4,5,13].

◼ ACTIVITIES

Cloned human nebulin fragments containing 6 to 17 of the 35-residue modules bind to F-actin with high affinity[8]. Nebulin also possesses binding sites for α-**actinin** and calmodulin[14,15]. Nebulin is a phosphoprotein[16].

◼ ANTIBODIES

Polyclonal and monoclonal antibodies to nebulin are generally crossreactive to nebulin isoforms from a variety of vertebrate species[2,4,5]. One monoclonal anti-nebulin is available from Sigma[11].

■ GENES

18 kB of human nebulin cDNA[6,7] and 2 kB of rabbit nebulin cDNA (EMBL X58122 and X58123) have been sequenced[7]. Human nebulin gene is localized to chromosome 2 q31-32[17,18].

■ REFERENCES

1. Wang, K. (1985) Cell Muscle Motil. 6, 315-369.
2. Wang, K. and Williamson, C.L. (1980) Proc. Natl. Acad. Sci. (USA) 77, 3254-3258.
3. Hu, D.H., Kimura, S. and Maruyama, K. (1986) J. Biochem. (Tokyo) 99, 1485-1492.
4. Wang, K. and Wright, J. (1988) J. Cell Biol. 107, 2199-2212.
5. Kruger, M., Wright, J. and Wang, K. (1991) J. Cell Biol. 115, 97-107.
6. Wang, K., Knipfer, M., Huang, Q.Q., Hsu, L., van Heerden, A., Browning, K., Quian, E. and Stedman, H. (1990) J. Cell Biol. 111, 428a.
7. Labiet, S., Gibson, T., Lakey, A., Leonard, K., Zeviani, M., Knight, P., Wardale, J. and Trinick, J. (1991) FEBS Lett. 282, 313-316.
8. Jin, J.-P. and Wang, K. (1991) J. Biol. Chem. 266, 21215-21223.
9. Pierobon-Bormioll, S., Betto, R. and Salviati, G. (1989) J. Musc. Res. Cell Motil. 10, 446-456.
10. Maruyama, K., Matsuno, A., Higuchi, H., Shimaoka, S., Kimura, S. and Shimizu, T. (1989) J. Muscle Res. Cell Motil. 10, 350-359.
11. Furst, D.O., Osborn, M. and Weber, K. (1989) J. of Cell Biol. 109, 517-527.
12. Bonilla, E., Miranda, A.F., Prelle, A., Salviati, G., Betto, R., Zeviani, M., Schon, E.A., DiMauro, S. and Rowland, L.P. (1988) Neurology 38, 1600-1603.
13. Wang, K. (1982) Meth. Enzymol. 85, 264-274.
14. Nave, R., Furst, D.O. and Weber, K. (1990) FEBS Lett. 269, 163-166.
15. Patel, K., Strong, P.N., Dubowitz, V. and Dunn, M.J. (1988) FEBS Lett. 234, 267-271.
16. Somerville, L. and Wang, K. (1988) Arch. Biochem. Biophys. 262, 118-129.
17. Stedman, H., Browning, K., Oliver, N., Oronzi-Scott, M., Fischbeck, K., Sarkar, S., Sylvester, J., Schmickel, R. and Wang, K. (1988) Genomics 2, 1-7.
18. Zeviani, M., Darras, B.T., Rizzuto, R., Salviati, G., Betto, R., Bonilla, E., Miranda, A.F., Du, J., Samitt, C., Dickson, G., Walsh, F.S., Dimauro, S., Francke, U. and Schon, E.A. (1988) Genomics 2, 249-256.

■ Kuan Wang:
Department of Chemistry and Biochemistry,
University of Texas,
Austin, TX, USA

Nuclear Actin Binding Protein (NAB)

Nuclear Actin Binding protein (NAB)[1] is a nuclear protein composed of two 34 kDa subunits that binds actin with a K_d of about 25 μM. Although relatively little is know about its function, it has a basic pI and binds DNA. It has been purified from Acanthamoeba nuclei and Xenopus oocyte nuclei[2] and localized to nuclei of human skeletal muscle, glial cells and retinal epithelial cells[3].

NAB was discovered via a fortuitous antigenic crossreactivity with *Acanthamoeba* **myosin-I**[4]. Monoclonal antibodies showed a nuclear labelling pattern that was not attributable to the 130 kDa myosin-I but rather to a smaller protein that was not a degradation product of myosin-I[4]. Initial studies comparing this protein to myosin-I showed little similarity other than the antigenic epitopes of nine monoclonal antibodies that all localize to a relatively small, unique region near the C-terminus of NAB. Some polyclonal antibodies to myosin-I also crossreact with NAB, however, two polyclonal antibodies generated against NAB do not react with myosin-I. Purified NAB has a molecular mass of about 65 kDa as calculated from its 4 nm Stokes' radius and sedimentation coefficient of 3.6 S. Comigration with a 34 kDa standard protein by SDS-PAGE suggests that the native form is a dimer. It binds **actin** in an ATP insensitive manner with a K_d of about 0.25 μM. DNA binding activity is also suggested by an overlay technique (southwestern blot[5]), although the specificity and significance of this interaction are not yet known. Monoclonal and polyclonal antibody immunoflourescence studies in *Acanthamoeba* show a diffuse nucleoplasmic staining pattern with some concentration around, but not within, the nucleolus of some cells (Figure A). Immunolocalization in a variety of human tissues, including skeletal muscle (Figure B) shows a similar pattern.

A protein similar to NAB was described in *Xenopus laevis*[2]. It was detected and purified by virtue of its ability to decrease the viscosity of actin gels in a Ca^{2+}-insensitive manner but its nuclear localization and physical properties suggest that it is the *Xenopus* version of *Acanthamoeba* NAB. The estimated native molecular weight, the Stokes' radius, the sedimentation coefficient and subunit composition and localization in the nucleus, all suggest that the proteins are related. Experiments specifically addressing the existence of shared epitopes between these two proteins have not yet been done.

■ PURIFICATION

NAB has been purified chromatographically from *Acanthamoeba* nuclei after salt extraction and ammonium sulphate precipitation. Good purification is

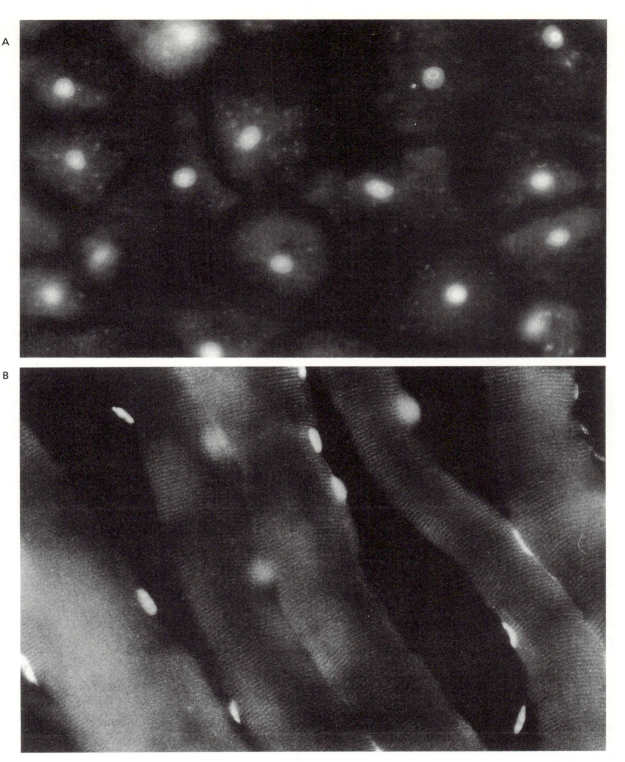

Figure. (A) Immunolocalization of NAB in *Acanthamoeba* nuclei (A) and human skeletal muscle nuclei (B). (A) Affinity purified polyclonal antibody JH-26 was used on acetone fixed cells and visualized with rhodamine conjugated goat anti-rabbit antibodies. (B) Monoclonal antibody M1.7 was used on frozen, acetone fixed human skeletal muscle and visualized with rhodamine conjugated goat anti-mouse antibodies.

obtained using hydroxylapatite, sizing with Sephacryl S-300, and enriching on Heparin-agarose. The *Xenopus* version of the protein was purified in a similar manner using DEAE and Mono-Q ion exchange, followed by hydroxylapatite chromatography.

ACTIVITIES

NAB binds actin with a K_d of about 25 µM. It also binds DNA in an as yet unspecified manner. The *Xenopus* NAB inhibits actin polymerization in a Ca^{2+}-insensitive fashion. The physiologic role of these proteins in the nucleus is not yet known.

ANTIBODIES

Nine monoclonal antibodies produced against *Acanthamoeba* myosin-I bind to NAB protein. One antibody (M1.5) binds NAB much more strongly than myosin-I. Two polyclonal antibodies have been produced against NAB and do not crossreact with myosin-I. Polyclonal antibodies to *Xenopus* NAB are also described[2]. Antibodies to both proteins show wide crossreactivity with other organisms.

GENES

No sequence data is available for *Acanthamoeba* NAB protein. A partial cDNA clone for *Xenopus* NAB has been isolated and sequenced[2]. There is some homology between this sequence and that of a similar sized actin capping protein (**small actin crosslinking protein**) from *Dictyostelium*.

REFERENCES

1. Rimm, D.L. and Pollard, T.D. (1989) J.Cell Biol. 109, 585-591.
2. Ankenbauer, T., Kleinschmidt, J.A., Walsh, M.J., Weiner, O.H. and Franke, WW. (1989) Nature 342, 822-825.
3. Madsen, J. personal communication.
4. Hagen, S.J., Kiehart, D.P., Kaiser, D.A. and Pollard, T.D. (1986) J. Cell Biol. 103, 2121-2128.
5. McKay, R.D.G. (1981) J. Mol. Biol. 145, 471-488.

■ *David L. Rimm:*
Dept. of Pathology,
Yale Medical School,
New Haven, CT, USA

Paramyosin

Paramyosin is a two-chain α-helical coiled-coil forming the core of myosin thick filaments in invertebrate muscles. It is present in especially large amounts in molluscan "catch" muscles. Paramyosin has a molecular weight of ~200,000, and is ~120 nm in length. Both the self-assembly of the molecule and its interactions with the rod region of myosin appear to be determined by long range repeats of charged residues, which are similar for the two molecules.

Genetic experiments in *C. elegans* indicate that paramyosin is probably essential for the functional organization of invertebrate thick filaments[1,2]. There is a rough correlation between thick filament length and paramyosin content[3]. Since tension development in a muscle is proportional to thick filament length, muscles with large amounts of paramyosin develop large tensions. These same muscles are specialized for catch contraction, and it has been suggested that paramyosin has a role in this mechanism[4,5]. There is also an inverse relation between speed of contraction and paramyosin content.

There is a close similarity in the fundamental design of various paramyosins (Figure a). All display the characteristic heptad repeat common to α-helical coiled-coil proteins[6], with a dominance of leucine residues on internal positions. The lengths of the helical regions are identical in the paramyosins thus far sequenced. Moreover, five interruptions in the heptad repeats ("skip" residues producing pitch alterations) occur in identical places; these positions are homologous to those in **myosin** rods and appear to be a conserved feature critical in assembly

(Figure b). Both the N- and C-termini have nonhelical extensions, but these differ in various paramyosins.

In addition to the short range repeats, paramyosin, as other fibrous proteins, is characterized by long range periodicities in its linear sequence related to its evolution and assembly. As in the case of myosin rod[7], there is a strong 28-residue repeat of alternating charge, which has probably evolved from gene duplication. Also, as in myosin rod, there is a longer related repeat of 196 residues (7x28)[8,9]. Such periodicities can account in part for the commonly occurring axial shifts between molecules (e.g. 14.5 nm, 29 nm, 43 nm 72 nm) found in both native filaments and *in vitro* paracrystals[8-11] (Figure c,d). Both the self-assembly and coassembly of myosin and paramyosin appear to be determined by surface charge interactions.

Paramyosin can be phosphorylated - generally on the serine residues - but the degree of phosphorylation and the location of the phosphorylation sites differ in different species. A recent report suggests that catch muscle paramyosin is more readily phosphorylated than the paramyosin of noncatch muscles[12].

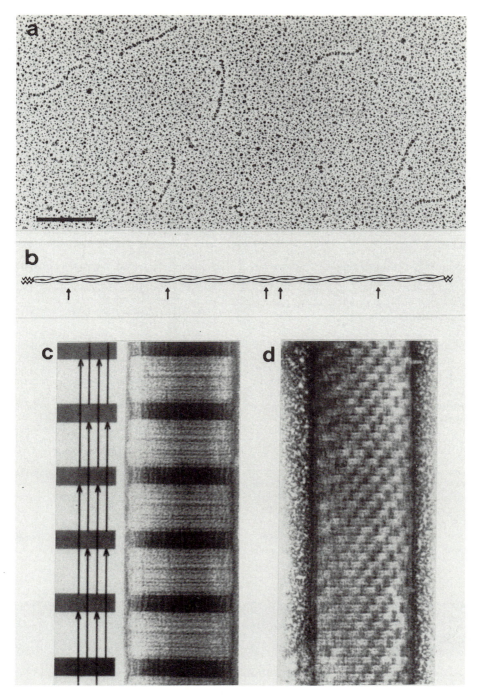

Figure. (a) Electron micrograph of rotary shadowed *Schistosome* paramyosin. Bar 100 nm. (b) Model of paramyosin derived from sequence studies. The two α-helices are parallel and in axial register. Breaks in the heptad repeats ("skips") along the rod are indicated by upward arrows. (c) 2-D representation of molecular packing in the so-called PI form of negatively stained paramyosin paracrystals showing a 72 nm repeat[10]. (d) Electron micrograph of a negatively stained thick filament from *Mercenaria* white muscle from which myosin had been removed. The underlying core of paramyosin shows a characteristic pattern of dark staining nodes that can be accounted for by axial shifts of subfilaments having the PI form[10,11] (shown in (c)).

People and animals infected with a variety of parasitic worms have strong antibody titers to paramyosin, which is under study as a possible vaccine. Serum from people infected with the parasite that causes river blindness, *Onchocerca volvulus*, has antibodies that preferentially recognize the N-terminus of the paramyosin[13]. In infections with *Taenia solium*, the pork tapeworm which causes cysticercosis, antigen-B is an immunodominant protein[14]. The amino acid sequence of this antigen is highly homologous to that of the paramyosin of the trematode *Schistosoma mansoni* so that antigen B is the paramyosin of *Taenia solium*[15] (and J.P. Laclette and C.B. Shoemaker, personal communication). It is of interest that paramyosin, an intracellular muscle protein, appears to be localized also in the tegument of certain platyhelminth parasites and may be a secretory product[14,16].

■ PURIFICATION

Paramyosin can be purified by exploiting its resistance to denaturation by organic solvents, such as ethanol, acetone, acid, or by its different solubility properties from myosin and actomyosin[17]. Proteolytic inhibitors as well as EDTA are added to minimize proteolysis.

■ ANTIBODIES

Polyclonal antisera against a variety of paramyosins are available and tend to react with paramyosins of different species. Monoclonal antibodies to the paramyosin of *C. elegans*[2,9,18], *Schistosoma mansoni*[19], and *Limulus polyphemus*[20] have been characterized.

■ GENES

DNA sequence analysis has been carried out on the paramyosin from the helminth nematode *C. elegans*[9] and the filarial nematodes *D. immitis* and *O. volvulus*[21]. These protein sequences are more than 90% identical[21]. The paramyosins of *S. mansoni*[15], *Drosophila melanogaster*[22] and *Taenia solium*[15] (and J.P. Laclette and C.B. Shoemaker, personal communication) have also been cloned and sequenced.

■ REFERENCES

1. Waterston, R.H., Fishpool, R.M. and Brenner, S. (1977) J. Mol. Biol. 117, 679-697.
2. Epstein, H.F., Ortiz, I. and Berliner, G.C. (1987) J. Musc. Res. Cell Motil. 8, 527-536.
3. Levine, R.J.C., Elfvin, M., Dewey, M.M. and Walcott, B. (1976) J. Cell Biol. 71, 273-279.
4. Achazi, R.K.A. (1979) Pfluegers Arch. 379, 197-201.
5. Cohen, C. (1982) Proc. Natl. Acad. Sci. (USA) 79, 3176-3178.
6. Cohen, C. and Parry, D.A.D. (1990) Proteins: Structure, Function and Genetics 7, 1-15.
7. McLachlan, A.D. and Karn, J. (1983) J. Mol. Biol. 164, 605-626.
8. Cohen, C., Lanar, D.E. and Parry, D.A.D. (1987) Bioscience Reports 7, 11-16.
9. Kagawa, H., Gengyo, K., McLachlan, A.D., Brenner, S. and Karn, J. (1989) J. Mol. Biol. 207, 311-333.
10. Cohen, C., Szent-Györgyi, A.G. and Kendrick-Jones, J. (1971) J. Mol. Biol. 56, 223-237.
11. Szent-Györgyi, A.G., Cohen, C. and Kendrick-Jones, J. (1971) J. Mol. Biol. 56, 239-258.
12. Watabe, S., Kantha, S.S. and Hashimoto, K. (1990) Comp. Biochem. Physiol. 96B, 81-88.
13. Steel, C., Limberger, R.J. and McReynolds, L.A. (1990) J. Immunol. 145, 3917-3923.
14. Laclette, J.P., Merchant, M. and Williams, K. (1987) J. Parasitol. 73, 121-129.
15. Laclette, J.P., Landa, A., Arcos, L., Willms, K., Davis, A.E. and Shoemaker, C.B. (1991) Mol. Biochem. Parasitol. 44, 287-295.
16. Matsumoto, Y., Perry, G., Levine, R.J.C., Blanton, R., Mahmoud, A.A.F. and Aikawa, M. (1988) Nature 333, 76-78.
17. Levine, R.J.C., Elfvin, M.J. and Sawyna, V. (1982) Methods Enzymol. 85, 149- 160.
18. Ardizzi, J.P. and Epstein, H.F. (1987) J. Cell Biol. 105, 2763-2770.
19. Pearce, E.J., James, S.L., Dalton, J., Barrall, A., Ramos, C., Strand, M. and Sher, A. (1986) J. Immunol. 137, 3593-3600.
20. Levine, R.J.C., Kensler, R.W. and Levitt, P. (1986) Biophys. J. 49, 135-138.
21. Limberger, R.J. and McReynolds, L.A. (1990) Mol. Biochem. Parasitol. 38, 271-280.
22. Becker, K.D. and Bernstein, S.I. (1990) J. Cell Biol. 111, 287a.

■ *Carolyn Cohen:*
Rosenstiel Basic Medical Sciences Research Center,
Brandeis University,
Waltham, MA, USA

Ponticulin

Ponticulin is an integral, actin binding, plasma membrane glycoprotein. Ponticulin binds directly to F-actin and also is involved in the nucleation of actin filament assembly at the cytoplasmic surfaces of plasma membranes.

Purified *Dictyostelium discoideum* ponticulin consists of six major 17 kDa isoforms with pI's ranging from 4.2 to 5.2 on 2-D SDS-PAGE[1,2]. In addition, minor polypeptides at 19 kDa (five isoforms, pI range 4.4 to 5.2) and 15 kDa (two isoforms, pI's 4.3 and 4.7) coisolate with ponticulin on F-**actin** affinity columns[2,3]. Because all of these polypeptides bind F-actin directly[2], ponticulin and the 19 and 15 kDa polypeptides may form a related family of integral membrane actin binding proteins. The 15 kDa polypeptide, in particular, is expected to be structurally similar to ponticulin because antibodies affinity purified against the 17 kDa ponticulin isoforms also recognize the 15 kDa iso-

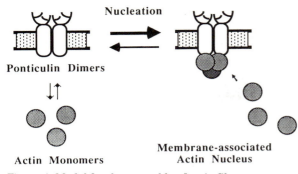

Ponticulin Dimers

Nucleation

Actin Monomers

Membrane-associated
Actin Nucleus

Figure 1. Model for the assembly of actin filaments on plasma membranes[5].

forms. Purified ponticulin in octylglucoside micelles has a sedimentation coefficient of 2.7 S and a Stokes radius of 3.6 nm, suggesting that this protein is an elongated monomer in detergent[4]. The quaternary organization of ponticulin in plasma membranes is unknown, but ponticulin has been postulated to be a dimer, or higher oligomer, because of its role in assembling and stabilizing actin nuclei at the membrane surface[5]. Ponticulin binding to F-actin is lowered in the presence of high salt concentrations and is eliminated by reduction of ponticulin with thiol reagents, suggesting that the actin binding site on ponticulin is composed of different local, probably electropositive, regions of primary sequence stabilized by disulphide bonds[2,3]. Ponticulin constitutes about 1% of the total *Dictyostelium* plasma membrane protein and, thus, is relatively abundant. Cell surface-labelling and concanavalin A binding studies show that ponticulin possesses a glycosylated extracellular domain as well as a cytoplasmic domain recognized by antibody adsorbed against intact cells[3].

Ponticulin appears to be a major link between the cytoskeleton and the plasma membrane in *Dicytostelium* amebae, the only system in which it has been investigated in depth so far (Figure 1). Monovalent antibody fragments directed against the cytoplasmic portion of ponticulin, but not those recognizing the external domain, block actin-membrane binding in sedimentation assays[3]. Removal of ponticulin from detergent extracts of plasma membranes correlates with the loss of membrane associated-actin nucleation activity which, in the presence of ponticulin, can be reconstituted upon dialysis or dilution of the detergent[6]. Furthermore, highly purified ponticulin reconstituted into vesicles containing *D. discoideum* lipids nucleates actin assembly in the same fashion as observed with plasma membranes[4,7].

The amount of ponticulin in the *D. discoideum* plasma membrane increases two to three-fold when amebae are forming aggregation streams, suggesting that the amount and/or intracellular localization of ponticulin is developmentally regulated[8]. By immunofluorescence microscopy (Figure 2), ponticulin in aggregating amebae is seen in regions of lateral cell-cell contacts and in arched regions of the plasma membrane reminiscent of early pseudopods[1]. Ponticulin also is present throughout the plasma membranes of both log-phase and aggregating amebae and in intracellular vesicles, including juxtanuclear vesicles associated with the Golgi complex[1]. A 17 kDa plasma membrane protein that specifically cross-reacts with antibodies affinity purified against *Dictyostelium* ponticulin has been found in human polymorphonuclear leukocytes, suggesting that ponticulin mediated linkages also are present in higher eukaryotic cells[1]. However, the tissue distribution of ponticulin cannot be determined until more strongly crossreactive antibodies are available.

■ PURIFICATION

Ponticulin can be purified by F-actin affinity chromatography of highly purified, octylglucoside solubilized plasma membranes[3]. Alternatively, ponticulin can be isolated by high salt extraction of detergent cytoskeletons, phase partitioning into Triton X-114 layers, and fractionation on F-actin columns[4,9].

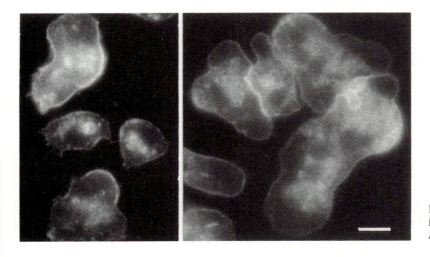

Figure 2. Immunofluorescence localization of ponticulin in aggregating *D. discoideum* amebae[1]. Bar 5 μm .

ACTIVITIES

Ponticulin binding to actin is monitored in cosedimentation binding assays[5,10] or in blot overlays with ^{125}I-labelled F-actin[2]. Ponticulin associated actin nucleating activity is measured by following the polymerization of pyrene-actin in the presence and absence of membrane vesicles containing ponticulin[7,11].

ANTIBODIES

Only one rabbit polyclonal antibody against ponticulin has been described[3]. This antibody also reacts with the actin binding 19 kDa and 15 kDa polypeptides that coisolate with ponticulin; probably only a few epitopes are recognized because almost all reactivity against all three proteins is lost after reduction of these proteins with thiol reagents[1]. Although this antibody is useful for immunoprecipitation, immunoblotting, and immunofluorescence of *Dictyostelium* ponticulin, its crossreactivity with vertebrate analogues is weak.

GENES

The cloning of *Dictyostelium* ponticulin is in progress[12].

REFERENCES

1. Wuestehube, L. J., Chia, C.P. and Luna, E.J. (1989) Cell Motil. Cytoskeleton 13, 245-263.
2. Chia, C.P., Hitt, A.L. and Luna, E.J. (1991) Cell Motil. Cytoskeleton 18, 164-179.
3. Wuestehube, L.J. and Luna, E.J. (1987) J. Cell Biol. 105, 1741-1751.
4. Chia, C.P., Shariff, A., Savage, S.A. and Luna, E.J. manuscript in preparation.
5. Schwartz, M.A. and Luna, E.J. (1988) J. Cell Biol. 107, 201-209.
6. Shariff, A. and Luna, E.J. (1990) J. Cell Biol. 110, 681-692.
7. Shariff, A., Chia, C.P., Savage, S. and Luna, E.J. (1990) J. Cell Biol. 111, 166a.
8. Ingalls, H.M., Barcelo, G., Wuestehube, L.J. and Luna, E.J. (1989) Differentiation 41, 87-98.
9. Chia, C.P., Savage, S.A., Hitt, A.L. and Luna, E.J. (1989) J. Cell Biol. 109, 267a.
10. Schwartz, M.A. and Luna, E.J. (1986) J. Cell Biol. 102, 2067-2075.
11. Cooper, J.A. and Pollard, T. D. (1982) Methods Enzymol. 85, 182-210.
12. Hitt, A.L. and Luna, E.J. work in progress.

Elizabeth J. Luna:
Worcester Foundation for Experimental Biology, Shrewsbury, MA, USA

Profilins

Profilin is a small, globular, cytoplasmic protein that binds actin monomers[1], associates with plasma membranes[2] and inhibits soluble phospholipase C (PLC) by binding to its substrate, phosphatidylinositol-4,5-bisphosphate (PIP_2)[3]. Profilin may provide a key link between transmembrane signalling and the actin cytoskeleton. Profilin null mutant Saccharomyces cerevisiae are conditionally lethal. They are very temperature sensitive, fail to divide properly, becoming quite enlarged, and show altered distributions of both chitin and actin[4].

Profilins have been purified from several sources including *Acanthamoeba castellanii*[5,6], *Saccharomyces cerevisiae*[4], various bovine tissues[1,7], *Tetrahymena*[8], rabbit alveolar macrophages[9], chicken skeletal muscle[10], *Physarum polycephalum*[11], two species of Echinoidea (*Clypeaster japonicus* and *Anthocidaris crassispina*)[11] and human platelets[12]. All known profilins consist of a single polypeptide chain of 124 to 140 amino acids. Profilins are among the most abundant cytoplasmic proteins, present at concentrations of 40 to 100 µM.

All well characterized profilins bind **actin** monomers with micromolar affinity[13,14], increase the rate of exchange of ATP bound to actin[15,16], inhibit actin monomer ATPase[17] and inhibit nucleation of actin polymerization[4,5,9]. The mechanism for inhibition of actin polymerization is thought to be more complex than simple sequestration of actin monomers, because the weak inhibition of elongation at the barbed end of actin filaments[5,17-19] can better be explained by low affinity binding of profilin at this site[20]. In cells, these properties of profilin should suppress spontaneous nucleation of actin filaments, but allow filaments

to elongate at the barbed end. However, given the relatively low affinity of profilin for actin and the two to six-fold molar excess of actin over profilin[12,13,18], it is unlikely that profilin alone can account for the high concentrations of unpolymerized actin found in the cytoplasm. Rather than simply acting stoichiometrically, profilin acts catalytically to increase the exchange of both nucleotide and divalent cation on actin monomers[12]. Although the significance of this observation is still being investigated, it may be the major mechanism of cytoskeletal regulation by profilin *in vivo*.

Profilin binds with micromolar affinity to small clusters of four to five molecules of phosphatidylinositol-4-monophosphate (PIP) and PIP_2. This interaction dissociates the profilactin complex[21,22], and inhibits the hydrolysis of PIP_2 by PLC. Tyrosine phosphorylation of PLC-γ from bovine brain by the activated EGF receptor will allow PLC-γ to overcome the profilin block and hydrolyze PIP_2 *in vitro*. This is the first reconstitution of a regulated signalling pathway[23]. Platelet profilin has a higher affinity for PIP_2 than for actin[3] and therefore might be largely bound to cell membranes in cells until the PLC becomes activated and the profilin block is overcome. The interaction of profilin with lipid is specific for PIP and PIP_2, other phospholipids tested did not interact significantly with profilin[3,24].

■ PURIFICATION

Profilin can be purified alone or as a profilin-actin complex. The original purification employed ammonium sulphate precipitation and chromatography on DEAE, hydroxylapatite and Sephadex G-75[5]. Poly-L-proline Sepharose affinity chromatography was first used by Lindberg et al.[25], to obtain profilactin complexes as well as pure profilin from calf spleen. Poly-L-proline chromatography yields in some cases profilactin complexes and in other cases pure profilin. An extract containing soluble proteins is passed through a column of poly-L-proline linked Sepharose. The column is washed with 3-4 M urea to elute weakly bound proteins, and profilin is eluted with 8 M urea[26]. The profilin is re-natured by dialysis into low salt buffer[26]. Profilin flows through DEAE-cellulose at pH 7.5, a step that can be used to clarify extracts before chromatography on poly-L-proline Sepharose. The profilin-I and profilin-II isoforms from *Acanthamoeba* can be separated on carboxymethylcellulose resin at pH 6.0[26]. Recombinant rat spleen profilin[27], as well as *Acanthamoeba* profilins I and II[28] have been expressed in *E. coli* and purified in active form.

3-D Structure: Crystals of *Acanthamoeba* profilin have the symmetry of the space group C2 with lattice constants a=110.4, b=31.7, c=33.5Å, b=112.2°. They diffract to at least 2.0Å resolution[29]. Profilactin crystals[30] have the space group $P2_12_12_1$ with a=38.6, b=71.3, c=187.0Å. The cell dimensions of these crystals are, however, highly variable and sensitive to such environmental factors as pH, temperature, ionic strength and composition of crystal bathing medium. The crystals are also affected by the nucleotide in the soaking medium (ATP or ADP) and seem to act as cooperative systems with the actin in a "zig-zag ribbon" network. Profilin makes extensive contacts with the larger of the two domains of actin, while DNase 1 contacts mainly the smaller domain of actin[31].

■ ACTIVITIES

Platelet profilin binds to monomeric platelet actin with an affinity of about 5 μM[12], and *Acanthamoeba* profilin binds to monomeric *Acanthamoeba* actin with an affinity of about 10 μM[5]. Profilin binds to clusters of four to five PIP_2 molecules in large unilamellar vesicles containing other lipids or to pure PIP_2 micelles. The following affinities for PIP_2 were estimated from inhibition of PLC activity and equilibrium gel filtration assays: Human platelet profilin K_d=0.1-1 μM; *Acanthamoeba* profilin-II K_d=5-10 μM; *Acanthamoeba* profilin-I K_d=50-100 μM. Poly-L-proline binds to profilin with approximately a 30 μM K_d and no more than ten consecutive proline residues are sufficient for binding in solution[32].

■ ANTIBODIES

Monoclonal antibodies are available that bind specifically to *Acanthamoeba* profilin-II or that bind to both profilin-I and profilin-II[26]. Polyclonal human platelet[26] and polyclonal yeast profilin[4] antibodies are also available.

■ GENES

Profilin cDNA's have been cloned from rat[27], mouse[33], yeast[34], human[35] and *Acanthamoeba*[36]. Disruption of the profilin gene in *Saccharomyces cerevisiae* is conditionally lethal[4,34].

■ REFERENCES

1. Carlsson, L., Nystrom, L.E., Sundkvist, I., Markey, F. and Lindberg, U. (1977) J. Mol. Biol. 115, 465-483.
2. Hartwig, J.H., Chambers, K.A., Hopica, K.L. and Kwiatkowski, D.J. (1989) J. Cell Biol. 109, 1571-1579.
3. Goldschmidt-Clermont, P.J., Machesky, L.M., Baldassare, J.J. and Pollard, T.D. (1990) Science 247, 1575-1578.
4. Haarer, B.K., Lillie, S.H., Adams, A.E.M., Magdolen, V., Bandlow, W. and Brown, S.S. (1990) J. Cell Biol. 110, 105-114.
5. Reichstein, E. and Korn, E.D. (1979) J. Biol. Chem. 254, 6174-6179.
6. Kaiser, D.A., Sato, M., Ebert, R.F. and Pollard, T.D. (1986) J. Cell Biol. 102, 221-226.
7. Kobayasha, R., Bradley, W.A. and Field, J.B. (1982) Anal. Biochem. 120, 106-110.
8. Edmatsu, M., Masafumi, H. and Watanabe, Y. (1990) Bioc. Biop. Res. 170, 957-962.
9. DiNubile, M.J. and Southwick, F.S. (1985) J. Biol. Chem. 12, 7402-7409.
10. Ooshima, S., Abe, H. and Obinata, T. (1989) J. Biochem. 105, 855-857.
11. Takagi, T., Mabuchi, I., Hosoya, H., Furuhashi, K. and Hatano, S. (1990) Eur. J. Biochem. 192, 777-781.
12. Goldschmidt-Clermont, P.J., Machesky, L.M., Doberstein, S.K. and Pollard, T.D. (1991) J. Cell Biol. 113, 1081-1089.
13. Tseng, P.C.-H., Runge, M.S., Cooper, J.A., Williams Jr., R.C. and Pollard, T.D. (1984) J. Cell Biol. 98, 214-221.
14. Pollard, T.D. and Cooper, J.A. (1986) Annu. Rev. Biochem. 55, 987-1035.
15. Mockrin, S.C. and Korn, E.D. (1980) Biochem. 19, 5359-5362.
16. Nishida, E. (1985) Biochem. 24, 1160-1164.
17. Tobacman, L.S. and Korn, E.D. (1982) J. Biol. Chem. 257, 4166-4170.
18. Tseng, P.C.-H. and Pollard, T.D. (1984) J. Cell Biol. 98, 214-221.
19. Tilney, L.G., Bonder, E.M., Colluccio, L.M. and Mooseker, M.S. (1983) J. Cell Biol. 97, 112-124.
20. Pollard, T.D. and Cooper, J.A. (1984) Biochem. 23, 6631-6641.
21. Lassing, I. and Lindberg, U. (1985) Nature 314, 472.
22. Lassing, I. and Lindberg, U. (1988) J. Cell Biochem. 37, 255.
23. Goldschmidt-Clermont, P.J., Kim, J.W., Machesky, L.M., Rhee, S.G. and Pollard, T.D. (1991) Science, 251, 1231-1233.
24. Machesky, L.M., Goldschmidt-Clermont, P.J. and Pollard, T.D. (1990) Cell Regul. 1, 937-950.
25. Lindberg, U., Schutt, C.E., Hellsten, E., Tjader, A.-C. and Hult, T. (1988) Biochim. Biophys. Acta 967, 391-400.
26. Kaiser, D.A., Goldschmidt-Clermont, P.J., Levine, B.A. and Pollard, T.D. (1989) Cell Motil. Cytoskel. 14, 251-262.
27. Babcock, G. and Rubenstein, P.A. (1989) Cell Motil. and Cytoskel. 14, 230-236.

28. Way, M., Machesky, L.M. and Pollard, T.D. unpublished results.
29. Magnus, K.A., Lattman, E.E., Sato, M. and Pollard, T.D. (1986) J. Biol. Chem. 28, 13360-13361.
30. Schutt, C.E., Lindberg, U., Myslik, J. and Strauss, N. (1989) J. Mol. Biol. 209, 735-746.
31. Kabsch, W., Mannherz, H.G., Suck, D., Pai, E.F. and Holmes, K.C. (1990) Nature 347, 37-44.
32. Machesky, L.M. and Pollard, T.D. manuscript in preparation.
33. Widada, J.S., Ferraz, C. and Liautard, J.-P. (1989) Nucleic Acids Res. 17, 2855.
34. Magdolen, V., Oeschner, U., Muller, G. and Bandlow, W. (1988) Mol. Cell Biol. 8, 5108-5115.

35. Kwiatkowski, D.J. and Bruns, G.A.P. (1988) J. Cell Biol. 12, 5910-5915.
36. Pollard, T.D. and Rimm, D.L. (1991) Cell Motil. Cytoskel. 20, 169-177.

■ *Laura M. Machesky and Thomas D. Pollard:*
Department of Cell Biology,
Johns Hopkins Medical School,
Baltimore, MD, USA

Protein 4.1

Erythroid protein 4.1 promotes complex formation between spectrin and actin and links the erythrocyte membrane skeleton to the overlying lipid bilayer. Nonerythroid 4.1 isoforms may perform similar integrating roles with the cytoskeleton and the mitotic apparatus.

Initially identified as one of the most abundant polypeptides of red cell membranes, protein 4.1 (80 kDa) plays a critical role in promoting associations between **spectrin** and **actin** *in vitro*[1]. It is one of the two most important proteins that link the spectrin-actin lattice to the overlying lipid bilayer[2].

A provisional domain map of erythroid 4.1 was proposed on the basis of differential proteolytic cleavage of interdomain segments[3] and is depicted in Figure 1. The domain responsible for promoting spectrin-actin interactions is a 67 amino acid segment[4] which contains sites for cAMP induced phosphorylation[5] and is adjacent to a domain that has sites for protein kinase c catalyzed phosphorylation[5]. The state of phosphorylation of 4.1 regulates both its capacity to stimulate spectrin-actin assembly and the attachment of 4.1 to the erythrocyte membrane[6,7]. The latter appears to be mediated by a specific interaction between an N-terminal segment of 4.1 and a glycophorin-PIP$_2$ complex[8].

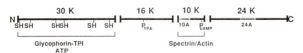

Figure 1. Domain map of erythroid 4.1.

Erythroid 4.1 contains O-linked N-acetylglucosamine residues covalently attached to amino acids located in the C-terminal half of the protein[9], but their functional significance remains undefined. Isolated erythroid 4.1 also binds **tubulin**[10] and **myosin**[11], suggesting that certain isoforms of 4.1 may play some role in integrating the membrane skeleton with the larger cytoskeleton.

cDNA clones for erythroid 4.1 have been isolated and sequenced and the complete primary structure determined[12]. Multiple transcripts of the 4.1 gene have been identified in reticulocytes which code for a family of closely related protein isoforms[13], one set contains a motif that codes for the spectrin-actin binding domain, that appears to be specific for cells of the erythroid lineage[14].

Avian erythrocytes contain multiple molecular variants of protein 4.1 that exhibit tissue-specific expression[15], and many 4.1 related proteins have been identified in other nucleated cells[16]. cDNA clones isolated from human lymphoid cell lines code for a variety of 4.1 isoforms that reflect tissue and functional specificity and appear to arise by alternative mRNA splicing[17]. Antisera raised against an N-terminal peptide present in lymphoid 4.1 isoforms, but absent in erythroid 4.1, react with a set of high molecular weight polypeptides by immunoblotting. These antisera localize to centrioles and interphase nuclei and membrane borders of contiguous cells when analyzed by indirect immunofluorescence (Figure 2). Protein 4.1 isoforms also appear to be associated with centrosomal regions of the mitotic apparatus[18].

The erythroid 4.1 gene, which codes for an 80 kDa protein, has been localized to the short arm (p) of chromosome one[19], and mutant forms of this gene have been described that cause a severe hemolytic anemia due to 4.1 deficiency[19].

■ PURIFICATION

Reasonable quantities of protein 4.1 can be isolated from human erythrocytes, using a high-salt extraction procedure followed by a one step ion exchange purification as described[20]. Since isolated 4.1 is easily degraded, care must be taken in preparing red cell ghost membranes free of leukocyte contamination, and buffers should contain proteolytic inhibitors. No reliable method has yet been described for the isolation of nonerythroid forms of 4.1.

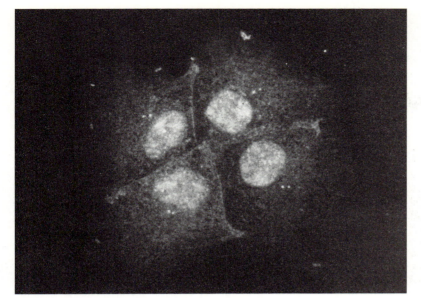

Figure 2. Immunofluorescence localization of anti-sera to nonerythroid
4.1 in cultured MDCK cells.

ACTIVITIES

Protein 4.1 promotes the association between erythroid
spectrin and actin. A simple sedimentation assay has been
described[1] which is also useful to assay the activities of
peptides derived from 4.1[4]. Erythroid 4.1 also binds to
liposomes reconstituted with glycophorin and PIP_2[8].

ANTIBODIES

Rabbit polyclonal antisera react with human erythroid 4.1
and crossreact with 4.1 preparations from a variety of
species. Some polyclonal antisera also react with nonerythroid forms, but strong reactivity to nonerythroid isoforms has been obtained only with polyclonal antisera
raised against synthetic peptides derived from 4.1 and to
fusion proteins produced in transfected *E. coli*[18].

GENES

Full length cDNA clones of erythroid 4.1 have been
derived from human reticulocytes[12] and clones coding for
larger and more diverse nonerythroid forms have been
isolated from human T cell leukemia libraries[17].

REFERENCES

1. Ungewickell, E., Bennett, P.M., Calvert, R., Ohanian, V. and
   Gratzer, W.B. (1979) Nature 280, 811-814.
2. Marchesi, V.T. (1985) Ann. Rev. Cell Biol. 1, 531-561.
3. Leto, T.L. and Marchesi, V.T. (1984) J. Biol. Chem. 259, 4603-
   4608.
4. Correas, I., Speicher, D.W. and Marchesi, V.T. (1986) J. Biol.
   Chem. 261, 13362-13366.
5. Horne, W.B., Leto, T.S. and Marchesi, V.T. (1985) J. Biol. Chem.
   260, 9073-9076.
6. Danilov, Y.N., Fennell, R., Ling, E. and Cohen, C.M. (1990) J.
   Biol. Chem. 265, 2556-2562.
7. Eder, P.S., Soong, C. and Tao, M. (1986) Biochem. 25, 1764-
   1770.
8. Anderson, R.A. and Marchesi, V.T. (1985) Nature 318, 295-298.
9. Holt, G.D., Haltiwanger, R.S., Torres, C. and Hart, G.W. (1987)
   J. Biol. Chem. 262, 14847-14850.
10. Correas, I. and Avila, J. (1988) Biochem. J. 255, 217-221.
11. Pasternack, G.R. and Racusen, R.H. (1989) Proc. Natl. Acad. Sci.
    86, 9712-9716.
12. Conboy, J., Kan, Y.W., Shohet, S.B. and Mohandas, N. (1986)
    Proc. Natl. Acad. Sci. USA 83, 9512-9516.
13. Conboy, J.G., Chan, J., Mohandas, N. and Kan, Y.W. (1988)
    Proc. Natl. Acad. Sci. USA 85, 9062-9065.
14. Tang, T.K., Leto, T.L., Correas, I., Alonso, M.A., Marchesi, V.T.
    and Benz Jr., E.J. (1988) Proc. Natl. Acad. Sci. 85, 3713-3717.
15. Granger, B.L. and Lazarides, E. (1984) Cell 37, 595-607.
16. Spiegel, J.E., Beardsley, D.S., Southwick, F.S. and Lux, S.E.
    (1984) J. Cell Biol. 99, 886-893.
17. Tang, T.K., Qin, Z., Leto, T.L., Marchesi, V.T. and Benz Jr., E.J.
    (1990) J. Cell Biol. 110, 617-624.
18. Marchesi, V.T., Tang, T.K., Huang, S.C., Benz Jr., E.J. manu-
    script in preparation.
19. Conboy, J., Mohandas, N., Tchernia, G. and Kan, Y.W. (1986)
    N. Engl. J. Med. 315, 680-685.
20. Tyler, J.M., Reinhardt, B.N. and Branton, D. (1980) J. Biol.
    Chem. 255, 7034-7039.

■ *V.T. Marchesi:*
Boyer Center for Molecular Medicine,
Yale University,
New Haven, CT 06510, USA

Radixin

Radixin is a barbed end actin capping protein which is concentrated at the undercoat of cell-to-cell adherens junctions during interphase[1] and at the cleavage furrow during cytokinesis[2]. Radixin probably plays a crucial role in the association of actin filaments with plasma membranes in an end-on fashion.

Radixin purified from the isolated adherens junctions of rat liver reveals an apparent molecular mass of 82 kDa on SDS-PAGE[2,3]; chick gizzard radixin has a slightly smaller apparent molecular mass[2]. Radixin shows a barbed end capping activity for F-**actin** *in vitro* in a Ca^{2+}-independent manner. Recent immunoblot analyses suggested that there may be at least three radixin isotypes in rat and mouse tissues.

Radixin was shown by immunofluorescence microscopy and immunoelectron microscopy to be localized at the cell-to-cell adherens junctions[1] (Figure 1). Biochemical analyses using isolated cell-to-cell adherens junctions from rat liver reveal that radixin is one of the major peripheral proteins of this type of junction, and not an integral membrane protein. Antibodies do not stain the cell-to-substrate adherens junctions (focal contacts). Radixin is also highly concentrated at the cleavage furrow during cytokinesis[1] (Figure 2). Radixin accumulates rapidly at the cleavage furrow at the onset of furrowing, continues to be concentrated there during anaphase and telophase, and is finally enriched at the midbody. Cells lacking both cell-to-cell and cell-to-substrate adherens junctions, for example myeloma cells also exhibited a clear accumulation of radixin at the cleavage furrow.

The most prominent feature shared by the cell-to-cell adherens junctions and the cleavage furrow is the tight association of actin filaments with plasma membranes. Considering that actin filaments are known to associate in general with plasma membrane at their barbed ends, this strongly suggests that radixin may be a key protein in the end-on association of actin filaments with plasma membranes.

■ PURIFICATION

Radixin can be purified from isolated cell-to-cell adherens junctions of rat or mouse liver cells[1] and from the crude membrane fraction of chick gizzard[2]. The purification scheme includes low salt extraction followed by DEAE-cellulose ion exchange, DNase 1-actin affinity, and carboxyl-cellulose ion exchange chromatographies.

■ ACTIVITIES

Substoichiometric concentrations of radixin largely inhibit actin filament assembly; when the molar ratio of radixin to G-actin is 1:1,000, the viscosity is reduced to 28% of the control value[1]. The barbed end capping activity of radixin has been demonstrated by direct electron microscopic examinations. This activity of radixin is Ca^{2+}-independent.

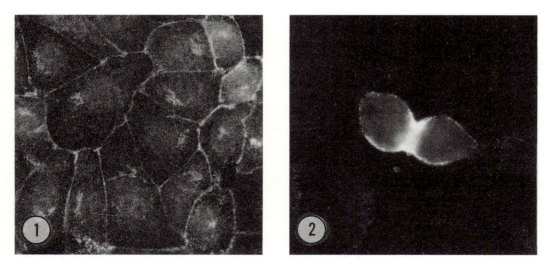

Figure 1. Immunofluorescence localization of radixin in cultured MDBK cells. x300.
Figure 2. Immunofluorescence localization of radixin in 3Y1 cells during cytokinesis. x700.

■ ANTIBODIES

Polyclonal antisera against rat radixin purified from the isolated cell-to-cell adherens junctions have been described[1]. A monoclonal antibody to chick radixin purified from gizzard has also been published[2]. Both generally crossreact with many species.

■ GENES

The sequence data of mouse radixin have been recently published[4] (GenBank EMBL DDBJ X60672). The composite cDNA is 4,241 nucleotides long and codes for a 583 amino acid polypeptide with a calculated molecular mass of 68.5 kDa. Mouse radixin shares 74.9% identity with mouse ezrin, which was reported to be a member of the band 4.1 family.

■ REFERENCES

1. Tsukita, Sa., Hieda, Y. and Tsukita, Sh. (1989) J. Cell Biol. 108, 2369-2382.
2. Sato, N., Yonemura, S., Obinata, T., Tsukita, Sa. and Tsukita, Sh. (1991) J. Cell Biol. 113, 321-330.
3. Tsukita, Sh. and Tsukita, Sa. (1989) J. Cell Biol. 108, 31-41.
4. Funayama, N., Nagafuchi, A., Sato, N., Tsukita, Sa. and Tsukita, Sh. (1991) J. Cell Biol. 115, 1039-1048.

■ *Sachiko Tsukita and Shoichiro Tsukita:*
Department of Information Physiology,
National Institute for Physiological Sciences,
Okazaki, Japan

Sarcomeric MM-Creatine Kinase

A significant fraction of the creatine kinase isoenzyme (MM-CK) in striated muscle is bound to the myofibrillar M-band region, where it contributes to the high electron density of the M-band[1]. In vitro this MM-CK is capable of regenerating the ADP produced by the Ca^{2+} stimulated myofibrillar myosin ATPase[2]. The binding of the creatine kinase is isoprotein specific and only the MM-CK dimer was shown to be capable of this interaction. The amino acid sequences responsible for this interaction are localized mainly in the C-terminal half of the M-CK subunit[3].

One of the components localized in the M-band region of sarcomeres is the MM-CK isoenzyme of creatine kinase, the other proteins so far identified are myomesin and M-protein. The enzyme bound in the M-band amounts to some 5-7% of the total CK activity in skeletal muscle[4]. The dimeric MM-CK isoenzyme is composed of two M-CK subunits with a molecular weight of 42,500 and is a member of the creatine kinase isoenzyme family. All members of this family catalyze the reaction to transphosphorylate the so-called "energy rich phosphate" from the phosphocreatine to ADP that is generated in many tissues with an intensive energy metabolism. On the other hand at least two mitochondrial CK subunitspecies (Mi-CK)[5] are involved in the conversion of the ATP generated in mitochondria to phosphocreatine and ADP. The two metabolic sites are linked in a phosphocreatine metabolic circuit[6].

During embryonic development brain CK (B-CK) transcripts and translation products (B-CK subunits) can be found in very young embryos of chicken and mouse in the presumptive nervous tissue and in the myotomal part of the somites. During muscle development the M-CK isoprotein gene is activated and after an isoenzyme transition becomes the only nonmitochondrial CK form of adult skeletal muscle[7,8]. Differentiating skeletal muscle cultures express both B-CK and M-CK subunits which combine to form BB-CK, MB-CK and MM-CK dimeric isoenzymes, however, the M-band bound CK consisted of MM-CK dimers[9]. In chicken heart cells, contrary to mammalian heart, there is no M-CK expression and no electron dense M-band can be observed, which has been shown in skeletal muscle and mammalian heart myofibrils to consist mainly of MM-CK[6]. However, the high molecular weight M-band proteins myomesin and M-protein are also present in sarcomeres devoid of an electron dense M-band. If cultured chicken heart cells were microinjected with syn-

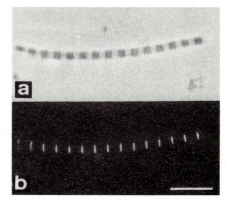

Figure. M-band staining of chicken myofibrils with antibody against M-CK. Myofibrils isolated from pectoral muscle were fixed in 3% paraformaldehyde in relaxing solution A, stained with culture supernatant of hybridoma cells secreting monoclonal antibody aC M-CK15 and RITC-labelled goat anti mouse secondary antibody[15].

thetic M-CK mRNA the resulting protein was shown to interact with the M-band. Chimeric B-CK/M-CK constructs retained part of this interactivity provided that the C-terminal half was of M-CK origin[3].

MM-CK localization in the M-band has been investigated with immunohistochemistry (Figure) and more exactly by the combination of electron microscopy and immunological methods[10]. Most of the electron density appears to be localized in the region of the 4 and 4' substriations of the M-band and seems to consist of MM-CK forming primary M-bridges[6]. The localization of M-CK bound to subcellular structures by immunofluorescence in cultured cells is always obscured by the excess of soluble enzyme which, however, can be removed by careful extraction with buffers containing nonionic detergents[3,9].

PURIFICATION

Since most of the enzymes are easily solubilized from muscle, tissue extraction of MM-CK does not pose a problem. For the preparation of homogeneous MM-CK, classical column chromatography has been used. Alternatively adsorption on antibody columns[11] or more recently affinity columns (Blue Sepharose, Pharmacia)[12] have all been shown to be very efficient in purification of creatine kinases.

ACTIVITIES

The enzymatic activity can be measured with different assay conditions either spectrophotometrically with a coupled assay system[11] or by the pH-stat method[2]. The property to associate with myofibrils, however, cannot be assayed in vitro and the system with cultured heart cells has to be utilized[3].

ANTIBODIES

The chicken CK antigens have been utilized to elicit antibodies in rabbits and goats, and polyclonal reagents were obtained that reacted isoprotein specifically[11]. Antibodies obtained against mammalian creatine kinases in rabbits and goats that inhibit the human MM-CK activity, are used in the diagnosis of human cardiac infarction. These antibodies are commercially available (Calbiochem, San Diego, USA; Merck, Darmstadt, Germany). More recently monoclonal antibodies have been generated[13,14] and were shown to be isoprotein specific like the mMaC MCK15, the antibody which was used to stain the myofibril shown in the Figure.

GENES

The cDNA sequences of many CK have been obtained and a complete set is available for the chicken and human isoproteins including M-CK, B-CK and Mi-CK[5,15,16].

These cDNAs were the basis for the construction of chimeric CKs utilized in the microinjection experiments described above[3]. In addition genomic CK-DNAs[17] have been characterized and the muscle specific gene M-CK has been shown to be under the control of the myogenic differentiation factor MyoD1 in the mouse[18]. The GenBank numbers for the M-CK cDNAs from some species are listed below: Chicken (M10012); dog (M11660); human (M14780); mouse (X03233); rabbit (K02831); rat (M10140).

REFERENCES

1. Turner, D.C., Wallimann, T. and Eppenberger, H.M. (1973) Proc. Natl. Acad. Sci. (USA) 70, 702-705.
2. Wallimann, T., Schlösser, T. and Eppenberger, H.M. (1984) J. Biol. Chem. 259, 5238-5246.
3. Schäfer, B.W. and Perriard, J.-C. (1988) J. Cell Biol. 106, 1161-1170.
4. Wallimann, T., Turner, D.C. and Eppenberger, H.M. (1977) J. Cell Biol. 75, 297-317.
5. Hossle, J.P., Schlegel, J., Wegmann, G., Wyss, M., Böhlen, P., Eppenberger, H.M., Wallimann, T. and Perriard, J.-C. (1988) Biochem. Biophys. Res. Com. 151, 408-416.
6. Wallimann, T. and Eppenberger, H.M. (1985) Cell Muscle Motil. (eds. Shaw, J.W.) pp.239-285.
7. Perriard, J.-C. (1979) J. Biol. Chem. 254, 7036-7041.
8. Perriard, J.-C., Hossle, J.P., Schafer, B.W., Soldati, T., Wegmann, G. and Wirz, T. (1989) in Cell. Mol. Biol. of Muscle Dev. (eds. L. Kedes and F. Stockdale) 545-554 (Edits: Alan R. Liss, Inc. New York, N.Y.).
9. Wallimann, T., Moser, H. and Eppenberger, H.M. (1983) J. Muscle Res. Cell Motil. 4, 429-441.
10. Wallimann, T., Doetschman, T.C. and Eppenberger, H.M. (1983) J. Cell Biol. 96, 1772-1779.
11. Perriard, J.-C., Caravatti, M., Perriard, E. and Eppenberger, H.M. (1978b) Arch. Biochem. Biophys. 191, 90-100.
12. Wallimann, T., Zurbriggen, B. and Eppenberger, H.M. (1985) Enzyme 33, 226-231.
13. Cerny, L. and Perriard, J.-C. (1991) in preparation.
14. Morris, G.E. (1982) FEBS Lett. 145, 163-168.
15. Babbitt, P.C., Kenyon, G.L., Kuntz, I.D., Cohen, F.E., Baxter, J.D., Benfield, P.A., Buskin, J.D., Gilbert, W.A., Hauschka, S.D., Hossle, J.P., Ordahl, C.P., Pearson, M.L., Perriard, J.C., Pickering, L.A., Putney, S.D., West, B.L. and Ziven, R.A. (1986) J. Protein Chem. 5, 1-14.
16. Haas, R.C. and Strauss, A.W. (1990) J. Biol. Chem. 265, 6921-6927.
17. Wirz, T., Brändle, U., Soldati, T., Hossle, J.P. and Perriard, J.-C. (1990) J. Biol. Chem. 265, 11656-11666.
18. Lassar, A.B., Buskin, J.N., Lockshon, D., Davis, R.L., Apone, S., Hauschka, S.D., Weintraub, H. (1989) Cell 58, 823-831.

Jean-Claude Perriard:
Institute for Cell Biology,
Swiss Federal Institute of Technology,
CH-8093 Zurich, Switzerland

Severin

Severin from Dictyostelium discoideum belongs to a family of proteins which in a Ca²⁺-dependent manner sever actin filaments, nucleate actin assembly and cap the fast growing end of actin filaments. The representatives of this family have molecular masses of approximately 40 kDa in lower eukaryotes and 80-90 kDa in higher eukaryotes.

Severin is an abundant 40 kDa cytoplasmic protein that is present throughout *Dictyostelium* development. It has been isolated based on its ability to inhibit F-**actin** sedimentability in a Ca²⁺-dependent manner[1]. Similar proteins in lower eukaryotes are **fragmin** from *Physarum polycephalum*[2] and a 45 kDa protein from sea urchin eggs[3], the corresponding proteins in higher eukaryotes are **gelsolin**[4,5] and **villin**[6,7].

All the actin modulating activities of severin are observed with these other proteins. A direct comparison of the functional properties of severin and gelsolin showed that both molecules are very similar and that the actin filament severing activity of severin, like that of gelsolin, is inhibited by polyphosphoinositides[8]. Severin does not however nucleate actin filament assembly as well as gelsolin. Comparison of the primary structure of severin[9], fragmin[10], gelsolin[11] and villin[12,13] confirmed the evolutionary relationship between these proteins and furthermore revealed their modular structure with fragmin and severin containing a threefold repeat and gelsolin and villin a sixfold repeat[14]. Expression of C-terminally truncated severin molecules in *Escherichia coli* and analysis of their activities led to the identification of domains responsible for severing, capping, Ca²⁺- and actin binding[15] (Figure).

Figure. Construction, expression and functional analysis of truncated severin derivatives. PCR-primers were used to design truncated cDS4-constructs coding for peptides that contain only severin-specific amino acids. Complete severin, the five constructs containing 277, 177, 151, 117 and 111 N-terminal amino acids, and the domain organization of severin are schematically presented. The table summarizes the activities of severin and severin derivatives.

Isolation of a *Dictyostelium* mutant deficient for severin showed that cell shape, cell motility and development are not altered compared to the parent strain[16]. This may indicate that severin is a component of a network of actin binding proteins whose functions are redundant[17].

■ PURIFICATION

Severin is purified from the cytoplasmic fraction of *Dictyostelium discoideum*. The flow-through of anion exchange chromatography is precipitated with a 60-80% ammonium sulphate cut and loaded onto a phosphocellulose column. Severin containing fractions are further purified on a hydroxylapatite column[1,15,18]. Activity is tested by low shear viscometry[19].

■ ACTIVITIES

Severin fragments actin filaments, nucleates actin assembly and caps the fast growing ends of actin filaments in a Ca^{2+}-dependent manner[1,15,20,21].

■ ANTIBODIES

Both mono- and polyclonal antibodies have been isolated. They recognize specifically *Dictyostelium* severin[9,16].

■ GENES

The sequence of a full length cDNA has been published[9] (GenBank/EMBL Data Bank accession number J03515).

■ REFERENCES

1. Brown, S.S., Yamamoto, K. and Spudich, J.A. (1982) J. Cell Biol. 93, 205-210.
2. Hasegawa, T., Takahashi, S., Hayashi, H. and Hatano, S. (1980) Biochemistry 19, 2677-2683.
3. Wang, L.-L. and Spudich, J.A. (1984) J. Cell Biol. 99, 844-857.
4. Norberg, R., Thorstensson, R., Utter, G. and Fagraeus, A. (1979) Eur. J. Biochem. 100, 575-583.
5. Yin, H.L. and Stossel, T.P. (1979) Nature 281, 583-586.
6. Bretscher, A. and Weber, K. (1979) Proc. Natl. Acad. Sci. (USA) 76, 2321-2325.
7. Matsudaira, P.T. and Burgess, D.R. (1979) J. Cell Biol. 83, 667-673.
8. Yin, H.L., Janmey, P.A. and Schleicher, M. (1990) FEBS Lett. 264, 78-80.
9. André, E., Lottspeich, F., Schleicher, M. and Noegel, A. (1988) J. Biol. Chem. 263, 722-727.
10. Ampe, C. and Vandekerckhove, J. (1987) EMBO J. 6, 4149-4157.
11. Kwiatkowski, D.J., Stossel, T.P., Orkin, S.H., Mole, J.E., Colten, H.R. and Yin, H.L. (1986) Nature 323, 455-458.
12. Arpin, M., Pringault, E., Finidori, J., Garcia, A., Jeltsch, J.-M., Vandekerckhove, J. and Louvard, D. (1988) J. Cell Biol. 107, 1759-1766.
13. Bazari, W.L., Matsudaira, P., Wallek, M., Smeal, T., Jakes, R. and Ahmed, Y. (1988) Proc. Natl. Acad. Sci. (USA) 85, 4986-4990.
14. Matsudaira, P. and Janmey, P. (1988) Cell 54, 139-140.
15. Eichinger, L., Noegel, A.A. and Schleicher, M. (1991) J. Cell Biol. 112, 665-676.
16. André, E., Brink, M., Gerisch, G., Isenberg, G., Noegel, A., Schleicher, M., Segall, J.E. and Wallraff, E. (1989) J. Cell Biol. 108, 985-995.
17. Bray, D. and Vasiliev, J. (1989) Nature 338, 203-204.
18. Schleicher, M., Gerisch, G. and Isenberg, G. (1984) EMBO J. 3, 2095-2100.
19. MacLean-Fletcher, S. and Pollard, T.D. (1980) J. Cell Biol. 85, 414-428.
20. Yamamoto, K., Pardee, J.D., Reidler, J., Stryer, L. and Spudich, J.A. (1982) J. Cell Biol. 95, 711-719.
21. Giffard, R.G., Weeds, A.G. and Spudich, J.A. (1984) J. Cell Biol. 98, 1796-1803.

■ *Angelika A. Noegel:*
Max-Planck-Institute for Biochemistry,
8033 Martinsried,
Fed. Rep. Germany

Small (Mr ≤ 35,000) Actin Crosslinking Proteins

Small actin crosslinking proteins are polypeptides with Mr ≤ 35,000 that crosslink actin filaments into either isotropic (gel) or anisotropic (bundle) arrays. These proteins are found in Dictyostelium and Acanthamoeba as well as in higher eukaryotes. Localization in the cortical cytoskeleton suggests a role in control of actin filament arrays at the cell surface.

The *Dictyostelium discoideum* 30 kDa protein was originally identified as an **actin** bundling protein whose actin crosslinking activity is inhibited at a free Ca^{2+} concentration $\geq 10^{-7}$ M[1]. The protein is monomeric with a molecular weight of 34,000 daltons[1-3]. The Stokes' radius of 2.7 nm allows calculation of an axial ratio of 3.5 and a length of 5.1 nm for the hydrated molecule[4]. The binding of Ca^{2+} induces a large decrease in the apparent affinity of

the protein for actin with consequent inhibition of the formation of crosslinked actin structures[1,5]. The presence of two EF hand Ca^{2+} binding sites in the cDNA sequence provides a structural basis for Ca^{2+} binding directly to the 30 kDa protein[6,7]. Maximal binding to actin is observed with one molecule of the 30 kDa protein bound per ten actin monomers in filaments[5]. The deduced amino acid sequence of the 30 kDa protein reveals only very limited

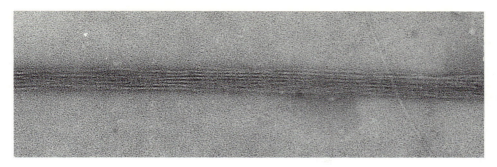

Figure. Reconstituted bundle formed in a mixture of rabbit skeletal muscle actin and the *Dictyostelium* 30 kDa protein.

sequence similarity to other known protein sequences[6]. A 27 kDa fragment of the protein, which lacks the N-terminal six amino acids and the C-terminal 7 kDa, exhibits Ca^{2+}-regulated actin crosslinking but not bundling activity[7]. The 30 kDa protein comprises 0.04% of the protein in vegetative cells[5], and exhibits a three fold decrease in abundance during development of *Dictyostelium* [4], due in part to lack of expression in prestalk cells[2]. The protein is selectively present in filopodia, as well as in the cortical cytoskeleton[2,5,8].

Proteins with antigenic similarity to the *Dictyostelium* 30 kDa protein, and with molecular masses of 34 kDa and 39 kDa have been identified in *Drosophila melanogaster* and *Schistosoma mansoni*, respectively[4]. Antigenic homologs with molecular weights of 34 kDa have also been identified in tissue culture cells of mammalian origin including mouse 3T3 and rat kidney fibro blasts, and are localized both at the cell surface and in stress fibres in these cultures[2]. Interestingly, the protein is more abundant in rat kidney fibroblasts following transformation with Kirsten Virus[2]. Antigenic homologs of the *Dictyostelium* 30 kDa protein have been identified in samples of human lung carcinoma in concentrations significantly higher than in normal human t issue[9]. The vertebrate homolog of the 34 kDa protein has been purified and shown to possess actin bundling activity[9].

A number of other small actin crosslinking proteins have been described, although their similarity to each other and to the *Dictyostelium* 30 kDa protein have not been tested in most cases. Four actin crosslinking proteins, the gelactins with subunits of 23, 28, 32 and 38 kDa, were purified from *Acanthamoeba castellanii*[10,11]. So far, neither the types of crosslinked actin structures formed, the intracellular localization, nor the potential derivation by proteolytic breakdown of high molecular weight proteins have been investigated. A 30 kDa actin bundling protein (p30b) was isolated from *Dictyostelium*, and shown to be distinct from the *Dictyostelium* 30 kDa protein using the criteria of peptide mapping, lack of antigenic cross-reactivity, and lack of regulation of the actin binding activity by Ca^{2+}[12]. A 36 kDa protein from *Physarum polycephalum* crosslinks actin filaments into aggregates in which the ability of the actin to interact with myosin is

dramatically reduced[13]. A unique ATP-sensitive actin gelling protein from *Acanthamoeba castellanii* is an ATP-sensitive protein that exists as a hexamer of six 35 kDa subunits[14]. Finally, a number of basic proteins of low molecular weight including ribonuclease, histones and lysozyme can crosslink actin filaments emphasizing the need for caution in extrapolating from biochemistry to physiology[15].

■ PURIFICATION

The *Dictyostelium* 30 kDa protein is purified by the sedimentation of an *in vitro* gelated and contracted extract and subsequent chromatography on DEAE-cellulose at pH 7.5, in which the 30 kDa protein is not bound and chromatography on hydroxylapatite[3]. An average preparation provides 0.85 mg of purified protein per 100 ml of packed cells (2-4 mg of total protein) and a 24% yield[3].

■ ACTIVITIES

The actin binding activity of *Dictyostelium* 30 kDa protein has been directly assessed by measurement of the cosedimentation of this protein with actin filaments[1,5]. The actin crosslinking activity has been measured using viscometry[1,5], light scattering[5,9], polarization microscopy[5] and electron microscopy[1,4,5]. Direct binding of Ca^{2+} to the 30 kDa protein has been qualitatively assessed using radiolabelled Ca^{2+} to probe electrophoretic blots[7].

■ ANTIBODIES

Affinity purified polyclonal antibodies reactive with the *Dictyostelium* 30 kDa protein crossreact with similar proteins from *Drosophila*, *Schistosoma*[4,5], mouse, rat and human[2,9]. Monoclonal antibodies reactive with the *Dictyostelium* 30 kDa protein are also available[7]. No antibodies to other small actin crosslinking proteins have been reported.

■ GENES

The cDNA for the *Dictyostelium* 30 kDa protein has been cloned and sequenced[6]. These data are available under

Gen Bank accession number M58022. The cDNA sequences of the other small actin crosslinking proteins have not been reported.

■ REFERENCES

1. Fechheimer, M. and Taylor, D.L. (1984) J. Biol. Chem. 259, 4514-4520.
2. Johns, J.A., Brock, A.M. and Pardee, J.D. (1988) Cell Motil. and Cytoskel. 9, 205-218.
3. Fechheimer, M. and Furukawa, R. (1991) Meth. in Enz. 196, 84-91.
4. Furukawa, R. and Fechheimer, M. (1990) Dev. Genetics 11, 362-368.
5. Fechheimer, M. (1987) J. Cell Biol. 104, 1539-1551.
6. Fechheimer, M., Murdock, D., Carney, M. and Gloves, C.V.C. (1991) J. Biol. Chem. 266, 2883-2889.
7. Fechheimer, M. and Furukawa, R. unpublished.
8. Fechheimer, M. and Taylor, D.L. (1987) Meth. in Cell Biol. 28, 179-190.
9. Personal communication from Dr. Joel Pardee.
10. Maruta, H. and Korn, E.D. (1977) J. Biol. Chem. 252, 399-402.
11. MacLean-Fletcher, S.D. and Pollard, T.D. (1980) J. Cell Biol. 85, 414-428.
12. Brown, S.S. (1985) Cell Motil. and Cytoskel. 5, 529-543.
13. Ogihara, S. and Tonomura, Y. (1982) J. Cell Biol. 93, 604-614.
14. Albanesi, J.P., Lynch, T.J., Fujisaki, H., Bowers, B. and Korn, E.D. (1987) J. Biol. Chem. 262, 3404-3408.
15. Griffith, L.M. and Pollard, T.D. (1982) J. Biol. Chem. 257, 9135-9142.

■ *Marcus Fechheimer:*
Department of Zoology,
University of Georgia
Athens, Georgia, USA

Spectrins (Fodrin)

Spectrin denotes a family of acidic, largely α-helical, high molecular weight, multifunctional, actin binding proteins usually found in association with the plasma membrane of mature cells[1-4]. These proteins exist as heterodimers (α,β) and as heterotetramers (α,β)$_2$ the latter being the most common. Larger oligomers also can form [e.g. (α,β)$_3$]. All spectrins share a repetitive 106 amino acid motif, with nonhomologous regions marking sites of functional specialization. Distant homology exists with α-actinin and dystrophin. The spectrins bestow mechanical stability on the membrane and link transmembrane proteins (and perhaps phospholipids) to the cytoskeleton. In nonerythroid cells, spectrins may be confined to distinct membrane domains where they are thought to mediate receptor organization and/or vesicle trafficking[3,4].

Based on their tissue of origin, disposition within the cell, and apparent role, at least three functional classes of spectrin are recognized: erythrocyte, intestinal brush border, and nonerythroid spectrin. The first spectrin to be isolated, and the one most divergent, is that from human erythrocyte ghosts. In humans this protein consists of an α-subunit of 280 kDa (2,429 residues)[5] and a β-subunit of 246 kDa (2,137 residues)[6]. Both subunits migrate anomalously on SDS-PAGE, with apparent sizes of 240 and 220 kDa respectively. Both subunits display a characteristic 106 residue repeat structure (Figure 1). The repeat-to-repeat identity is 10-30%. *In vitro* reconstitution studies and direct visualization *in situ* indicate that spectrin and **actin** form an anastomosing and stoichiometric planar array at the cytoplasmic face of the erythrocyte membrane. This array is linked to the anion transporter (band 3) and to other integral membrane glycoproteins by ankyrin and **protein 4.1** (Figure 2). Formation of the spectrin-actin lattice requires competent spectrin self-association and spectrin-F-actin binding. Several hemolytic diseases with

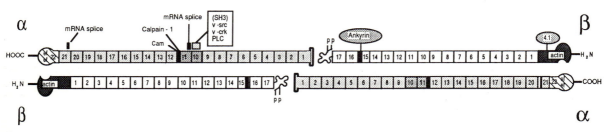

Figure 1. Schematic diagram of nonerythroid spectrin tetramer structure. The two subunits are arranged antiparallel to each other, and each is composed of multiple, approximately 106 residue, repeats. Nonhomologous segments are marked in black or shaded. The sites of major binding proteins are as indicated, and two regions of alternative mRNA transcription of α-fodrin are indicated[11]. The region of *src* homology in the 10th unit is shown.

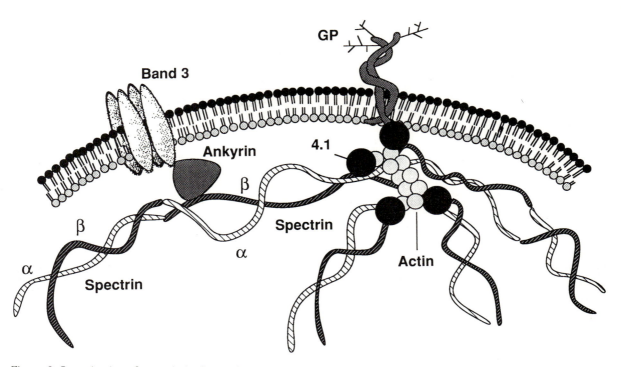

Figure 2. Organization of spectrin in the erythrocyte membrane cytoskeleton. A two dimensional planar lattice is formed by the ability of spectrin to self-associate at the amino terminal end of the α-subunit, and bind actin at the amino terminal end of the β-subunit. This array is linked to integral membrane proteins by means of ankyrin and protein 4.1.

enhanced erythrocyte fragility and/or shape abnormalities arise from point mutations or deletions in spectrin that alter its self-association or binding to protein 4.1 and actin[7] (Figure 1).

A second functional class is exemplifyed by terminal web spectrin of avian intestinal brush borders, called *TW260/240*[2,8]. This spectrin shares a common α-subunit (in avians) with other nonerythroid spectrins, but has a unique β-subunit (estimated 260 kDa by SDS-PAGE). This protein crosslinks microvillar core actin bundles but does not intimately associate with the plasma membrane, presumably due to the inability of the (TW260) β-subunit to bind either ankyrin or protein 4.1[2]. A similar terminal web-specific isoform of spectrin is not present in mammals[2].

The third and most general class of spectrins are the nonerythroid spectrins. Many trivial names for these exist, including fodrin, calspectin, brain actin binding protein (BABP), calmodulin binding protein, and α,γ-spectrin. Most workers prefer either *fodrin*[9] or *nonerythroid spectrin*. Human α-fodrin has a molecular weight of 284 kDa (2,472 residues)[10,11], and is 54% identical to human erythrocyte α-spectrin, 63% identical to *Drosophila* α-spectrin[12], and 96% identical to chicken α-fodrin[5,10]. In mammals, at least three isoforms of α-fodrin arise by alternative mRNA splicing[11]. Since antibodies raised to the nonerythroid spectrins typically display broad species crossreactivity, the pressure to preserve nonerythroid

spectrin structure during evolution must be strong. This same level of conservation is not enjoyed by the erythroid α-spectrins, or by the β-spectrins. In addition to the variant β-spectrin of TW260, other β-isoforms arise from multiple genes or by alternative mRNA processing[2,6,13,14]. One of these, *Drosophila* (β$_H$), is unusual since its size (430 kDa) is similar to dystrophin[13]; another is an erythroid like β-spectrin that appears not to exist in association with an α-subunit but is localized with clustered acetylcholine receptors in myocytes[15].

The spectrins interact cooperatively with several ligands, and are the target of several posttranslational regulatory events, including covalent phosphorylation, Ca^{2+} and calmodulin binding, and proteolysis[16].

■ PURIFICATION

Erythrocyte spectrin is prepared by extraction of fresh erythrocyte ghosts at low ionic strength with EDTA, followed by gel filtration in isotonic buffers on a large pore Sepharose (CL-2B or 4B) column[17]. Nonerythroid spectrin (fodrin) is prepared from brain membranes by extraction with either high[8] or low[18] ionic strength buffer, followed by ion exchange chromatography. The high-salt procedure yields fodrin of greater purity. Recombinant spectrin and fodrin peptides which retain functional activity have also been prepared[19].

ACTIVITIES

Spectrin has no intrinsic enzymatic activity. It associates *in vitro* with itself, ankyrin, F-actin, protein 4.1, **adducin**, calmodulin, **calpain-I**, and possibly other cytoskeletal proteins[1-3]. Its ability to self-associate is most easily detected by nondenaturing PAGE[20]. Ankyrin, adducin, and F-actin binding are best demonstrated by sedimentation velocity[1,21]. In these assays, the binding of spectrin to actin can be stimulated by protein 4.1 and inhibited under some conditions by Ca^{2+} and/or calmodulin[1,2,21].

ANTIBODIES

Polyclonal and monoclonal antibodies to a variety of spectrins have been reported[2,8,9,14], some of these are available from Chemicon Inc. Most erythrocyte spectrin antibodies react poorly, at best with the nonerythroid spectrins, and vice versa. However, most nonerythroid α-spectrin antibodies react well across species lines.

GENES

Full length cDNA clones have been identified and sequenced for human erythrocyte α-spectrin (GenBank M61877)[5], human erythrocyte β-spectrin (GenBank J05500)[6], chicken nonerythroid α-spectrin[10] (EMBL X14519, X13701), human brain and fibroblast α-spectrin (fodrin) (GenBank J05243)[11], and *Drosophila* α-spectrin[12] (GenBank M26400). Several reports of partial sequences have also appeared, as cited in the above references.

REFERENCES

1. Bennett, V. (1989) Biochim. Biophys. Acta 988, 107-121.
2. Coleman, T.R., Fishkind, D.J., Mooseker, M.S. and Morrow, J.S.(1989) Cell Motil.12, 225-247.
3. Morrow, J.S. (1989) Curr. Opin. Cell Biol. 1, 23-29.
4. Rodriques-Boulan, E. and Nelson,W.J. (1989) Science. 245,718-725.
5. Sahr, K.E., Laurila, P., Kotula, L., Scarpa, A.L., Coupal, E., Leto, T.L., Linnenbach, A.J., Winkelmann, J.C., Speicher, D.W., Marchesi, V.T., Curtis, P.J. and Forget, B.G. (1990) J. Biol. Chem. 265, 4434-4443.
6. Winkelmann, J.C., Chang, J.G., Tse, W.T., Scarpa, A.L., Marchesi, V.T. and Forget, B.G. (1990) J. Biol. Chem. 265, 11827-11832.
7. Marchesi, S.L. (1989) In "Red Blood Cell Membranes" (Agre, P. and Parker, J., eds.) Marchel Dekker, New York, pp. 77-85.
8. Glenney, J.R., Glenney, P., Osborn, M. and Weber, K. (1982) Cell 28,843-854.
9. Levine, J. and Willard, M. (1981) J. Cell Biol. 90,631-643.
10. Wasenius, V.M., Suraste, M., Salven, P., Eramaa, M., Hoem, L., Lehto, V.P. (1989) J. Cell Biol. 108, 79-93.
11. Moon, R.T. and McMahon, A.P. (1990) J. Biol. Chem. 265, 4427-4433.
12. Dubreuil, R.R., Byers, T.J., Sillman, A.L., Bar-Zvi, D., Goldstein, L.S. and Branton, D. (1989) J. Cell Biol. 109, 2197-2205.
13. Dubreuil, R.R., Byers, T.J., Stewart, C.T. and Kiehart, D.P. (1990) J. Cell Biol. 111, 1849-1858.
14. Zagon, I.S., Higbee, R., Riederer, B.M. and Goodman, S.R. (1986) J. Neurosci. 6, 2977-2986.
15. Bloch, R.J. and Morrow, J.S. (1989) J. Cell Biol. 108,481-493.
16. Mische and Morrow (1988) Protoplasma 145, 167-175.
17. Bennett, V. (1983) Meth Enzymol. 96, 313-324.
18. Davis, J. and Bennett, V. (1983) J. Biol. Chem. 258, 7757-7766.
19. Kennedy, S.P., Warren, S.L., Forget, B.G. and Morrow, J.S. (1991) J.Cell Biol. 115, 267-277.
20. Morrow, J.S. and Haigh Jr., W.B. (1983) Meth. Enzymol. 96, 298-304.
21. Harris, A.S. and Morrow, J.S. (1990) Proc. Natl. Acad. Sci. (USA) 87, 3009-3013.

Jon S. Morrow:
Department of Pathology,
Yale Medical School,
New Haven, CT, USA

Tenuin

Tenuin is a high molecular weight protein localized at the undercoat of adherens junctions and at microfilament bundles such a stress fibres[1]. It cannot bind to actin filaments in vitro. Tenuin probably functions as a key protein responsible for the maintenance of the structural integrity of actin filament bundles in nonmuscle cells.

Tenuin reveals an apparent molecular mass of 400 kDa by SDS-PAGE[1]. In the low angle rotary shadowing electron microscopy, the tenuin molecules look like slender rods about 400 nm long (Figure 1). Tenuin molecules do not bind to **actin** filaments *in vitro*. This molecule was first identified as a major constituent of the undercoat of cell-to-cell adherens junctions isolated from rat liver. Immunoblot analysis with anti-tenuin mAb showed that this protein occurs in various types of tissues including liver, intestine, brain, skeletal and cardiac muscle.

Immunofluorescence microscopy and immunoelectron microscopy revealed that this protein is distributed not only at the undercoat of adherens junctions but also along actin bundles associated with the junction in nonmuscle cells: stress fibres in cultured fibroblasts and

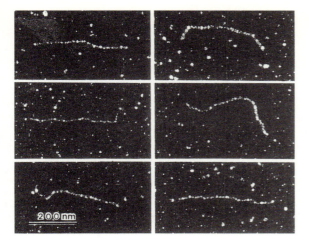

Figure 1. Rotary shadow electron micrograph of tenuin (x70,000).

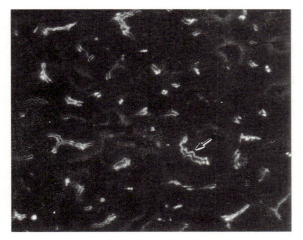

Figure 2. Immunofluorescence localization of tenuin in rat liver (x200).

circumferential bundles in epithelial cells[1] (Figure 2). Taken together, tenuin may play a crucial role in forming and maintaining actin filament bundles in nonmuscle cells.

■ PURIFICATION

Tenuin can be purified from isolated cell-to-cell adherens junctions of rat liver[1,2]. Tenuin is effectively extracted from isolated adherens junctions with a low salt solution at pH 9.2. After DEAE ion exchange column chromatography, tenuin is finally purified by gel filtration through sepharose CL-4B column.

■ ACTIVITIES

Tenuin is concentrated in actin bundles, but it cannot bind to actin filaments *in vitro*[1].

■ ANTIBODIES

A monoclonal antibody against rat liver tenuin is published[1]. This monoclonal antibody crossreacts with mouse, bovine, human and chick tenuin.

■ GENES

No sequence data of tenuin have yet been published.

■ REFERENCES

1. Tsukita, Sh., Itoh, M. and Tsukita, Sa. (1989) J. Cell Biol. 109, 2905-2915.
2. Tsukita, Sa. and Tsukita, Sh. (1989) J. Cell Biol. 108, 31-41.

■ Shoichiro Tsukita and Sachiko Tsukita:
Department of Information Physiology,
National Institute for Physiological Sciences,
Okazaki, Japan

Thymosin β4 (Tβ4)

Thymosin β4 (Tβ4; previously called "Fx" as isolated from platelet supernatants[1]) is a 5 kDa peptide which is widely distributed in vertebrate cells. Tβ4 binds to and sequesters actin monomers, but does not cut or cap actin filaments. Tβ4 is the major actin sequestering protein in human platelets.

Thymosin β4 is a single polypeptide of 43 residues[2,3] (Figure 1). It is unusually rich in polar amino acids, and has a pI of 5.1[2]. A recent study by NMR shows that the free peptide is largely unfolded in aqueous solution; in helix stabilizing solvents such as 60% trifluoroethanol, Tβ4 folds into two short α-helical domains (boldface in Figure 1) connected by a flexible nonhelical region[4]. Tβ4 binds to G-**actin** to form a complex which can be identified by its characteristic mobility in nondenaturing SDS-PAGE[1]. The electrophoretic mobility of the complex is slightly higher than that of G-actin alone, while the mobility of free Tβ4 despite its small mass and high negative charge, is consid-

| Ac | S | D | K | **P** | **D5** | **M** | **A** | **E** | I | **E10** | **K** | **F** | **D** | **K** | **S15** |

| | **K** | L | K | K | T20 | E | T | Q | E | K25 | N | P | L | **P** | **S30** |

| | **K** | **E** | **T** | **I** | **E35** | **Q** | **E** | **K** | **Q** | **A40** | G | E | S | | |

Figure 1. Primary sequence of thymosin β4/Fx. Residues which fold into α-helical segments in 60% trifluoroethanol are indicated in boldface. (Redrawn from Safer et al.[3].)

Figure 2.

Tropomyosin	194	K	L	K	E	A	E	T	R	A	E	203
			I	I	I		:	I	I			
Thymosin β4	16	K	L	K	K	T	E	T	Q	E	K	25
			I	I		:	I	I		:	I	
Actobindin	14	A	L	K	H	A	E	T	V	D	K	23
			I	I		:	I	I		:	I	
Actobindin	50	E	L	K	H	A	E	T	V	D	K	59
			I	I		:	I	I				
Myosin heavy chain	839	L	L	K	S	A	E	T	E	K	E	848
			I	I				I	:			
α–actinin	89	C	L	K	H	I	E	S	H	G	V	98

Figure 2. Partial sequence comparison of thymosin β4 with other actin binding proteins: equine platelet tropomyosin[19], *Acanthamoeba* actobindin[20], rat skeletal myosin heavy chain[21], and *Dictyostelium* α-actinin[22]. Identical residues are indicated by vertical lines, conservative substitutions by paired closed circles. (Reprinted, with permission, from Safer et al.[3].)

erably lower than that of G-actin or the actin-Tβ4 complex (Safer, unpublished observations). These data suggest that free Tβ4 has a high frictional coefficient, consistent with an unfolded conformation, and may form a compact structure when it binds to actin. The sequence of Tβ4 shows a short region of ten residues with some similarities to other actin binding proteins (including **actobindin**, **α-actinin**, **myosin** heavy chain and **tropomyosin**); while suggestive, these similarities do not represent statistically significant homologies and the region of the sequence is far too small to provide a complete protein-protein interface (Figure 2).

Thymosin β4 has been detected in a wide range of vertebrate species[5], and a closely related peptide, thymosin β4[Xen], has been found in *Xenopus* tissues and oocytes[6]. The amount of Tβ4 has been measured in numerous mammalian tissues[7], cells[8], and cell lines[9]. Several other peptides closely homologous to Tβ4 have been reported whose functions are unknown[10]. The concentration in 60-day old rat tissues varied from 452 mg/g (wet weight) for spleen to 11 mg/g for muscle[7]. Tβ4 was measured at 6.9-31.7 fg/cell in human platelets[8]; estimating platelet cell volume as seven fl the mean concentration is about 600 μM. Similarly it is estimated at about 200 μM in polymorphonuclear leukocytes. These measurements indicate that Tβ4 is by far the most abundant actin sequestering protein in these cells[1]. In cultured cells the levels ranged from 1.75 in macrophages and 1.63 in fibroblasts to <0.003 μg/mg protein for mouse DS19 erythroleukemia cells[9].

■ PURIFICATION

Several chromatographic procedures have been reported for the purification of Tβ4[2,7,1]. We have found it most convenient to extract cells or tissues with cold 0.5 M perchloric acid[7]; after neutralization with KOH, the clarified, dialyzed extract is applied to column of DEAE-cellulose in 25 mM NH_4HCO_3, washed, and eluted with a step of 0.25 M NH_4HCO_3. Final purification is accomplished by preparative HPLC[1].

■ ACTIVITIES

Tβ4 forms a 1:1 complex with α-actin and inhibits salt induced polymerization, but does not cut or cap actin filaments[1]. The K_d of the actin-Tβ4 complex, measured in 0.1 M KCl, 2 mM $MgCl_2$, pH7.5, is 2.1 μM for muscle actin and 0.7 μM for platelet actin[11], which is consistent with the observation[1] that the bulk of the unpolymerized actin in resting platelets is complexed with Tβ4. The stability of the actin-Tβ4 complex is not affected by Ca^{2+} at physiological concentrations[11]. Formation of the actin-Tβ4 complex inhibits ATP exchange by actin monomers (Dr. P. Goldschmidt-Clermont, personal communication). Preliminary data indicate that **profilin**, **vitamin D-binding protein** (VDBP), and DNAse 1 all compete with Tβ4 for binding to actin, suggesting that the binding site for Tβ4 may overlap both the DNAse 1 binding site and the profilin/VDBP binding site on the actin monomer (Goldschmidt-Clermont, personal communication; Safer and Nachmias, unpublished data). Stimulation of platelets by thrombin results in a rapid (30 sec) decrease in the level of actin-Tβ4 complex, paralleling the increase in the concentration of F-actin (Nachmias and Golla, unpublished data). A number of hormonal and immunomodulatory effects have previously been reported for Tβ4[12]; these several extracellular function(s) of Tβ4 *in vivo* have not yet been resolved[12].

■ ANTIBODIES

Antibodies against Tβ4 have been reported[13-15]; in one study[15] they were found to recognize two separate epitopes containing residues 1-8 and 22-32. Monoclonal antibodies have been raised in our laboratory using Tβ4 coupled to hemocyanin.

■ GENES

Full-length cDNA sequences have been reported for rat[16] and human[17] Tβ4. Hybridization analysis using a rat cDNA probe has shown that the Tβ4 gene is expressed in a wide range of rat tissues and lymphoid cells[18], confirming the HPLC data for the peptide itself[7,8,9].

■ REFERENCES

1. Safer, D., Golla, R. and Nachmias, V.T. (1990) Proc. Natl. Acad. Sci. (USA) 87, 2536-2540.
2. Low, T.L.K., Hu, S.-K. and Goldstein, A.L. (1981) Proc. Natl. Acad. Sci. (USA) 78, 1162-1166.
3. Safer, D., Elzinga, M. and Nachmias, V.T. (1991) J. Biol. Chem. 266, 4029-4032.
4. Zarbock, J., Oschkinat, H., Hannappel, E., Kalbacher, H., Voelter, W. and Holak, T.A. (1990) Biochem 29, 7814-7821.
5. Erickson-Viitanen, S., Ruggieri, S., Natalini, P. and Horecker, B.L. (1983) Arch. Biochem. Biophys. 221, 570-576.
6. Hannappel, E., Kalbacher, H. and Voelter, W. (1988) Arch. Biochem. Biophys. 260, 546-551.
7. Hannappel, E. (1986) Analt. Biochem. 156, 390-396.
8. Hannappel, E. and Van Kampen, M. (1987) J. Chromatog. 397, 279-285.
9. Goodall, G.J., Morgan, J.I. and Horecker, B.L. (1983) Arch. Biochem. Biophys. 221, 598-601.
10. Horecker, B.L., Erickson-Viitanen, S. and Hannappel, E. (1985) Methods in Enzymol. 116, 265-269.
11. Weber, A., Pennise, C., Pring, M., Safer, D. and Nachmias, V.T. (1990) J. Cell Biol. 111, 161a and personal communication.
12. Spangelo, B.L, Hall, N.R. and Goldstein, A.L. (1987) Ann. N.Y. Acad. Sci. 496, 196-204.
13. Naylor, P.H., McClure, J.E., Spangelo, B.L., Low, T.L.K. and Goldstein, A.L. (1984) Immunopharmacol. 7, 9-16.
14. Dalakas, M.C. and Trapp, B.O. (1986) Ann. Neurol. 19, 349-355.
15. Goodall, G.J., Hempstead, J.L. and Morgan, J.I. (1983) H. Immunol. 131, 821-825.
16. Goodall, G.J., Richardson, M., Furuichi, Y., Wodnar-Filipowicz, A. and Horecker, B.L. (1985) Arch. Biochem. Biophys. 236, 445-447.
17. Gondo, H., Kudo, J., White, J.W., Barr, C., Selvanayagam, P. and Saunders, G.F. (1987) J. Immunol. 139, 3840-3848.
18. Gomez-Marquez, J., Dosil, M., Segade, F., Bustelo, X.R., Pichel, J.G., Dominguez, F. and Freire, M. (1989) J. Immunol. 143, 2740-2744.
19. Lewis, W.G., Cote, G.P., Mak, A.S. and Smillie, L.B. (1983) FEBS Lett. 156, 269-273.
20. Vandekerckhove, J., Van Damme, J., Vancompernolle, K., Bubb, M.R., Lambooy, P.K. and Korn, E.D. (1990) J. Biol. Chem. 265, 12801-12805.
21. Strehler, E.E., Strehler-Page, M.A., Perriard, J.C., Periasamy, M. and Nadal-Ginard, B. (1986) J. Mol. Biol. 190, 291-299.
22. Noegel, A., Witke, W. and Schleicher, M. (1987) FEBS Lett. 221, 391-396.

■ *Daniel Safer and Vivianne T. Nachmias:*
Department of Anatomy, School of Medicine,
University of Pennsylvania,
Philadelphia, PA 19104-6058, USA

Titin

Titin, the largest polypeptide so far identified, is an extraordinarily long and slender elastic protein, abundant in the sarcomere of striated muscles. It forms an elastic filamentous matrix in the sarcomere that provides structural continuity and elastic restoring force. In developing muscles, titin molecules may act as a template/scaffold in myofibrillogenesis.

Titin (also called connectin) is the third most abundant protein in skeletal and cardiac muscle sarcomere, comprising ~10% of myofibrillar protein[1-4]. Titin is a family of giant polypeptides, 2,000-3,500 kDa. The expression of titin size variants or isoforms appears to be development-, tissue- and species-dependent[1-6]. On high resolution, low porosity SDS gels, titin appears as either a singlet (T1: 2,800-3,200 kDa) or a doublet (T1 and T2: 2,100-2,400 kDa)[1,2,5-7]. Native T2, presumably a proteolytic product of the longer T1 polypeptide, has been extensively characterized. In rotary shadowed (Figure) or negatively stained preparations, T2 appears as extremely long and flexible strand (~1 μm long, 3-4 nm in diameter) with or without a globular head[8-11]. T2 may self-associate, via its globular

Figure. Electron micrograph of rotary-shadowed native titin (T2). Circles indicate beaded morphology.

head, to dimers or higher oligomers[11]. Frequently it entangles to form bundles or meshes of filamentous strands[8]. Titin is a multidomain protein, rich in β-sheet structure and consisting of two types of 100 residue repeats that are related to **fibronectin** type III domain and C2 set of immunoglobulin domain[12].

In adult sarcomeres, each titin molecule spans from the M-line to the Z-line[13,14], with the globular head binding near the M-line[11] and its C-terminus near the Z-line[15]. Titin binds to and is rendered inextensible by thick filaments along most of its length[1,2,6,13,14]. The remaining segment serves as an elastic spring that connects the ends of thick filaments to the Z-line[1,2,6,13,14]. The reversible extension of this titin segment generates resting tension[6]. In developing skeletal and cardiac muscles, titin participates in myofibrillogenesis by serving as a template or scaffold in the assembly of individual A-bands and I-Z-I bundles[16-19].

Titin is degraded somewhat in muscle tissues of Duchenne and Fukuyama muscular dystrophy patients[20,21]. Auto-antibodies to titin are found in the sera of patients with *myasthenia gravis* and *thyoma*[21].

In invertebrate skeletal and flight muscles, a family of 800-1,200 kDa proteins (twitchin[22], projectin[23-26] and mini-titin[27]) share the sequence motifs, secondary structure, filamentous morphology and physical properties of vertebrate titin. In insect asynchronous flight muscles, mini-titin (0.26 μm in length) appears to constitute the elastic connecting filaments that link the Z disc to the A band, somewhere along the length of thick filaments or near the M-line[24,26,27]. In contrast, these proteins are found only in the **myosin**-containing A-band in insect synchronous leg muscles[22,24,26,27], suggesting a distinct function. Nematode twitchin may be involved in the regulation of the contraction relaxation cycle[22].

■ PURIFICATION

Titin polypeptides can be readily purified from SDS-solubilized myofibrils or muscle tissues by a novel procedure that utilizes NaCl to precipitate titin preferentially[7]. T1 and T2 are then partially resolved in large-pore gel filtration columns (Sephacryl-S1000 or Bio Gel-A150M) in the presence of SDS[7]. Native T2 is purified from fresh muscles by high salt extraction, followed by ion exchange or

hydroxylapatite columns[8-11]. These T2 preparations can be resolved into monomers and oligomers by gel filtration chromatography[8-11]. Purification of native T1 has recently been accomplished[28]. In most purification procedures, the inclusion of protease inhibitors is essential for reproducible results. Similar protocols are applicable to mini-titins from insect muscles[26,27].

■ ACTIVITIES

In solution of low ionic strength, titin binds and aggregates myosin rods, entangles actin filaments and slightly enhances actomyosin ATPase[29]. It also binds to AMP-deaminase and **C-protein**[30], **myomesin** and **M-protein**[16]. Nematode twitchin may be a myosin light chain kinase[22].

■ ANTIBODIES

Polyclonal and monoclonal antibodies are generally cross-reactive to titin isoforms from a wide range of tissues and species[1,2]. Monoclonal antibodies directed to A-I junction titin epitopes are available from Sigma Chemical Co. and from Developmental Hybridoma Bank (The Johns Hopkins University). Antibodies to twitchin, projectin and mini-titin crossreact with each other in most invertebrate flight and skeletal muscles[23-27]. Certain antibodies crossreact with vertebrate titin[26]. Anti-titin is useful in distinguishing tumors of muscle tissue[31].

■ GENES

DNA sequences of rabbit skeletal titin (EMBL X17329 and X17330)[12], nematode twitchin (EMBL X15423)[22], Lethocerous projectin[24], and *Drosophila* projectin (GenBank M73433-M73435)[25] have been reported. The human titin gene localizes in chromosome 2q13-35[12] and the *Drosophila* projectin localizes on the fourth chromosome at 102C/D[25].

■ REFERENCES

1.Wang, K (1985) Cell Muscle Motil. 6, 315-369.
2.Maruyama, K. (1986) Int. Rev. Cyto. 104, 81-114.

3. Trinick, J. (1991) Curr. Opin. Cell Biol. 3, 112-119.
4. Fulton, A.B. and Isaacs, W.B. (1991) Bioessays 13, 157-161.
5. Hu, D.H., Kimura, S. and Maruyama, K. (1986). J. Biochem. (Tokyo) 99, 1485-1492.
6. Wang, K., McCarter, R., Wright, J., Beverly, J. and Ramirez-Mitchell, R. (1991) Proc. Natl. Acad. Sci. (USA) 88, 7101-7105.
7. Wang, K. (1982) Meth. Enzymol. 85, 264-274.
8. Wang, K., Ramirez-Mitchell, R. and Palter, D. (1984) Proc. Nat. Acad. Sci. (USA) 81, 3685-3689.
9. Trinick, J., Knight, P. and Whiting, A. (1984) J. Mol. Biol. 180, 264-274.
10. Maruyama, K., Kimura, S., Yoshidomi, H., Sawada, H. and Kukuchi, M. (1984) J. Cell Biol. 109, 1423-1493.
11. Nave, R., Furst, D.O. and Weber, K. (1989) J. Cell Biol. 109, 2177-2187.
12. Labeit, S., Barlow, D.P., Gautel, M., Gibson, T. Holt, J., Hsieh, C.L., Francke, U., Leonard, K., Wardale, J., Whiting, A. and Trinick, J. (1990) Nature 345, 273-276.
13. Itoh, J., Suzuki, T., Kimura, S., Ohashi, K., Higuchi, H., Sawada, H., Shimizu, T., Shibata, M. and Mauryama, K. (1988) J. Biochem. (Tokyo) 104, 504-508.
14. Furst, D.O., Osborn, M., Nave, R. and Weber, K. (1988) J. Cell Biol. 106, 1563-1572.
15. Wang, S.M., Jun, M.C. and Jeng, C.J. (1991) Biochem. Biophys. Res. Commun. 176, 189-193.
16. Furst, D.O., Osborn, M. and Weber, K. (1989) J. Cell Biol. 109, 517-527.
17. Isaacs, W.B., Kim, I.S., Struve, A. and Fulton, A.B. (1989) J. Cell Biol. 109, 2189-2195.
18. Wang, S., Greaser, M., Schultz, E., Bulinski, J., Lin, J.J.-C. and Lessard, J.L. (1988) J. Cell Biol. 107, 1075-1083.
19. Tokuyasu, K.T. and Maher, P.A. (1987) J. Cell Biol. 105, 2781-2793.
20. Matsumura, K., Shirnizu, T., Nonaka, I. and Mannen, T. (1989) J. Neurol. Sci. 93, 147-156.
21. Matsumura, K., Shimizu, T., Sunado, Y., Mannen, T., Nonaka, I., Kimura, S. and Maruyama, K. (1990) J. Neurol. Sci. 98, 155-196.
22. Benian, G.M., Kiff, J.E., Neckelmann, N., Noerman, D.G. and Waterston, R.H. (1989) Nature 342, 45-50.
23. Saide, J.D. (1981) J. Mol. Biol. 153, 661-679.
24. Lakey, A., Ferguson, C., Labeit, S., Reedy, M., Larkins, A., Butcher, G., Leonard, K. and Bullard, B. (1990) EMBO J. 9, 3459-3467.
25. Ayne-Southgate, A., Vigoreaux, J., Benian, G. and Pardue, M.L. (1991) Proc. Natl. Acad. Sci. (USA) 88, 7973-7977.
26. Hu, D.H., Matsuno, A., Terakado, K., Matsuura, T., Kimura, S. and Maruyama, K. (1990) J. Muscle Res. Cell Motil. 11, 497-511.
27. Nave, R. and Weber, K. (1990) J. Cell Sci. 95, 535-554.
28. Maruyama, T., Nakauchi, Y., Kimura, S. and Maruyama, K. (1989) J. Biochem. (Tokyo) 105, 323-326.
29. Kimura, S., Maruyama, K. and Huang, Y.P. (1984) J. Biochem. (Tokyo) 96, 499-506.
30. Koretz, J., Irving, T. and Wang, K. (1992) Biochem. Biophys. Res. Commun. in press.
31. Osborn, M., Hill, C., Aetmannsberger, M. and Weber, K. (1986) Lab. Invest. 55, 101-108.

■ Kuan Wang:
Department of Chemistry and Biochemistry,
University of Texas,
Austin, TX, USA

Tropomodulin

Tropomodulin (Tmod) is a tropomyosin binding protein that binds to one of the ends of tropomyosin and weakens tropomyosin-actin interactions. Tmod is a component of the human erythrocyte membrane skeleton and is also present in a variety of nonerythroid cells and tissues. The cDNA derived amino acid sequence of Tmod reveals that this protein is not related to any previously described tropomyosin binding proteins.

Tropomodulin (Tmod) is a M_r 43,000 **tropomyosin** binding protein that has been purified from the human erythrocyte membrane skeleton[1]. A combination of rotary shadowing electron microscopy[2], chemical crosslinking, nondenaturing gel electrophoresis and Scatchard analysis of Tmod-tropomyosin binding[3] indicate that Tmod probably exists as dimers and tetramers in solution. Tmod binds to one of the ends of erythrocyte tropomyosin (Figure 1) and inhibits tropomyosin binding to F-**actin** without itself binding to actin[2]. Based on its ability to abolish cooperative binding of tropomyosin to F-actin[2], Tmod is proposed to weaken tropomyosin-actin interactions by blocking head-to-tail association of tropomyosin molecules along the actin filament (Figure 2). Tmod binding to tropomyosins is isoform specific, since Scatchard analysis of solid phase binding assays show that binding of brain, platelet and skeletal muscle

tropomyosins to Tmod is characterized by decreases in affinity and/or saturation binding capacity in comparison to erythrocyte tropomyosin[3].

Isolation and sequencing of the cDNA for Tmod predicts a protein of 359 amino acids with a calculated molecular weight of 40,600 and a pI of 4.8[4]. Secondary structure predictions indicate that Tmod is predominantly α-helical with little or no β-sheet. The tropomyosin binding domain has been mapped to the N-terminal region containing residues 39 to 138, but appears to require the C-terminal region for correct folding and expression of its tropomyosin-binding activity. The Tmod sequence has no significant homology with any known proteins in the database, including tropomyosin binding proteins such as actin, **caldesmon**, **troponin**-I or troponin-T. However, crossreactivity of Tmod antibodies with striated muscle troponin I[2] suggests that Tmod and troponin I

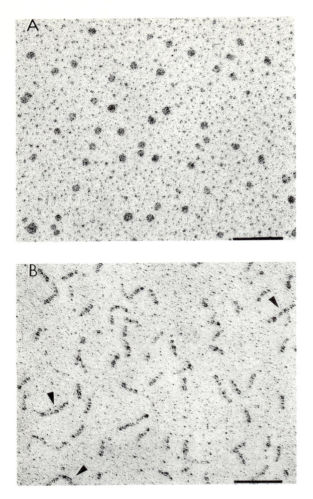

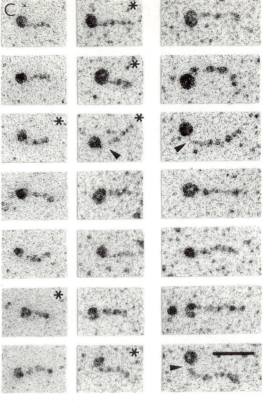

Figure 1. Electron micrographs of tungsten/carbon rotary shadowed images of (A) Tmod, (B) tropomyosin, (C) Tmod-tropomyosin complexes. A,B, bar = 100 nm; C, bar = 50 nm. (Reproduced with permission from ref. 2).

may share common features of secondary or tertiary structure.

Immunoreactive Tmod polypeptides are present in a variety of nonerythroid cells and tissues, including striated muscle, brain, eye lens, neutrophils and endothelial cells[2]. Immunolocalization of Tmod in stretched and resting length myofibrils from rat psoas muscle shows that Tmod is associated with the free (pointed) ends of the thin filaments in the muscle sarcomeres[5]. We hypothesize that Tmod binds to the terminal tropomyosin molecules on the thin filaments and acts as a tropomyosin capping protein, preventing additional tropomyosin molecules from binding to the distal portion of the filament. Since actin filaments without tropomyosin are susceptible to the action of actin depolymerizing or severing proteins, Tmod could function with tropomyosin to regulate actin filament length at the pointed end in the muscle sarcomere or in the erythrocyte membrane skeleton.

■ PURIFICATION

Tmod has so far only been purified from the human erythrocyte membrane[1,2]. It is extracted from the Triton insoluble fraction (membrane skeleton) by 1 M NaBr and purified to 95% homogeneity by anion exchange chromatography on DEAE-cellulose followed by gel filtration and sucrose gradient sedimentation.

■ ACTIVITIES

The tropomyosin binding activity of Tmod can be assayed by quantitative analysis of ^{125}I-tropomyosin binding to nitrocellulose dots to which purified Tmod has been adsorbed[1,3]. The ability of Tmod to inhibit tropomyosin-actin interactions can be assayed by measuring the effect of purified Tmod on the amount of ^{125}I-tropomyosin sedimenting with F-actin at 100,000xg[2].

■ ANTIBODIES

Polyclonal antibodies to purified human erythrocyte Tmod have been generated in rabbits[2], and work well in

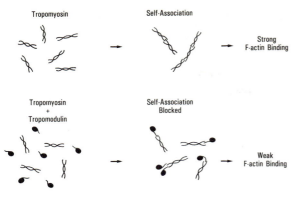

Figure 2. Model to explain how Tmod inhibits tropomyosin binding to F-actin. Binding to F-actin of individual tropomyosin molecules with six actin binding sites would be expected to be considerably weaker than binding of end-to-end tropomyosin dimers or trimers which would have 12 or 18 actin binding sites, respectively.

immunoblotting[2] or immunoprecipitation assays (V. Fowler, unpublished data), as well as in immunofluorescence on erythrocyte membranes, lens fibre cells, and striated muscle.

■ GENES

cDNAs for the Tmod have been isolated and sequenced from human foetal liver and reticulocyte libraries (GenBank M 77016)[4]. The Tmod gene is located on human chromosome 9 at bands q22.2 - q22.3[6].

■ REFERENCES

1. Fowler, V.M. (1987) J. Biol. Chem. 262, 12792-12800.
2. Fowler, V.M. (1990) J. Cell Biol. 111, 471-482.
3. Sussman, M.A. and Fowler, V.M. Eur. J. Biol. Chem. submitted.
4. Sung, L.A., Fowler, V.M., Lambert, K., Sussman, M.A., Karr, D. and Chien, S. (1992) J. Biol. Chem. 267, 2616-2621.
5. Sussman, M.A., Daniels, M.P., Flucher, B.E. and Fowler, V.M. (1991) J. Cell Biol. 115, 179a.
6. Lin, C.C., Sung, L.A., Fan, Y.S., Lambert, K., Fowler, V.M. and Chien, S. (1991) Cytogenet. and Cell Genet. Proc. Eleventh Int. Workshop on Human Gene Mapping.

■ *Velia Fowler:*
Department of Molecular Biology,
Research Institute of Scripps Clinic,
La Jolla, CA, USA

Tropomyosins

Tropomyosins (TM) are found in virtually all eukaryotic cells in association with actin[1,2]. In skeletal and cardiac muscles, TM together with the troponin complex, is a component of the regulatory mechanism by which the interaction of F-actin and myosin is controlled by the level of Calcium[3]. In smooth muscle and nonmuscle cells in which the troponin complex is absent, its role is less clear. However, recent descriptions of its interaction with caldesmon, calponin and tropomodulin are suggestive of an important regulatory role in these tissues (see 4-6 and references therein).

Tropomyosin (TM) molecules exist as dimers of two identical or similar α-helical polypeptide chains arranged in parallel and in-register as a coiled-coil. In the case of rabbit skeletal TM the overall length has been estimated as 410 ± 4Å. The structure is stabilized by interactions at the face of the two helices contributed by nonpolar side chains at positions *a* and *d* of the pseudoheptapeptide repeat (residues *a* to *g*) of the amino acid sequence. Residues at other positions are predominantly, but not exclusively, ionic or polar. Further stabilization probably also occurs through ionic interactions between acidic residues at position *e* of one helix and basic residues at position *g* of the other[1,7]. Two broad classes of tropomyosin isoforms, long and short, have been identified in vertebrates. The polypeptide chains of the former, present in striated and smooth muscles, are 281-285 amino residues in length and span seven **actin** monomers on each strand of F-actin (14 on both strands). The shorter form, present in nonmuscle cells, spans only six actins (or 12 on both strands) and are 245-251 amino acids in length. Both long and short forms can be present in nonmuscle cells[8]. A pattern of 14 (or 12) amino acid repeats (each 19-2/3 residues) in rabbit skeletal α- and β-TM (or in β-TM in platelets) has been correlated with F-actin monomer binding[9-11]. Tropomyosin molecules polymerize in a head-to-tail manner by an overlap at their ends of 8-11 amino acid residues. Removal of 11 residues at the C-terminus eliminates polymerization and markedly reduces cooperative binding to F-actin[12]. Binding is restored by adding the **troponin** complex[13]. Removal of ten residues at the N-terminus also virtually eliminates actin binding[14]. The propensity for head-to-tail polymerization of TMs is in the order: smooth muscle > skeletal muscle > platelets. These differences have been correlated with altered amino acid

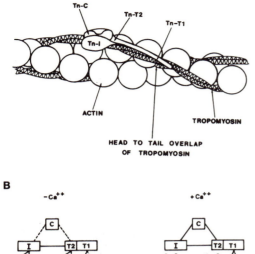

A

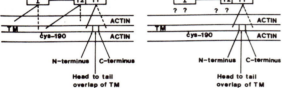

B

Figure. (A) Arrangement of actin, TM and troponin (Tn) components in thin filaments of striated muscle. (B) Schematic representation and effects of Ca^{2+} on interactions among thin filament proteins. Figure taken from Heeley *et al.* (1987) J. Biol. Chem. 262, 9971-9978.

sequences at the N-terminus (in the case of platelet) and at the C-terminus (for skeletal) when compared with smooth muscle TM[15].

In striated muscles the highly asymmetric troponin-T component, represented by its fragments T1 (residues 1-158) and T2 (residues 159-259), interacts with TM both at the head-to-tail overlap (T1) and in a region about 13-15 nm from its C-terminal end (T2)[16,17]. Troponin-I and -C are located in this latter region through interactions with the T2 region of troponin-T. Troponin-I probably also interacts with TM in this region (see Figure). A number of these interactions are sensitive to Ca^{2+}-concentration. Smooth muscle TM, platelet TM and nonpolymerizable TM (whose C-terminal amino acid sequences are different or partially eliminated) interact only weakly with the T1 fragment of troponin-T[18].

All TMs are probably acetylated at their N-terminus. The lack of acetylation of recombinant TM expressed in *E. coli* affects head-to-tail polymerization and binding to F-actin[19]. Skeletal muscle and cardiac TMs, but not chicken gizzard or equine platelet TMs, are partially phosphorylated at their penultimate C-terminal serine residue. Head-to-tail polymerization, interaction with troponin and ATPase activities of reconstituted TM-troponin-acto-myosin S1 are altered[20].

■ PURIFICATION

Protocols[21,22] for the purification of TM take advantage of its solubility at neutral pH or at pH 2.0, its insolubility at its isoelectric point (~pH 4.2), the reversible nature of its denaturation at elevated temperatures, precipitation with ammonium sulphate and chromatographic behavior on anion exchangers or hydroxylapatite. Its reduced mobility in SDS-PAGE in the presence of 6 M urea is useful for identification purposes.

■ ACTIVITIES

Biological assays include measurements of its inhibitory or activating effects on nonregulated or regulated acto-myosin ATPase systems[23-25].

■ ANTIBODIES

Monoclonal antibodies against TM have been employed (see references in **26**).

■ GENES

Partial or complete sequences for TM genes have been reported for human, rat, chicken, quail and *Drosophila*. The diversity of TM isoforms arises from multiple genes (four identified in humans), by alternative transcriptional promotors and by alternative splicing of primary gene transcripts. A single gene can encode only one isoform or as many as nine [including striated muscle, smooth muscle and nonmuscle; both long and short (see **27,28** and references therein)]. Alternatively spliced exons correspond approximately to the TM domains implicated in actin interaction, head-to-tail polymerization and troponin binding (or lack thereof)[8].

■ REFERENCES

1. Smillie, L.B. (1979) Trends in Biochem. Sci. 4, 151-154.
2. Côté, G.P. (1983) Mol. Cell. Biochem. 57, 127-146.
3. Leavis, P. and Gergely, J. (1984) Crit. Rev. Biochem. 16, 235-305.
4. Lehman, W., Moody, C. and Craig, R. (1990) Ann. N.Y. Acad. Sci. 599, 75-84.
5. Winder, S.J. and Walsh, M.P. (1990) J. Biol. Chem. 265, 10148-10155.
6. Fowler, V.M. (1990) J. Cell. Biol. 111, 471-482.
7. McLaughlan, A.D. and Stewart, M. (1975) J. Mol. Biol. 98, 293-304.
8. MacLeod, A.R. (1987) Bioessays 6, 208-212.
9. McLaughlan, A.D. and Stewart, M. (1976) J. Mol. Biol. 103, 271-298.
10. Mak, A.S., Smillie, L.B. and Stewart, G.R. (1980) J. Biol. Chem. 255, 3647-3655.
11. Hitchcock-DeGregori, S.E. and Varnell, T.A. (1990) J. Mol. Biol. 214, 885-896.
12. Mak, A.S. and Smillie, L.B. (1981) Biochem. Biophys. Res. Commun. 101, 208-214.
13. Mak, A.S., Golosinska, K. and Smillie, L.B. (1983) J. Biol. Chem. 258, 14330-14334.

14. Cho, Y.-J., Liu, J. and Hitchcock-DeGregori, S.E. (1990) J. Biol. Chem. 265, 538-545.

15. Sanders, C. and Smillie, L.B. (1985) J. Biol. Chem. 260, 7264-7275.

16. Mak, A.S. and Smillie, L.B. (1981) J. Mol. Biol. 149, 541-550.

17. White, S.P., Cohen, C. and Phillips Jr., G.N. (1987) Nature 325, 826-828.

18. Pearlstone, J.R. and Smillie, L.B. (1982) J. Biol. Chem. 257, 10587-10592.

19. Hitchcock-DeGregori, S.E. and Heald, R.W. (1987) J. Biol. Chem. 262, 9730-9735.

20. Heeley, D.H., Watson, M.H., Mak, A.S., Dubord, P. and Smillie, L.B. (1989) J. Biol. Chem. 264, 2424-2430.

21. Smillie, L.B. (1982) in Methods in Enzymology (ed. S.P. Colowick and N.B. Kaplan), Vol. 85, 234-241, Academic Press, N.Y.

22. Côté, G.P. and Smillie, L.B. (1981) J. Biol. Chem. 256, 11004-11010.

23. Lehrer, S.S. and Morris, E.P. (1982) J. Biol. Chem. 257, 8073-8080.

24. Williams, D.L., Jr., Greene, L.E. and Eisenberg, E. (1984) Biochemistry 23, 4150-4155.

25. Heeley, D.H., Smillie, L.B. and Lohmeier-Vogel, E.M. (1989) Biochem. J. 258, 831-836.

26. Wang, S.-M., Wang, S.-H., Lin, J.-L. and Lin, J.-C. (1990) J. Muscle Res. Cell Motil. 11, 191-202.

27. Lees-Miller, J.P., Yan, A. and Helfman, D.M. (1990) J. Mol. Biol. 213, 399-405.

28. Lees-Miller, J.P., Goodwin, L.O. and Helfman, D.M. (1990) Mol. Cell. Biol. 10, 1729-1742.

■ Lawrence B. Smillie:
MRC Group in Protein Structure and Function,
Department of Biochemistry,
University of Alberta, Edmonton,
Canada T6G 2H7

Troponins

Troponin[1-3] (Tn) is a complex of three subunits: troponin C (TnC), troponin I (Tn) and troponin T (TnT); together with tropomyosin (TM) it constitutes the Ca^{2+}-dependent regulatory machinery localized in the actin filaments of striated muscle. The binding of Ca^{2+} - released in vivo from the sarcoplasmic reticulum upon stimulation of the muscle - to TnC induces a conformational change in that subunit, and the effect is transmitted to the other thin filament components. This results in the activation of the actomyosin ATPase and, in vivo, force development and shortening of muscle fibres.

TnC, the Ca^{2+}-binding subunit, consists of a single polypeptide chain containing 159 residues, with a mass of 17,846 Da (parameters given here are those of fast rabbit skeletal muscle constituents). The structure (Figure 1) emerging from recent X-ray crystallographic studies shows two globular domains connected by a single nine turn α-helix[4,5]. In each domain there are two divalent cation binding sites, the two N-terminal ones being specific for Ca^{2+} and the two C-terminal ones being able to bind both Ca^{2+} and Mg^{2+}. Binding constants for Ca^{2+} are 3×10^5 M^{-1} and 2×10^6 M^{-1} at the N- and C-terminal sites, respectively[7]; that for Mg^{2+} at the C-terminal sites is 2×10^3 M^{-1}. Ca^{2+} binding to the N-terminal sites triggers actomyosin activation, while the metal bound at the Ca^{2+}-Mg^{2+} sites appears to play a structure stabilizing role. The four metal binding sites are homologous to the two sites found in parvalbumin, a sarcoplasmic Ca^{2+}-binding protein; they are known as EF-hand or helix-loop-helix structures[6]. The Ca^{2+}-binding loop and a residue in the C-terminally adjacent helix provide seven oxygen ligands resulting in pentagonal bipyramidal Ca^{2+}-coordination[8]. One of the ligands may be a water molecule. Site I in cardiac and slow skeletal muscle TnC contains amino acid replacements that inactivate it as a Ca^{2+}-binding site. The number of functional binding sites may be less in some lower species[2,9]. For a collection of TnC amino acid sequences see reference 10.

TnI, the inhibitory subunit of troponin, contains 179 amino acids and has a mass of 20,864 Da. Although TnI by itself inhibits actomyosin ATPase, Ca^{2+}-dependent regulatory activity requires TnC, TnT and **tropomyosin**. A highly basic peptide (residues 96-116) corresponding to one of the interaction sites with TnC has inhibitory activity.

TnT, the TM binding subunit, contains 259 amino acid residues and has a mass of 30,503 Da. Its N-terminal region has high α-helical content and interacts with the C-terminal portion of one TM molecule and with the N-terminus of the next, overlapping, TM. The C-terminal segment forms a globular structure, interacting with the region of TM that contains Cys-190 and with TnC and TnI.

The dumbbell shaped structure of TnC revealed by X-ray diffraction may not be that prevailing in solution or in the intact thin filaments[11,12]. The short stretch of TnI corresponding to the inhibitory peptide interacts with both the N- and the C-terminal domain of TnC[13]. Differences in the structure of the apo form of the N-terminal domain and the Ca^{2+}-bound C-terminal domain revealed by X-ray studies have led to the suggestion that the conformational changes, accompanying Ca^{2+}-binding to the N-terminal Ca^{2+}-specific sites, involve a change in the angle between pairs of helical segments leading to the opening up of a hydrophobic patch[14] (Figure 2). Binding of TnI at this exposed site appears to be a key step in regulation. Introduction by protein engineering of an -S-S- link, or by manipulation of charge interaction in the N-terminal domain leads to alteration in Ca^{2+}-binding and reduced or

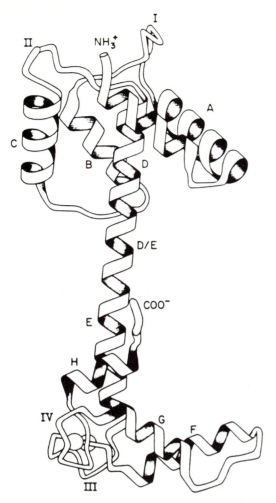

Figure l. Ribbon model derived from X-ray diffraction of skeletal muscle TnC crystals. Helices are labelled A through H; Ca^{2+}-binding domains are labelled I through IV (from ref. 4; used by permission).

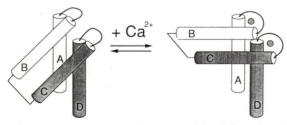

Figure 2. Schematic representation of the Ca^{2+}- induced triggering conformational transition in the N-terminal domain of TnC (based on refs. 14 and 39). The hydrophobic cavity exposed in the Ca^{2+}-bound form provides an interaction site with TnI. A similar site is located in the C-terminal domain. Both sites appear to interact with the inhibitory segment of TnI.

abolished regulatory function, presumably by interfering with the above conformational change[15,16]. Another established step in the regulatory mechanism is a movement of TnI away from **actin**; this has been deduced from binding and electron microscopic studies and has recently been directly demonstrated by resonance energy transfer[17]. Site directed mutants have also been used to study the Ca^{2+} binding sites[8,18,19]; mutants in the central helix have so far shown little effect on function[20,21].

The subunits of troponin exist in various isoforms to be found in muscle fibres of different types, some isoforms being controlled by the developmental stage of the muscle[22]. Some of these forms are products of distinct genes while others result from alternative splicing of mRNA[23-26].

■ PURIFICATION

The three subunits of TnC can be purified from extracts of ether dried muscle powder by combination of ammonium sulphate precipitation, DEAE and CM-Sephadex chromatography in 6 M urea and EDTA[27]. Reconstitution of the complex from the isolated subunits is done by combining them in urea followed by dialysis.

■ ACTIVITIES

Ca^{2+}-binding to the Tn complex or to TnC is determined by equilibrium dialysis with the use of $^{45}Ca^{2+}$ and Ca^{2+}-EGTA buffers[6], or spectrophotometrically with suitable Ca^{2+}-indicators[9,28]. The activity of Tn reconstituted from native, chemically modified, or genetically engineered subunits can be determined by the activation of actomyosin ATPase as a function of Ca^{2+}-concentration or by utilizing *in vitro* motility systems[29]; purified **myosin**, or its active fragments, can be used in the presence of actin-TM. TnC can be tested in myofibrils (ATPase) or glycerol extracted fibres (ATP-induced tension) after selective removal of the endogenous TnC[30,31].

■ ANTIBODIES

In early work polyclonal antibodies were used to distinguish subunit isoforms[1,2]. Recently monoclonal antibodies have been used to study TnI and TnT isoforms occurring in cardiac, slow and fast muscle[32-34].

■ GENES

TnC[35,36] (EMBL GenBank Y00760), (EMBL/GenBank Y03462)[37], TnI[25] (EMBL/GenBank X54163)[38], and TnT[23] genes have been cloned and sequenced. TnC has been expressed in *E. coli*[18,21,35,36].

■ REFERENCES

1. Leavis, P.C. and Gergely, J. (1984) CRC Crit. Rev. Biochem. 16, 235-305.
2. Ohtsuki, I., Maruyama, K. and Ebashi, S. (1986) Adv. Protein Chem. 38, 1-67.

3. Zot, A.S. and Potter, J.D. (1987) Annu. Rev. Biophys. Biophys. Chem. 16, 535-559.
4. Herzberg, O. and James, M.N.G. (1985) Nature 313, 653-659.
5. Sundaralingam, M., Bergstrom, R., Strasburg, G., Rao, S. and Roychowdhury, P. (1985) Science 227, 945-948.
6. Kretsinger, R.H. and Nockolds, C.E. (1973) J. Biol. Chem. 248, 3313-3326.
7. Potter, J.D. and Gergely, J. (1975) J. Biol. Chem. 250, 4628-4633.
8. Strynadka, N.C.J. and James, M.N.G. (1989) Annu. Rev. Biochem. 58, 951-998.
9. Collins, J.H., Theibert, J.L., François, J.M., Ashley, C.C. and Potter, J.D. (1991) Biochemistry 30, 702-707.
10. Collins, J.H. (1991) J. Muscle Res. Cell Motil. 12, 3-25.
11. Heidorn, D.B. and Trewhella, J. (1988) Biochemistry 27, 909-914.
12. Wang, C-L.A., Zhan, Q., Tao, T. and Gergely, J. (1987) J. Biol. Chem. 262, 9636-9640.
13. Leszyk, J., Grabarek, Z., Gergely, J. and Collins, J.H. (1990) Biochemistry 29, 299-304.
14. Herzberg, O., Moult, J. and James, M.N.G. (1986) J. Biol. Chem. 261, 2638-2644.
15. Fujimori, K., Sorenson, M., Herzberg, O., Moult, J. and Reinach, F.C. (1990) Nature 345, 182-184.
16. Grabarek, Z., Tan, R.-Y., Wang, J., Tao, T. and Gergely, J. (1990) Nature 345, 132-135.
17. Tao, T., Gong, R.-J. and Leavis, P.C. (1990) Science 247, 1339-1341.
18. Putkey, J.A., Sweeney, H.L. and Campbell, S.T. (1989) J. Biol. Chem. 264, 12370-12378.
19. Sweeney, H.L., Brito, R.M.M., Rosevear, P.R. and Putkey, J.A. (1990) Proc. Natl. Acad. Sci. (USA) 87, 9538-9542.
20. Sheng, Z., Strauss, W.L., François, J.M. and Potter, J.D. (1990) J. Biol. Chem. 265, 21554-21560.
21. Xu, G.-Q. and Hitchcock-DeGregori, S.E. (1988) J. Biol. Chem. 263, 13962-13969.
22. Pette, D. and Staron, R.S. (1990) Rev. Physiol. Biochem. Pharmacol. 116, 1-76.
23. Breitbart, R.E., Nguyen, H.T., Medford, R.M., Destree, A.T., Mahdavi, V. and Nadal-Ginard, B. (1985) Cell 41, 67-82.
24. Maisonpierre, P.C., Hastings, K.E.M. and Emerson, C.P. (1987) Methods Enzymol. 139, 326-337.
25. Vallins, W.J., Brand, N.J., Dabhade, N., Butler-Browne, G., Yacoub, M.H. and Barton, P.J.R. (1990) FEBS Lett. 270, 57-61.
26. Bucher, E.A., Maisonpierre, P.C., Konieczny, S.F. and Emerson, C.P. (1988) Molecular and Cell Biol. 10, 4134-4142.
27. Potter, J.D. (1982) Methods Enzymol. 85, 241-263.
28. Ogawa, I.Y. (1985) J. Biochem. (Tokyo) 97, 1011-1032.
29. Honda, H. and Asakura, S. (1989) J. Mol. Biol. 205, 677-684.
30. Cox, J.A., Coste, M. and Stein, E.A. (1981) Biochem. J. 195, 205-211.
31. Zot, H.G. and Potter, J.D. (1982) J. Biol. Chem. 257, 7678-7683.
32. Sabry, M.A. and Dhoot, G.K. (1989) J. Muscle Res. Cell Motil. 10, 85-91.
33. Bird, I.M., Dhoot, G.K. and Wilkinson, J.M. (1985) Eur. J. Biochem. 150, 517-525.
34. Ogasawara, Y., Komiya, T. and Obinata, T. (1987) J. Biochem. (Tokyo) 102, 25-30.
35. Reinach, F.C. and Karlsson, R. (1988) J. Biol. Chem. 263, 2371-2376.
36. Chen, Q., Taljanidisz, J., Sarkar, S., Tao, T. and Gergely, J. (1988) FEBS Lett. 228, 22-26.
37. Zot, A.S., Potter, J.D. and Strauss, W.C. (1987) J. Biol. Chem. 262, 15418-15421.
38. Koppe, R.I., Hallauer, P.L., Karpati, G. and Hastings, K.E.M. (1989) J. Biol. Chem. 264, 14327-14333.
39. Richardson, J.S. and Richardson, D.C. (1988) Proteins 4, 229-239.

■ John Gergely, Zenon Grabarek, and Terence Tao:
Department of Muscle Research,
Boston Biomedical Research Institute,
Boston, MA 02114, USA

Villin

Villin[1-3], is one of the major structural proteins which are associated with the actin cytoskeleton of the intestinal or renal epithelial cell brush border microvilli. Villin belongs to a family of actin binding proteins which share sequence homologies and which have similar effects on actin polymerization in vitro. It is most likely involved in the assembly of the brush border cytoskeleton[4].

Villin isolated from intestinal epithelial cells of different species is an acidic polypeptide with an apparent molecular weight of 95,000, occurring under monomeric form. Two villin isoforms have been detected, the origin of which remains unknown. Villin's primary structure is organized in two large duplicated domains, the core, followed by a unique C-terminal domain, the headpiece. Each half of the core presents the same motif constituted by four areas, one of which is present only once in each domain, while the three others are identical and repeated. The organization of the duplicated half is highly conserved throughout evolution of animal species. The same organization is found in **gelsolin**, present in mammalia, and fragmin and **severin**, present in lower eukaryotes[5]. In contrast, the C-terminal headpiece domain is a structural feature which is villin specific. Villin contains at least two **actin** binding sites, one of which is Ca^{2+}-dependent and located on the core domain, while the second is Ca^{2+}-independent and situated on the headpiece domain[6]. Three distinct Ca^{2+}-binding sites have been determined for villin. Binding of Ca^{2+} to that located in the headpiece domain induces a conformational change of the core part of the molecule and thereby may induce a change in the molecule's biochemical behaviour[7].

Villin is a marker for a few organs in adults and in embryos, as well as a differentiation marker for epithelial

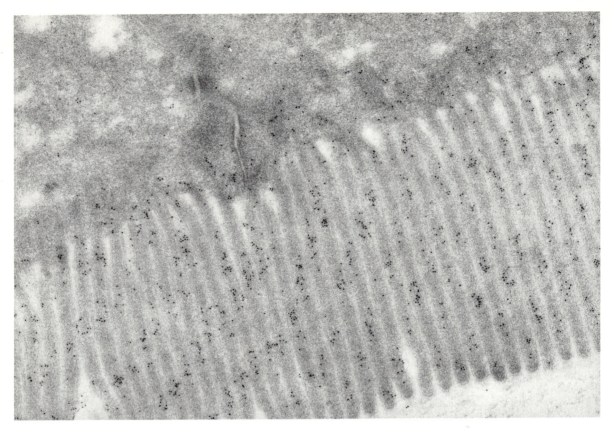

Figure 1. The electron transmission microscopy micrograph shows the brush border of a differentiated enterocyte from human small intestine. Cells have been labelled for villin by the immunogold technique using a villin specific antiserum. Villin is distributed along the microvilli and the rootlets of the brush border.

cells displaying a brush border[3,8] (Figure 1). Using immunocytochemical techniques, villin has been detected only in a subpopulation of cells from the gastrointestinal and the urogenital tract. Villin is present in epithelial cells developing a brush border and in those related to the former by their embryonic origin (e.g. cells of the biliar ductus)[9]. In adults villin synthesis increases during the differentiation process of the enterocyte, which takes place during the migration of the enterocyte from the crypt to the tip of the intestinal villus[9]. During early embryonic development in chicken and mouse, villin is detectable in endodermic cells of the primitive gut which are precursors of the adult intestine[10,11]. Recent experiments confirm the idea that villin participates in the assembly of the intestinal brush border cytoskeleton. Results obtained in an *in vitro* reconstitution experiment show that villin and **fimbrin,** another major actin binding protein of the brush border, dictate the structural organization of the microvilli actin bundle. Addition of these two proteins to pure F-actin results in the formation of actin bundles constituted of oriented microfilaments which exhibit almost the same periodical arrangement as those of the brush

border[12]. In a different approach the effect of villin on the structural organization of living cells was investigated by transfection of the human villin gene in a cell line (CV1 cells) which does not produce villin and does not form a brush border. In these cells, large amounts of villin induce the growth of microvilli (Figure 2) and a reorganization of actin microfilaments. Concomitant with the appearance of actin bundles supporting the plasma membrane of the microvilli, the stressfibres are disrupted. The formation of microvilli is impaired when villin is truncated at its C-terminal headpiece domain, suggesting the functional importance for this domain[13]. These findings show that villin alone is able to modify cell morphology and indicate that at least some of the villin *in vitro* activities on actin polymerization and organization may occur *in vivo*.

Since villin expression is maintained in neoplasic tissues, villin is a useful marker for primary tumours or metastases deriving from tissues which normally express villin[9]. Its presence in the blood of patients presenting colorectal carcinomas makes it a diagnostic adjunct for the detection of these cancers[14].

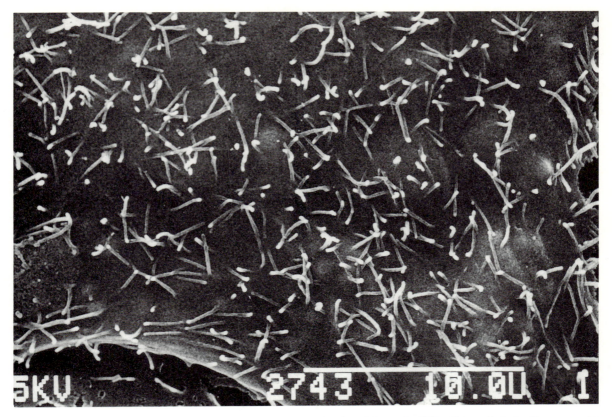

Figure 2. Scanning electron microscopy micrograph showing the dorsal face of a CVI cell transfected with a recombinant DNA encoding human villin. The cell is covered by numerous long microvilli, the growth of which has been induced by villin.

■ PURIFICATION

Villin is purified from isolated brush borders of chicken or pig intestine by affinity chromatography on immobilized DNAse 1, followed by anion exchange columns[15,16].

■ ACTIVITIES

Villin is a Ca^{2+}-dependent actin polymerization modulating protein, *in vitro*[2,4,17]. At high Ca^{2+}-concentrations (0.2-1 mM) villin induces depolymerization of preformed actin microfilaments by severing the filament. The severing activity of villin is inhibited in the presence of PIP_2. At intermediate Ca^{2+}-concentrations (25 μM-100 μM) villin caps the fast growing end of actin microfilaments and thereby prevents elongation. At the same Ca^{2+}-concentration villin nucleates microfilament growth when added to actin monomers. In the absence of Ca^{2+}, villin has no effect on actin polymerization, but it bundles actin microfilaments, an activity which is in comparison to the other actin binding proteins of the villin specific family. **Tropomyosin** I, an actin binding protein present in the terminal web of the brush border, competes with villin for the same binding site on F-actin and thereby inhibits actin bundling by villin. Functional domains of villin have been mapped *in vitro* by testing actin polymerization modulating activities of proteolytic villin fragments. The villin core retains the Ca^{2+}-dependent severing and capping activities but does not induce bundling of actin microfilaments. The headpiece domain binds to F-actin in a Ca^{2+}-independent manner but has no activity on actin polymerization or microfilament bundling.

■ ANTIBODIES

Polyclonal sera have been raised against chicken and porcine villin isolated from intestine epithelia as well as against synthetic peptides containing sequences from the N- or C-termini of chicken villin[1,9,10,18]. Monoclonal antibodies against porcine villin which recognize the headpiece domain have been obtained[19]. In general, anti-villin antibodies exhibit good crossspecies reactivity.

■ GENES

Villin has been cloned and sequenced from human[5] (EMBL/GenBank Data libraries X 12901) and chicken[20]

(EMBL/GenBank Data libraries J03781) intestinal epithelial cells. The villin gene is located on chromosome 2q35-q36 in man and on chromosome 1 in mouse, near the loci of the genes encoding the fast skeletal muscle isoform of the **myosin I** light chain and the isocitrate dehydrogenase[21]. In contrast to chicken and rat villin which is encoded by one mRNA, two human villin mRNAs which differ in their 3'-noncoding regions have been detected[22], the biological significance of these two mRNAs is unknown.

■ REFERENCES

1. Bretscher, A. and Weber, K. (1979) Proc. Natl. Acad. Sci. (USA) 76, 2321-2325.
2. Mooseker, M.S. (1985) Ann. Rev. Cell Biol. 1, 209-241.
3. Louvard, D. (1989) Curr.Opin. in Cell Biol. 1, 51-57.
4. Friederich, E., Pringault, E., Arpin, M. and Louvard, D. (1990) Bioessays 12, 403-408.
5. Arpin, M., Pringault, E., Finidori, J., Garcia, A., Jeltsch, J.-M., Vankekerckhove, J. and Louvard, D. (1988) J. Cell Biol. 107, 1759-1766.
6. Janmey, P.A. and Matsudaira, P.T. (1988) J. Biol. Chem. 263, 16738-16743.
7. Hesterberg, L.K. and Weber, K. (1983) J. Biol. Chem. 258, 359-364.
8. Coudrier, E., Kerjaschki, D. and Louvard, D. (1988) Kidney Int. 34, 309-320.
9. Robine, S., Huet, C., Moll, R., Sahuquillo-Merino, C., Coudrier, E., Zweibaum, A. and Louvard, D. (1985) Proc. Natl. Acad. Sci. (USA) 82, 8488-8492.
10. Shibayama, T., Carobni, J.M. and Mooseker, M.S. (1987) J. Cell Biol. 105, 335-344.
11. Maunoury, R., Robine, S., Pringault, E., Huet, C., Guénet, J.L., Gaillard, J.A. and Louvard, D. (1988) EMBO J. 7, 3321-3329.
12. Coluccio, L.M. and Bretscher, A. (1989) J. Cell Biol. 108, 495-502.
13. Friederich, E., Huet, C., Arpin, M. and Louvard, D. (1989) Cell 58, 461-475.
14. Dudouet, B., Jacob, L., Beuzeboc, P., Magdelenat, H., Robine, S., Chapuis, Y., Christoforov, B., Cremer, G.A., Pouillard, P., Bonnichon, P., Pinon, F., Salmon, R.J., Pointereau-Bellanger, J., Maunoury, M.T. and Louvard, D. (1990) Cancer Res. 50, 438-443.
15. Glenney, J.R. and Weber, K. (1981) Proc. Natl. Acad. Sci. (USA) 78, 2810-2814.
16. Coleman, T.R. and Mooseker, M.S. (1985) J. Cell Biol. 101, 1850-1857.
17. Vandekerckhove, J. (1990) Curr. Opin. in Cell Biol. 2, 41-56.
18. Matsudaira, P., Jakes, R., Cameron, L., Atheron, E. (1985) Proc. Natl. Acad. Sci. (USA) 82, 6788-6792.
19. Dudouet, B., Robine, S., Huet, C., Sahuquillo-Merino, C., Blair, L., Coudrier, E. and Louvard, D. (1987) J. Cell Biol. 105, 359-369.
20. Bazari, W.L., Matsudaira, P., Wallek, M., Smeal, T., Jakes, R. and Ahmed, Y. (1988) Proc. Natl. Acad. Sci. (USA) 85, 4986-4990.
21. Rousseau-Merck, M.F., Simon-Chozottes, D., Arpin, M., Pringault, E., Louvard, D., Guénet, J.L. and Berger, R. (1988) Genetics.
22. Pringault, E., Arpin, M., Garcia, A., Finidori, J. and Louvard, D. (1986) EMBO J. 5, 3119-3124.

■ *Evelyne Friederich and Daniel Louvard:*
Institut Pasteur, Department of Molecular Biology,
25, rue du Docteur Roux,
Paris 75015, France

Vitamin D Binding/Gc Protein (DBP/Gc)

Vitamin D binding/Gc protein (DBP/Gc)[1,2] is an extracellular protein found in the plasma of vertebrate animals. DBP/Gc binds to monomeric actin[3,4] with nanomolar K_d[5,6]. The biological roles of DBP/Gc include the transport of vitamin D in the extracellular space[1]. DBP/Gc also participates to the process of sequestration of actin and its clearance from the extracellular compartment where actin leaks from damaged cells[7-10].

Vitamin D binding/Gc protein is a single polypeptide of ~55 kDa[11,12]. The genes for albumin (ALB), α-fetoprotein (AFP) and DBP/Gc belong to a family which originates from the triplication of a common ancestor nearly one billion years ago[13-16]. All three genes are located on the same human chromosome 4[14] and rat chromosome 14[13]. Although DBP/Gc is the smallest protein of the family, the size of the DBP/Gc gene is twice the size of the genes for ALB and AFP, mainly as a result of the presence of a very large first intron[16]. The single sterol binding site, which is specific to DBP/Gc, is located at the N-terminus of the protein in a region encoded by the first two exons[16]. The **actin** binding site, which is not found on ALB and AFP, is located near the C-terminus and is encoded by exon 10[16]. ALB, AFP and DBP/Gc are expressed in liver. Preserved sequence motifs in the pro-moter element of these genes are probably responsible for the common tissue specificity of their expression and their common enhanced expression in response to certain stimuli[16].

Smithies discovered this major (5-10 µM) highly polymorphic serum glycoprotein and named it the group specific component or Gc-globulin[2]. Three major alleles result from glycosylation of a single polypeptide: Gc_{1Fast}, Gc_{1Slow} (the glycan moeity represent ~1% of the molecular mass) and unglycosylated Gc_2[17,18]. DBP/Gc appears in the human population as three genetic phenotypes: DBP/Gc1, DBP/Gc2 and DBP/Gc1-2[17,19]. Each phenotype can be readily identified by its electrophoretic pattern: two major bands for DBP/Gc1 (pI: ~4.9 and ~4.95), one band for DBP/Gc2 (pI: ~5.0) and three bands for DBP/Gc1-2[20]. Interestingly, DBP/Gc⁰ homozygote was never detected in spite of extensive worldwide studies[21,22]. Together with the widespread occurrence of DBP/Gc in vertebrate animals[23], these data suggest that the null-mutation might be lethal and that DBP/Gc encodes a function necessary to life.

■ PURIFICATION

Several methods have been described for the purification of DBP/Gc from plasma. Rapid methods use pseudoaffinity matrices like Cibacron Blue 3-GA[24,25], or affinity columns containing immobilized sterols[26], antibodies[27] or G-actin[28].

■ ACTIVITIES

DBP/Gc is the major carrier in plasma for vitamin D metabolites[1], and binds to[2] $25(OH)D_3$ with a K_d of 50 nM. Given the concentration of DBP/Gc and vitamin D metabolites in plasma, 90-99% of DBP/Gc is unbound[29].

DBP/Gc binds with high affinity (1-10 nM K_d) to monomeric actin with a stoichiometry of one to one. DBP/Gc binds to proteolytic fragments containing the C-terminus end of actin, suggesting that this region corresponds to the binding site for DBP/Gc[30]. Competition *in vitro* actin between DBP/Gc and **profilin** (which binds the C-terminus of actin) for binding to actin also indicates that the binding site for DBP/Gc is located near the C-terminus of actin[30,31]. The stability of the complex between DBP/Gc and actin allows to separate DBP/Gc bound to actin or actin-DNaseI complex, from free DBP/Gc, using gel filtration[3,4] or electrophoretic methods such as isoelectric focusing[7] and PAGE in nondenaturating conditions[5,32]. *In vivo*, interaction of DBP/Gc with actin is believed to result from the leakage of actin from cells during necrosis and apoptosis[7,33-35]. Together with plasma **gelsolin**[36], DBP/Gc constitutes a mechanism whereby actin released from cells can be depolymerized and sequestered as monomers unable to polymerize[8-10,37]. The potential threat of saturating this system has been documented[38,39].

DBP/Gc interact with the membrane of certain cells[40,41]. Although a specific binding site has not yet been identified, it might involve membrane lipids[42], and the extracellular domain of membrane receptors[43], and seems to bind DBP/Gc with a rather slow off-rate[44].

DBP/Gc binds to C5a and thereby may increase neutrophils chemotaxis[45].

■ ANTIBODIES

Antisera against DBP/Gc are commercially available. Monoclonal antibodies against DBP/Gc have been produced[41].

■ GENES

Full length cDNAs for human[14,15] and rat[13,16] DBP/Gc have been published (GenBank/EMBL M60197-M60206).

■ REFERENCES

1. Smithies, O. (1955) Biochem. J. 61, 629-641.
2. Daiger, S.P., Schanfield, M.S. and Cavalli-Sforza, L.L. (1975) Proc. Natl. Acad. Sci. (USA) 72, 2076-2080.
3. Van Baelen, H., Bouillon, R. and De Moor, P. (1980) J. Biol. Chem. 255, 2270-2272.
4. Haddad, J.G. (1982) Arch. Biochem. Biophys. 213, 538-544.
5. Goldschmidt-Clermont, P.J., Galbraith, R.M., Emerson, D.L., Marsot, F., Nel, A.E. and Arnaud, P. (1985) Biochem. J. 228, 471-477.
6. McLeod, J.F. and Cooke, N.E. (1989) J. Biol. Chem. 264, 1260-1267.
7. Emerson, D.L., Arnaud, P. and Galbraith, R.M. (1983) Am. J. Reprod. Immunol. 4, 185-189.
8. Lind, S.E., Smith, D.B., Janmey, P.A. and Stossel, T.P. (1986) J. Clin. Invest. 78, 736-742.
9. Harper, K.D., McLeod, J.F., Kowalski, M.A. and Haddad, J.G. (1987) J. Clin. Invest. 79, 1365-1370.
10. Goldschmidt-Clermont, P.J., Van Baelen, H., Bouillon, R., Shook, T.E., Williams, M.H., Nel, A.E. and Galbraith, R.M. (1988) J. Clin. Invest. 81, 1519-1527.
11. Haddad, J.G. and Walgate, J. (1976) J. Biol. Chem. 251, 4803-4809.
12. Bouillon, R., Van Baelen, H., Rombauts, W. and De Moor, P. (1978) J. Biol. Chem. 253, 4426-4431.
13. Cooke, N.E. and David, E.V. (1985) J. Clin. Invest. 76, 2420-2424.
14. Yang, F., Brune, J.L., Naylor, S.L., Cupples, R.L., Naberhaus, K.H. and Bowman, B.H. (1985) Proc. Natl. Acad. Sci. (USA) 82, 7994-7998.
15. Cooke, N.E. (1986) J. Biol. Chem. 261, 3441-3450.
16. Ray, K., Wang, X., Zhao, M. and Cooke, N.E. (1991) J. Biol. Chem. 266, 6221-6229.
17. Svasti, J., Kurosky, A., Bennett, A. and Bowman, B.H. (1979) Biochemistry 18, 1611-1617.
18. Viau, M., Constans, J., Debray, H. and Montreuil, J. (1983) Biochem. Biophys. Res. Commun. 117, 324-331.
19. Cleve, H. and Bearn, A. (1962) Prog. Med. Genet. 2, 64-82.
20. Constans, J. and Viau, M. (1977) Science 198, 1070-1071.
21. Cleve, H. (1973) Isr. J. Med. Sci. 9, 1133-1146.
22. Mikkelson, M., Jacobsen, P. and Henningsen, K. (1977) Hum. Hered. 27, 105-107.
23. Bouillon, R., Van Kerkhove, P. and De Moor, P. (1976) Calc. Tiss. Res. (suppl.) 21, 172-176.
24. Chapuis-Cellier, C., Gianazza, E. and Arnaud, P. (1982) Biochem. Biophys. Acta 709, 353-357.
25. Miribel, L., Goldschmidt-Clermont, P.J., Galbraith, R.M. and Arnaud, P. (1986) J. Chromatography 363, 448-455.
26. Haddad, J.G., Abrams, J. and Walgate, J. (1981) Metab. Bone Dis. Rel. Res. 3, 43-46.
27. Bouillon, R. and Van Baelen, H. (1982) In Vitamin D (Norman, A., Schaefer, K., Herrath, D., Grigoleit, H., eds.) pp. 1181-1186, Walter de Gruyter, Berlin.
28. Haddad, J.G., Kowalski, M.A. and Sanger, J.W. (1984) Biochem. J. 218, 805-810.
29. Cooke, N.E. and Haddad, J.G. (1989) Endocrine Rev. 10, 294-307.
30. Goldschmidt-Clermont, P.J., Van Alstyne, E.L., Day, J.R., Emerson, D.L., Nel, A.E., Lazarchick, J. and Galbraith, R.M. (1986) Biochemistry 35, 6467-6472.
31. McLeod, J.F., Kowalski, M.A. and Haddad, J.G. (1989) J. Biol. Chem. 264, 1260-1267.
32. Safer, D. (1989) Anal. Biochem. 178, 32-37.
33. Lee, W.M., Emerson, D.L., Werner, P.A.M., Arnaud, P., Goldschmidt-Clermont, P.J. and Galbraith, R.M. (1985) Hepatology 5, 271-275.

34. Young, W.O., Goldschmidt-Clermont, P.J., Emerson, D.L., Lee, W.M., Jollow, D.J., and Galbraith, R.M. (1987) J. Lab. Clin. Med. 110, 83-90.
35. Smith, D.B., Janmey, P.A. and Lind, S.E. (1988) Am. J. Pathol. 130, 261-267.
36. Chaponier, C., Borgia, R., Rungger-Brandle, E., Weil, R. and Gabbiani, G. (1979) Experientia (Basel) 35, 1039-1040.
37. Janmey, P.A., Stossel, T.P. and Lind, S.E. (1986) Biochem. Biophys. Res. Commun. 136, 72-79.
38. Goldschmidt-Clermont, P.J., Lee, W.M. and Galbraith, R.M. (1988) Gastroenterology 94, 1454-1458.
39. Haddad, J.G., Harper, K.D., Guoth, M., Pietra, G.G. and Sanger, S.W. (1990) Proc. Natl. Acad. Sci. (USA) 87, 1381-1385.
40. Petrini, M., Emerson, D.L. and Galbraith, R.M. (1983) Nature, 306, 73-74.
41. McLeod, J.F., Kowalski, M.A. and Haddad, J.G. (1986) Endocrinology 119, 77-83.
42. Williams, M.H., Van Alstyne, E.L. and Galbraith, R.M. (1988) Biochem. Biophys. Res. Commun. 153, 1019-1024.
43. Petrini, M., Galbraith, R.M., Emerson, D.L., Nel, A.E. and Arnaud, P. (1985) J. Biol. Chem. 260, 1804-1810.
44. Guoth, M., Murgia, A., Smith, R.M., Prystowsky, M.B., Cooke, N.E. and Haddad, J.G.(1990) Endocrinology 127, 2313-2321.
45. Kew, R.R. and Webster, R.O. (1988) J. Clin. Invest. 82, 364-369.

■ *Pascal J. Goldschmidt-Clermont:*
Division of Cardiology, Department of Medicine,
Johns Hopkins University,
School of Medicine,
Baltimore, Maryland 21205, USA

25 kDa Inhibitor of Actin Polymerization (25 kDa IAP)

The 25 kDa IAP is a low molecular weight heat shock protein which inhibits actin polymerization and is constitutively expressed in muscle tissues[1,2]. The protein was originally detected in the vinculin rich fraction from turkey and chicken smooth muscle and is, at least in part, responsible for the effect on actin viscosity originally attributed to vinculin[3,4].

The 25 kDa IAP migrates on SDS-PAGE as a single polypeptide of ~25 kDa and tends to form disulphide linked dimers upon storage. On gel filtration columns[5] it migrates as a ~150 kDa molecule suggesting that in its native form the 25 kDa IAP is an oligomer (probably a hexamer). This notion is also supported by electron microscopy. This protein is apparently distinct from **tensin**[5] and its relationships to **insertin**[6] are still unclear. Partial protein sequencing followed by cloning and complete cDNA sequencing[2] indicated that the 25 kDa IAP is a 193 amino acid polypeptide. Analysis of the 25 kDa IAP sequence indicated a high degree of homology (67% identity, 80% similarity) to the human 25 kDa heat shock protein[7]. The notion that the 25 kDa IAP is, in fact, a heat shock protein was then confirmed by exposing chick fibroblasts to a single or double heat shock, resulting in ~15 fold increase in the level of the protein[2].

In smooth muscle cells the 25 kDa IAP shows an overall distribution similar to that of **actin**. In cultured cardiac myocytes the protein is particularly enriched along myofibrils. In nonheat shocked fibroblast, the protein is hardly detectable, while after exposure to elevated temperature (45°C) it appears in the form of cytoplasmic granules and later appears to be associated with mitochondria. High resolution data on its distribution are not yet available.

■ ACTIVITIES

The 25 kDa IAP, when added to F-actin, causes a dramatic decrease in low shear viscosity[1]. In addition, it clearly inhibits actin polymerization, as determined fluorimetrically using pyrenyl-actin polymerization assay. The protein does not affect the lag time of polymerization and apparently increases the critical actin concentration under physiological salt conditions. Its exact mode of action is still unclear, yet it behaves as a nonnucleating, barbed end capping protein. The 25 kDa IAP also induces disassembly of F-actin, either free or crosslinked with α-**actinin**[1].

■ GENES

The complete 25 kDa IAP sequence has been deduced from cDNA (EMBL/GenBank/DDBJ X59541).

■ REFERENCES

1. Miron, T., Wilchek, M. and Geiger, B. (1988) Eur. J. Biochem. 178, 543-553.
2. Miron, T., Vancompernolle, K., Vandekerckhove, J., Wilchek, M. and Geiger, B. (1991) J. Cell Biol. 114, 255-261.
3. Jockusch, B.M. and Isenberg, G. (1981) Proc. Natl. Acad. Sci. (USA) 78, 3005-3009.
4. Wilkins, J.A. and Lin, S. (1982) Cell 28, 83-90.

5. Wilkins, J.A., Risinger, M.A. and Lin, S. (1986) J. Cell Biol. 103, 1483-1494.
6. Ruhnau, K., Gaertner, A. and Wegner, A. (1989) J. Mol. Biol. 210, 141.
7. Hickey, E., Brandon, S.E., Potter, R., Stein, G., Stein, J. and Weber, L.A. (1986) Nucl. Acids Res. 14, 4127-4145.

■ Benjamin Geiger:
Department of Chemical Immunology
The Weizmann Institute of Science,
Rehovot, Israel

The 43 kDa Protein (43 K)

The 43 kDa protein is a synaptic peripheral protein that codistributes with the nicotinic acetylcholine receptor (AChR) in electrocyte postsynaptic membranes and at the neuromuscular junction. It has been proposed to play an important role in the genesis and the stabilization of the AChR-rich postsynaptic domains of the motor endplate.

First reported in electric organ as the major companion protein of AChR during postsynaptic membrane purification[1], the 43 kDa protein (43 K) is also present at the vertebrate (rodent[2], frog[3]) neuromuscular junction. In SDS-PAGE, it migrates, at 43 kDa, between **actin** and **muscle creatine kinase**. It is a peripheral[4] protein tightly associated to the AChR rich membrane, approximately equimolar with AChR[5], and it can be extracted from these membranes by alkaline pH[4], detergent[6] or anhydrides[7].

The sequence[8] of the *Torpedo* 43 K protein does not share homologies with any known protein and is characterized by its high content of cysteine for an intracellular protein. The spacing of cysteines at the C-terminal domain is similar to that observed at the regulatory domain of protein kinase C. The 43 K protein contains phosphorylation sites and a cAMP-dependent kinase site (KRSS) near the C-terminus. The C-terminus is a Y-V sequence, related to the long mRNA transcript[9] (two 43 K cDNA differing by an extension at the 3' end have been reported in electric organ). The deduced amino acid sequence shows a myristoylation[9] site at the N-terminus. The protein is effectively myristoylated[10,11] and phosphorylated[12] in electrocyte postsynaptic membranes and phosphorylatable[12,13] on serine[12] residues. The three to five isovariants[13,14] (pI 6.4-7.8) are possibly related to different states of phosphorylation[13]. The protein is highly conserved especially in its N- and C-terminus[15,16].

Cytosolic and membrane associated pools of 43 K[17,18] coexist in the electrocyte throughout its development. The amount of membrane bound 43 K increases abruptly upon synapse maturation such that the molar ratio of the two forms is 1:1 in the adult electric organ[18]. Detected in the cytoplasm of early embryonic electric tissue[17,19],

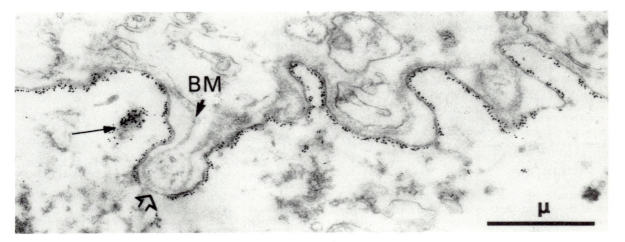

Figure. Immunoelectron micrograph showing the cytoplasmic face of the *Torpedo* electrocyte decorated with an anti-43 K protein antibody (immunogold staining). The distribution of 43 K protein is coextensive with that of AChR. The aggregate of gold particles (arrow) corresponds to a tangential section of an invagination of the postsynaptic membrane. Note the absence of 43 K staining in the bottom of a fold which is devoided of AChR (open arrow). BM: basement membrane (taken from ref. 22).

the 43 K protein is restricted to the postsynaptic domain in the adult electrocyte, where it accumulates at the cytoplasmic face[20-23,24] in a coextensive distribution[17,19,22,23] with the AChR (Figure). This coextensive distribution is also observed in vertebrate neuromuscular junctions (rodents[2], batracians[3]) and at AChR clusters on rat[25] and Xenopus[26,27] muscle cultures. Close proximity of 43 K protein and AChR has been suggested by images of 2-D crystal arrays of tubular AChR rich postsynaptic membrane vesicles[28], as well as by the chemically induced crosslinking of 43 K protein to the β-subunit of the AChR[29]. Conflicting results on the coordination of regulation of 43 K protein and AChR subunit transcripts have been reported[16,30,31].

Removal of the 43 K protein from electrocyte postsynaptic membranes[4] or cultured rat myotubes[25] is accompanied by a significant increase in the susceptibility of AChR to heat denaturation[32] and proteolytic degradation[33] and in their translational[34] and rotational[35] mobility within the membrane plane. Furthermore, clusters of AChR can be induced in quail fibroblasts[36], stably transfected with the AChR, upon transient expression of recombinant 43 K protein and in Xenopus oocytes[37] injected with AChR and 43 K protein transcripts. The receptor properties (kinetics of binding of acetylcholine, ion-gated channel opening, binding of local anesthetics) however are not altered[4] by removal of the 43 K protein.

The 43 K protein thus is not required for the function of AChR as a ligand gated channel but is suspected to play a role in the induction of AChR clustering in neuromuscular junction and the maintenance and stabilization of the postsynaptic structures. Present in close association with the AChR, it might serve as an intermediate piece between the postsynaptic membrane and the cytoskeleton.

■ PURIFICATION

The electrocyte is the only cell rich in 43 K protein. Purification[8,12] has been achieved by elution of the protein from SDS gels of electrocyte postsynaptic membranes or of their detergent or alkaline extracts (the 43 K protein aggregates at neutral pH).

■ ACTIVITIES

No biological assays are available. Estimation of 43 K can be achieved by immunochemical methods (Elisa[5,18], immunoprecipitation[30]).

■ ANTIBODIES

Monoclonal[3,22,24,26], polyclonal[2,10,17] and anti-peptide[17,18] antibodies against electrocyte 43 K have been described. Some of them crossreact with mouse[10], rat[2,3], frog[3] and Xenopus[26] 43 K protein.

■ GENES

cDNA[9,16] coding for two 43 kDa proteins which differ by an extension of 23 amino acids at the C-terminus have been described for electric organ (GenBank J02952, J02953). The complete 43 K gene sequences for mouse muscle[15] (GenBank J03962) and mouse muscle BC3H1 cell lines[15] are also available and there are partial sequences for Xenopus[16] 43 K.

■ REFERENCES

1. Sobel, A., Weber, M. and Changeux, J.P. (1977) Eur. J. Biochem. 80, 215-244.
2. Froehner, S.C., Gulbandsen, V., Hyman, C., Jeng, A.Y., Neubig, R.R. and Cohen, J.B. (1981) Proc. Natl. Acad. Sci. (USA) 78, 5230-5234.
3. Froehner, S.C. (1984) J. Cell Biol. 99, 88-96.
4. Neubig, R.R., Krodel, E.K., Boyd, N.D. and Cohen, J.B. (1979) Proc. Natl. Acad. Sci. (USA) 76, 690-694.
5. La Rochelle, W.J. and Froehner, S.C. (1986) J. Biol. Chem. 261, 5270-5274.
6. Elliott, J., Blanchard, S.G., Wu, W., Miller, J., Strader, C.D., Hartig, P., Moore, H.P., Racs, J. and Raftery, M.A. (1980) Biochem. J. 185, 667-677.
7. Eriksson, H., Liljeqvist, G. and Heilbronn, E. (1983) Biochim. Biophys. Acta 728, 449-454.
8. Carr, C., McCourt, D. and Cohen, J.B. (1987) Biochemistry 26, 7090-7102.
9. Frail, D.E., Mudd, J., Shah, V., Carr, C., Cohen, J.B. and Merlie, J.P. (1987) Proc. Natl. Acad. Sci. (USA) 84, 6302-6306.
10. Musil, L.S., Carrr, C., Cohen, J.B. and Merlie, J.P. (1988) J. Cell Biol. 107, 1113-1121.
11. Carr, C., Tyler, A.N. and Cohen, J.B. (1989) FEBS Lett. 243, 65-69.
12. Hill, J.A., Nghiêm, H.O. and Changeux, J.P. (1991) Biochemistry 30, 5579-5585.
13. Saitoh, T. and Changeux, J.P. (1980) Eur. J. Biochem. 105, 51-62.
14. Gysin, R., Wirt, M. and Flanagan, S.D. (1981) J. Biol. Chem. 256, 11373-11376.
15. Frail, D.E., McLaughlin, L.L., Mudd, J. and Merlie, J.P. (1988) J. Biol. Chem. 263, 15602-15607.
16. Baldwin, T.J., Theriot, J.A., Yoshihara, C.M. and Burden, S.J. (1988) Development 104, 557-564.
17. Kordeli, E., Cartaud, J., Nghiêm, H.O., Devillers-Thiéry, A. and Changeux, J.P. (1989) J. Cell Biol. 108, 127-139.
18. Nghiêm, H.O., Hill, J. and Changeux, J.P. (1991) In: "The living cell in its 4th dimension". American Institute of Physics editors, pp. 416-429. Nghiêm, H.O., Hill, J. and Changeux, J.P. Development, in press.
19. La Rochelle, W.J., Witzemann, V., Fiedler, W. and Froehner, S.C. (1990) J. Neurosci. 10, 3460-3467.
20. Wennogle, L.P. and Changeux, J.P. (1980) Eur. J. Biochem. 106, 381-393.
21. Saint-John, P.A., Froehner, S.C., Goodenough, D.A. and Cohen, J.B. (1982) J. Cell Biol. 92, 333-342.
22. Nghiêm, H.O., Cartaud, J., Dubreuil, C., Kordeli, C., Buttin, G. and Changeux, J.P. (1983) Proc. Natl. Acad. Sci. (USA) 80, 6403-6407.
23. Sealok, R., Wray, B.E. and Froehner, S.C. (1984) J. Cell Biol. 98, 2239-2244.
24. Bridgman, P.C., Carr, C., Pedersen, S.E. and Cohen, J.B. (1987) J. Cell Biol. 105, 1829-1846.
25. Block, R.J. and Froehner, S.C. (1987) J. Cell Biol. 104, 645-654.
26. Burden, S.J. (1985) Proc. Natl. Acad. Sci. (USA) 82, 8270-8273.
27. Peng, B. and Froehner, S.C. (1985) J. Cell Biol. 100, 1698-1705.
28. Toyoshima, C. and Unwin, N. (1988) Nature 336, 247-250.
29. Burden, S.J., Depalma, R.L. and Gottesman, G.C. (1983) Cell 35, 687-692.

30. Frail, D.E., Musil, S., Buonanno, A. and Merlie, J.P. (1989) Neuron 2, 1077-1086.
31. Froehner, S.C. (1989) FEBS Lett. 249,229-233.
32. Saitoh, T., Wennogle, L.P. and Changeux, J.P. (1980) FEBS Lett. 108, 489-494.
33. Klymkowsky, M.W., Heuser, J.E. and Stroud, R.M. (1980) J. Cell Biol. 85, 823-838.
34. Lo, M.M.S., Garland, P.B., Lamprecht, J. and Barnard, E.A. (1980) FEBS Lett. 111, 407-412.
35. Rousselet, A., Cartaud, J., Devaux, P.F. and Changeux, J.P. (1982) EMBO J. 1, 439-445.
36. Phillips, W.D., Kopta, C., Blount, P., Gardner, P.D., Steinbach, J.H. and Merlie, J.P. (1991) Science 251, 568-570.
37. Froehner, S.C., Luetje, C.W., Scotland, P.B. and Patrick, J. (1990) Neuron 5, 403-410.

■ *Hoàng-Oanh Nghiêm and Jean-Pierre Changeux:*
Institut Pasteur, Neurobiologie Moléculaire,
25 rue du Dr. Roux,
75015 Paris,
France

2

Tubulin and Associated Proteins

Microtubules assembled from isolated centrosomes in vitro. Polymerized microtubules are labeled with antibodies against tubulin.

(Courtesy of Dr Steve Doxsey, University of California, San Francisco.)

Tubulin and Associated Proteins

Microtubules, assembled primarily from heterodimers of α- and β-tubulin and a collection of nontubulin associated proteins (Microtubule Associated Proteins or MAPs), are hollow, 25 nm cylindrical polymers whose in vitro assembly from subunits is spontaneous, requires GTP (which is hydrolyzed to GDP just after assembly), and is driven by elevated temperature and reversed by cooling. With few exceptions, microtubules in vivo are comprised of what appear in cross-section to be 13 longitudinal protofilaments, although the lattice can also be indexed by three or five start helices. The lattice is inherently polar, so that the two ends of a microtubule are different and display different kinetics of subunit addition and loss. Arbitrarily defining the end that grows faster as the plus end, microtubules in cells are usually organized in a polar fashion with the plus ends pointing toward the extremities of the cell. In the axons of neurons, for example, essentially all microtubules have their plus ends extending toward the tip of the axon. Spontaneous assembly cannot produce such polarity, but in vivo most (perhaps all) microtubules grow from specific microtubule organizing centres (often called MTOCs). Among the known nucleation sites are the basal bodies that lie at the base of each flagellum and the centriole-containing centrosomes that are attached to the nuclear membrane during interphase and become the spindle poles during mitosis.

Microtubule assembly *in vivo* from tubulin and associated proteins is a dynamic process. At one level this is obvious since the interphase array of microtubules radiating from a nuclear bound MTOC is disassembled as the cell enters mitosis and spindle microtubules are formed from the liberated subunits. What came unexpectedly was the discovery both *in vivo* and *in vitro* that even when assembly has reached an apparent steady state, individual microtubules remain very dynamic, a property termed dynamic instability[1]. Microtubules coexist in growing and shrinking populations which interconvert rather infrequently. These states seem to reflect properties of the ends of the lattice and it is likely that cleavage of the GTP bound to the terminal subunit plays an important role in governing the transition between growing and shrinking. Photobleaching experiments following injection of fluorescently labelled **tubulins** as well as *in vitro* experiments using interphase or mitotic extracts from *Xenopus* eggs have revealed that microtubules turnover much more rapidly in mitotic extracts[2-4]. This cell cycle dependent modulation in dynamics could involve capping of the polymer (either at ends or at positions along the length) during interphase, or even stimulation of the rate of GTP hydrolysis during mitosis. While except for the **STOP** proteins there are no candidate capping factors, components that bind to ends or laterally to the wall of microtubules must be present (for example, when microtubules attach to kinetochores).

■ WHAT ROLES DO MICROTUBULES PERFORM IN CELLS?

Microtubules are used both for specialized roles and for two general functions that are essential to all higher eukaryotes. They were first identified in one of the specialized roles; microtubules comprise the most prominent component of eukaryotic cilia and flagella and are used there to produce flagellar beating. The molecular basis for flagellar motility is now known: a motor protein (**dynein**) attaches to and powers sliding between flagellar microtubules that are held in a highly ordered array. This in turn causes a series of local bendings of the flagellum that together comprise the motile stroke.

The most obvious general use of microtubules is as the primary structural component of the mitotic spindle. During mitosis microtubules are responsible for attaching to each duplicated chromosome pair, for translocating each pair to the mitotic midplate, and finally (during anaphase) for the translocation of one copy of each chromosome to each of the two spindle poles prior to cell division. Given such a role, it seems certain that mitotic microtubules are essential for all dividing cells (genetics has proven this true in yeast[5]).

A second, but less appreciated, general function of microtubules is in organizing the cytoplasm. In concert with **actin** filaments and intermediate filaments, cytoskeletal microtubules establish and maintain the overall internal architecture of the cytoplasm and in so doing comprise a major determinant of overall cell shape. The evidence that microtubule integrity is required for maintenance of normal cell architecture is overwhelming: disruption of microtubules with any of a series of agents results in gross changes in cell shape and in re-distribution of cytoplasmic components. One such rearrangement that has long been known is the collapse of intermediate filament arrays upon perturbation of microtubules, but this is only the tip of the iceberg. A direct role for cytoplasmic microtubules in the distribution of organelles was first demonstrated in neurons, where the use of video enhanced microscopy showed that the oriented microtubule arrays in axons serve as the tracks along which vesicles and cell organelles (such as mitochondria) are translocated from the cell centre to the periphery and back again[6,7]. This is all but certain to be true in most other cells as well. Microtubules and associated microtubule motor proteins (the **kinesin** and **cytoplasmic dynein** families) are not only responsible for the targeted delivery of vesicles, but also for the establishment and maintainance of the localization and organization of the endo-plasmic reticulum[8,9].

COMPLEXITY OF TUBULIN SUBUNITS

The variety of cellular events in which microtubules are involved encouraged early speculation that functional differences between microtubules might be the result of the assembly of different tubulin gene products. Thus was born the multitubulin hypothesis that envisaged individual tubulin genes encoding functionally divergent polypeptides. Despite evidence from unicellular eukaryotes that single tubulin gene products are sufficient for the construction of all the essential microtubules[5], molecular genetic approaches quickly lead to the discovery that in most eukaryotes small multigene families encode both α- and β-polypeptides. Work principally from Cowan's group and my own have established that in vertebrates approximately six functional α-tubulin and a corresponding number of β-tubulin gene products have persisted during evolution. Within each family, individual subunits diverge from each other (both within and across species) at less than 10% of amino acid positions[10]. The greatest divergence among the tubulin molecules is localized to a variable domain comprised of the C-terminal 15 residues. The variable domains of β-tubulin are nearly absolutely conserved in the subunits utilized in the same cell types in different species, and thus define five evolutionarily conserved isotypes[11].

These highly acidic, C-terminal variable domain sequences found on both α-tubulin and β-tubulin contain at least a portion of the site of interaction with several MAPs. For example, microtubules cleaved with proteases near the C-termini of both α- and β-tubulin fail to interact productively with dynein or **MAP2**[12]. Although the MAP binding site on tubulin is unlikely to be limited to the variable domain[13], the influence of the variable regions raises the possibility of preferential interaction of some MAPs with specific tubulin isotypes.

MULTIPLE TUBULIN GENES SOMETIMES ENCODE FUNCTIONALLY DIVERGENT SUBUNITS

Do the different tubulin gene products affect the function of the microtubules into which they are assembled? In lower eukaryotes, genetic evidence has clearly shown functional redundancy between isotypes. For example, the two α-tubulin polypeptides in yeast share only 80% sequence identity but can substitute for each other[14,15]. Similar functional redundancy has been shown for the two β-tubulins in Aspergillus[16]. Also consistent with the interchangeability of isoforms, use of isotype specific antibodies has clearly revealed that all vertebrate microtubules examined are constructed of copolymers of available isotypes[17,18].

However, in vitro and in vivo biochemical differences that could be of functional importance have been documented for some β-tubulin isotypes. Recent genetic evidence from Raff's group has proven that at least one Drosophila isotype can poison the function of another[19]. Normally, the β2-gene product is the major β-tubulin utilized for the construction of most classes of microtubules during spermiogenesis. By introducing into the Drosophila genome an additional gene whose transcriptional control elements were derived from the β2-gene, but whose encoded polypeptide is the wild type product of the β3-gene, Hoyle and Raff determined that when β3 is coexpressed in the male germ line at a level above 20% of the wild type level of β2, it acts in a dominant way to disrupt normal axoneme assembly[19]. These results establish that β2 and β3 have different intrinsic functional capabilities. Whether such differences derive from inherent assembly properties of each isotype or from differential binding to, or interaction with, MAPs remains to be determined.

THREE UNUSUAL MODIFICATIONS OF TUBULINS: BUT OF WHAT SIGNIFICANCE?

Beyond the genetic complexity of tubulins, two modifications of α-tubulin and a third, unprecedented modification of both α and β, are now known. The α-tubulin specific modifications are reversible acetylation of lysine at position 40 and removal of the C-terminal tyrosine and its enzymatic readdition. Both detyrosinated and acetylated tubulins are usually enriched in the subpopulation of microtubules that are more stable, but the functional significance remains unproven since it is also clear that neither modification directly causes stabilization[20].

The final modification is addition of one to five glutamic acid residues to the gamma carboxyl group of a specific glutamic acid residue that lies in the C-terminal variable region of some α[21] and β[22] tubulins. This modification (thus far found only in brain tubulin and sufficient to account for the long-known heterogeneity [>20 isoelectric forms] in brain tubulins) requires a change in the way we envision neuronal microtubules, since these polyglutamyl residues result in a negatively charged "bottlebrush" structure protruding from the microtubule surface. It is not yet known whether polyglutamylated microtubules are present in axons, dendrites and/or neuronal cell bodies. Unknown too is what influence this modification has on assembly properties or MAP binding.

DEFINING MAPS

Microtubules isolated from cell extracts by multiple cycles of assembly/disassembly and differential centrifugation yield a final microtubule preparation of which about 80% is tubulin, while the remaining 20% is a variety of non-tubulin, MAP proteins whose complexity and composition depends greatly on the choice of starting tissue and organism. From this emerged the first operational definition of what constitutes a MAP: a protein that copurifies with tubulin during repeated cycles of assembly. Interest in MAPs was stimulated when it was found that tubulin depleted in MAPs assembled poorly (if at all) *in vitro*, but assembly could be fully restored by readdition of MAPs. This finding led to two primary criteria for defining a true MAP. The first was the ability to stimulate microtubule nucleation and elongation from purified tubulin *in vitro* (as initially documented for the neuronal MAPs **tau** and MAP2). The second required that a MAP quantitatively fractionate with microtubules during repeated assembly cycles.

But both criteria were always unsatisfactory, not only because an alarming list of other agents could supplant the assembly inducing properties (DEAE-dextran, dimethylsulfoxide, Mg^{2+}, glycerol), but also because the requirement for stimulating both nucleation and elongation ignored that *in vivo* these two steps are likely to be quite distinct (with nucleation the province of specific MTOCs). That this unnecessarily restricted focus on assembly diverted attention from other potential MAP functions is now most clearly seen with the discovery of motor proteins such as kinesin which, although among the most interesting microtubule related components, would fail to qualify under this early definition of MAPs.

The original MAPs isolated were from mammalian neurons and named according to the three major size classes of polypeptides: MAP1 (>250 kDa), MAP2 (~200 kDa) and tau (50-70 kDa). MAP1 is something of a misnomer in that the size class to which it refers contains at least three polypeptides (**MAP1A, MAP1B**, and MAP1C) which are unrelated to each other. Ironically, although the MAP1 components have been the least studied, MAP1C is the only one of these neuronal MAPs with an identified function. This MAP is the long anticipated cytoplasmic dynein[23], a molecule which powers transport along microtubules of components in a direction from the nerve tip back toward the cell body. Cytoplasmic dynein (MAP1C) is not restricted to neurons, however, and may serve similar transport functions in other cell events, including during mitosis.

Interest in **tau** was renewed when it was recognized by several groups to be a major component of the abnormal, intracellular tangles of filaments that accumulate in the brains of Alzheimer's patients. Always distressingly heterogeneous in size, the various taus were initially feared to be proteolytic products of the larger MAPs. This is not the case; rather, a single tau gene produces multiple polypeptides through alternative RNA splicing. At least four of the 14 exons of the >63 kB bovine tau gene may be included or deleted during splicing and two different 5' splice donors are utilized in removing the final intron[24], resulting in a minimum of 32 potential tau polypeptide products.

The determination of the primary structures of both tau[25] and MAP2[26] revealed a common feature: both proteins carry a similar set of three (sometimes four) imperfect, 18 amino acid repeats that comprise a portion, but probably not the entirety, of the tubulin binding domain. Although this raised the possibility that this repeat is a general feature of microtubule binding proteins, no such domain is present on kinesin[27], its family members, or MAP1B[28]. In the case of MAP1B, a basic domain containing multiple copies of the short motif KKEE or KKE$_V^\downarrow$ appears to represent the tubulin binding domain.

■ AN ALTERNATIVE DEFINITION OF MAPS: COMPONENTS SELECTIVELY ASSOCIATED WITH MICROTUBULES

Viewing the weaknesses in the initial definition of MAPs and the potential that important microtubule related components would be overlooked by focusing exclusively on *in vitro* coassembly, Solomon and coworkers[29] proposed an alternative definition: MAPs are proteins that are attached to microtubules *in vivo*. By identifying proteins retained only in microtubule containing cytoskeletons, but not in cytoskeletons prepared under conditions that did not preserve microtubules, additional MAPs (such as the **chartins**) were identified. Although it has not been generally adopted for biochemical approaches, this method (and underlying MAP definition) is in fact universally used in conjunction with immunofluorescence microscopy to define components that are microtubule related.

■ γ-TUBULIN: A NEW TUBULIN FAMILY MEMBER ASSOCIATED WITH MICROTUBULE NUCLEATION SITES

A genetic approach by Oakley and colleagues has identified an unexpected microtubule component. Beginning with an initial temperature sensitive mutation in the major β-tubulin gene of *Aspergillus*, the mipA gene was isolated by its ability to suppress growth arrest at the nonpermissive temperature[30]. However, the sequence of this putative MAP gene revealed not a new MAP protein, but rather a new tubulin that was neither α-tubulin nor β-tubulin, but about equally similar to both. This novel subunit, now named γ-**tubulin**, has been shown in *Aspergillus* to be an essential gene, since its disruption causes a reduction in the number and length of cytoplasmic microtubules and a virtual absence of the mitotic apparatus[31]. *Drosophila* and mammalian homologues have now been cloned revealing ~65% sequence conservation with fungal γ-tubulin[32,33]. Present at a level of about 5% of β-tubulin[32], antibody localization reveals that this novel subunit is not assembled randomly into microtubules, but is restricted to MTOCs (centrosomes of the spindle pole bodies). The obvious implication is that γ-tubulin is a critical component for microtubule nucleation.

■ WHAT IS KNOWN ABOUT THE IN VIVO PROPERTIES OF THE MAPS?

Immunolocalization of MAP2 and tau gave the first clue as to what properties these MAPs might contribute. Expressed together within most neurons, they localize to separate subcellular compartments. Tau is restricted to axons, the thin tubes that constitute the conducting unit of the neuron. MAP2 is found largely in dendrites, the arborized extensions of the cell body that serve as the neuron's chief signal receptor apparatus. How this segregation is achieved is unresolved, particularly since they share a common tubulin binding domain. Microinjection of biotinylated MAP2 into primary cultures of neurons has revealed that MAP2 can be transported into axons as well as dendrites. The odd thing is that in axons MAP2 barely binds to microtubules and is rapidly degraded[34].

What tau and MAP2 do has been addressed in three ways: First, for tau, microinjection into fibroblasts (which do not express their own tau genes) demonstrated that tau both induced additional tubulin assembly and stabilized microtubules against depolymerization, without an obvious change in filament organization[35]. The second method used DNA transfection to express a single tau cDNA in fibroblastic cells. In both transient transfections[36,37] and in stable lines[36], tau accumulation lead to a dramatic reorganization of microtubules into bundles and an increase in total tubulin content. Equally striking bundling was observed when MAP2 was expressed by transfection[37]. A third method using antisense approaches has demonstrated that both tau and MAP2 appear to be required for initial neurite growth. Antisense oligonucleotides added to the culture medium inhibit the synthesis of tau in primary cultures of rat cerebellar neurons and also suppress elaboration of a stable axon-like neurite[38]. Similarly, stable integration of a gene expressing an RNA that is complementary to the mRNA encoding MAP2 blocks neurite extension in a cell line that can otherwise be induced to extend neurites[39].

That these MAPs can induce microtubule bundling *in vivo* is certain, but neither the mechanism of bundling nor what this means concerning the *in vivo* roles of the MAPs is settled. An initial proposal that bundling was the result of MAP dimerization through a C-terminal domain has been disproven[40], but it is not known whether MAPs directly crosslink adjacent microtubules, whether an as yet undefined cellular factor forms crossbridges between microtubule bound MAPs or whether MAPs simply stabilize normally dynamic microtubules thereby allowing them time to bundle through an inherent microtubule-microtubule affinity[41].

Bundling of a qualitatively different sort has been found for a newly identified MAP named **dynamin**. This 100 kDa polypeptide was identified by Shpetner and Vallee[42] as one of three polypeptides that bind to microtubules in a nucleotide dependent manner, the other two being the force producing translocators cytoplasmic dynein and kinesin. Released from microtubules in the presence of GTP and AMP-PNP, purified dynamin induces hexagonally packed bundles of microtubules spaced 13 nm apart. The binding is extraordinarily cooperative, with portions of a microtubule covered with dynamin, while adjacent domains are completely unbound. Further, in the presence of ATP, the bundles fragment and elongate, indicating dynamin-induced sliding between microtubules. Still, the *in vivo* role of dynamin remains obscure, particularly since in neurons there is no known homologue of the close packed bundling seen *in vitro*. Moreover, antibody localization shows that it is neither tightly associated with microtubules in either the cell bodies or neurites in one cultured cell model[43]. Nonetheless, sequence data[44] have revealed dynamin to be a novel GTP binding protein with a homologue in yeast that is essential for protein sorting. Further, rat dynamin is a 70% identical with a *Drosophila* protein, mutations in which block endocytosis and lead to a depletion in synaptic vesicles[45]. It seems very likely, but as yet unproven, that dynamin provides the motor for some forms of vesicular transport, particularly during endocytosis.

■ HOW CAN THE IN VIVO ROLES OF THE MAPS BE DETERMINED?

As indicated above and in the accompanying summaries, except for the neuronal motor proteins kinesin and cytoplasmic dynein (which participate anterograde and retrograde axonal transport, respectively), the *in vivo* roles of the MAPs remain unproven. This is, in fact, also true for the cytoplasmic dyneins and kinesin family members in nonneuronal cells. How can functions be established? Microinjection either of purified proteins or transfection of the corresponding genes may reveal a portion of the answer, as it has already for MAP2 and tau. Other obvious approaches include microinjection of antibodies that inhibit function. This is particularly attractive for members of the kinesin and dynein families where, despite their presence in a variety of cells, defining in which cellular event(s) a specific motor protein is required is unlikely to yield to biochemical or cytolocalization approaches.

Beyond this, other genetic approaches seem the most promising. An early example has already emerged using an antisense oligonucleotide to block expression of tau in cultured neurons. In the absence of tau, neurite extension was inhibited, providing strong evidence that in these cells tau accumulation is required for this event[46]. Classical genetic methods may also produce some definitive answers (and identify new MAPs) in those lower eukaryotes where genetics can be exploited. This has already been demonstrated by the discovery of γ-tubulin (in *Aspergillus*) and the **kinesin related proteins** (in yeast)[47] and nimC (in *Aspergillus*)[48].

■ FUTURE DIRECTIONS FOR MICROTUBULES: NUCLEATION, TERMINATION, CELL CYCLE AND DYNAMICS

Several areas seem to this observer to be the obvious outstanding problems concerning microtubule assembly and function.

How is the control of microtubule nucleation established both in normally cycling cells and in specialized microtubular arrays (e.g. in axons)? What are the constitutents of such nucleation sites and what are the cell cycle-dependent modifications that alter the activity of such sites?

Once nucleated, what specifies changes in microtubule assembly dynamics through the cell cycle? It seems likely that MAPs are involved in this, but which ones and how?

How do microtubules terminate? There must be specific attachment sites for microtubules at kinetochores, but what are the key MAP proteins? How is it that such kinetochore linkages are assembled/activated by the cell cycle clock? Are there other microtubule termination (or capping) sites, for example, that attach microtubules to the plasma membrane?

Given the bias of how MAPs have been traditionally identified, it seems likely that new approaches (e.g. pursuit of extragenic suppressors of tubulin mutations) will provide novel insights into these processes. Indeed, the fact that γ-tubulin escaped detection by a phalanx of cell biological and biochemical approaches is an unmistakable warning that the identification of key players in microtubule nucleation, assembly, and termination still lies ahead.

■ REFERENCES

1. Mitchison, T. and Kirschner, M.W. (1984) Nature 312, 237-242.
2. Salmon, E.D., Leslie, R.J., Karow, W.M., McIntosh, J.R. and Saxton, R.J. (1984) J. Cell Biol. 99, 2165-2174.
3. Belmont, L.D., Hyman, A.A., Sawin, K.E. and Mitchison, T.J. (1990) Cell 62, 579-589.
4. Verde, F., Labbe, J.-C., Doree, M. and Karsenti, E. (1990) Nature 343, 233-238.
5. Neff, N.F., Thomas, J.H, Grisafi, P. and Botstein, D. (1983) Cell 33, 211-219.
6. Allen, R.D., Weiss, D.G., Hayden, J.H., Grown, D.T., Fujiwake, H. and Simpson, M. (1985) J. Cell Biol. 100, 1736-17522.
7. Vale, R.D, Schnapp, B.J., Reese, T.S. and Sheetz, M.P. (1985) Cell 40, 449-454.
8. Dabora, S.L. and Sheetz, M.P. (1988) Cell 54, 27-35.
9. Lee, C, Ferguson, M. and Chen, L.B. (1989) J. Cell Biol. 109, 2045-2055.
10. Sullivan, K.F. (1988) Ann. Rev. Cell Biol. 4, 687-716.
11. Sullivan, K.F. and Cleveland, D.W. (1986) Proc. Natl. Acad. Sci. 83, 5327-5331.
12. Paschal, B.M., Obar, R.A. and Vallee, R.B. (1990) Nature 342, 569-572.
13. Cleveland, D.W., Joshi, H. and Murphy, D.B. (1990) Nature 344, 389.
14. Adachi, Y., Toda, T., Niwa, O. and Yanagida, M. (1986) Mol. Cell. Biol. 6, 2168- 2178.
15. Shatz, P.J., Soloman, F. and Botstein, D. (1986) Mol. Cell. Biol. 6, 3722-3733.
16. May, G.S. (1989) J. Cell Biol. 109, 2267-2274.
17. Lewis, S.A., Gu, W. and Cowan, N.J. (1987) Cell 49, 539-548.
18. Lopata, M.A. and Cleveland, D.W. (1987) J. Cell Biol. 105, 1707-1720.
19. Hoyle, H.D. and Raff, E.C. (1990) J. Cell Biol. 111, 1009-1026.
20. Khawaja, S., Gundersen, G.G. and Bulinski, J.C. (1988) J. Cell Biol. 106, 141-149.
21. Edde, B., Rossier, J., Le Caer, J.-P., Desbruyeres, E., Gros, F. and Denoulet, P. (1990) Science 247, 74-77.
22. Alexander, J.E., Hunt, D.F., Lee, M.K., Shabanowitz, J., Michel, H., Berlin, S.C., MacDonald, T.L, Sundberg, R.J., Rebhun, L.I. and Frankfurter, A. (1991) Proc. Natl. Acad. Sci. 88, 4685-4689.
23. Paschal, B.M. and Vallee, R.B. (1987) Nature 330, 181-183.
24. Himmler, A. (1989) Mol. Cell. Biol. 9, 1389-1396.
25. Lee, G, Cowan, N. and Kirschner, M. (1988) Science 239, 285-288.
26. Lewis, S.A., Wang, D. and Cowan, N.J. (1988) Science 242, 936-939.
27. Yang, J.T., Laymon, R.A. and Goldstein, L.S.B. (1989) Cell 56, 879-889.
28. Noble, M., Lewis, S.A. and Cowan, N.J. (1989) J. Cell Biol. 109, 3367-3376.
29. Solomon, F., Magendantz, M. and Salzman, A. (1979) Cell 18, 431-438.
30. Oakley, C.E. and Oakley, B.R. (1989) Nature 338, 662-664.
31. Oakley, B.R., Oakley, C.E.and Jung, M.K. (1990) Cell 61, 1289-1301.
32. Zheng, Y., Jung, M.K. and Oakley, B.R. (1991) Cell 65, 817-824.
33. Stearns, T., Evans, L. and Kirschner, M.W. (1991) Cell 65, 825-836.
34. Okabe, S. and Hirokawa, N. (1989) Proc. Natl. Acad. Sci. 86, 4127-4131.
35. Drubin, D.G. and Kirschner, M.W. (1986) J. Cell Biol. 103, 2739-2746.
36. Kanai, Y., Takemura, R., Oshima, T., Mori, H., Ihara, Y., Masashi, Y., Masaki, T. and Hirokawa, N. (1989) J. Cell Biol. 109, 1173-1184.
37. Lewis, S.A., Ivanov, I.E., Lee, G.-H. and Cowan, N.J. (1989) Nature 342, 498-505.
38. Caceres, A. and Kosik, K.S. (1990) Nature 343, 461-463.
39. Dinsmore, J.H. and Solomon, F. (1991) Cell 64, 817-826.
40. Lewis, S.A. and Cowan, N.J. (1990) Nature 345, 674.
41. Chapin, S.J., Bulinski, J.C. and Gundersen, G.G. (1991) Nature 349,24.
42. Shpetner, H.S. and Vallee, R.B. (1989) Cell 59, 421-432.
43. Scaife, R. and Margolis, R. (1990) J. Cell Biol. 111, 3023-3033.
44. Obar, R.A., Collins, C.A., Hammarback, J.A., Shpetner, H.S. and Vallee, R.B. (1990) Nature 347, 256-261.
45. Chen, M.S., Obar, R.A., Schroeder, C.C., Austin, T.W., Poodry, C.A., Wadsworth, S.C. and Vallee, R.B. (1991) Nature 351, 583-586.
46. Caceres, A. and Kosik, K.S. (1990) Nature 343, 461-463.
47. Mulah, P.B. and Rose, M.D. (1990) Cell 60, 1029-1041.
48. Enos, A.P. and Morris, N.R. (1990) Cell 60, 1019-1027.

■ *Don W. Cleveland:*
Department of Biological Chemistry,
Johns Hopkins University, School of Medicine,
Baltimore, MD 21205, USA

Chartins

The chartins are a complex family of proteins originally identified as microtubule components by their association with isolated microtubules from cultured cells. Chartins of different primary structure are present in different cell types. Those forms of chartins which are more heavily phosphorylated are preferentially associated with assembled microtubules in vivo.

Chartins were first identified as microtubule associated proteins based on analyses of microtubules isolated from cultured cells after nonionic detergent extraction[1,2]. In this assay, microtubule associated proteins were defined as those proteins which cofractionated with assembled **tubulin**, and so were present in detergent extracted cytoskeletons if, and only if, the microtubules were preserved[2,3]. In parallel preparations from cells preincubated with microtubule depolymerizing drugs, or extracted in buffers which depolymerize microtubules, neither assembled tubulin nor the associated proteins are present. The relatively minor associated proteins can be enriched by *in vitro* coassembly of radiolabelled fractions with carrier calf brain tubulin, or identified directly by analysis of extracts on 2-D gels[4]. Among the proteins in this category is the 69 kDa chartin which is found in rodent and avian cells, but not primate cells[4]. The chartins are components of both interphase and mitotic microtubules[5]. Chartins fulfill two other criteria for association with microtubules: coassembly *in vitro* and colocalization *in situ* with microtubules. First, one of the bovine brain proteins which coassembles to constant specific activity, and copurifies with **tau** on phosphocellulose columns, is homologous to the 69 kDa chartin, as demonstrated by peptide mapping and antibody crossreactivity[4,6]. Despite similarity in apparent molecular weights, chartins are distinct from taus by the same criteria[4,6]. Second, immunofluorescence microscopy with affinity purified antibodies against chartins demonstrates that the antigen is colinear with microtubules in cultured cells[6].

In rodent cells of neural origin (neuroblastoma, glioma, and PC12 cells) as well as brain, three distinct bands of chartins are found with apparent molecular masses of 69, 72 and 80 kDa[7,8]. Proteins with similar sizes and properties are present in cultured neurons[9]. In nonneural cells, the 80 kDa chartins are not detected. All three size classes can be resolved on 2-D gels into polypeptides of varying isoelectric point and apparent size. Complete tryptic peptide maps demonstrate that the 69 kDa and 80 kDa proteins are present as a set of structurally related polypeptides, and that those two proteins are closely related to one another: all of the 35 methionine containing tryptic peptides yielded by the 69 kDa protein are among the 38 peptides derived from the 80 kDa protein. The sequence contained in the three peptides unique to the 80 kDa chartin may participate in neural specific functions.

The fractionation of the isoelectric forms of chartins suggest that covalent modification may correlate with function. The differences in pI are generated at least in part by increasing phosphorylation, as assessed by $^{32}P/^{35}S$ ratios. At least five serines and threonines, but no tyrosines, can be phosphorylated per molecule. The detergent extraction assay permits separation of the assembled and unassembled microtubule components. Resolution of the chartins in those extracts demonstrates that the more heavily phosphorylated forms are restricted to the assembled pool, and are not detected in the unassembled pool[7]. The relative abundance of the more acidic, more heavily phosphorylated forms of chartins increases in PC12 cells during long term NGF treatment[10]; the proportion of assembled tubulin increases also with similar kinetics. Drugs that affect adenylate cyclase, and therefore increase cAMP levels, cause a depression of neurite outgrowth in PC12 cells. Under the same conditions, there is a decrease in the extent of phosphorylation of chartins[11]. Taken together, these results suggest a correlation between extent of chartin phosphorylation and distribution between the assembled and unassembled pools of microtubule components. It is not known if the covalent modification regulates the assembly states of the chartins, or is a consequence of assembly[7,12].

■ PURIFICATION AND ACTIVITIES

The chartins have not been purified significantly, and no *in vitro* activities have been identified.

■ ANTIBODIES

Polyclonal antibodies raised against both calf brain and neuroblastoma chartins bind to bands which have the appropriate size in western blots, and which fractionate with assembled microtubules[6].

■ GENES

There is no information available about the genes encoding chartins.

■ REFERENCES

1. Osborn, M. and Weber, K. (1977) Cell 12, 561-571.
2. Solomon, F., Magendantz, M. and Salzman, A. (1979) Cell 18, 431-438.
3. Solomon, F. (1986) Methods Enzymol. 134, 139-147.
4. Duerr, A., Pallas, D. and Solomon, F. (1981) Cell 24, 203-211.
5. Zieve, G. and Solomon, F. (1982) Cell, 28, 233-242.

6. Magendantz, M. and Solomon, F. (1985) Proc. Natl. Acad. Sci. (USA) 82, 6581-6585.
7. Pallas, D. and Solomon, F. (1982) Cell 30, 407-414.
8. Zieve, G. and Solomon, F. (1984) Mol. Cell. Biol. 4, 371-374.
9. Peng, I., Binder, L.I. and Black, M.M. (1985) Brain Research, 361, 200-211.
10. Black, M.M., Aletta, J.M. and Greene, L.A. (1986) J. Cell Biol. 103, 545-557.
11. Greene, L.A., Drexler, S.A., Connolly, J.L., Rukenstein, A. and Green, S.H. (1986) J. Cell Biol. 103, 1967-1978.

12. Aletta, J. and Greene, L. (1987) J. Cell Biol., 105, 277-290.

■ *Frank Solomon:*
Department of Biology,
Massachusetts Institute of Technology
Cambridge, MA, USA

MAP1A

Microtubule associated protein 1A (MAP1A) is a high molecular weight protein made up of four distinct polypeptide chains, and copurifies with microtubules out of extracts prepared from brain, and selected other tissue and cell sources. When present, MAP1A is distributed along the lengths of microtubules and is believed to contribute to their stability.

The development of methods for purifying microtubules from brain extracts by repetitive cycles of assembly and disassembly led to the discovery of several microtubule associated proteins (MAPs) that copurify with **tubulin**, the major component. Among the first and most conspicuous of the brain MAPs to have been described were high molecular weight species[1] which could be resolved into two electrophoretically distinct groups, called MAP1 and **MAP2**[2]. Subsequently, MAP1 was resolved into three separate protein species, MAPs 1A, 1B and 1C, distinguished from one another on the basis of electrophoretic, biochemical and immunological criteria[3,4]. MAP1A, which has also been termed MAP1[5] or MAP1.1[6], is usually more abundant in microtubules purified from adult mammalian brain than either **MAP1B** (also called MAP5[7] or MAP1x[8]) or MAP1C, the last of which is now known to be the heavy chain of brain **cytoplasmic dynein**[9].

Protein preparations consisting solely of MAP1A and MAP1B have been obtained by conventional fractionation methods, and found to contain copurifying low molecular weight polypeptides[10-12]. Later immunoprecipitation studies indicated that these represent light chains for both MAP1A and MAP1B, and that the native MAP1A molecule probably contains a single copy of each of four distinct subunit polypeptides: an ~350 kDa heavy chain and light chains of ~30 kDa, ~28 kDa and ~18 kDa[13]. The microtubule binding domain of MAP1A is included within a complex formed by the three light chains and an ~60 kDa region located at one end of the heavy chain[13]. The remainder of MAP1A forms an arm-like projection on the outer wall of the microtubule[10,13,14]. A molecular model which summarizes these features of MAP1A is presented in Figure 1. Examination of immunopurified MAP1A by low angle rotary shadowing electron microscopy indicated that the molecule is a long, thin rod, 100-150 nm long[14].

Within the nervous system, MAP1A is found in axons, dendrites and neuronal cell bodies (Figure 2), as well as in many glial cells[3,5]. Other tissues that contain appreciable levels of MAP1A include anterior pituitary[15] and adrenal medulla[16], and the protein is readily detectable in many types of cultured mammalian cells[17]. Both immunofluorescence[17] and immunoelectron[14] microscopy have demonstrated that MAP1A is associated with microtubules in cells. Curiously, however, the MAP1A in retinal ganglion neurons is transported toward axon terminals at a rate ~10-fold faster than that observed for tubulin[18], implying that the association of MAP1A with microtubules is not static.

MAP1A is believed to be involved in stabilizing microtubules and long cytoplasmic processes that they occupy. This hypothesis is based on the developmental pattern of MAP1A expression in the brain, and the fact that microtubules are required for the growth and maintenance of axons and dendrites. In the rat brain, MAP1A rises ~6-fold in abundance from the day of birth to 15 days after birth[13,19], a pattern which is opposite to that found for MAP1B[13]. During this period of brain development, many axons and dendrites convert from a growth phase to a more stable, mature state, suggesting that MAP1A and MAP1B serve complementary roles *in vivo*. For example,

Figure 1. A model for the structure of MAP1A. The protein is a long, thin, rod-shaped molecule containing an ~350 kDa heavy chain and three distinct light chains (LC-1 ~30 kDa, LC-2 ~28 kDa and LC-3 ~18 kDa). The microtubule binding domain (shaded portion) is located within a region that contains the three light chains and an ~60 kDa segment of the heavy chain. The remainder of the heavy chain (unshaded portion) forms an arm-like structure that protrudes forms the microtubule wall. (This illustration was adapted from Figure 15 in Schoenfeld et al.[13]).

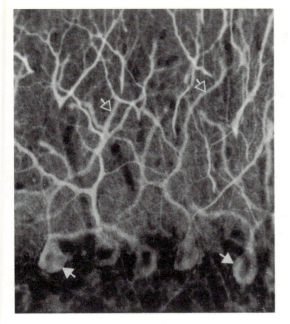

Fig 2. The distribution of MAP1A in the adult rat cerebellum. Prominently labeled structures in this immunofluorescence image include the cell bodies (closed arrows) and dendrites (open arrows) of Purkinje neurons.

perhaps MAP1B stimulates tubulin assembly during early brain development in a manner that promotes neuronal process outgrowth. By contrast, maybe MAP1A, which gradually replaces much of the MAP1B as development proceeds, stabilizes fully differentiated axons and dendrites by increasing the stability of their resident microtubules.

Phosphorylated MAP1A has been found in brain[20], where it is apparently restricted to mature axons[13]. The function of this posttranslational modification is not known, but EGF and insulin stimulate MAP1A phosphorylation in cultured cells[21].

Antibodies that react with MAP1A stain two types of inclusions characteristic of Alzheimer's disease and other neurodegenerative disorders. Hirano bodies were labeled by one monoclonal antibody specific for MAP1A, and by another monoclonal that recognizes a phosphorylated epitope common to MAP1A, MAP1B and two neurofilament subunit polypeptides[22]. Neurofibrillary tangles were stained by two additional monoclonal antibodies directed against phosphorylated epitopes common to neurofilament subunits and high molecular weight MAPs, apparently including MAP1A[23]. In contrast, other monoclonals that react with total or phosphorylated MAP1A have not stained neurofibrillary tangles[22]. MAP1A may thus be a component of Hirano bodies, its presence in neurofibrillary tangles is debatable, and its precise role in neuropathological diseases remains to be determined.

■ PURIFICATION

Using isolated brain microtubules as starting material, mixtures of MAP1A and MAP1B uncontaminated by other proteins can be obtained by several methods. Typically, MAPs are separated from the bulk of the tubulin, and are then resolved by multiple chromatographic steps. One method sequentially employs DEAE-Sephadex, hydroxylapatite and gel filtration columns[24]. Another procedure, which begins with MAPs obtained from brain white matter rather than whole brain, relies upon the successive use of gel filtration and CM-Sephadex columns[10]. An alternative method entails washing isolated brain microtubules with poly-(L-aspartic acid) to solubilize selectively MAP1A and MAP1B, which are then copurified using successive phosphocellulose and gel filtration columns[25]. The only method that has been reported to yield MAP1A free of other proteins involves passing brain cytosol over an anti-MAP1A antibody affinity column[14].

■ ACTIVITIES

Purified mixtures of MAP1A and MAP1B have been reported to stimulate tubulin assembly[10,25,26]. In addition, binding to preformed, taxol stabilized microtubules has been reported for MAP1A/MAP1B preparations that either contain or lack the accompanying light chains[10]. Pure MAP1A, prepared by antibody affinity chromatography, also binds to taxol stabilized microtubules[14].

■ ANTIBODIES

Several monoclonal anti-MAP1A antibodies that are suitable for immunofluorescence and immunoelectron microscopy, immunoprecipitation and immunoblotting have been described[4,5,14]. Commercial sources for some of these antibodies include the Amersham Corporation[4] and Sigma Chemical Company[5].

■ GENES

Recently, a set of cDNAs encoding MAP1A were isolated and characterized[27]. The sequence demonstrates that MAP1A is structurally related to MAP1b.

■ REFERENCES

1. Murphy, D.B. and Borisy, G.G. (1975) Proc. Natl. Acad. Sci. (USA) 72, 2696-2700.
2. Sloboda, R.D., Rudolph, S.A., Rosenbaum, J.L. and Greengard, P. (1975) Proc. Nat. Acad. Sci. (USA) 72, 177-181.
3. Bloom, G.S., Schoenfeld, T.A. and Vallee, R.B. (1984) J. Cell Biol. 98, 320-330.
4. Bloom, G.S., Luca, F.C. and Vallee, R.B. (1985) Proc. Nat. Acad. Sci. (USA) 82, 5405-5408.
5. Huber, G. and Matus, A. (1984) J. Neurosci. 4, 151-160.
6. Greene, L.A., Liem, R.K.H. and Shelanski, M.L. (1983) J. Cell Biol. 96, 76-83.
7. Riederer, B., Cohen, R. and Matus, A. (1986) J. Neurocytol. 15, 763-775.

8. Calvert, R.A., Woodhams, P.L. and Anderton, B.H. (1987) Neurosci. 23, 131-141.

9. Vallee, R.B., Wall, J.S., Paschal, B.M. and Shpetner, H.S. (1988) Nature 332, 561-563.

10. Vallee, R.B. and Davis, S.E. (1983) Proc. Nat. Acad. Sci. (USA) 80, 1342-1346.

11. Kuznetsov, S.A., Rodionov, V.I., Nadezhdina, E.S., Murphy, D.B. and Gelfand, V.I. (1986) J. Cell Biol. 102, 1060-1066.

12. Kuznetsov, S.A. and Gelfand, V.I. (1987) FEBS Lett. 212, 145-148.

13. Schoenfeld, T.A., McKerracher, L., Obar, R. and Vallee, R.B. (1989) J. Neurosci. 9, 1712-1730.

14. Shiomura, Y. and Hirokawa, N. (1987) J. Neurosci. 7, 1461-1469.

15. Bloom, G.S., Luca, F.C. and Vallee, R.B. (1985) Biochem. 24, 4185-4191.

16. Kotani, S., Murofushi, H., Maekawa, S., Sato, C. and Sakai, H. (1986) Eur. J. Biochem. 156, 23-29.

17. Bloom, G.S., Luca, F.C. and Vallee, R.B. (1984) J. Cell Biol. 98, 331-340.

18. Nixon, R.A., Fischer, I. and Lewis, S.E. (1990) J. Cell Biol. 110, 437-448.

19. Riederer, B. and Matus, A. (1985) Proc. Nat. Acad. Sci. (USA) 82, 6006-6009.

20. Luca, F.C., Bloom, G.S. and Vallee, R.B. (1986) Proc. Nat. Acad. Sci. (USA) 83, 1006-1010.

21. Erickson, A.K., Ray, L.B. and Sturgill, T.W. (1990) Biochem. Biophys. Res. Commun. 166, 827-832.

22. Peterson, C., Kress, Y., Vallee, R. and Goldman, J.E. (1988) Acta Neuropathol. 77, 168-174.

23. Ksiezak-Reding, H. and Yen, S.-H. (1987) J. Neurochem. 48, 455-462.

24. Kuznetsov, S.A., Rodionov, V.I., Gelfand, V.I. and Rosenblat, V.A. (1981) FEBS Lett. 135, 237-240.

25. Fujii, T., Nakamura, A., Ogoma, Y., Kondo, Y. and Arai, T. (1990) Anal. Biochem. 184, 268-273.

26. Kuznetsov, S.A., Rodionov, V.I., Gelfand, V.I. and Rosenblat, V.A. (1981) FEBS Lett. 135, 241-244.

27. Langkopf, A. Hammerback, J.A., Muller, R., Vallee, R.B. and Garner, C.C. (1992) J. Biol. Chem. 267, 16561-16566.

■ George S. Bloom:
Department of Cell Biology and Neuroscience,
University of Texas,
Southwestern Medical Center,
5323 Harry Hines Blvd.,
Dallas, TX 75235, USA

MAP1B/MAP5

Microtubule associated protein 1B/5 (MAP1B/MAP5) is a heat labile high molecular weight microtubule associated protein. Its function is not known, but its very early expression during nervous system development is suggestive of a role in neuronal morphogenesis.

This protein was originally described under a variety of names of which two, MAP1B and MAP5, are currently in use. MAP1B/MAP5 binds to microtubules during successive cycles of repolymerization *in vitro* (although more weakly than **MAP2** and **tau** protein), and immunoaffinity purified MAP1B/MAP5 promotes the polymerization of **tubulin** *in vitro*[1,2]. MAP1B/MAP5 has been shown by rotary shadowing electron microscopy to be a long rod approximately 10x200 nm with a small sphere at one end[3]. In axons and dendrites it has been localized to fine filamentous crossbridges between microtubules, and it forms crossbridges in microtubule pellets[3]. Microtubule binding is mediated by a 120 kDa protease resistant fragment (probably corresponding to the spherical portion seen by EM) which also binds two low molecular weight proteins termed MAP1 light chains I (34 kDa) and III (18 kDa)[4-6]. The contribution of the light chains to microtubule binding is unknown. It has been reported[7] that light chain I is derived proteolytically from the C-terminal end of the encoded MAP1B/MAP5 protein. MAP1B/MAP5 contains two repeat motifs[8]: the first is a highly basic region near the N-terminus including multiple copies of the form KKEE or KKE(I,V) that has been implicated in microtubule binding *in vitro* and *in vivo*[8]. The second is a set of 12 imperfect repeats of 15 amino acids each near the C-terminus whose function is unknown. Neither set of repeats is similar to those that mediate the microtubule binding of MAP2 and tau[8].

The expression of MAP1B/MAP5 is very closely correlated with neuronal morphogenesis. In developing neurons, MAP1B/MAP5 is detectable before any other known MAPs when the first nascent axonal process emerges from the cell body[9]. The expression and phosphorylation of MAP1B/MAP5 increases during neurite outgrowth in PC12 cells[10,11], and phosphorylated forms of the protein are abundant in developing axons *in vivo*[3,6]. As a substrate MAP1B/MAP5 is best phosphorylated by casein kinase II[12]. The close correlation of MAP1B/MAP5 with neuronal morphogenesis is shown by its continued high level of expression in retinal photoreceptors, the olfactory system and the amphibian retinotectal system[9,13], all sites at which neuronal differentiation continues in the adult brain.

■ PURIFICATION

Preparations of microtubule associated proteins are made by repolymerizing brain microtubules *in vitro* followed by ion exchange chromatography on phosphocellulose or DEAE-Sepharose. Further purification of MAP1B/MAP5 has been achieved by immunoprecipitation and immunoaffinity chromatography, but yields are very limited[2,3].

■ ACTIVITIES

MAP1B/MAP5 promotes tubulin polymerization *in vitro*[2].

ANTIBODIES

A large number of monoclonal antibodies to MAP1B/MAP5 have been described[3,6,14], some of which are specific to phosphorylated epitopes. Three of them are commercially available from Sigma (M4403, M1406) and from Amersham (RPN.1193)[6]. Rabbit antisera raised against two bacterially expressed protein each containing one of the two repeat regions have also been described[8].

GENES

The full length mouse cDNA has been assembled and sequenced[8] (GenBank HX 51396) and several partial rat cDNAs have been isolated[15-17], one of which has been used to map the MAP1B/MAP5 gene to chromosome 13 in the rat[15].

REFERENCES

1. Bloom, G.S., Luca, F.C. and Vallee, R.B. (1985) Proc. Natl. Acad. Sci. (USA) 82, 5404-5408.
2. Riederer, B., Cohen, R. and Matus, A. (1986) J. Neurocytol. 15, 763-775.
3. Sato-Yoshitake, R., Shiomura, Y., Miyasaka, H. and Hirokawa, N. (1989) Neuron 3, 229-238.
4. Kuznetsov, S.A. and Gelfand, V.I. (1987) FEBS Lett. 212, 145-148.
5. Kuznetsov, S.A., Rodionov, V.L., Nadezhdina, E.S., Murphy, D.B. and Gelfand, V.I. (1986) J. Cell Biol. 102, 1060-1066.
6. Schoenfeld, T.A., McKerracher, L., Obar, R. and Vallee, R.B. (1989) J. Neurosci. 9, 1712-1730.
7. Hammarback, J.A., Obar, R.A., Hughes, S.M. and Vallee, R.B. (1991) Neuron 7, 129-139.
8. Noble, M., Lewis, S.A. and Cowan, N.J. (1989) J. Cell Biol. 109, 3367-3376.
9. Matus, A. (1988) Ann. Rev. Neurosci. 11, 29-44.
10. Drubin, D.G., Feinstein, S., Shooter, E. and Kirschner, M. (1985) J. Cell Biol. 101, 1799-1807.
11. Aletta, J.M., Lewis, S.A., Cowan, N.J. and Greene, L.A. (1988) J. Cell Biol. 106, 1573-1581.
12. Diaz-Nido, J., Serrano, L., Mendez, E. and Avila, J. (1988) J. Cell Biol. 106, 2057-2065.
13. Viereck, C., Tucker, R.P. and Matus, A. (1989) Neurosci. 9, 3547-3557.
14. Garner, C.C., Matus, A., Anderton, B. and Calvert, R. (1989) Mol. Brain Res. 5, 85-92.
15. Garner, C.C., Garner, A., Huber, G., Kozak, C. and Matus, A. (1990) J. Neurochem. 55, 146-154.
16. Safaei, R. and Fisher, I. (1989) J. Neurochem. 52, 1871-1879.
17. Kirsch, J., Littauer, U.Z., Schmitt, B., Prior, P., Thomas, L. and Betz, H. (1990) FEBS Lett. 262, 259-262.

■ Nicholas J. Cowan:
Biochemistry Department,
New York University Medical School,
New York, USA
■ Andrew Matus,
Friedrich Miescher-Institut,
Basel, Switzerland

MAP2

Microtubule associated protein 2 (MAP2) is found specifically in neuronal cells and consists of three components whose expression is regulated both temporally and at the subcellular level. MAP2 binds to microtubules and promotes their assembly from purified tubulin. The biological role of MAP2 is unknown, but may in part be to regulate microtubule dynamics in neuronal processes.

Purified MAP2 consists of three components[1-3], with the following molecular masses by SDS-PAGE: MAP2a (280 kDa), MAP2b (270 kDa) and MAP2c (70 kDa). Peptide mapping and monoclonal antibody reactivity suggest that MAP2a and MAP2b are closely related[2], but because the exact relationship between the two is unknown we use the term MAP2 to refer to the high molecular weight form MAP2a/b. MAP2 and MAP2c arise by alternative splicing of transcripts of a single gene[4,5]. MAP2c mRNA encodes a polypeptide lacking amino acids 152-1514 of the MAP2 sequence[6]. The high and low molecular weight components are differentially expressed. MAP2 is present at very low levels in embryonic and newborn brain but becomes the major MAP in adult brain[3]. In contrast, MAP2c is the major MAP in embryonic brain but is expressed at very low levels in the adult, persisting in areas where neuronal growth continues[5]. MAP2c is present in axons whereas MAP2 is not[7]. This different localization could result in

part from the segregation of mRNAs; MAP2 mRNA is found in dendrites and cell bodies[8] while MAP2c mRNA is found only in cell bodies[7]. In addition, MAP2 is selectively degraded in axons[9].

MAP2 has two domains[10], a 200 kDa N-terminal projecting arm that forms regularly spaced 100 nm lateral projections from the microtubule surface, and a C-terminal microtubule binding domain. The protease sensitive site that separates the two domains has been mapped to near amino acid 1625[11]. The binding domain of both high and low molecular weight MAP2 contains a series of three imperfect 18 amino acid repeats each separated by 13-14 amino acids, very similar to those found in **tau**[12]. The three repeat sequences of MAP2 act in conjunction with adjacent sequences to affect microtubule binding affinity[13]. In vitro, a synthetic peptide corresponding to the second 18 amino acid repeat is capable of promoting microtubule nucleation and elongation[11]. However, this peptide is only

MAP 2   111

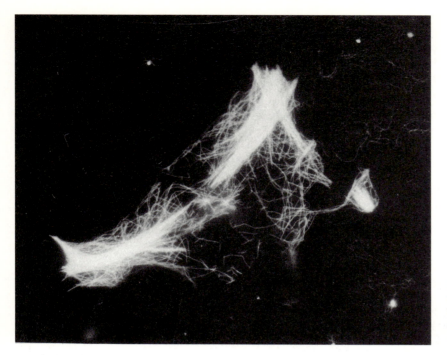

Figure. MAP2 bundled micro-tubules in transfected cells are stable in the presence of the microtubule depolymerizing drug nocodazole.

effective at concentrations 100-fold higher than intact MAP2. Both high and low molecular weight MAP2 bind RII, the regulatory subunit of cAMP dependent kinase. The RII binding domain has been localized to the N-terminus (amino acids 83-113)[14].

MAP2 has been shown by electron microscopy to be a component of the crossbridges between microtubules and neurofilaments[15]. However, binding of MAP2 to neurofilaments seems only to occur in the absence of microtubules[16]. MAP2 is associated with **actin** in the dendritic spines which lack microtubules[17]. When expressed in cultured cells by transfection, MAP2 reorganizes the interphase microtubules into a static network of bundles which are stable to cold and microtubule depolymerizing drugs[13] (Figure). Concomitantly, there is an approximate two-fold increase in the level of microtubule polymer. These changes depend solely on the microtubule binding domain. MAP2 can be phosphorylated by both cAMP dependent and cAMP independent routes[17]. Since extensive MAP2 phosphorylation reduces its ability to stimulate microtubule assembly and to bind to preformed microtubules[17], this may be a means whereby MAP2 can modulate neuronal plasticity.

■ PURIFICATION

MAP2 and its associated cAMP dependent protein kinase can be purified by cycles of microtubule polymerization followed by chromatography on DEAE-Sephadex and chromatography on Biogel α-15m[18]. Alternatively MAP2 can be purified by exploiting its resistance to heat denaturation[19,20].

■ ACTIVITIES

Purified MAP2 promotes the assembly of purified **tubulin** *in vitro*[19,21].

■ ANTIBODIES

Commercially available (Sigma) mouse monoclonal antibodies to MAP2[13] react with MAP2a/b/c in human, rat, mouse, bovine, chicken and quail tissues and cells by immunoblotting and immunofluorescence. Numerous polyclonal MAP2 antibodies have been generated, see for example references **4** and **14**.

■ GENES

The following MAP2 cDNAs exist: high molecular weight mouse full length cDNA (GenBank M21041)[12], rat full length cDNA (EMBL X51842)[6], human partial cDNA (GenBank M25668)[22]; MAP2c; rat full length (EMBL X171682)[7].

■ REFERENCES

1. Murphy, D.B. and Borisy, G.G. (1975) Proc. Natl. Acad. Sci. (USA) 72, 2696-2700.
2. Herrman, H., Dalton, J.M. and Wiche, G. (1985) J. Biol. Chem. 260, 5797-5803.
3. Riederer, B. and Matus, A. (1985) Proc. Natl. Acad. Sci. (USA) 82, 6006-6009.
4. Lewis, S.A. Villasante, A., Sherline, P. and Cowan, N.J. (1986) J. Cell Biol. 102, 2098-2105.

5. Garner, C.C. and Matus, A. (1988) J. Cell Biol. 106, 779-784.
6. Kindler, S., Schwanke, B., Schulz, B. and Garner, C.C. (1990) Nucl. Acids. Res. 18, 2822.
7. Papandrikopoulou, A., Doll, T., Tucker, R.P., Garner, C.C. and Matus, A. (1989) Nature 340, 650-652.
8. Garner, C.C., Tucker, R.P. and Matus, A. (1988) Nature 336, 674-677.
9. Okabe, S. and Hirokawa, N. (1989) Proc. Natl. Acad. Sci. (USA) 86, 4127-4131.
10. Vallee, R.B. (1980) Proc. Natl. Acad. Sci. (USA) 77, 3206-3210.
11. Joly, J.C., Flynn, G. and Purich, D.L. (1989) J. Cell Biol. 109, 2289-2294.
12. Lewis, S.A., Wang, D. and Cowan, N.J. (1988) Science 242, 936-939.
13. Lewis, S.A., Ivanov, I.E., Lee, G.-H. and Cowan, N.J. (1989) Nature 342, 498-505 (1990) ibid 345, 674.
14. Rubino, H.M., Dammerman, M., Sahfit-Zagardo, B. and Erlichman, J. (1989) Neuron 3, 631-638.
15. Hirokawa, N., Hisanaga, S.-I. and Shoimura, Y. (1988) J. Neurosci. 8, 2769-2779.
16. Glynn, G., Joly, J.C. and Purich, D.L. (1987) Biochem. Biophys. Res. Commun. 148, 1453-1459.
17. Olmsted, J.B. (1986) Ann. Rev. Cell Biol. 2, 421-457.
18. Vallee, R.B. (1984) Methods Enzymol. 134, 89-104.
19. Herzog, W. and Weber, K. (1978) Eur. J. Biochem. 92, 1-8.
20. Sloboda, R.D. and Rosenbaum, J.L. (1982) Meth. Enzymol. 85, 409-416.
21. Sandoval, I.V. and Vanderkerchove, J. (1981) J. Biol. Chem. 256, 8795-8800.
22. Kosik, K.S., Orecchio, L.D., Bakalis, S., Duffy, L. and Neve, R.L. (1988) J. Neurochem. 51, 587-589.

■ *Paul D. Walden, Sally A. Lewis and Nicholas J. Cowan:*
Department of Biochemistry
NYU Medical Center,
New York, USA.

MAP3

Microtubule associated protein 3 (MAP3)[1] is a large microtubule associated protein, found in various tissues where it is present in cells of marked morphological asymmetry[2]. In addition MAP3 seems to occur at sites where there are high concentrations of intermediate filaments, such as in neurofilament rich axons in the brain[1] and in vimentin rich podocytes of kidney epithelia[2], suggesting a possible dual association with the microtubular and intermediate filament systems of the cytoskeleton. MAP3 expression is developmentally regulated in the brain, there being far higher levels of MAP3 in developing embryos than in the adult.

Purified MAP3 appears on SDS-PAGE as pair of protein bands, MAP3a and MAP3b, of apparent molecular mass of 180 kDa[1]. Despite the occurence of MAP3 in conjunction with intermediate filaments it does not appear to be strongly bound to them, because during biochemical purification from brain all the MAP3 coassembles with **tubulin** polymers and none is found in the intermediate filament (neurofilament and glial filament)[1] fraction. A functional association of MAP3 with tubulin is suggested by its ability to stimulate tubulin polymerization *in vitro*[3]. However, the specificity of this effect is unclear because many neuronal MAPs exhibit this property. More interesting is the influence of monoclonal antibodies to MAP3 on microtubule assembly *in vitro*[3]. Addition of pure anti-MAP3 immunoglobulin leads to inhibition of microtubule polymerization and results in the formation of abnormally short microtubules.

Several MAPs with M_r's around 200,000 have been described. Among these the one with the closest similarity to MAP3 is the heat-stable 190 kDa **MAP4** (MAP-U)[4], which like MAP3 is present in a variety of tissues including brain. However, the two differ in tissue distribution; MAP3 is prominent in kidney (Figure) and present at trace levels in adrenal gland[2] whereas the opposite is true for MAP-U[4].

■ PURIFICATION

Preparations of microtubule associated proteins enriched in MAP3 are made by repolymerizing brain microtubules *in vitro*[5] followed by ion exchange chromatography on phosphocellulose[6] or DEAE-Sepharose[7]. Further purification of MAP3 can be achieved by heat treatment of the microtubules, which precipitates tubulin and most of the MAPs[7,8] leaving a supernatent containing **MAP2**, **tau** and MAP3[2]. MAP3 can be resolved from the other components by gel filtration; FPLC separation using a Superose-6B column is a convenient means of achieving this (B. Brugg and A. Matus, unpublished observations). The low abundance of MAP3, even in the richest source, developing brain, presents a difficulty for purification. For this reason, we have assayed the contribution of MAP3 to brain microtubule assembly by immunodepletion of brain MAP preparations using monoclonal antibody[3].

■ ACTIVITIES

Like several other MAPs, including MAP2 and tau, MAP3 promotes tubulin polymerization *in vitro*[3].

MAP 3 113

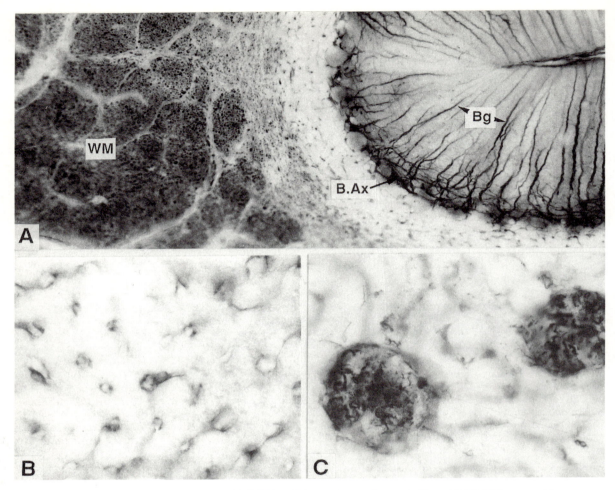

Figure. Immunohistochemical localization of MAP3 in adult rat tissues: In cerebellum (A) MAP3 is present in Bergman glial cell processes of the cortex (Bg) and in neurofilament rich axons of basket cells (B.Ax) and myelinated axons in the white matter (WM). In the liver (B) MAP3 is found in Kupfer cells and in the kidney (C) in podocytes of the gromeruli.

■ ANTIBODIES

Published studies on MAP3 have used a mouse mono-clonal antibody (1991), clone "N"[1]. This antibody reacts with both MAP3a and MAP3b by immunoblotting of brain microtubules as well as some additional lower molecular weight bands in preparations from other tissues[2].

■ GENES

cDNA clones have not yet been obtained.

■ REFERENCES

1. Huber, G., Alaimo Beuret, D. and Matus, A. (1985) J. Cell Biol. 100, 496-507.
2. Huber, G. and Mauts, A. (1990) J. Cell Sci. 95, 237-246.
3. Huber, G., Pehling, G. and Matus, A. (1986) J. Biol. Chem. 261, 2270-2273.
4. Aizawa, H., Kawasaki, H., Murofushi, H., Kotani, S., Suzuki, K. and Sakai, H. (1989) J. Biol. Chem. 264, 5885-5890.
5. Karr, T.L., White, H.D. and Purich, D.L. (1979) J. Biol. Chem. 254, 6107-6111.
6. Weingarten, M., Lockwood, A., Hwo, S.-Y. and Kirschner, M. (1975) Proc. Natl. Acad. Sci. (USA) 72, 1858-1862.
7. Herzog, W. and Weber, K. (1978) Eur. J. Biochem. 92, 1-8.
8. Fellous, A., Francon, J., Lennon, A.M. and Nunez, J. (1977) Eur. J. Biochem. 78, 167-174.

■ Gerda Huber,
  Pharmaceutical Research,
  Hoffmann-La Roche,
  Basel
■ Andrew Matus
  Friedrich Miescher Institute,
  Basel, Switzerland.

MAP4 (MAP-U)

MAP4 is a family of microtubule associated proteins that is distributed ubiquitously in tissues and across mammalian species. MAP4 promotes tubulin polymerization and binds to reconstituted microtubules in vitro. The functions of these MAPs are not yet known.

MAP4 describes a family of microtubule associated proteins that have an apparent molecular mass of ~200 kDa on SDS gels, and are thermostable. Recent molecular cloning data have demonstrated that the bovine 190 kDa MAP (MAP-U[1]), originally purified from adrenal cortex[2,3], the HeLa 210 kDa MAP[4,5] and mouse MAP4[4] are all members of the MAP4 family. These MAPs exist as complexes of related polypeptides[6,7], and the mouse isoforms have been found to vary with tissue origin[8] and during development[7,9]. This MAP has a long flexible rod-like structure with a contour length of 100 nm[2], and like **MAP2** and **tau**, consists of a microtubule binding domain and a projection domain extending from the microtubule wall[2,10].

MAP4 is encoded by a single gene[4] expressing multiple mRNAs[1,4]. Comparison of sequences for human[4], mouse MAP4[4] and the bovine MAP-U/MAP4[1] show ~80% overall homology at the amino acid level between these three species[4]. The molecular weight of MAP4 derived from sequence analysis is ~40% less than that derived from SDS-PAGE analyses. Cross-species sequence comparisons[4] and biochemical data obtained with synthetic peptides[11] and genetically engineered constructs[12] have defined several domains for MAP4. The protein consists of an acidic N-terminal region that corresponds to the projection domain, a C-terminal basic region that corresponds to the microtubule binding domain, and a very short hydrophobic acidic tail[1,4,12]. The major acidic domain of the protein has no sequence homology with other proteins, including MAP2[13] or tau[14].

Two distinctive features of the acidic domain are a highly conserved N-terminus[4] and a series of degenerate 14-mer repeats, with consensus sequence KD(M/V)XLPXETEVALA[1,4,5]. The number of repeats varies with species, bovine having 18[1], mouse having 19[4], and human having 26[4,5]. The basic domain is comprised of a conserved proline-rich region[4], a region rich in serine and proline containing conserved phosphorylation sites[4], and a series of imperfect 18-mer repeats (consensus VXSKXGSXXNIKHXPGGG) characteristic of the microtubule-binding domains of MAP2 and tau[1,4]. The number of repeats and presence of intervening sequences varies between and within the three species[4]. Synthetic peptides corresponding to the first repeat found in the bovine MAP4 compete directly for the binding of tau to microtubules[11], and a bacterially expressed fragment of this region[12] induces polymerization of tubulin into morphologically normal microtubules. However, the affinity of the expressed fragment[12] and especially of the synthetic peptides[11] to microtubules is lower than that of intact bovine MAP4. An expressed fragment of the proline-rich domain[1] (which includes both the conserved proline region and the serine, proline rich region[4]) has a higher affinity for microtubules, but induces bundles of microtubules with nonuniform diameter[12]. It has been suggested that this region may function as an "enhancer" of the repeat domain and thus bind MAP4 to microtubules with high affinity.

Phosphorylation with protein kinase C[15], cdc2 kinase[16] or MAP2 kinase[17] suppresses the activity of bovine MAP4 to induce microtubule assembly, suggesting that assembly promoting activity is controlled by a phosphorylation-dephosphorylation system. Phosphate is incorporated specifically into the proline-rich region by the former two protein kinases[15,16], a domain containing phosphorylation sites that are highly conserved in position for all three species sequenced[4].

MAP4 promotes the assembly of microtubules in vitro[2,17,18], and is induced during neurite differentiation in vivo[9,19]. Immunological data have shown that MAP4 is widely distributed in cells[7,20] and tissues[8,21] of various origin, and is present on both interphase and mitotic microtubule arrays[7,20,22]. In brain, MAP4 is restricted to glial cells[21,23]; this contrasts with MAP2 and tau, which are localized to neurons. Fluorescent MAP4 injected into cells localized to both interphase and mitotic mirotubules[24]. Photobleaching analyses on injected MAP4 have demonstrated that this protein turns over 4-5 times more rapidly than **tubulin** in interphase cells, and that dynamics change as a function of the cell cycle[24]. Early work with monoclonal antibodies demonstrated that an epitope of the human MAP4 is present only in mitotic cells[25]. Phosphorylation of MAP4 has been demonstrated to occur as a function of the cell cycle[26], and a cdc2-like kinase activity that may be associated with this activity recently identified in isolated mitotic spindles from CHO cells[27].

■ PURIFICATION

Using cultured cells, MAP4 can be enriched by cycles of microtubule polymerization[19,28] and further purified by boiling[8]. In brain tissue, MAP4 is a minor component, and is difficult to purify free from MAP2, which has similar fractionation properties[29]. MAP4 can be purified from thermostable preparations of tissues lacking MAP2 (e.g. heart, lung) by selective rebinding onto purified microtubules stabilized with taxol[24]. MAP4 has been purified from bovine adrenal cortex by heat treatment, DEAE-cellulose, hydrophobic chromatography and cosedimentation with reconstituted microtubules followed by heat treatment to remove denatured tubulin[2].

MAP 4 (MAP-U)   115

■ ACTIVITIES

MAP4 promotes the assembly of microtubules *in vitro*[2,18]. MAP4 does not crosslink **actin**, as found for MAP2 and tau, has no effect on the apparent viscosity of F-actin as measured by falling ball viscometry, and does not cosediment with F-actin filaments[2]. *In vivo* functions are unknown, although recent studies have suggested that MAP4 associated with the mitotic spindle is preferentially phosphorylated by a cdc2-like enzyme[27].

■ ANTIBODIES

The first polyclonal antibodies raised to the HeLa[20] or mouse MAP4[7,8] were found to be highly species specific, but more recent antibodies to human[29] and rat[30-32] MAP4 have shown greater species crossreactivity. A number of monoclonal antibodies to mouse MAP4[29] and to human MAP4[25,33] have been produced. Some of the polyclonal antibodies raised against a synthetic polypeptide with amino acid sequence of the first microtubule binding repeat of bovine MAP4[11] crossreact with tau and MAP2 (Murofushi, unpublished results).

■ GENES

cDNA sequences and derived complete open reading frames for bovine MAPU/MAP4 (GenBank J05557)[1], human (GenBank M64571)[4] and mouse (GenBank M72414)[4] MAP4 have been published. A partial amino acid sequence for human MAP4 has also appeared[5].

■ REFERENCES

1. Aizawa, H., Emori, Y., Murofushi, H., Kawasaki, H., Sakai, H. and Suzuki, K. (1990) J. Biol. Chem. 265, 13849-13855.
2. Murofushi, H., Kotani, S., Aizawa, H., Hisanaga, S., Hirokawa, N. and Sakai, H. (1986) J. Cell Biol. 103, 1911-1919.
3. Kotani, S., Murofushi, H., Maskawa, S., Sato, C. and Sakai, H. (1986) Eur. J. Biochem. 156, 23-29.
4. West, R.R., Tenbarge, K.M. and Olmsted, J.B. (1991) J. Biol. Chem. 266, 21888- 21896.
5. Chapin, S.J. and Bulinski, J.C. (1991) J. Cell Sci. 98, 27-36.
6. Bulinski, J.C. and Borisy, G.G. (1980) J. Cell Biol. 87, 792-801.
7. Olmsted, J.B., Asnes, C.F., Parysek, L.M., Lyon, H.D. and Kidder, G.M. (1986) Ann. NY Acad. Sci. 466, 292-305.
8. Parysek, L.M., Asnes, C.F. and Olmsted, J.B. (1984) J. Cell Biol. 99, 1309-1315.
9. Olmsted, J.B., Cox, J.V., Asnes, C.F., Parysek, L.M. and Lyon, H.D. (1984) J. Cell Biol. 99 (Suppl.), 28s-32s.
10. Aizawa, H., Murofushi, H., Kotani, S., Hisanaga, S., Hirokawa, N. and Sakai, H. (1987) J. Biol. Chem. 262, 3782-3787.
11. Aizawa, H., Kawasaki, H., Murofushi, H., Kotani, S., Suzuki, K. and Sakai, H. (1989) J. Biol. Chem. 264, 5885-5890.
12. Aizawa, H., Emori, Y., Mori, A., Murofushi, H., Sakai, H. and Suzuki, K. (1991) J. Biol. Chem. 266, 9841-9846.
13. Lewis, S.A., Wang, D. and Cowan, N.J. (1988) Science 242, 936-939.
14. Lee, G., Cowan, N.J. and Kirschner, M. (1988) Science 239, 285-288.
15. Mori, A., Aizawa, H., Saldo, T.C., Kawasaki, H., Mizuno, K., Murofushi, H., Suzuki, K. and Sakai, H. (1991) Biochemistry 30l, 9341-9346.
16. Aizawa, H., Kamijo, M., Ohba, Y., Mori, A., Okuhara, K., Kawasaki, H., Murofushi, H., Suzuki, K. and Yasuda, H. (1991) Biochem. Biophys. Res. Commun. 179, 1620-1626.
17. Hoshi, M., Ohta, K., Gotoh, Y., Mori, A., Murofushi, H., Sakai, H. and Nishida, E. (1992) Eur. J. Biochem. 203, 43-52.
18. Bulinski, J.C. and Borisy, G.G. (1980) J. Biol. Chem. 255, 11570-11576.
19. Olmsted, J.B. and Lyon, H.D. (1981) J. Biol. Chem. 256, 3507-3511.
20. Bulinski, J.C. and Borisy, G.G. (1980) J. Cell Biol. 87, 802-808.
21. Parysek, L.M., Wolosewick, J.J. and Olmsted, J.B. (1984) J. Cell Biol. 99, 2287-2296.
22. Kotani, S., Murofushi, H., Maekawa, S., Aizawa, H., Kaji, K. and Sakai, H. (1987) Cell Struct. Funct. 12, 1-9.
23. Parysek, L.M., DelCerro, M. and Olmsted, J.B. (1985) Neuroscience 15, 869-876.
24. Olmsted, J.B., Stemple, D.L., Saxton, W.M., Neighbors, B.W. and McIntosh, J.R. (1989) J. Cell Biol. 109, 211-223.
25. Izant, J.G., Weatherbee, J.A. and McIntosh, J.R. (1983) J. Cell Biol. 96, 424-434.
26. Vandre, D.D., Centonze, V.E., Peloquin, J., Tombes, R.M. and Borisy, G.G.(1991) J. Cell Sci. 98, 577-588.
27. Tombes, R.M., Peloquin, J.G. and Borisy, G.G. (1991) Cell Reg. 2, 861-874.
28. Bulinski, J.C. (1986) Methods Enzymol. 134, 147-156.
29. Olmsted, J.B., unpublished results.
30. Kotani, S., Murofushi, H., Maekawa, S., Aizawa, H. and Sakai, H. (1988) J. Biol. Chem. 263, 5385-5389.
31. Murofushi, H., Kotani, S., Aizawa, H., Maekawa, S. and Sakai, H. (1987) J. Biochem. 102, 1101-1112.
32. Murofushi, H., Suzuki, M., Sakai, H. and Kobayashi, S. (1989) Cell Tissue Res. 255, 315-322.
33. Izant, J., Weatherbee, J. and McIntosh, J.R. (1982) Nature 295, 248-250.

■ *J.B. Olmsted:*
Department of Biology,
University of Rochester,
Rochester, NY 14627, USA
■ *Hiromu Murofushi:*
Department of Biophysics and Biochemistry,
Faculty of Science,
University of Tokyo,
Tokyo, Japan

MARPs

MARP-1 and MARP-2 (MARP: Microtubule Associated Repetitive Protein) are two closely related, abundant microtubule associated proteins from the membrane skeleton of the parasitic hemoflagellate Trypanosoma brucei. They are heat-stable, high molecular weight proteins which contain large numbers of conserved 38 amino acid repeat units. Their functions most likely involve stabilizing the microtubules and mediating their interaction with the overlying cell membrane.

The cytoskeleton of the parasitic protozoon *Trypanosoma brucei* essentially consists of a membrane skeleton formed by a helical array of closely spaced membrane associated microtubules. This membrane skeleton encloses the entire cell body[1,2]. It may constitute the only skeletal structure of the cell body since no transcellular structural elements such as **actin** filaments or intermediate filaments have been observed so far. The closely packed microtubules of the membrane skeleton are extensively crosslinked among each other, and they are tightly connected to the overlying cell membrane[3,4]. They are highly resistant to the action of microtubule disrupting drugs, and they are very stable upon extraction of the cell[5]. This unusual property is most likely due to the extensive crosslinking by microtubule associated proteins.

Immunofluorescence and immunogold electron microscopy have demonstrated that the MARPs are located all along the microtubules of the membrane skeleton, but that they are absent from the microtubules of the flagellar axoneme[6]. Recent evidence from quick-freeze, deep-etch immunogold electron microscopy indicates that they are located exclusively at the domain of the microtubular surface which is apposed to the cell membrane, and that they are absent from the surface domains exposed to the cytoplasm[7].

MARP-1[6] and MARP-2 both consist of a short N-terminus, followed by a large number (>50) of tandemly arranged, highly conserved (within each protein species) 38 amino acid repeat units. The 38 amino acid repeat unit of MARP-1 represents a novel microtubule binding motif[8]. The sequence similarity between the repeat units of MARP-1 and MARP-2 is 50%. Both proteins end with a C-terminus of approximately 250 amino acids length, whose sequence is almost identical between the two (M. Affolter and I. Roditi, unpublished observations). The C-termini of MARP-1 and MARP-2 contain hydrophobic domains which may be involved in membrane contacts. On the other hand, transfection experiments have recently established that the C-termini also contain microtubule binding domains (A. Hemphill, unpublished). No sequence similarities to other known proteins or to the microtubule binding domains of other microtubule associated proteins have been observed.

Homologues of MARP-1 and MARP-2 have been detected by hybridization or antibody techniques in most trypanosomatids analyzed so far. No related proteins or DNA sequences were detected in any other organism.

■ PURIFICATION

The MARPs can be purified by heat treatment of a high-salt extract from isolated cytoskeletons, ammonium sulphate fractionation of the heat resistant proteins, followed by gel filtration chromatography[8]. The molecular mass of MARP-1, as determined by analytical ultracentrifugation, is about~ 320 kDa (U. Aebi, personal communication).

■ ACTIVITIES

No activities of the MARPs are currently known, though their predominant function may be that of microtubule stabilizers.

■ ANTIBODIES

Polyclonal rat antibodies have been raised against biochemically purified MARPs and against β-galactosidase-fusion proteins carrying MARP-1 and MARP-2 repeats, respectively.

■ GENES

The N- and C-termini of MARP-1 and MARP-2, as well as several repeats from MARP-2 have recently been sequenced (M. Affolter and I. Roditi, unpublished). The sequence of several MARP-1 repeats[6] is available in GenEMBL under the accession number M20569.

■ REFERENCES

1. Seebeck, T., Hemphill, A. and Lawson, D. (1990) Parasitol. Today 6, 49-52.
2. Sherwin, T. and Gull, K. (1989) Cell 57, 211-221.
3. Souto-Padron, T., De Souza, W. and Heuser, J.E. (1984) J. Cell Sci. 69, 167-178.
4. Hemphill, A., Lawson, D. and Seebeck, T. (1991) J. Parasitol. 77, 603-612.
5. Schneider, A., Sherwin, T., Sasse, R., Russell, D., Gull, K. and Seebeck, T. (1987) J. Cell Biol. 104, 431-438.
6. Schneider, A., Hemphill, A., Wyler, T. and Seebeck, T. (1988) Science 241, 459-462.
7. Hemphill, A., Seebeck, T. and Lawson, D. (1991) J. Struct. Biol. 107, 211-220.
8. Hemphill, A., Affolter, M. and Seebeck, T. (1992) J. Cell Biol. 117, 95-103.

■ *Marianne Affolter, Andrew Hemphill and Thomas Seebeck:*
Institute for General Microbiology,
CH-3012 Bern, Switzerland

Pericentrin

Pericentrin is a conserved protein that is an integral component of centrosomes and other microtubule organizing centers (MTOCs). It appears to be involved in organizing the microtubule spindle during mitosis and meiosis.

Pericentrin[1-3] was originally identified by screening λgt11 cDNA libraries with human autoimmune sera that had previously been shown to recognize highly conserved centrosomal antigens[2,4-6]. The deduced amino acid sequence from a pericentrin cDNA predicts a polypeptide that is largely α-helical. The nucleotide and amino acid sequences bear no strong homology to any known gene or protein. Two rare mRNAs of 7.5 and 9 kB have been identified; the significance of the two mRNAs is unknown.

Pericentrin is a rare 200-220 kDa protein that is localized exclusively to the centrosome (Figure) in several mammalian species, *Xenopus*, and *Drosophila*. It is located in the area of the centrosome involved in nucleating the growth of new microtubules[7], the pericentriolar material (hence the name pericentrin). The protein is also present in MTOCs other than centrosomes, such as the acentriolar meiotic spindle poles of mouse oocytes[2].

Pericentrin undergoes cell cycle fluctuations that are concomitant with changes in the microtubule nucleating capacity of the centrosome[2]. The amount of centrosome-associated protein is highest in pro- and metaphase and lowest in late ana- and telophase. At all stages of the cell cycle pericentrin's association with the centrosome is resistant to high salt and detergent extraction indicating that it is an integral part of the organelle.

Pericentrin plays a role in the organization of the microtubule spindle during meiosis and mitosis. These conclusions are drawn from experiments in which affinity-purified anti-pericentrin antibodies were injected into mouse oocytes and *Xenopus* embryos. In mouse oocytes the antibody disrupted meiotic spindle formation and in *Xenopus* embryos mitotic cell division was arrested. Preliminary results in cell-free *Xenopus* extracts[3,8] suggested that the antibody inhibited the ability of microtubule nucleating components to assemble into functional nucleating centers.

■ PURIFICATION

Pericentrin copurifies with centrosomes isolated from cultured cells as described by Mitchison and Kirschner[9]. A detailed protocol for its isolation has not yet been developed.

■ ACTIVITIES

None determined.

■ ANTIBODIES

Several polyclonal and monoclonal antibodies have been generated against a bacterially expressed pericentrin

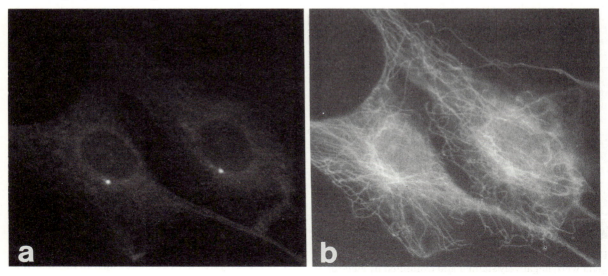

Figure. Immunofluorescence localization of pericentrin in Rat 1 cells using an affinity purified antiserum raised against a bacterial fusion protein. The staining is confined to one or two bright, perinuclear foci (a) at the site of microtubule convergence (b). [(a) Anti-pericentrin, (b) anti-tubulin].

fusion protein. The polyclonal antisera react with mammals[1,2], *Xenopus*[2], and *Drosophila*, while the monoclonal antibodies react only with mouse and rat.

GENES

A cDNA encoding 95% of the pericentrin molecule has been cloned and sequenced from mouse[1,2]. A human homologue has been identified.

REFERENCES

1. Doxsey, S.J., Calarco, P., Evans, L., Stein, P., Siebert, P. and Kirschner, M. (1990) J. Cell Biol. 111, 179a.
2. Doxsey, S.J., Stein, P. Calarco, P., Evans, L.,. and Kirschner, M. Submitted.
3. Doxsey, S.J. and Kirschner, M. (1991) J. Cell Biol. 111, 172a.
4. Tuffanelli, D., McKeon, F., Kleinsmith, D.M., Burham, T.K. and M. Kirschner, M. (1983) Arch. Dermatol. 119, 560-566.
5. Calarco-Gillam, P. D., Siebert, M.C., Hubble, R., Mitchison, T. and Kirschner, M.(1983) Cell 35, 621-629.
6. Clayton, L., Black, C.M. and Lloyd, C. (1985) J. Cell Biol. 101, 319-325.
7. Brinkley, B.R. (1985) Ann. Rev. Cell Biol. 1, 145-172.
8. Murray, A.W. and Kirschner, M.W. (1989) Nature 339, 275-280.
9. Mitchison, T. and Kirschner, M. (1984) Nature 312, 232-237.

■ *Stephen J. Doxsey and Marc Kirschner:*
Department of Biochemistry and Biophysics,
University of California,
San Francisco, CA, USA

Radial Spoke Proteins

The radial spoke is a multisubunit structure found in eukaryotic cilia and flagella. Its apparent role is to regulate the bending pattern to produce an asymmetric, ciliary-type beat.

Radial spoke proteins (RSPs) 1 through 17 are defined as the set of flagellar polypeptides that are absent from the flagella of a *Chlamydomonas* mutant lacking radial spokes (as determined by electron microscopy). They range in molecular mass from 34 to 124 kDa[1]. The loci encoding several of these proteins have been identified by genetic and biochemical studies[2,3].

Mutants lacking the head portion (Figure) are deficient in RSP1, 4, 6, 9, and 10[1]. Biochemical removal of spoke-heads removes these same polypeptides plus RSP5, suggesting that RSP5 might be located at the stalk/head junction. The remaining 11 polypeptides are located in the stalk[1]. The absence of any RSPs in the flagella of a mutant that does not produce RSP3[4] indicates that this polypeptide might mediate the binding of the stalk to the microtubule. Proteins 2, 3, 5, 13 and 17 are phosphorylated. In a mutant cell deficient for this modification,

Figure. Chlamydomonas radial spokes. Wild type flagella were negatively stained and examined by electron microscopy. The radial spoke stalks extend upward from the outer doublet microtubules and terminate in a globular head. (Reprinted from reference 11.)

spoke assembly is incomplete and the flagella are paralyzed, indicating a role for phosphorylation in assembly or function of the spokes[5].

In vitro studies on isolated axonemes suggested that interactions between the radial spokes and projections from the central pair microtubules are required for the conversion of microtubule sliding into bending movements[6]. Reversion analysis of radial spoke mutants indicated that a symmetric bending pattern is possible in the absence of radial spokes, and the function of spokes is to convert this into the asymmetric pattern used in forward swimming[7]. The studies of revertants suggest the existence of a complex mechanism in which the spoke/central pair interactions are involved in the regulation of **dynein** arm activity[8].

■ PURIFICATION

A spokehead fraction (RSP1,4,5,6,9,10) can be obtained by low-salt dialysis of axonemes[1]. Individual polypeptides were isolated from 2-D polyacrylamide gels for production of antisera (see below).

■ ACTIVITIES

Biochemical activities unknown.

■ ANTIBODIES

Polyclonal antisera have been raised against the following RSPs after isolation from 2-D polyacrylamide gels: stalk proteins 3[9], 2 and 7 (unpublished), and head proteins 1,5,6[10],9 and 10 (unpublished).

GENES

Genomic clones have been isolated for RSP1, 6[10], 3[9], 2 and 4 (unpublished). The protein deduced from the sequence of the RSP3 gene[9] (EMBL X14549) does not share homology with other proteins, including several known MAPs. This gene has been used to transform *pf-14* (a paralyzed mutant that lacks RSP3[4]) to wild type[11].

Sequence analysis of the genes for head proteins 4 and 6 confirms that these genes are very closely linked[5] and shows them to be homologous to each other (manuscript in preparation). The protein sequences are rich in proline and would be expected to have extended conformations.

REFERENCES

1. Piperno, G., Huang, B., Ramanis, Z. and Luck, D.J.L. (1981) J. Cell Biol. 88, 73-79.
2. Luck, D.J. (1984) J. Cell Biol. 98, 789-794.
3. Huang, B.P.H. (1986) Int. Rev. Cytol. 99, 181-215.
4. Luck, D., Piperno, G., Ramanis, Z. and Huang, B. (1977) Proc. Natl. Acad. Sci. (USA) 74, 3456-3460.
5. Huang, B., Piperno, G., Ramanis, Z. and Luck, D.J.L. (1981) J. Cell Biol. 88, 80-88.
6. Witman, G.B., Plummer, J. and Sander, G. (1978) J. Cell Biol. 76, 729-747.
7. Brokaw, C.J., Luck, D.J.L. and Huang, B. (1982) J. Cell Biol. 92, 722-732.
8. Huang, B., Ramanis, Z. and Luck, D.J.L. (1982) Cell 28, 115-124.
9. Williams, B.D., Velleca, M.A., Curry, A.M. and Rosenbaum, J.L. (1989) J. Cell. Biol. 109, 235-245.
10. Williams, B.D., Mitchell, D.R. and Rosenbaum, J.L. (1986) J. Cell. Biol. 103, 1-11.
11. Diener, D.R., Curry, A.M., Johnson, K.A., Williams, B.D., Lefebvre, P.A., Kindle, K.L. and Rosenbaum, J.L. (1990) Proc. Natl. Acad. Sci. (USA) 87, 5739-5743.

Alice M. Curry and Joel L. Rosenbaum:
Department of Biology,
Yale University
New Haven, CT, USA

Sea Urchin MAPs and Microtubule Motors

Several microtubule associated proteins (MAPs); including motors copurify with microtubules prepared from sea urchin egg or embryo cytosol. Although the functions of these proteins have not been directly demonstrated, they are likely to play important roles in the activities of microtubules during the multiple cell divisions that occur during the cleavage stage of embryogenesis.

Microtubules prepared from sea urchin egg or embryo cytosol[1-3] contain numerous projections visible by electron microscopy, plus several copurifying microtubule associated proteins (MAPs), including 80 kDa MAP and 235, 205, 150, 52, 50 and 37 kDa polypeptides, as well as copurifying ATPase activity[2], arising from (at least) two 20 S **dynein** isoforms, a 10 S ATPase and **kinesin**.

The major MAP that copurifies with sea urchin cytoplasmic microtubules is a protein of 75-80 kDa[1,4]. Current data generally fail to distinguish whether there is a single MAP or a group of distinct proteins in this size range. Monoclonal antibodies to a 77 kDa MAP stain mitotic spindle fibres in embryos as well as microtubules in interphase sea urchin coelomocytes[5] (Figure). A 75 kDa protein named buttonin forms a coat of hexagonally packed globular structures on the surfaces of microtubules and increases the rate and extent of bovine brain microtubule assembly *in vitro*[4]. An 80 kDa protein coextracts and coassembles with **tubulin** from mitotic spindles[6], and 2-D gel electrophoresis of microtubules assembled from such spindle tubulin reveals one major and several minor polypeptides in the general area of 80 kDa[7].

In addition to the major 80 kDa MAP(s), sea urchin egg microtubules contain polypeptides of 235, 205, 150, and 37 kDa which behave as MAPs during microtubule isolation[1]. Monoclonal antibodies to these polypeptides stain mitotic spindles in sea urchin blastomeres[1] and microtubule like structures in coelomocytes[8]. Additionally, a 50 kDa acidic MAP[9] and a 52 kDa basic MAP[7] have been identified as components of detergent extracted mitotic cytoskeletons.

Egg cytoplasm contains also at least two 20 S dynein isoforms whose heavy chains are susceptible to vanadate sensitized UV cleavage[10]. A soluble "latent activity" isoform that crossreacts with flagellar dynein antibodies can be purified from microtubule depleted cytosol by chromatographic procedures and is thought to be a ciliary precursor, whereas a second isoform (HMr3) cosediments with microtubules in an ATP-sensitive manner, and may be a bona fide **cytoplasmic dynein**[10]. Dynein-like proteins have been localized to mitotic spindles and cortices of dividing sea urchin eggs, but the significance of these findings are unclear.

Kinesin from sea urchin eggs and embryos[11] can be purified by microtubule affinity binding or monoclonal antibody immunoadsorption, and consists of two heavy chains (130 kDa on SDS-PAGE) plus a light chain doublet (78 and 84 kDa)[12]. The heavy chain sequence that has been deduced from cloned cDNAs predicts a 117 kDa product, organized into head, stalk and tail domains[13]. Sea urchin

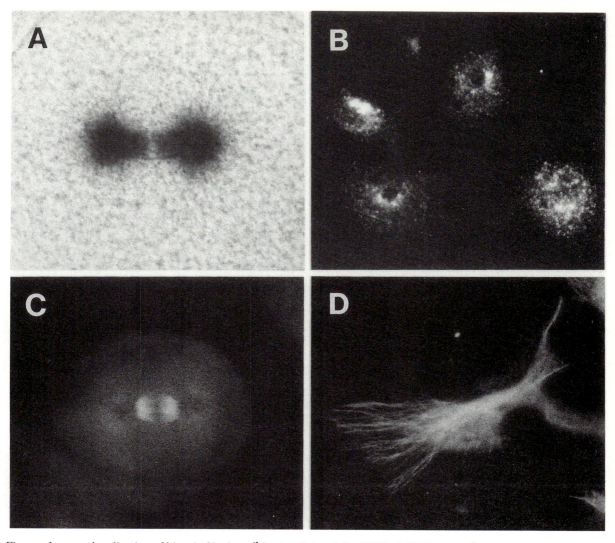

Figure. Immunolocalization of kinesin [(A,B) see[15] for details] and the 77 kDa MAP [(C,D) see[5] for details] to sea urchin blastomere mitotic spindles (A,C) and to vesicle-like structures (B) or microtubule arrays (D) in coelomocytes.

kinesin uses ATP to move microtubules at a V_{max} of 0.6 µm/sec, and the nucleotide specificity and inhibitor sensitivity of this motility have been quantitated[14]. Kinesin localizes to microtubule-membrane arrays in early, but not late, cleavage-stage sea urchin embryos, and to vesicle-like structures in coelomocytes[13]. Peptide antibodies to conserved regions of the kinesin motor domain react with presumptive kinesin related polypeptides of 170, 95 and 85 kDa in sea urchin egg microtubules.

Sea urchin egg cytoplasmic microtubules also contain a 10 S micotubule activated ATPase distinct from cytoplasmic dynein and kinesin, which may represent a motor[15].

■ PURIFICATION

Cytoplasmic microtubules are easily prepared in yields of 2-5 mg per 10 ml packed eggs/embryos, by differential centrifugation of cytosol treated with taxol[1,2] or with dimethyl sulfoxide[3]. The microtubules can be washed, subjected to cycles of assembly and disassembly[3] and treated with elevated ionic strength buffers or nucleotides to elute MAPs or motors. Purification protocols have been described for kinesin[12], cytoplasmic dyneins[10], the 50 kDa MAP[9], and the 75 kDa MAP buttonin[4].

ACTIVITIES

Sea urchin egg kinesin and flagellar dynein (but so far not the egg dyneins), when bound to coverslips, stimulate the movement of MAP-free microtubules in the presence of nucleotide triphosphate and Mg^{2+}[14,16]. The 75 kDa MAP (buttonin) stimulates the assembly of microtubules when present in relatively high molar ratio to tubulin[4].

ANTIBODIES

Several mono- and polyclonal antibodies to specific domains on the kinesin heavy chain have been prepared and characterized[13,17,18]. Three monoclonals bind to the kinesin heads and block kinesin-driven motility[17,18]. Vallee and Bloom have raised monoclonal antibodies to MAPs[1,5]. Polyclonal antibodies to the 50 kDa MAP[9] and kinesin[17] are available.

GENES

The coding sequence for the sea urchin kinesin heavy chain has been determined[13], and is available from EMBL/GenBank/DDBL under accession number X56844.

REFERENCES

1. Vallee, R.B. and Bloom, G.S. (1983) Proc. Natl. Acad. Sci. (USA) 80, 6259-6263.
2. Scholey, J.M., Neighbors, B., McIntosh, J.R. and Salmon, E.D. (1984) J. Biol. Chem. 259, 6516-6525.
3. Suprenant, K.A. and Marsh, J.C. (1987) J. Cell Sci. 87, 71-84.
4. Hirokawa, N. and Hisanaga, S.I. (1987) J. Cell Biol. 104, 1553-1561.
5. Bloom, G.S., Luca, F.C., Collins, C.A. and Vallee, R.B. (1985) Cell Motil. 5, 431-446.
6. Keller, T.C.S. and Rebhun, L.I. (1982) J. Cell Biol. 93, 788-796.
7. Leslie, R.J. and Wilson, L. (1989) Analytical Biochem. 181, 51-58.
8. Bloom, G.S., Luca, F.C., Collins, C.A. and Vallee, R.B. (1986) Ann. N.Y. Acad. Sci. 466, 328-339.
9. Raymond, M.N., Foucault, G., Renner, M. and Pudles, J. (1987) Eur. J. Cell Biol. 45, 302-310.
10. Porter, M.E., Grissom, P.M., Scholey, J.M., Salmon, E.D. and McIntosh, J.R. (1988) J. Biol. Chem. 263, 6759-6771.
11. Scholey, J.M., Porter, M.E., Grissom, P. and McIntosh, J.R. (1985) Nature 318, 483-486.
12. Buster, D. and Scholey, J.M. (1991) J. Cell Sci., 14s, 109-115.
13. Wright, B.D., Henson, J.H., Wedaman, K.P., Willy, P.J., Morand, J.N. and Scholey, J.M. (1991) J. Cell Biol. 113, 817-833.
14. Cohn, S.A., Ingold, A.L. and Scholey, J.M. (1989) J. Biol. Chem. 264, 4290-4297.
15. Collins, C.A. and Vallee, R.B. (1986) Proc. Natl. Acad. Sci. (USA) 83, 4799-4803.
16. Paschal, B.M., King, S.M., Moss, A.G., Collins, C.A., Vallee, R.B. and Witman, G.B. (1987) Nature 330, 672-674.
17. Ingold, A.L., Cohn, S.A. and Scholey, J.M. (1988) J. Cell Biol. 107, 2657-2667.
18. Scholey, J.M., Heuser, J., Yang, J.T. and Goldstein, L.S.B. (1989) Nature 338, 355-357.

■ *Jonathan M. Scholey and Roger J. Leslie:*
Department of Zoology,
Univ. of California,
Davis, CA 95616-8755, USA

STOPs

STOPs (Stable Tubulin Only Polypeptides) constitute a family of Ca^{2+}-Calmodulin regulated, microtubule stabilizing proteins that are responsible for the formation of cold stable microtubules in various tissue extracts. These proteins colocalize with stable microtubules in vivo.

The best characterized STOP was isolated from rat brain crude extracts as a 145 kDa polypeptide[1] ($STOP_{145}$). Purified $STOP_{145}$ has an apparent Stokes radius corresponding to a globular protein of 500 kDa[1]. It is not known whether it binds to microtubules as a monomer or a multimeric complex.

Other forms of STOPs have been isolated from bovine brain[2]. Immunoblot analysis suggests a strong interspecies variability in the apparent molecular weight of neuronal STOPs. The existence of proteins with STOP activity has been demonstrated in nonneuronal tissues, but the physical properties of the corresponding proteins have not yet been determined[3]. $STOP_{145}$ is localized to the stable microtubule subpopulation of mitotic spindles in dorsal root ganglion cells in primary culture (Figure)[4]. This observation correlates with the distribution suggested in a model detailing possible STOP involvement in chromosome movement[5]. STOP must also have a function as a MAP in the nonmitotic neuronal tissue from which it is derived.

The exact physiological role of STOPs is presently unknown.

PURIFICATION

Highly purified $STOP_{145}$ can be obtained from rat brain cold stable microtubules by a two step procedure involving a DEAE-column and a calmodulin affinity chromatography step[1]. Alternatively it can be purified directly from a rat brain high speed supernatant using S-Sepharose or Heparin-Sepharose columns, to which the protein binds, followed by an immunoaffinity step[3].

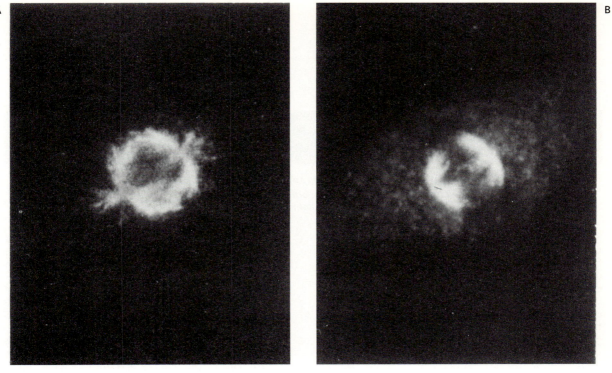

Figure. Immunolocalization of STOP$_{145}$ in mitotic dorsal root ganglion cells. The cell is double stained with an anti-tubulin antibody (A), and an anti-STOP$_{145}$ antibody (B)[4].

ACTIVITIES

STOP protein activity can be quantitated by a filter binding assay. These assays use (^{3}H)-GTP labelled microtubules which are diluted into destabilizing conditions in the presence of added STOPs, after which residual polymers, protected against disassembly, are quantitated after trapping on glass fibre filters[3,4,6]. Using the dilution assay, it is possible to detect microtubule stabilizing proteins solubilized by high salt or chaotropic agents[3,4,6]. Such solubilization procedures are often necessary for purified STOPs, which are often insoluble in standard buffers.

Microtubule bound STOPs may slide along the polymer[7]. At a molar ratio of 1/10 to **tubulin**, they make microtubules extremely resistant to Ca^{2+}, drugs or to cold temperature[8]. Their effect is abolished by Ca^{2+}-Calmodulin concentrations of 10^{-6} - 10^{-5}M[9].

ANTIBODIES

Three mouse monoclonal antibodies recognizing separate epitopes on the 145 kDa neuronal STOP are available: Map 296, Mab 175 and Mab 378. Mab 296 and Mab 378 are specific to rat neuronal tissues. Mab 175 crossreacts with neuronal STOPs from other mammalian species. None of these antibodies yields a clear crossreaction with nonneuronal tissue proteins. Mab 296 has been successfully used in immunofluorescence studies of rat dorsal root ganglion explant cells[4]. Mab 296 and Mab 175 have been used for immunoaffinity purification of rat[3,4] and bovine[2] STOPs.

GENES

No cDNA clones have been isolated so far.

REFERENCES

1. Margolis, R.L., Rauch, C.T. and Job, D. (1986) Proc. Natl. Acad. Sci. (USA) 83, 639-643.
2. Pirollet, F., Job D. and Margolis, R.L. (1989) J. Cell. Bio. 109, 80A.
3. Pirollet, F., Rauch, C.T., Job, D. and Margolis, R.L. (1989) Biochemistry 28, 835-842.
4. Margolis, R.L., Rauch, C.T., Pirollet, F. and Job, D. (1990) EMBO J. 9, 4095-4102.
5. Garel, J.R., Job, D. and Margolis, R.L. (1987) Proc. Natl. Acad. Sci. (USA) 84, 3599-3603.
6. Job, D., Pabion, M. and Margolis, R.L. (1985) J. Cell Biol. 101, 1680-1689.
7. Pabion, M., Job, D. and Margolis, R.L. (1984) Biochemistry 23, 6642-6648.
8. Job, D., Rauch, C.T. and Margolis, R.L. (1987) Biochem. Biophys. Res. Commun. 148, 429-434.
9. Job, D., Fischer, E.H. and Margolis, R.L. (1981) Proc. Natl. Acad. Sci. (USA) 78, n° 8, 4679-4682.

■ Didier Job:
Département de Biologie Moléculaire et Structurale
and INSERM U 244,
Centre d'Etudes Nucléaires Grenoble,
France

■ Robert L. Margolis:
The Fred Hutchinson Cancer Research Center,
Seattle, Washington, USA

Syncolin

Syncolin[1] is a 280 kDa microtubule associated protein (MAP) identified in chicken erythrocytes. It colocalizes with marginal band microtubules and possesses microtubule bundling activity. It is likely that syncolin is related to a protein previously suggested to play a role in the crosslinking of amphibian and avian erythrocyte microtubules[2-4].

Immunofluorescence microscopy of chicken bone marrow cells indicates that syncolin is expressed at all stages of erythrocyte differentiation (Figure). In early erythroblasts its distribution is diffuse throughout the cytoplasm, but it becomes restricted to marginal band microtubules at later stages. Syncolin's association with the marginal band is dependent on the integrity of microtubules, as shown by temperature dependent de- and repolymerization of marginal band microtubules[1].

Syncolin shares epitopes with brain **MAP2** and both proteins have the same apparent molecular weights (~280,000) when analyzed by SDS-PAGE. However, striking differences in ultrastructure and in several biochemical characteristics suggest that syncolin and MAP2 are distinct proteins[1].

Electron microscopy of negatively stained or shadowed syncolin molecules reveals a ring-like or globular structure with a diameter of ~13 nm, in contrast to filamentous MAPs, such as **tau** and MAP2[1,5]. Considering its dimensions and its sedimentation coefficient of 22-27 S, it is likely that syncolin exists in solutions as a globular particle of ~1,000 kDa. Syncolin cosediments with microtubules assembled *in vitro*, and it is displaced from the polymer by salt; at saturation the molar ratio of syncolin to **tubulin** dimers is 1:12. Purified syncolin induces aggregation into bundles of taxol polymerized brain or erythrocyte tubulin; up to 15 microtubules could be observed in these bundles by electron microscopy. Syncolin may function as a crosslinker and stabilizer of marginal band microtubules in nucleated (chicken) erythrocytes.

■ PURIFICATION

The protocol for purifying syncolin from chicken erythrocytes involves cosedimentation with taxol assembled microtubules, release from the polymer at low temperature, sucrose gradient centrifugation, and chromatography on hydroxylapatite and DEAE-cellulose columns[1]. Preparations of purified syncolin contain tightly associated proteins of 50-60 kDa.

■ ACTIVITIES

Syncolin bundles erythrocyte and brain microtubules *in vitro*, but does not promote their assembly. In solid-phase binding assays, the protein binds to nitrocellulose immobilized α- and β-tubulin from brain, but not to immobilized erythrocyte tubulins. On the other hand, native soluble (oligomeric) erythrocyte tubulin binds to immobilized syncolin[5]. The purified protein is significantly phosphorylated by protein kinase C, and poorly by protein kinase A.

■ ANTIBODIES

A polyclonal serum and affinity-purified antibodies against gel-purified chicken syncolin are available[1]. An

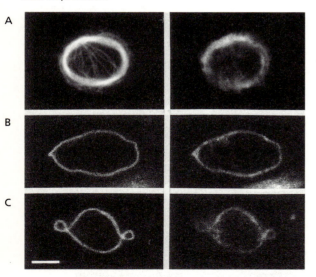

Figure. Localization of syncolin in chicken erythrocytes at late developmental stages using double immunofluorescence microscopy, tubulin (left), syncolin (right). (A) Polychromatophilic erythroblasts, (B, C) mature erythrocytes with normal and supertwisted marginal bands. Bar 5 µm.

antiserum raised to hog brain MAP2[6] shows crossreactivity with chicken erythrocyte syncolin[1].

■ GENES

So far, no sequence information is available.

■ REFERENCES

1. Feick, P., Foisner, R. and Wiche, G. (1991) J. Cell Biol. 112, 689-699.
2. Sloboda, R.D. and Dickerson, K. (1980) J. Cell Biol. 87, 170-179.
3. Centoze, V.E. and Sloboda, R.D. (1986) Exp. Cell Res. 167, 471-483.
4. Centoze, V.E., Ruben, G.C. and Sloboda, R.D. (1985) Eur. J. Cell Biol. 39, 190-197.
5. Wiche, G., Oberkanins, C. and Himmler, A. (1991) Int. Rev. Cytol. 124, 217-273.
6. Wiche, G., Briones, E., Hirt, H., Krepler, R., Artlieb, U. and Denk, H. (1983) EMBO J. 2, 1915-1920.

■ Gerhard Wiche:
Institute of Biochemistry,
University of Vienna,
1090 Vienna,
Austria

Tau

Tau is a complex family of microtubule associated proteins (MAPs) originally purified based on their promotion of microtubule assembly from purified tubulin. Found specifically in neuronal cells along the length of microtubules, tau most likely functions by regulating microtubule dynamics during neurite outgrowth.

Purified tau from mammalian brain consists of four to six discrete isoforms ranging in apparent molecular mass on SDS-PAGE from 35-65 kDa[1]. This heterogeneity in tau is generated at least in part by alternate splicing of exons from a single large gene[2]. During brain development, the expression of tau isoforms changes[3,4]. All of the isoforms contain a cluster of three or four imperfect 18 amino acid repeats spaced 13-14 amino acids apart and found within the C-terminal half of tau[5,6]. These repeats which are also found in **MAP2**[7], **MAP4/MAPU**[8,9] are thought to be the core of the microtubule binding domain of tau with sequences outside this region also affecting the binding affinity[10]. Synthetic peptides corresponding to the first or third repeats promote microtubule assembly[11] and short fragments of tau derived from the expression of truncated cDNAs containing the full or partial repeat region bind to microtubules[5,10,12]. Purified tau protein strongly promotes the nucleation of microtubule assembly and the elongation of existing microtubules at 100-fold lower concentrations than that required with the repeat peptides[11]. Phosphorylation at multiple sites has profound structural effects leading to a stiffening and elongation of the already long and rather flexible molecule[13]. A decrease in the ability of tau to promote microtubule assembly accompanies this structural alteration by phosphorylation[14].

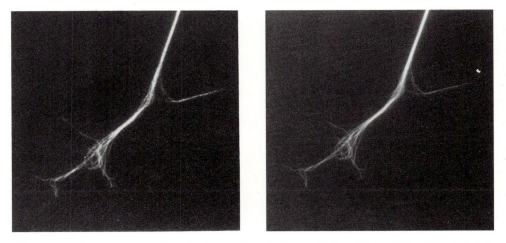

Figure: Double immunofluorescent staining of a differentiated PC-12 cell. Tau (7A5 rabbit anti-tau) staining appears on the left and tubulin (DM1α mouse anti-tubulin) on the right.

Tau is localized along the length of microtubules in PC12 cells which normally express tau[15] (Figure) or when introduced into cell lines by transfection[16] or microinjection[17]. In mammalian brain neurons, tau is specifically localized along microtubules in the axon and excluded from cell bodies and dendritic processes[18]. Tau is also a major component of the neuritic tangles associated with Alzheimers disease[19]. Introduction of tau into cells leads to an increase in the total levels of microtubule polymer and renders the microtubule array resistant to depolymerization by nocodazole and cold[16,17]. Expression in PC12 cells correlates closely with neurite outgrowth and the growth of microtubules[20]. Anti-sense oligonucleotides to tau blocked neurite outgrowth in rat cerebellar neurons, suggesting that expression of tau is essential for this morphogenetic event[21].

■ PURIFICATION

Tau can be purified by cycles of microtubule polymerization followed by chromatography on phosphocellulose, ammonium sulphate precipitation, and chromatography on hydroxylapatite[1]. Alternately, tau can be purified by exploiting its resistance to denaturation by heat or perchloric acid[22].

■ ACTIVITIES

Tau promotes the nucleation and elongation of microtubules[23]. Microtubule growth was assayed by turbidity or with electron micrographs of fixed specimens.

■ ANTIBODIES

Rabbit polyclonal antisera against bovine tau recognizes tau from many vertebrates[3]. Rabbit antisera against specific peptides from tau have been published[2]. A large number of monoclonal antibodies to tau have also been described[18,24].

■ GENES

Full length cDNAs for human (GenBank J03778)[25], mouse (M18775,M18776)[6], bovine[5] and rat[26] tau are published.

■ REFERENCES

1. Cleveland, D. W., Hwo, S.-Y. and Kirschner, M.W. (1977) J. Mol. Biol. 116, 207-225.
2. Himmler, A. (1988) Mol. Cell. Biol. 1389-1396.
3. Drubin, D. G., Caput, D. and Kirschner, M.W. (1984) J. Cell. Biol. 98, 1090-1097.
4. Francon, J., Lennon, A.M., Fellous, A., Mareck, A., Pierre, M. and Nunez, J. (1982) Eur. J. Biochem. 129, 465-471.
5. Himmler, A., Drechsel, D., Kirschner, M.W. and Martin, D.J. (1988) Mol. Cell. Biol. 9, 1381-1388.
6. Lee, G., Cowan, N. and Kirschner, M. (1988) Science 239, 285-288.
7. Lewis, S. A., Wang, D.H. and Cowan, N.J. (1988) Science 242, 936-939.
8. Chapin, S.J. and Bulinski, J.C. (1991) J. Cell Sci. 92, 27-36.
9. Aizawa, H., Kawasaki, H., Murofushi, H., Kotani, S., Suzuki, K. and Sakai, H. (1989) J. Biol. Chem. 264, 5885-5890.
10. Butner, K. A. and Kirschner, M. W. (1991) J. Cell Biol. 115, 717-730.
11. Ennulat, D. J., Liem, R.K.H., Hashim, G.A. and Shelanski, M.L. (1989) J. Biol. Chem. 264, 5327-5330.
12. Aizawa, H., Kawasaki, H., Murofushi, H., Kotani, S., Suzuki, K. and Sakai, H. (1988) J. Biol. Chem. 263, 7703-7707.
13. Hagestedt, T., Lichtenberg, B., Wille, H., Mandelkow, E.-M. and Mandelkow, E. (1989) J. Cell Biol. 109, 1643-1651.
14. Lindwall, G. and Cole, D. R. (1985) J. Biol. Chem. 259, 12241-12245.
15. Drubin, D., Kobayashi, S. and Kirschner, M.W. (1986) Ann. NY Acad. Sci. 466, 257-268.
16. Kanai, Y., Takemura, R., Oshima, T., Mori, H., Ihara, Y., Yanagisawa, M., Masaki, T. and Hirokawa, N. (1989) J. Cell Biol. 109, 1173-1184.
17. Drubin, D. G. and Kirschner, M. W. (1986) J. Cell. Biol. 103, 2739-2746.
18. Binder, L. I., Frankfurter, A. and Rebhun, L.I. (1985) J. Cell Biol. 101, 1371-1378.
19. Kosik, K. (1990) Curr. Op. in Cell Biol. 2, 101-104.
20. Drubin, D. G., Feinstein, S.C., Shooter, E.M. and Kirschner, M.W. (1985) J. Cell Biol. 101, 1799-1807.
21. Caceres, A. and Kosik, K. S. (1990) Nature 343, 461-463.
22. Drubin, D. and Kirschner, M. W. (1986) Methods Enzymol. 134, 156-170.
23. Witman, G. B., Cleveland, D.W., Weingarten, M.D. and Kirschner, M.W. (1976) Proc. Natl. Acad. Sci. 73, 4070-4074.
24. Kosik, K. S., Orrechio, L.D., Binder, L., Trojanowski, J.Q., Lee, V.M.-Y. and Lee, G. (1988) Neuron 1, 817-825.
25. Goedert, M., Spillantini, M.G., Potier, M.C., Ulrich, J. and Crowther, R.A. (1989) EMBO J. 8, 393-399.
26. Kosik, K. S., Orecchio, L.D., Bakalis, S. and Neve, R.L. (1989) Neuron 2, 1389-1397.

■ *David N. Drechsel and Marc Kirschner:*
Department of Biochemistry and Biophysics,
UCSF
San Francisco, CA, USA

α/β-Tubulin

Tubulin is the subunit protein of microtubules, one of the major components of the eukaryotic cytoskeleton. Microtubules are hollow cylinders formed by the self assembly of tubulin. They bind a variety of microtubule associated proteins (MAPs) and serve as tracks for microtubule dependent motor proteins. The main functions of microtubules include chromosome segregation in mitosis, intracellular transport, ciliary and flagellar bending, or providing static support[1,2].

Tubulin has a molecular mass of ~50 kDa and contains ~450 amino acid residues with some variation depending on the isoform. It is a highly conserved protein. Tubulin isotypes fall into two main classes, α- and β-tubulin, each of which can be further subclassified[3]. A third class (γ-**tubulin**) has recently been reported[4]. The α/β-heterodimer is the effective microtubule forming subunit. Each heterodimer binds two molecules of GTP. One of them is bound nonexchangeably to α-tubulin (GTP_n), the other binds exchangeably to β-tubulin (GTP_e) and is hydrolyzed during microtubule assembly. Tubulin consists of (at least) two domains, separated by a protease sensitive region around residue 300. The N-terminal domain binds GTP, it shows a limited sequence homology with other G-proteins; the C-terminal domain binds MAPs, and both domains are involved in self-assembly[5].

Tubulin can be posttranslationally modified in several ways[6,7]: phosphorylation of Ser-444 of β-tubulin[8], acetylation of Lys-40 of α-tubulin[9], reversible removal of the C-terminal tyrosine of α-tubulin (involving a **carboxypeptidase** and a tubulin-specific **tyrosine ligase**[10]), and glutamylation of Glu-445 of α-tubulin[11]. The functions of these modifications are still a matter of debate. "Old" microtubules, i.e. those that turn over slowly, tend to be acetylated (α-tubulin) and detyrosinated (α-tubulin), but this may be explained simply because the enzymes have different activities with microtubules and depolymerized tubulin[12].

The synthesis of tubulin is autoregulated by the pool of unpolymerized tubulin. When this pool is lowered by microtubule assembly, tubulin synthesis is increased, and vice versa. The regulation is achieved during translation by an interaction of the first four residues of nascent β-tubulin chain (MREI) with tubulin which leads to the degradation of the polyribosome bound mRNA[13,14].

Assembly of microtubules (Figure 1) *in vivo* is nucleated mainly on microtubule organizing centres (MTOC), e.g. the region around centrosomes or basal bodies[15]; *in vitro* microtubules can grow either from preformed "seeds" (e.g. from isolated axonemes (MTOC)), or nucleate and grow spontaneously (homogeneous nucleation). Tubulin assembly is promoted by MAPs, GTP, Mg^{2+}, elevated temperature (37°C) and the drug taxol. Disassembly is promoted by Ca^{2+}, GDP, low temperature (4°C), or drugs such as colchicine, vinblastine, griseofulvin and nocodazole. Because tubulins from different sources have different sensitivities to certain drugs, microtubule poisons can be used in chemotherapy of cancer, as herbicides or fungicides[16].

The self-assembly of tubulin leads to microtubules and several related polymorphic forms[17]. Their common feature is that they are built from protofilaments, strings of alternating α- and β-tubulin, spaced 4 nm apart and pointing in the same direction. A microtubule of about 25 nm diameter contains 13 protofilaments juxtaposed with a spacing of about 5 nm and staggered by about 0.9 nm (Figure 2-4). The structure of the tubulin core of microtubules is similar in most cells, while the nature and the composition of MAPs attached to the microtubule surface is variable, depending on cell type and function. However, microtubules with other protofilament numbers also exist (12-16), as well as polymorphic forms such as incomplete microtubule walls ("sheets" or "C-tubules"), protofilament spirals (e.g. "vinblastine spirals"), protofilaments coiled into 36 nm rings (e.g. at low temperature), and others. Microtubules can associate into doublets (as in cilia and flagella) and triplets (as in basal bodies and centro-

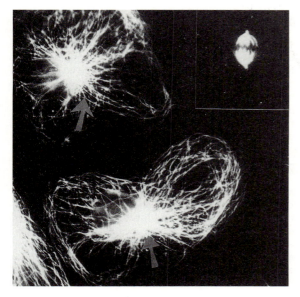

Figure 1. Microtubule network of a PtK2 cell, shown by indirect immunofluorescence during interphase and mitosis (insert upper right). Magnification X700. (From Füchtbauer et al. EMBO J. 4, 2807 (1985))

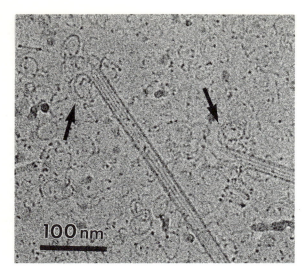

Figure 2. Cryoelectron micrograph of unstained microtubules embedded in vitrified ice, obtained during the disassembly phase. Note the curling and breaking of protofilaments into oligomers at the microtubule ends. (From Mandelkow et al. J. Cell Biol. 114 (1991))

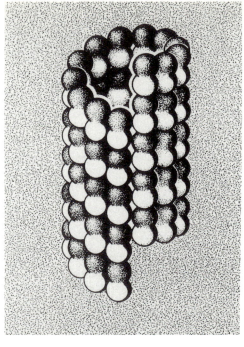

Figure 3. Surface lattice of microtubules, showing α-tubulin (open spheres) and β-tubulin (stippled). (From Mandelkow et al. J. Cell Biol. 102, 1067 (1986))

somes). Microtubules with added incomplete microtubule walls ("hooks" in cross-section) can be used as indicators of polarity[18].

Microtubules show the unique properties of dynamic instability, autonomous oscillations and spatial pattern formation: even after assembly to an overall steady state, individual microtubules continue to switch between phases of disassembly and reassembly[19]. When the phases are synchronized, the solution oscillates and is capable of forming various dissipative structures[20]. The energy of the dynamic behaviour is derived from the GTP hydrolysis during assembly[21]. Assembly takes place mainly by endwise addition of tubulin heterodimers, disassembly by oligomers which subsequently disintegrate into dimers (Figure 2). Nucleotide exchange on β-tubulin is possible only on dimers, but not on microtubule or oligomers.

■ PURIFICATION

One distinguishes between "tubulin" in the strict sense (i.e. α/β-heterodimers) and "microtubule protein" (i.e. mixture of tubulin and MAPs). Microtubule protein can be prepared from cell extracts (usually brain tissue) by repeated cycles of temperature and GTP dependent assembly and disassembly and intervening centrifugation steps. Tubulin can be purified from microtubule protein by ion exchange chromatography (e.g. phosphocellulose, DEAE-chromatography). Commonly used buffers are PIPES, MES, glutamate, or phosphate. In addition one requires Mg^{2+}, GTP (cofactors for assembly), and EGTA (to remove Ca^{2+})[22].

■ ACTIVITIES

The functional state of tubulin can be assessed by drug binding (e.g. colchicine which binds to nonpolymerized tubulin), or by measuring microtubule assembly. Assembly can be monitored by light scattering or turbidity, electron microscopy, viscosimetry, X-ray scattering, or video-enhanced microscopy; the latter allows direct visualization of individual microtubules and their dynamics in vitro[23] and living cells[24].

■ ANTIBODIES

Many groups have produced antibodies against tubulin, e.g. polyclonal, monoclonal, or peptide antibodies. The C-terminal regions of α- and β-tubulin tend to have the highest antigenicity. Some antibodies have become widely used: YOL1-34 and YL1/2[25], and 1A2[26] (monoclonal, against the C-terminal region of α-tubulin; YL1/2 and 1A2 against tyrosylated α-tubulin; available from Serolab and Sigma, respectively); peptide antibodies distinguishing between tyrosylated and detyrosylated α-tubulin[26,27]; the monoclonal antibodies DM1A and DM1B distinguishing between α- and β-tubulin[28] (both binding sites near C-terminus; available from Amersham and Sigma), and a monoclonal antibody against acetylated α-tubulin[9]; these

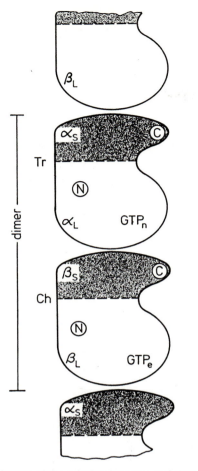

Figure 4. Model of the tubulin domains and their interactions along a protofilament. Both α- and β-tubulin consist of a large N-terminal and a smaller C-terminal domain (stippled). They can be separated by limited proteolysis with trypsin (Tr, for α-tubulin) or chymotrypsin (Ch, for β-tubulin). The N-terminal domains bind GTP and have limited sequence homology with other GTP binding proteins (Linse & Mandelkow, J. Biol. Chem. 263, 15205, 1988). The intra-dimer bond is formed by the large N-terminal domain of α-tubulin (with bound GTP_n) and the small C-terminal domain of β-tubulin. Conversely, the inter-dimer bond leading to protofilament elongation is formed by the small C-terminal domain of α-tubulin and the N-terminal domain of β-tubulin; the latter carries the exchangeable GTP_e which is hydrolysed upon assembly. From Kirchner and Mandelkow, EMBO J. 9, 2397 (1985).

antibodies have permitted the mapping of the distribution not only of microtubules, but also of their posttranslational modifications. A series of isotype-specific antibodies was used by Cowan, Cleveland and coworkers to show that cellular microtubules are usually copolymers of different isotypes[29,30]. Anti-idiotypic antibodies against

tubulin peptides have revealed their interaction sites with MAPs[31].

■ GENES

The isotypes of α- and β-tubulin are usually generated by different genes. Many genes have now been characterized from a variety of sources. The SWISSPROT databank contains more than 80 tubulin sequences, identified by the file initials TBA (for α-tubulins) or TBB (for β). Vertebrates contain about six gene products for α- and β-tubulin, several of which can occur in a single cell type[32,33]. The multiplicity suggests either different functions for different tubulins (multitubulin hypothesis), developmental regulation, or simply redundancy. There is some evidence for developmental regulation of isotype expression, but not for functional specialization[3].

■ REFERENCES

1. Kirschner, M.W. and Weber, K., eds. (1989) Curr. Opin. Cell Biol. 1, 1-160.
2. Kirschner, M.W. and Weber, K., eds. (1990) Curr. Opin. Cell Biol. 2, 1-168.
3. Sullivan, K.F. (1988) Ann. Rev. Cell Biol. 4, 687-716.
4. Oakley, C.E. and Oakley, B.R. (1989) Nature 338, 662-664.
5. Mandelkow, E. and Mandelkow, E.-M. (1989) Cell Movement 2, 23-45 (F.D. Warner and J.R. McIntosh, eds.) A.R. Liss, New York.
6. Greer, K. and Rosenbaum, J.L. (1989) Cell Movement 2, 47-66. (F.D. Warner and J.R. McIntosh, eds). A.R. Liss, New York.
7. Joshi, H.C. and Cleveland, D.W. (1990) Cell Motil. Cytoskel. 16, 159-163.
8. Luduena, R.F., Zimmermann, H.P. and Little, M. (1988) FEBS Lett. 230, 142-146.
9. LeDizet, M. and Pierno, G. (1987) Proc. Natl. Acad. Sci. (USA) 84, 5720-5724.
10. Wehland, J. and Weber, K. (1987) J. Cell Biol. 104, 1059-1067.
11. Eddé, B., Rossier, J., Le Caer, J.-P. Desbruyeres, E., Gros, F. and Denoulet, P. (1990) Science (Wash.) 247, 83-85.
12. Webster, D., Gundersen, G., Bulinski, J. and Borisy, G. (1987) Proc. Natl. Acad. Sci. (USA) 84, 9040-9044.
13. Cleveland, D.W. (1989) Int. Rev. Cytol. 115, 139-170.
14. Cleveland, D.W. (1989) Cell Motil. Cytoskel. 14, 147-155.
15. Brinkley, B.R. (1985) Ann. Rev. Cell Biol. 1, 145-172.
16. Dustin, P. (1984) Springer Verlag, Heidelberg.
17. Amos, L.A. (1982) Electron Microscopy of Proteins 3, 207-250. Harris, J.R. ed. Academic Press, London.
18. Heidemann, S.R. and McIntosh, J.R. (1980) Nature 286, 517-519.
19. Mitchison, T.J. (1988) Ann. Rev. Cell Biol. 4, 527-549.
20. Mandelkow, E., Mandelkow, E.-M., Hotani, H., Hess, B. and Müller, S.C. (1989) Science (Wash.) 246, 1291-1293.
21. Carlier, M.F. (1989) Int. Rev. Cytol. 115, 139-170.
22. Vallee, R.B. (1986) Meth. Enzymol. 134, 1-748.
23. Walker, R., O'Brien, E., Pryer, N., Soboeiro, M., Voter, W., Erickson, H. and Salmon, E. (1988) J. Cell Biol. 107, 1437-1448.
24. Gorbsky, G.J. and Borisy, G.G. (1989) Methods Cell Biol. 29, 175-193.
25. Kilmartin, J.V., Wright, B. and Milstein, C. (1982) J. Cell Biol. 93, 576-582.
26. Kreis, T.E. (1986) EMBO J. 5, 931-941.
27. Bulinski, J.C. (1986) Int. Rev. Cytol. 103, 281-302.

28. Blose, S.H., Meltzer, D.I. and Feramisco, J.R. (1984) J. Cell Biol. 98, 847-858.
29. Lopata, M.A. and Cleveland, D.W. (1987) J. Cell Biol. 105, 1707-1720.
30. Gu, W., Lewis, S. and Cowan, N. (1988) J. Cell Biol. 106, 2011-2022.
31. Rivas, C.I., Vera, J.C. and Maccioni, R. (1988) Proc. Natl. Acad. Sci. (USA) 85, 6092-6096.
32. Villasante, A., Wang, D., Dobner, P., Dolph, P., Lewis, S. and Cowan, N.J. (1986) Molec. Cell. Biol. 6, 2409-2419.
33. Monteiro, M. and Cleveland, D. (1988) J. Mol. Biol. 199, 439-446.

■ Eva-Maria Mandelkow and Eckhard Mandelkow:
Max-Planck-Unit for Structural Molecular Biology,
Hamburg,
Germany

γ-Tubulin

γ-tubulin is related to α- and β-tubulin and is, thus, a member of the tubulin superfamily of proteins. It is associated with microtubule organizing centres (MTOCs) and is apparently required for the assembly of MTOCs in vivo.

γ-tubulin was identified by molecular genetic means in the filamentous fungus *Aspergillus nidulans*[1]. γ-tubulin genes or cDNAs have subsequently been cloned from several organisms including *Drosophila melanogaster* and *Homo sapiens*[2], *Schizosaccharomyces pombe*[3,4] and *Xenopus laevis*[4]. γ-tubulin has not yet been purified in a functional form. The amino acid sequences predicted by γ-tubulin genes and cDNAs give some information, however, on the physical characteristics of the protein. The predicted mass of γ-tubulins identified to date are between 50-53 kDa and their predicted pI's are between 5.5-5.9. In *Aspergillus nidulans* the mass of the protein as determined by immunoblotting of SDS-polyacrylamide gels is similar to the predicted mass of the protein[5]. The γ-tubulins identified to date share 66% or greater amino acid identity and γ-tubulin, thus, is highly conserved. γ-tubulins share approximately 28-32% identity with α-**tubulins** from various organisms and 32-36% identity with β-tubulins[1] (B.R. Oakley, unpublished). Some regions (including regions thought to be involved in GTP binding) are highly conserved among α-, β- and γ-tubulins and the similarities among these proteins leave no doubt that they are all members of the same superfamily of proteins[1].

γ-tubulin is located at the spindle pole bodies of *Aspergillus nidulans*[5] and *Schizosaccharomyces pombe*[3,6] and at the centrosomes of human[2,7], *Drosophila*[7] and mouse[7], and *Xenopus*[4] cells. γ-tubulin may, thus, be a universal component of microtubule organizing centres. In *A. nidulans*, disruption of the γ-tubulin gene causes a strong inhibition of nuclear division, a weaker inhibition of nuclear migration and the elimination of all mitotic and most cytoplasmic microtubules[5]. Earlier genetic analyses[8] showed that the product of the *mipA* gene of *A. nidulans*, which we now know to be γ-tubulin[1], interacts specifically, probably physically, with β-tubulin. These data have led Oakley et al.[5] to hypothesize that γ-tubulin nucleates microtubule assembly from spindle pole bodies and establishes the polarity of microtubules that assemble from spindle pole bodies.

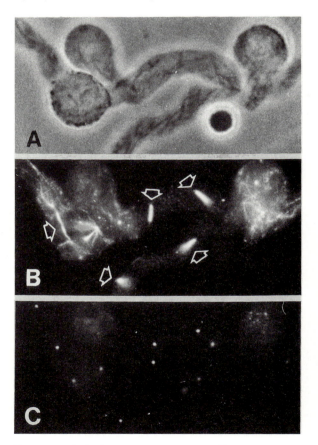

Figure. γ-tubulin localization at the spindle pole bodies of *Aspergillus nidulans*. (A) Phase-contrast of an *A. nidulans* hypha containing mitotic nuclei. (B) A fluorescence micrograph of the same field. Mitotic spindles (arrows) are stained with an antibody against α-tubulin. (C) A fluorescence micrograph of the same field showing anti-γ-tubulin staining. γ-tubulin is located at the spindle-pole bodies at the ends of the mitotic spindles. Bar 10 μm. (Reproduced with permission from reference 5).

PURIFICATION

γ-tubulin has not yet been purified in a functional form.

ACTIVITIES

Little is known as yet. Sequence similarities suggest that γ-tubulin binds GTP[1]. Gene disruption experiments suggest it is required for microtubule assembly *in vivo*[5].

ANTIBODIES

Polyclonal antisera against an *A. nidulans* γ-tubulin fusion protein have been produced[5]. The antisera are crossreactive with γ-tubulins of human[2], mouse[7] and *Drosophila* cells[7] but require affinity purification before they are useful. Polyclonal antisera against fusion proteins of portions of γ-tubulin genes of *S. pombe* and *X. laevis* and against a peptide conserved among γ-tubulins. These antisera have been affinity purified before use. The purified antibodies are crossreactive to γ-tubulins of a variety of animal cells.

GENES

γ-tubulin genes and/or cDNAs have been cloned and sequenced from *A. nidulans*[1], *D. melanogaster*[2] (GenBank M61765), *Homo sapiens*[2] (GenBank M61764), *Schizosaccharomyces pombe*[3,4] and *Xenopus laevis*[4].

REFERENCES

1. Oakley, C.E. and Oakley, B.R. (1989) Nature 338, 662-664.
2. Tanaka, H., Zheng, Y., Oakley, C.E., Jung, M.K. and Oakley, B.R. (1990) J. Cell Biol. 111, 412a.
3. Horio, T., Uzawa, S., Jung, M.K., Oakley, B.R., Tanaka, K. and Yanagida, M. (1991) J. Cell Sci. 99, 693-700.
4. Stearns, T., Evans, L. and Kirschner, M. (1991) Cell 65, 825-836.
5. Oakley, B.R., Oakley, C.E., Yoon, Y. and Jung, M.K. (1990) Cell 61, 1289-1301.
6. Masuda, H., Sevik, M. and Cande, W.Z. (1990) J. Cell Biol. 111, 182a.
7. Zheng, Y., Jung, M.K. and Oakley, B.R. (1991) Cell 65, 817-823.
8. Weil, C.F., Oakley, C.E. and Oakley, B.R. (1986) Mol. Cell. Biol. 6, 2963-2968.

■ *Berl R. Oakley:*
Department of Molecular Genetics,
Ohio State University,
Columbus, Ohio, USA

Tubulin Tyrosine Ligase (TTL) and Tubulin Carboxypeptidase (TCP)

The reversible detyrosination/tyrosination of α-tubulin is the best characterized post-translational modification of tubulin[1,2]. Although encoded by the mRNA, the C-terminal tyrosine of α-tubulin shows a high turnover in various cells and tissues. While the removal of the C-terminal tyrosine by the tubulin carboxypeptidase (TCP) is not well understood, the addition of the tyrosine by the tubulin tyrosine ligase (TTL) is well characterized. The cyclic detyrosination/tyrosination of α-tubulin may serve to differentiate subclasses of microtubules for specific activities.

TTL activity has been found in extracts of various vertebrate tissues and cells, in several invertebrates and in protozoa[1]. TTL is most commonly isolated from mammalian brain, it is a globular monomer containing a single polypeptide chain of 43 kDa. Gel filtration experiments have shown it forms a one to one complex with the α/β-**tubulin** dimer[3-5]. In addition to its catalytic interaction site with the C-terminus of α-tubulin, TTL has a second binding site on β-tubulin, which explains the high specificity of the enzyme[6]. α/β-tubulin is the only known substrate for TTL, which does not act *in vitro* on peptides spanning the C-terminal residues of detyrosinated α-tubulin or on denatured α/β-tubulin. Thus, binding of TTL to β-tubulin should be the prerequisite to enzymatic action at the C-terminus of α-tubulin within the same tubulin dimer. In mammalian tissues, TTL activity is highest in nervous and muscle tissues, especially during early developmental stages.

TCP is readily found in brain extracts and is distinct from pancreatic carboxypeptidase A which also can utilize α-tubulin as a substrate[7,8]. The molecular weight of TCP, 45 kDa, is much lower than the 90 kDa originally suggested[9]. TTL and TCP, reveal different specificities in utilizing microtubules and soluble tubulin as substrates. In contrast to TCP, which acts on microtubules, TTL prefers the tubulin dimer as a substrate. The availability of antibodies specific for either tyrosinated or detyrosinated α-tubulin provides a powerful method to determine the extent of α-tubulin tyrosination in cells and tissues. The rat monoclonal antibody YL1/2 raised against yeast tubulin[10] reacts specifically with tyrosinated α-tubulin[11]. Several polyclonal and monoclonal antibodies specific for tyrosinated and detyrosinated α-tubulin have been elicited with synthetic peptides spanning the C-terminal sequences of the two tubulin forms[12-14].

Although detyrosination is correlated with microtubule

stability[1,2], it is now clear that the function of tubulin detyrosination is not to stabilize microtubules directly[15]. Instead it may serve to further differentiate stable microtubules for specific activities such as the interaction with specific microtubule associated proteins, cell organelles or other structures.

■ PURIFICATION

TTL is routinely purified from brain extracts by conventional DEAE-cellulose chromatography followed by ATP and/or tubulin affinity chromatographical procedures, that yield an essential homogeneous enzyme[3-5,16]. Alternatively TTL can easily be purified in an active form by immunoaffinity chromatography[5,16]. TCP can be partially purified from brain extracts by ammonium sulphate precipitation and chromatography on CM-Sephadex.

■ ACTIVITIES

TTL catalyzes the addition of tyrosine to the C-terminal glutamic acid of detyrosinated α-tubulin. The enzyme is highly specific for tubulin and requires Mg^{2+}, K^+ and ATP. Even though tyrosine is the preferred substrate, TTL also catalyzes the addition of phenylalanine and 3,4-dihydroxyphenylalanine (DOPA) to detyrosinated tubulin in vitro [1]. TTL activity is assayed by measuring the incorporation of [^{3}H]-tyrosine into microtubule protein or purified tubulin.

TCP catalyzes the release of the C-terminal tyrosine from tyrosinated α-tubulin[7,8]. TCP activity is assayed by measuring the release of [^{3}H]-tyrosine from tyrosinated tubulin which has been recharged with [^{3}H]-tyrosine using TTL.

■ ANTIBODIES

Rabbit polyclonal sera against TTL from porcine brain have been described[6] as well as two monoclonal antibodies of which one inhibits TTL activity[5,6]. Antibodies against TCP have not been described.

■ GENES

No sequences for TTL or TCP are available.

■ REFERENCES

1. Barra, H.S., Arce, C.A. and Argarana, C.E. (1988) Mol. Neurobiol. 2, 133-153.
2. Greer, K. and Rosenbaum, J.L. (1989) in Cell Movement 2, (Warner, F.D. ed.) Alan R. Liss, Inc, pp. 47-66.
3. Raybin, D. and Flavin, M. (1977) Biochemistry 16, 2189-2194.
4. Murofushi, H. (1980) J. Biochemistry 87, 979-984.
5. Schroeder, H.C., Wehland, J. and Weber, K. (1985) J. Cell Biol. 100, 276-281.
6. Wehland, J. and Weber, K. (1987) J. Cell Biol. 104, 1059-1067.
7. Argarana, C.E., Barra, H.S. and Caputto, R. (1980) J. Neurochem. 34, 114-138.
8. Kumar, N. and Flavin, M. (1981) J. Biol. Chem. 256, 7678-7686.
9. Webster, D.R. and Bulinski, J.C. (1990) J. Cell Biol. 111, 174a.
10. Kilmartin, J.V., Wright, B. and Milstein, C. (1982) J. Cell Biol. 93, 576-582.
11. Wehland, J., Willingham, M.C. and Sandoval, I.V. (1983) J. Cell Biol. 97, 1467-1475.
12. Gundersen, G.G., Kalnoski, M.H. and Bulinski, J.C. (1984) Cell 38, 779-789.
13. Kreis, T.E. (1987) EMBO J. 6, 2597-2606.
14. Wehland, J. and Weber, K. (1987) J. Cell Sci. 88, 185-203.
15. Webster, D.R., Wehland, J., Weber, K. and Borisy, G.G. (1990) J. Cell Biol. 111, 113-122.
16. Wehland, J., Schroeder, H.C. and Weber, K. (1986) Methods Enzym. 134, 170-179.

■ *Juergen Wehland:*
National Research Centre for Biotechnology,
D-3300 Braunschweig
Germany

X-MAP (*Xenopus*)

X-MAP (Xenopus microtubule assembly protein)[1] is a high molecular weight microtubule associated protein (MAP), originally identified in Xenopus eggs by its ability to promote microtubule assembly in vitro. Found predominantly in oocytes, eggs and early embryos, X-MAP may regulate microtubule assembly during these important stages of development.

X-MAP is a 215 kDa protein, the purification of which was originally based on its promotion of centrosome nucleated microtubule assembly[1,2]. Gel filtration and ultracentrifugation indicate that purified X-MAP is an elongate monomer, with a sedimentation coefficient of approximately 5 S. The X-MAP polypeptide is very susceptible to proteolysis by Ca^{2+}-activated proteases (**calpains**) in Xenopus eggs, being sequentially cleaved into fragments of 160, 130, and 60 kDa. The generation of similar sized fragments by other proteases, including trypsin, suggests that X-MAP consists of several domains linked by protease sensitive hinges. Neither the X-MAP activity nor X-MAP protein exhibit heat stability.

Functional analysis of microtubule assembly from egg

extracts suggests that X-MAP is the major microtubule assembly promoting activity in eggs. Immunoblot analysis using X-MAP antisera indicates that it accumulates late in oogenesis (Dumont stages IV-VI), remains constant in amount (relative to **tubulin**) through early embryogenesis, and then decreases dramatically at tailbud stages (Gard, unpublished observation). Minor amounts of X-MAP are found in adult brain and testis[2].

X-MAP is a cell cycle dependent phosphoprotein, with peak incorporation of $^{32}PO_4$ during meiotic and mitotic metaphase[2]. Phosphorylation occurs on multiple sites, containing both serine and threonine residues (Muller and Gard, unpublished observations). No phosphotyrosine has yet been detected.

■ PURIFICATION

X-MAP can be purified from *Xenopus* eggs to 90-95% homogeneity through conventional biochemical techniques employing phosphocellulose chromatography, ammonium sulphate fractionation, FPLC anion exchange chromatography, and HPLC size exclusion chromatography[2]. X-MAP containing fractions are identified by a functional assay (microtubule assembly promotion) or by immunoblotting using specific antibodies against X-MAP.

■ ACTIVITIES

X-MAP promotes *in vitro* microtubule assembly from centrosomes or axonemes. Using axoneme nucleated assembly, it has been shown that X-MAP preferentially promotes assembly of the microtubule plus-end[2], though the basis for this specificity is unknown. X-MAP binds to taxol stabilized microtubules *in vitro*.

■ ANTIBODIES

Rabbit polyclonal antibodies to X-MAP show no crossreactivity with tubulin, **MAP1**, **MAP2**, or **Tau** proteins from *Xenopus*, bovine, or mouse brain. The two antisera currently available have been used to identify X-MAP by immunoblotting and immunoprecipitation, but do not recognize the protein in samples prepared for immunofluorescence. Peptide analysis suggests that the antisera generated to date recognize a limited number of spatially restricted epitopes on the polypeptide.

■ GENES

cDNA clones for X-MAP have not been obtained.

■ REFERENCES

1. Gard, D.L. and Kirschner, M.W. (1987) J. Cell Biol. 105, 2203-2215.
2. Gard, D.L. and Kirschner, M.W. (1987) J. Cell Biol. 105, 2191-2201.

■ *David L. Gard:*
Department of Biology, University of Utah,
Salt Lake City, Utah
USA

205K MAP (*Drosophila*)

The 205 kDa microtubule associated protein (205K MAP) is one of the principal MAPs in Drosophila[1]. It is present in several isoforms and it is thermostable.

205K MAP was originally isolated from *Drosophila* as a protein of ~200 kDa apparent molecular mass judged by SDS-PAGE. The predicted molecular weight based on the amino acid sequence of 205K MAP is 130 kDa, substantially smaller than the size obtained by SDS-PAGE.

A plot of the charge distribution of amino acids in this protein reveals a three domain structure. The N-terminal 762 amino acids form a region with a pI of 4.4. The next region (amino acid residues 763-1102) is very basic with a predicted pI of 12.9. The last domain containing the 61 C-terminal amino acids has a predicted pI of 4.4. This organization is very similar to that of **MAP2**. Yet at the sequence level, 205K MAP and MAP2 share no homology. Thus far, only one domain within 205K MAP has been associated with a particular function. Using truncated forms of *in vitro* expressed 205K MAP, the microtubule binding site has been localized to a 232 amino acid region within the basic domain[2]. The 205K MAP microtubule binding site has no sequence similarity to any of the other microtubule binding sites that have been identified thus far (**kinesin**[3]; **tau**[4]; MAP2[5]; mammalian **MAP1B**[6]).

Several potential phosphorylation sites are also located within the basic domain. Currently there is no information about the usage of these phosphorylation sites *in vivo*. A phosphorylated 205K MAP species has been immunoprecipitated from labelled S2 cells (P. Pesavento, I. Irminger-Finger, and L. Goldstein, unpublished).

Deletion and hypomorphic mutants have been made for the 205K MAP gene *in vivo*[7]. Intriguingly, animals completely lacking this gene are viable and have no obvious phenotype. This observation suggests an extant protein(s) that is able to substitute for 205K MAP. Alternatively, the 205K MAP has an inessential function.

PURIFICATION

205K MAP was originally isolated from the *Drosophila* cultured cell line, Schneider S2, using a taxol dependent microtubule polymerization procedure[1]. Associated proteins were released from microtubules with 0.4 M NaCl and heated to 100°C. After this treatment, two groups of soluble proteins (four species at 205 kDa and one at 150 kDa) were detected by SDS-PAGE and Coomassie Blue staining. The 205 kDa species were isolated from the gel and used for antibody production.

ANTIBODIES

Antibodies against the 205K MAP antibodies were raised against SDS-PAGE purified antigen[1]. These include a polyclonal rabbit antiserum and a mouse monoclonal IgM antibody. Both antibodies give intense staining of mitotic spindles and cytoplasmic microtubules in methanol but not formaldehyde fixed S2 cells. The affinity purified rabbit anti-205K MAP antibodies recognize a doublet of 205 kDa as well as a 100 kDa protein by immunoblotting of whole adult fly, adult head, and embryo homogenates[7]. The 100 kDa species is not encoded by the 205K MAP locus since it is present in adult animals that are completely deficient for the 205K MAP gene. These antibodies have been used to isolate the 205K MAP coding sequence from a λgt11 expression library[1] which allowed the production of antisera directed against different regions of the protein. An antiserum raised against a fusion protein containing the C-terminal domain of 205K MAP was found to crossreact with HeLa and mouse **MAP4/UMAP**. This crossreactivity is, however, difficult to explain since human and mouse MAP4/UMAP share no detectable sequence similarity with the 205K MAP[9]

GENES

Embryonic cDNA clones have been isolated and characterized[2]. DNA sequence analysis of one of these clones has revealed that the 205K MAP shares no detectable sequence similarity with any of the MAPs that have been sequenced to date.

The 205K MAP DNA sequence is available from EMBL GenBank DDBJ under accession number X54061.

REFERENCES

1. Goldstein, L.S.B., Laymon, R.A. and McIntosh, J.R. (1986) J. Cell Biol. 107, 2076-2087.
2. Irminger-Finger, I., Laymon, R.A. and Goldstein, L.S.B. (1990) J. Cell Biol. 111, 2563-2572.
3. Yang, J.T., Laymon, R.A. and Goldstein, L.S.B. (1989) Cell 56, 879-889.
4. Himmler, A., Prechsel, D., Kirschner, M.W. and Martin, D.W. (1989) Mol. Cell Biol. 9, 1381-1388.
5. Lewis, S.A., Wang, D. and Cowan, N.J. (1988) Science (Wash. D.C.) 242, 936-939.
6. Noble, M., Lewis, S.A. and Cowan, N.J. (1989) J. Cell Biol. 109, 3367-3376.
7. Pereira, A.J., Doshen, J., Tanaka, E. and Goldstein, L.S.B. (1992) J. Cell Biol. 116, 377-383.
8. West, R.R., Tenbarge, K.M., Gorman, M., Goldstein, L.S.B. and Olmsted, J.B. (1988) J. Cell Biol. 107, 460a.
9. West, R.R., Tenbarge, K.M. and Olmsted, J.B. (1991) J. Biol. Chem. 266, 21886-21896.

■ *Andrea Pereira and Lawrence S.B. Goldstein:*
Department of Cellular and Developmental Biology,
Harvard University,
Cambridge, MA, USA

3

The Intermediate Filaments

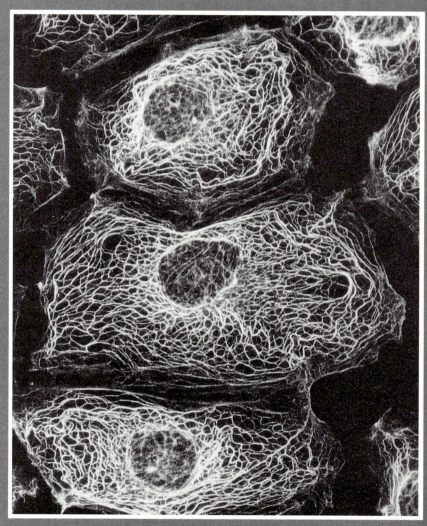

Cytokeratin filaments in PtK$_2$ cells. Immunofluorescence labeling of PtK$_2$ cells with a polyclonal antibody against cytokeratin.

(Courtesy of Dr Werner W. Franke, DKFZ, Heidelberg). For further details see 'Cytokeratins' by W.W. Franke and J. Kartenbeck.

The Intermediate Filaments and Associated Proteins

The morphological appearance of the eukaryotic cell (its shape and its internal architecture) is a major criterion in distinguishing the numerous diverse cells and tissues. Important architectural elements in cell type-specific organization are the filaments of the "cytoskeleton". This collective term includes all structures resistant to cell lysis and extraction with nondenaturing detergents, which as "cell residues" often still maintain typical features of the specific cell-architecture. Amongst the major categories of cytoskeletal filaments the intermediate-sized filaments ("intermediate filaments"; IF[1]) of 7-14 nm diameter are distinguished from the ~5 nm actin containing microfilaments on the one hand, and the thick myosin filaments or the 20-25 nm microtubules on the other hand. IF are also practically "insoluble" in high salt buffers, and refold spontaneously and rapidly in vitro and in near-physiological buffers from the denatured state to α-helical coiled-coils. These coiled-coils then self-assemble to 2-3 nm double coiled-coils ("tetramers"), to long protofilaments of 2-3 nm diameter and finally into IF that are essentially indistinguishable from those formed in the living cell[2-8]. The typical appearance of IF, which tend to fasciate into bundles and are often attached to the plaques of desmosomes, is shown in Figure 1, in comparison with other filamentous structures of the cytoplasm and the extracellular matrix.

The IF structures occurring in different kinds of cells are not formed by the same protein(s) but by different members of a large multigene family, the "IF proteins", whose members are differentially expressed in patterns specific for a given cell type or pathway of differentiation[5,9-14]. Hence, the specific cellular IF protein complement is characteristic of that cell type.

In human cells and tissues, for example, the products of more than 45 different IF protein genes have been identified, including at least three different genes encoding **lamins**, i.e. "karyoskeletal" proteins forming the intranuclear lamina. The members of the IF protein family can be divided into subgroups according to their sequence homologies, immunological crossreactivities, compatibilities of self-assembly, and with respect to their characteristic occurrence in particular cell types (Table 1). The by far largest and most complex class of IF proteins are the **cytokeratins**, the hallmarks of epithelial differentiation, which fall into two subfamilies, the type I and type II cytokeratins. In contrast to the other IF proteins which are able to assemble into homopolymeric IF, the cytokeratins are obligatory heteropolymers in which the IF subunit tetramers are made up of two type I and two type II chains.

■ THE COMMON MOLECULAR STRUCTURE

The various IF proteins differ considerably in molecular size and electric charge, but have a common basic molecular arrangement (Figure 2). The central element is a core segment, the "rod", of 309-331 amino acids, which is characterized by a predominance of amino acids favouring α-helical conformation and a certain arrangement of units ("heptades") of hydrophobic amino acids that is known for its tendency to assume a two-chain coiled-coil α-helical configuration[7,8,15,16]. The α-helical character of the rod may be shortly interrupted at one or two sites and therefore divided into two or three subdomains (coils 1A, 1B and 2 in Figure 2).

The rod domain has been relatively well conserved between the diverse IF proteins during evolution, not only in its size and structural character, but also in terms of amino acid sequence homology. Even distantly related members of the IF protein family share sequence similarity in the rod, most conspicuously at the edges of the individual subdomains (in addition, certain small conserved sequence motifs may also be identified in other regions of some IF proteins). Cytoplasmic IF proteins of certain invertebrates as well as nuclear lamins of both vertebrates and invertebrates contain an additional insert of 42 residues in coil 1B and therefore form a slightly longer rod (Figure 2)[17]. Therefore, these molecules may represent the type of IF protein that arose early in evolution.

Studies of IF assembly *in vitro*, using purified proteins, and *in vivo*, using cells transfected with cloned DNA molecules encoding a specific IF polypeptide, have made clear that an intact rod domain is essential for IF formation[18-23]. Deletions of rod segments or even certain point mutations in the rod result in the appearance of aberrant structures, often aggregate formation, and this even occurs when only a small proportion of the total cellular protein is altered[18,20-26]. From the observation that the introduction of such a rod-defective epidermal cytokeratin gene into transgenic mice resulted in a severe skin disease[27] it was suggested that the inheritable human blistering disease, *Epidermolysis bullosa simplex* (EBS), might reflect a genetic defect in the IF rod domain. Soon thereafter, this hypothesis was confirmed by the demonstration of distinct mutations in the rod domains of cytokeratins 5 or 14 of EBS patients[28-30].

In contrast, the domains flanking the rod, i.e. the N-terminal "head" and the C-terminal "tail", do not exhibit a constitutive α-helical coiled-coil forming character and vary greatly in size and amino acid sequence (Figure 2). For example, the size of the tail domain can vary from practically nonexistent, as in cytokeratin 19 of amphibia and mammals[31-33], to the tail of mammalian **neurofilament triplet protein** NF-H comprising as many as 607 amino acids[34] (Figure 2). It is widely assumed that these

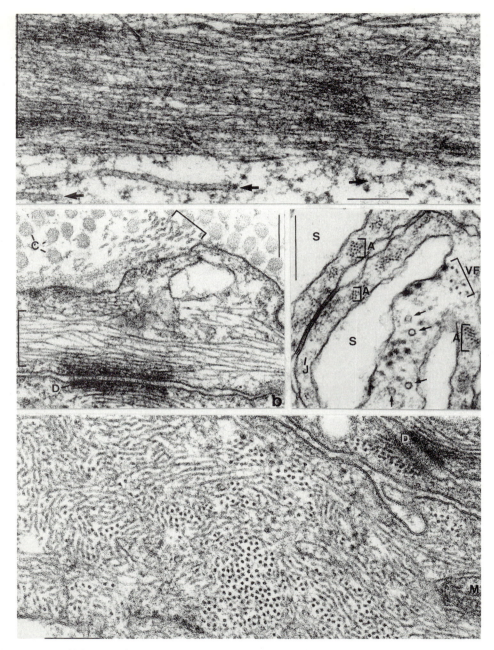

Figure 1. Appearance of IF (here of the vimentin type) as seen in the electron microscope (ultrathin sections; for cells and tissues, fixation and other methods see references 3 and 13): (a) Section through a human meningioma cell in culture, showing a fleece of loosely arranged ~10 nm vimentin IF (top, demarcated by bracket in the upper right corner) is seen next to a bundle ("cable") of ~5 nm actin microfilaments (demarcated by bracket on the left margin) and several microtubules of diameter ~20 nm (denoted by arrowheads). (b) Section through the periphery of two adjacent cells in bovine meningeal tissue, showing vimentin IF in bundles laterally associated with a desmosome (D), to be compared with bundles of elastin-containing "microfibrils" (bracket in the top portion) and collagen fibrils (some are denoted "C") in the extracellular matrix. (c) Peripheral parts of three adjacent Sertoli cells of rat testis, showing cross-sections through vimentin IF arranged in a bundle (VF) or as an individual IF (VF), in comparison with bundles of actin microfilaments (some are denoted "A") and with microtubules (arrowheads). S, spaces of endomembrane cisternae; J, intercellular junction of the gap junction type. (d) Abundance of vimentin IF in the cytoplasm of bovine meningeal cells in the tissue, some showing attachment to a desmosome (D). M, mitochondrion, (Bars 0.2 µm).

THE FAMILY OF INTERMEDIATE FILAMENT (IF) PROTEINS

IF Protein Class	M_r*	Type of Sequence Homology	Structure and Typical Appearance+
1. Cytokeratins			
Type I (acidic) cytokeratins	40 - 64	I	IF; epithelia
Type II (basic) cytokeratins	52 - 68	II	
2. Non-cytokeratins			
A.			
Vimentin	~55	III	IF; mesenchymal cells
Desmin	~53		IF; muscle cells
Glial filament protein	50 - 52		IF; astroglial cells
Peripherin	~54		IF; various neuronal cells
B.			
Neurofilament polypeptides		IV	IF; neurons
NF-L	~68		
NF-M	100 - 110		
NF-H	110 - 130		
α-Internexin	~66		IF; immature neurons
β-Internexin	~70		IF; neurons
C.			
Nestin	~240	VI	IF, CNS stem cells
3. Lamins		V	Nuclear lamina
Type A	62 - 72		many, but not all cells
Type B	65 - 68		all cell types

* Relative molecular mass of the diverse polypeptides in units of 10^3

+ The specific IF type may also occur in certain other cell types, often in coexistence with other IF proteins.

CNS, central nervous system

parts of the molecule are not integral components of the coiled-coil protofilament axis but protrude laterally, and may thus be involved in interactions with other cytoplasmic components, including cell type-specific ones.

Early observations that IF rod portions alone, obtained by proteolytic cleavage, are not capable of assembly into IF[6,35] have been confirmed by experiments using IF proteins truncated by recombinant DNA techniques[20,22,36]. In similar experiments it has also been shown that certain portions of the head domain are important for IF formation *in vitro* and *in vivo*[6,18,21-24,26,27,37]. For example, in scanning analyses an evolutionarily conserved nonapeptide with a central diarginine has been identified in the type III subgroup of IF proteins (Table 1) as a motif important for proper IF assembly[37].

In contrast, the tail domain has been shown to be dispensable for IF assembly under a wide range of conditions. The mere existence of the practically tail-less type I cytokeratin 19 as a major protein of IF of various epithelial cells[31-33] shows that a tail domain is not necessary for incorporation into the IF structure. Moreover, cytokeratin 19 as well as other type I cytokeratins, when tail-deleted, were clearly able to coassemble with appropriate intact type II cytokeratins[18,22,24], and normal-looking IF can also be formed from combinations in which both the type I and the type II cytokeratin partner are totally devoid of a tail domain[36,38]. Similarly, tail-deleted **vimentin** can also assemble into normal-looking IF, at least *in vitro*[39]. Interestingly, however, the normal restriction of IF to the cytoplasm is often relaxed for such tail-deficient IF pro-

teins, and the IF they form can accumulate in the nucleus[36,39]. This indicates that interactions of the tail domain with yet unknown cytoplasmic elements may contribute to IF topogenesis.

■ ALTERNATIVE STRUCTURAL STATES

Although IF proteins are known for their tendency to assume the IF organization they can also occur in other structural forms, i.e. in non-IF states.

Particularly spectacular are the transient changes of the IF cytoskeleton during mitosis in various lines of cultured cells as well as in certain normal or malignant tissues: Large parts or practically all of the IF disappear during mitosis, and the IF protein appears in large, mostly spheroidal aggregates of granulofibrillar texture in which the typical IF structure is no longer resolved. This was first reported for cytokeratins[40-43] but subsequently also noted for IF containing vimentin and/or **desmin**[44-46]. At the end of mitosis, extended bundles of typical IF can reform from such aggregates, often within minutes, indicating a precise and efficient control of this reversible structural transformation.

However, the existence of such spheroidal aggregates containing IF protein material is not limited to mitotic cells. Large cytoplasmic bodies containing IF protein occur in different kinds of interphase cells of lower vertebrates, particularly in certain developmental stages of amphibia ("figures of Eberth") and in fishes (for references see **47**).

Aggregates with a granulofibrillar texture, closely resembling the cytoplasmic spheroid bodies observed in living cells, were also observed after *in vitro* assembly of IF proteins that had been mutated in the head domain[37].

Large, spheroidal aggregates of densely packed, homogeneous cytokeratin material have been noticed in epidermal keratinocytes of EBS patients[27-30,48]. Similarly dense cytoplasmic aggregates have been described for cytoplasmic assemblies of partly deleted desmin in transfected cells[49].

A different form of large aggregates of haphazardly arranged, somewhat thicker, shorter and fuzzy-contoured filaments that contain cytokeratins appears in diseased hepatocytes ("Mallory bodies"), where they are characteristic of advanced stages of certain forms of toxically induced hepatitis, including human alcoholic hepatitis[50,51]. To what extent these diverse structural changes are - or can be - regulated or whether they reflect irreversible pathologically induced alterations remains to be examined.

A form of unusually thick (20-30 nm diameter), smoothly contoured filamentous structure assembled by IF proteins has been observed *in vitro* in the presence of elevated (~5 mM) Ca^{2+} concentrations[52] but its biological significance is not clear.

■ REGULATION OF IF AND IF PROTEIN ASSEMBLY

Structural changes of the IF cytoskeleton such as the various rearrangements observed during mitosis, with intact looking IF[53] or with transient IF disassembly and aggregate

MOLECULAR DOMAINS

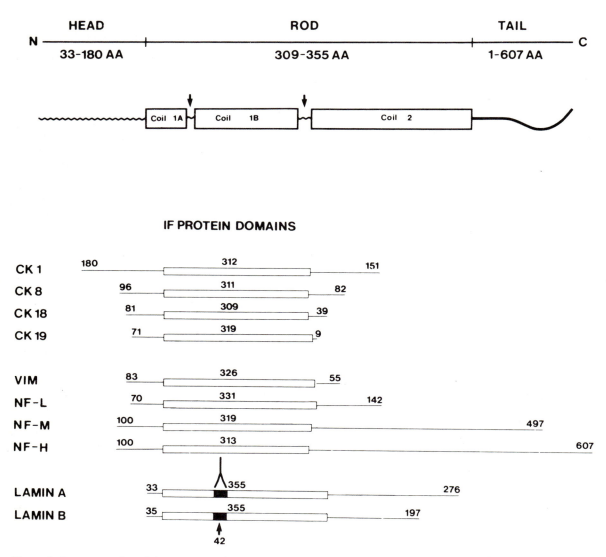

Figure 2. Representation of the major conformational domains in IF proteins in a general scheme (top) and for several kinds of IF proteins (bottom part). N, N-terminus; C, C-terminus; AA, amino acids. The numbers give the amino acid residues in the specific domain. The α-helical rod domain is represented by a rectangle(s), the individual subdomains (coils 1A, 1B and 2) are assumed to be separated by short nonhelical interruptions (denoted by downward arrowheads). The head and tail domains are denoted by thin lines (differently arranged in the top scheme). CK, cytokeratins (human); VIM, vimentin; NF, neurofilament proteins; L, M and H, "light", "middle" and "heavy" NF polypeptides. Note in lamins A and B the 42 AA insertion in the rod.

formation[40-46], have been found to correlate with certain forms of phosphorylation of the IF molecules involved[45,46,54,55]. In addition, phosphorylation of purified IF or IF proteins by certain kinases *in vitro* favours IF protein disassembly, and the crucial phosphorylated hydrox-yamino acids are located in the head domain[45,46,55-59]. In this respect, the regulation of the perimitotic plastic modulation or disassembly of the IF proteins seems to be similar to the regulation of lamin disassembly and reassembly during mitosis[60,61].

In addition to these effects of posttranslational modification on IF structure itself, both the intracellular display of IF and the distribution in general of IF protein may also be regulated by certain such modifications. For example, phosphorylation has been reported to affect the interaction of vimentin IF with other cytoplasmic components such as **plectin**[62], and the adenoviral protein E1B has been described as an example of a protein that induces IF disruption and promotes IF protein disassembly[63].

In all these regulated assembly-disassembly reactions it is not yet clear whether the disassembly proceeds to an equilibrium with the actual basic subunit of exchange. *In vitro* studies have shown that distinct soluble subunits with properties indicative of the tetrameric state ("double coiled-coil") can be induced to assemble rapidly and nearcompletely into IF[52,64-67]. The occurrence of a pool of such distinct subunits in the living cell has been indicated by the observations of monodisperse forms of similarly sized soluble IF protein complexes in "supernatant fractions" obtained after cell lysis, although at very low concentrations (e.g. [68,69]), and by studies of the kinetics of incorporation of solubilized IF proteins into IF following their microinjection into living cells[70,71].

■ SYNTHESIS OF IF PROTEINS AND CELL TYPE DIFFERENTIATION

The general patterns of the differential expression of the genes encoding the various IF proteins during embryogenesis as well as in adult vertebrates, and in certain pathogenic processes, including tumour formation and metastasis, have been established[5,6,8-14] (Table 1). As a result of the relatively high degree of maintenance of the expression patterns specific for certain cell types during malignant transformation, antibodies to the various IF proteins have become valuable probes for immunohistochemical "cell typing" in tumour diagnosis, including cases of metastatic tumours in which the site of origin of the primary tumour is unknown[13,72].

In certain cell types, particularly during embryogenesis and in tumours, additional IF protein genes can be expressed, often spontaneously. These are not part of the "normal" complement of the specific cell type, for example cytokeratins 8 and 18 in certain smooth muscle cells (for review see [73]), or glial filaments in certain epithelial cells of mammary gland[74]. Detailed biochemical analyses of certain cultured cell lines have disclosed that such "dysregulations" of the normally inactivated synthesis of IF proteins can occur at different levels of regulation, transcriptional, translational and at the level of protein interactions, because an individual cytokeratin chain may be unstable unless properly complexed with a complementary type cytokeratin[75,76] (see there for further references).

■ POSSIBLE FUNCTIONS

The function(s) of IF and IF proteins, in general, or in any specific cell, is still unclear; no distinct cellular function(s) can be ascribed to them. Clearly IF and IF proteins are not ubiquitous cellular elements and therefore unlikely to serve "housekeeping" cell functions common to different cells. Moreover, the various proteins of this multigene family are synthesized in different combinations, each being specific for a certain cell differentiation program, and - even more perplexingly - the type of IF protein present in a given kind of cell may differ from one species to another (for review see [73]). Findings that several cultured lines and tumours totally lack IF proteins (e.g. [77-80]) and that immunoprecipitation of IF by antibodies injected into living cells does not result in detectable damage or functional disturbance (e.g. [81-84]) has further increased the problem of assigning specific cell functions to these proteins. Obviously, certain differentiation processes can take place even in the absence of any IF. For example, the targeted inactivation of a specific IF protein gene by homologous recombination techniques has yielded a negative result[85]: Mouse embryo cells in which the gene encoding cytokeratin 8, the only type II cytokeratin expressed, has been disrupted do not contain any detectable IF but are able to differentiate to a well organized polar epithelium.

Experiments in which the IF system has been disturbed by induced expression of "ectopic" ("foreign") IF protein(s) or by deregulation, either by microinjection of protein[70] or of mRNA encoding foreign protein(s)[37,86,87], or by transfecting cultured cells or animals with IF protein genes brought under control of appropriate promoters[18-27,36,37,88-97] have also failed to reveal special IF functions. In most of these cases, the foreign or overexpressed IF protein(s) were well accommodated in the specific host cell or animal. No conspicuous effects on cell appearance or function were noted, except for the example of cataract formation in eye lenses that were loaded with overexpressed vimentin, desmin or neurofilament protein[92,96,97]. Even when ectopically synthesized epidermal cytokeratins were expressed in mammary gland-derived cells in great excess over the endogenous cytokeratins, the resulting permanent cell lines were perfectly viable and did not show major shape changes[91].

In contrast, functional disturbances have been noticed in some experiments in which animals had been transfected with recombinant genes experimentally mutated in the rod domain to interfere with orderly IF formation. The expression of such "interference constructs" in transgenic mice (see above) leads to dramatic changes in the IF system such as aggregate formations that correlate with severe defects of mechanical stability and intercellular coherence[18,20,27,29], similar to *Epidermolysis bullosa simplex* (EBS)[28-30]. However, such consequences have not been observed in other cells. For example, when rod-defective desmin genes were introduced into fibroblast or myoblast cell cultures, IF aggregate formation and extensive disorganization of the desmin/vimentin IF system also occurred but did not noticeably interfere with cell function and sarcomeric differentiation[98].

These first series of experiments aiming at the elucidation of the functions of IF and IF proteins have at least made clear that there is no general answer to this question but that it has to be asked and studied specifically for each cell type and differentiation step.

■ REFERENCES

1. Ishikawa, H., Bischoff,R. and Holtzer, H. (1968) J. Cell Biol. 38, 538-555.
2. Lee, L.D. and Baden, H.P. (1976) Nature 264, 377-379.
3. Steinert, P.M., Idler, W.W. and Zimmermann, S.B. (1976) J. Mol. Biol. 108, 547-567.
4. Renner, W., Franke, W.W., Schmid, E., Geisler, N., Weber, K. and Mandelkow, E. (1981) J. Mol. Biol. 149, 285-306.
5. Franke, W.W., Schmid, E., Schiller, D.L., Winter, S., Jarasch, E.-D., Moll, R., Denk, H., Jackson, B.W. and Illmensee, K. (1982) Cold Spring Harbor Symp. Quant. Biol. 46, 431-453.
6. Traub, P. (1985) Intermediate Filaments. A Review. Springer Verlag, Berlin, pp. 1-266.
7. Aebi, U.M., Häner, M., Troncoso, J., Eichner, R. and Engel, A. (1988) Protoplasma 145, 73-81.
8. Steinert, P.M. and Roop, D.R. (1988) Annu. Rev. Biochem. 57, 593-625.
9. Franke, W.W., Schmid, E., Osborn, M. and Weber, K. (1978) Proc. Natl. Acad. Sci. 75, 5034-5038.
10. Franke, W.W., Weber, K., Osborn, M., Schmid, E. and Freudenstein, C. (1978) Exp. Cell Res. 116, 419-445.
11. Franke, W.W., Appelhans, B., Schmid, E., Freudenstein, C., Osborn, M. and Weber, K. (1979) Differentiation 15, 7-25.
12. Moll, R., Franke, W.W., Schiller, D., Geiger, B. and Krepler, R. (1982) Cell 31, 11-24.
13. Osborn, M. and Weber, K. (1983) Lab. Invest. 48, 372-394.
14. Fuchs, E., Tyner, A.L., Giudice, G.J., Marchuk, D., RayChaudhury, A. and Rosenberg, M. (1987) Curr. Top. Devel. Biol. 22, 5-34.
15. Crewther, W.G., Dowling, L.M., Steinert, P.M. and Parry, D.A.D. (1983) Int. J. Biol. Macromol. 5, 267-274.
16. Conway, J.F. and Parry, D.A.D. (1990) Int. J. Biol. Macromol. 12, 328-334.
17. Dodemont, H., Riemer, D. and Weber, K. (1990) EMBO J. 9, 4083-4094.
18. Albers, K. and Fuchs, E. (1987) J. Cell Biol. 105, 791-806.
19. Van den Heuvel, R.M.M., van Eys, G.J.J.M., Ramaekers, F.C.S., Quax, W.J., Vree Egberts, W.T.M., Schaart, G., Cuypers, H.T.M. and Bloemendal, H. (1987) J. Cell Sci. 88, 475-482.
20. Coulombe, P.A., Chan, Y.-M., Albers, K. and Fuchs, E. (1990) J. Cell Biol. 111, 3049-3064.
21. Gill, S.R., Wong, P.C., Monteiro, M.J. and Cleveland, D.W. (1990) J. Cell Biol. 111, 2005-2019.
22. Lu, X. and Lane, E.B. (1990) Cell 62, 681-696.
23. Raats, J.M.H., Pieper, F.R., Vree Egberts, W.T.M., Verrijp, K.N., Ramaekers, F.C.S. and Bloemendal, H. (1990) J. Cell Biol. 111, 1971-1985.
24. Albers, K. and Fuchs, E. (1989) J. Cell Biol. 108, 1477-1493.
25. Trevor, K.T. (1990) New Biol. 2, 1004-1014.
26. Wong, P.C. and Cleveland, D.W. (1990) J. Cell Biol. 111, 1987-2004.
27. Vassar, R., Coulombe, P.A., Degenstein, L., Albers, K. and Fuchs, E. (1991) Cell 64, 365-380.
28. Bonifas, J.M., Rothman, A.L. and Epstein, E.H. (1991) Science 254, 1202-1205.
29. Coulombe, P.A., Hutton, M.E., Letai, A., Hebert, A., Paller, A.S. and Fuchs, E. (1991) Cell 66, 1301-1311.
30. Ishida-Yamamoto, A., McGrath, J.A., Chapman, S.J., Leigh, I.M., Lane, E.B. and Eady, R.J.A. (1991) J. Invest. Dermatol. 97, 959-968.
31. Bader, B.L., Magin, T.M., Hatzfeld, M. and Franke, W.W. (1986) EMBO J. 5, 1865-1875.
32. Bader, B.L., Jahn, L. and Franke, W.W. (1988) Eur. J. Cell Biol. 47, 300-319.
33. Stasiak, P.C., Purkis, P.E., Leigh, I.M. and Lane, E.B. (1989) J. Invest. Dermatol. 92, 707-716.
34. Lees, J.F., Shneidman, P.S., Skuntz, S.F., Carden, M.J. and Lazzarini, R.A. (1988) EMBO J. 7, 1947-1955.
35. Geisler, N., Kaufmann, E. and Weber, K. (1982) Cell 30, 277-286.
36. Bader, B.L., Magin, T.M., Freudenmann, M., Stumpp, S. and Franke, W.W. (1991) J. Cell Biol. 115, 1293-1307.
37. Herrmann, H., Hofmann, I. and Franke, W.W. (1992) J. Mol. Biol. 223, 637-650.
38. Hatzfeld, M. and Weber, K. (1990) J. Cell Sci. 97, 317-324.
39. Eckelt, A., Herrmann, H. and Franke,W.W. (1992) Eur. J. Cell Biol. in press.
40. Franke, W.W., Schmid, E., Grund, C. and Geiger, B. (1982) Cell 30, 103-113.
41. Lane, E.B., Goodman, S.L. and Tejdosiewicz, L.K. (1982) EMBO J. 11, 1365-1372.
42. Brown, D.T., Anderton, B.H. and Wylie, C.C. (1983) Int. J. Cancer 32, 163-169.
43. Geiger, B., Kreis, T.E., Gigi, O., Schmid, E., Mittnacht, S., Jorcano, J.L., von Bassewitz, D.B. and Franke, W.W. (1984) In: Cancer Cells 1. The Transformed Phenotype. Cold Spring Harbor Laboratory, Cold Spring Harbor. pp. 201-215.
44. Franke, W.W., Grund, C., Kuhn, C., Lehto, V.-P. and Virtanen, I. (1984) Exp. Cell Res. 154, 567-580.
45. Chou, Y.-H., Rosevear, E. and Goldman, R.D. (1989) Proc. Natl. Acad. Sci. (USA) 86, 1885-1889.
46. Chou, Y.-H., Bischoff, J.F., Beach, D. and Goldman, R.D. (1990) Cell 62, 1063-1071.
47. Jahn, L., Fouquet, B., Rohe, K. and Franke, W.W. (1987) Differentiation 36, 234-254.
48. Anton-Lamprecht, I. (1983) J. Invest. Dermatol. (Suppl.) 81, 149s-156s.
49. Raats, J.M.H., Henderik, J.B.J., Verdihk, M., van Oort, F.L.G., Gerards, W.L.H., Ramaekers, F.C.S. and Bloemendal, H. (1991) Eur. J. Cell Biol. 56, 84-103.
50. Denk, H., Franke, W.W., Eckerstorfer, R., Schmid, E. and Kerjaschki, D. (1979) Proc. Natl. Acad. Sci. (USA) 76, 4112-4116.
51. Denk, H., Lackinger, E. and Vennigerholz, F. (1986) Prog. Liver Dis. 8, 237-251.
52. Hofmann, I., Herrmann, H. and Franke, W.W. (1991) Eur. J. Cell Biol. 115, 1293-1307.
53. Aubin, J.E., Osborn, M., Franke, W.W. and Weber, K. (1980) Exp. Cell Res. 129, 149-165.
54. Evans, R.M. and Fink, L.M. (1982) Cell 39, 43-52.
55. Evans, R.M. (1988) Eur. J. Cell Biol. 45, 152-160.
56. Inagaki, M., Nishi, Y., Nishizawa, K., Matsuyama, M. and Sato, C. (1987) Nature 328, 649-652.
57. Inagaki, M., Gonda, Y., Matsuyama, M., Nishizawa, K., Nishi, Y. and Sato, C. (1988) J. Biol. Chem. 263, 5970-5978.
58. Inagaki, M., Gonda, Y., Nishizawa, K., Kitamura, S., Sato, C., Ando, S., Tanabe, K., Kikuchi, K., Tsuiki, S. and Nishi, Y. (1990) J. Biol. Chem. 265, 4722-4729.
59. Geisler, N. and Weber, K. (1988) EMBO J. 7, 15-20.
60. Heald, R. and McKeon, F. (1990) Cell 61, 579-589.
61. Heitlinger, E., Peter, M., Häner, M., Lustig, A., Aebi, U. and Nigg, E.A. (1991) J. Cell Biol. 113, 485-495.
62. Foisner, R., Traub, P. and Wiche, G. (1991) Proc. Natl. Acad. Sci. (USA) 88, 3812-3816.
63. White, E. and Cipriani, R. (1990) Mol. Cell Biol. 10, 120-130.
64. Rueger, D.C., Huston, J.S., Dahl, D. and Bignami, A. (1979) J. Mol. Biol. 135, 53-68.
65. Zackroff, R.V. and Goldman, R.D. (1979) Proc. Natl. Acad. Sci. (USA) 76, 6226-6230.
66. Huiatt, T.W., Robson, R.M., Arakawa, N. and Stromer, M.H. (1980) J. Biol. Chem. 255, 6981-6989.
67. Ip, W., Hartzer, M.K., Pang, Y.-Y.S., Pang, R.M. and Robson, R.M. (1985) J. Mol. Biol. 183, 365-375.

68. Soellner, P., Quinlan, R.A. and Franke, W.W. (1985) Proc. Natl. Acad. Sci. (USA) 82, 7929-7933.
69. Franke, W.W., Winter, S., Schmid, E., Soellner, P., Haemmerling, G. and Achtstaetter, T. (1987) Exp. Cell Res. 173, 17-37.
70. Vikstrom, K.L., Borisy, G.G. and Goldman, R.D. (1989) Proc. Natl. Acad. Sci. (USA) 86, 549-553.
71. Miller, R.K., Vikstrom, K. and Goldman, R.D. (1991) J. Cell Biol. 113, 843-855.
72. Osborn, M. and Weber, K. eds. (1989) Cytoskeletal Proteins in Tumour Diagnosis. Curr. Comm. Mol. Biol. Cold Spring Harbor Laboratory, Cold Spring Harbor, pp. 1-244.
73. Franke, W.W., Jahn, L. and Knapp, A.C. (1989) In: Cytoskeletal Proteins in Tumour Diagnosis. Curr. Comm. Mol. Biol. Cold Spring Harbor Laboratory, Cold Spring Harbor, pp. 151-172.
74. Gould, V.E., Koukoulis, G.K., Jansson, D.S., Nagle, R.B., Franke, W.W. and Moll, R. (1990) Am. J. Pathol. 137, 1143-1156.
75. Knapp, A.C. and Franke, W.W. (1989) Cell 59, 67-79.
76. Knapp, A.C., Bosch, F.X., Hergt, M., Kuhn, C., Winter-Simanowski, S., Schmid, E., Regauer, S., Bartek, J. and Franke, W.W. (1989) Differentiation 42, 81-102.
77. Venetianer, A., Schiller, D.L., Magin, T. and Franke, W.W. (1983) Nature 305, 730-733.
78. Giese, G. and Traub, P. (1986) Eur. J. Cell Biol. 40, 266-274.
79. Hedberg, K.K. and Chen, L.B. (1986) Exp. Cell Res. 163, 590-517.
80. Lilienbaum, A., Legagneux, V., Portier, M.-M., Dellagi, K. and Paulin, P. (1986) EMBO J. 11, 2809-2814.
81. Gawlitta, W., Osborn, M. and Weber, K. (1981) Eur. J. Cell Biol. 26, 83-90.
82. Klymkowsky, M.W. (1981) Nature 291, 249-251.
83. Klymkowsky, M.W., Bachant, J.B. and Domingo, A. (1989) Cell Motil. Cytoskel. 14, 209-331.
84. Lin, J.J.-C. and Feramisco, J.F. (1981) Cell 24, 185-193.
85. Baribault, H. and Oshima, R.G. (1991) J. Cell Biol. 115, 1675-1684.
86. Kreis, T.E., Geiger, B., Schmid, E., Jorcano, J.-L. and Franke, W.W. (1983) Cell 32, 1125-1137.
87. Franke, W.W., Schmid, E., Mittnacht, S., Grund, C. and Jorcano, J.L. (1984) Cell 36, 813-825.
88. Giudice, G.J. and Fuchs, E. (1987) Cell 48, 453-463.
89. Domenjoud, L., Jorcano, J.L., Breuer, B. and Alonso, A. (1988) Exp. Cell Res. 179, 352-361.
90. Kulesh, D.A. and Oshima, R.G. (1988) Mol. Cell. Biol. 8, 1540-1550.
91. Blessing, M., Jorcano, J.L. and Franke, W.W. (1989) EMBO J. 8, 117-126.
92. Capetanaki, Y., Smitz, S. and Heath, J.P. (1989) J. Cell Biol. 109, 1653-1664.
93. Kulesh, D.A., Ceceña, G., Darmon, Y.M., Vasseur, M. and Oshima, R.G. (1989) Mol. Cell. Biol. 9, 1553-1565.
94. Lersch, R., Stellmach, V., Stocks, C., Giudice, G. and Fuchs, E. (1989) Mol. Cell. Biol. 9, 3685-3697.
95. Pieper, F.R., Slobbe, R.L., Ramaekers, F.C.S., Cuypers, H.T. and Bloemendal, H. (1987) EMBO J. 6, 3611-3618.
96. Dunia, I., Pieper, F., Manenti, S., van de Kemp, A., Devilliers, G., Benedetti, E.L. and Bloemendal, H. (1990) Eur. J. Cell Biol. 53, 59-74.
97. Monteiro, M.J., Hoffman, P.N., Gearhart, J.D. and Cleveland, D.W. (1990) J. Cell Biol. 111, 1543-1557.
98. Schultheiss, T., Lin, Z., Ishikawa, H., Zamir, I., Stoeckert, C.J. and Holtzer, H. (1991) J. Cell Biol. 114, 953-966.

■ Werner W. Franke:
Institute of Cell and Tumour Biology,
German Cancer Research Centre,
6900 Heidelberg,
Germany

Cytokeratins

Of the various intermediate filament (IF) proteins the cytokeratins*, the hallmark IF proteins of epithelial differentiations, are the most diversified and complex subgroup.

The human genome contains at least 20 different genes encoding epithelial cytokeratin polypeptides, plus another ten cytokeratin polypeptides characteristic of hair- and nail-forming cells ("trichocytic" or "hard" cytokeratins), and these are expressed in different patterns in the various epithelial and trichocytic cell types (Figure 1 and Table)[1-10]. The cytokeratin gene complexity in other vertebrate species, from mammals to amphibia and fish, is similar[11,12]. The individual cytokeratin polypeptides differ remarkably in molecular weight (from ~40,000 to ~68,000) and pI (ranging, in high urea concentrations, from 4.8 to ~8.0). They can be divided into two subfamilies, the more acidic type I and the more basic type II cytokeratins (Figure 1), although some type II cytokeratins of lower vertebrates have a negative charge similar to type I cytokeratins[12]. The two subfamilies, however, share only <30% amino acid sequence identity in their best conserved domain, the α-helical "rod"[5,9]. The head and the tail domains (see introduction on **Intermediate Filaments and Associated Proteins**) differ markedly between the various cytokeratin polypeptides, even of the same subfamily, although certain motifs can be found in subgroups such as glycine-rich tri- or tetrapeptide repeats in the heads and tails of several cytokeratins that occur in stratified squamous epithelia.

The subunit ("building block") of cytokeratin IF is a tetramer, i.e. a double coiled-coil comprising two polypeptides of each type I and type II and, in contrast to the other IF protein classes, the cytokeratins are obligatory heteropolymers based on the formation of IF subunits from stoichiometric numbers of type I and type II cytokeratins[9,13]. The organization of the primary subunit is still disputed but most recent evidence speaks for a predominance of a heterodimeric coiled-coil[13-16] although homotypic coiled-coils can also exist[14-17]. The heterotypic cytokeratin subunits associate into typical IF which in turn fasciate into bundles (fibrils). The IF packing density within these bundles may be increased by certain cytokeratin-associated proteins, of which one, **filaggrin**, has been identified in upper layers of epidermis[18]. The typical light microscopic appearance of cytokeratin IF bundles in epithelial cells, as visualized by immunofluorescence microscopy of human hepatocytes, is shown in Figure 2, both in a monolayer culture of hepatoma cells (a) and in liver tissue (b). Many of the cytokeratin IF bundles are anchored at desmosomes, resulting in higher order arrays of corresponding cytokeratin fibril patterns in adjacent cells and hence throughout the tissue[3,4].

During development cytokeratins are the earliest IF proteins expressed, and both ecto- and endoderm are characterized by typical polarized epithelial cells with extended arrays of IF containing cytokeratins 8 and 18 (in some cases also with cytokeratin 19) that are mostly attached to desmosomes[19,20] (for amphibian see also references **21,22**). Subsequently, cytokeratin synthesis ceases in some cell types, mostly nonepithelial tissues, but is maintained in the various epithelia with specific patterns of expression (Figure 2). For example, one-layered ("simple") epithelia are distinguished in their cytokeratin pattern (cytokeratins 7, 8, 18, 19 and 20) from typical stratified epithelia, and certain types of epithelial cells can be characterized by their cytokeratin patterns[3,7,23]. In addition, different cytokeratin genes can be expressed in different layers and cell types of stratified epithelia and in the trichocytes of hair- or nail-forming follicles[5,6,9,23-26].

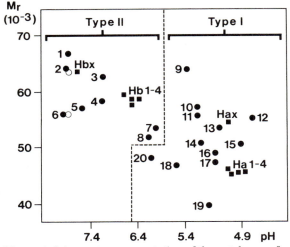

Figure 1. Schematic representation of the catalogue of human cytokeratins. The epithelial (circles, cytokeratin 1-cytokeratin 20) and trichocytic (squares *Ha 1-4*, *Hax*, *HB 1-4*, *Hbx*) cytokeratin polypeptides are arranged according to their migration in 2-D gel electrophoresis: abscissa, pI in the presence of 9.5 M urea; ordinate, M_r value in SDS-PAGE. Open circles indicate the existence of a second kind of cytokeratin 2 (one typically expressed in epidermis, the other in gingiva and hard palate epithelium) and possibly also cytokeratin 6.

*The broad term "cytokeratins" is used here in preference over "keratins", because the latter was originally used in biochemistry for the complex of different filamentous and amorphous proteins present in hair, wool, quill, and upper epidermis of skin and includes so-called β-keratins and certain "high-sulphur content" proteins that are chemically unrelated to the IF-forming a-helical components.

TABLE CYTOKERATIN POLYPEPTIDE PATTERNS OF EPITHELIAL TISSUES[1] OBTAINED BY TWO-DIMENSIONAL ELECTROPHORESIS AND/OR IMMUNOHISTOCHEMISTRY

Cytokeratin No.[2]	Stratified epithelium type														Simple epithelium type				
	Type II						Type I								Type II		Type I		
	1	2	3	4	5	6	9	10/11	12	13	14	15	16	17	7	8	18	19	20
Stratified squamous epithelia																			
Epidermis (hairy skin)	●	·			●			●			●	·							
Epidermis of palm and sole	●	·			●	●	●	●			●	·							
Non-cornifying strat. squam. epith.					●	·				●		·	·	·				·	
Corneal epithelium			●		●				●										
Simple epithelia																			
Secretory epithelial cells[3]																●	●		
Intestinal/gastric faveolar epith.																●	●	●	●
Ductal epith., endocervix, endometr.																●	●	●	
Mesothelium						·										●	●	●	
Urothelium				·		·				●					·	●	●	●	·
Mammary gland ducts						●					●	·			●	●	●	●	
Respiratory epithelium						●	·							●	●	·	●	●	
Molecular weight (x 10⁻³)	68	65,5	63	59	58	56	64	56	55	54	50	50	48	46	54	52,5	45	40	46

[1] No separation according to individual cell types; size of dot indicates relative abundance class.
[2] According to Moll et al. [7,8]
[3] E.G., hepatocytes, pancreatic acinar cells. Proximal and distal tubular cells of the kidney show the same pattern.

The observation that cytokeratin synthesis is generally maintained in malignantly transformed cells, even in morphologically altered ones that have lost their epithelial structure, has made cytokeratin antibodies valuable tools in the diagnosis of epithelium derived tumours, notably carcinomas[7,26]. In conjunction with desmosomal proteins, cytokeratins provide the best biochemical markers of epithelial character and hence for the identification of carcinomas and their metastases[7,26,27].

In addition to epithelia, cytokeratins can also be synthesized in specific nonepithelial cells, particularly in some lower vertebrates[12,28,29], in embryonic tissues and in certain tumours or cell culture lines, either in a general "homogeneous" pattern or restricted to individual cells or cell clusters occurring at different frequencies[30,31].

■ PURIFICATION

Most protocols for cytokeratin enrichment and purification start from cytoskeletal fractions obtained after extractions of other materials with nondenaturing detergents, combined with low and high salt buffers. These steps are followed by denaturation in high urea and/or guanidinium hydrochloride solution, ion exchange chromatography and often also preparative gel electrophoresis[2,3,32,33]. In addition, methods for producing individual cytokeratin polypeptides, wild-type or mutated, from recombinant DNA clones in *Escherichia coli* have been described[15,34].

■ ANTIBODIES

As a result of the widespread use of cytokeratin antibodies in pathology, particularly the detection and classification of carcinomas, numerous cytokeratin antibodies are offered commercially by more than a dozen companies. Some of these antibodies react with several cytokeratins, others are specific for individual cytokeratin polypeptides.

■ GENES

For most cytokeratins of human, bovine, rodent or amphibian (*Xenopus laevis*) cDNA clones have been isolated and characterized, many cytokeratin genes have been sequenced and their chromosomal localization determined.

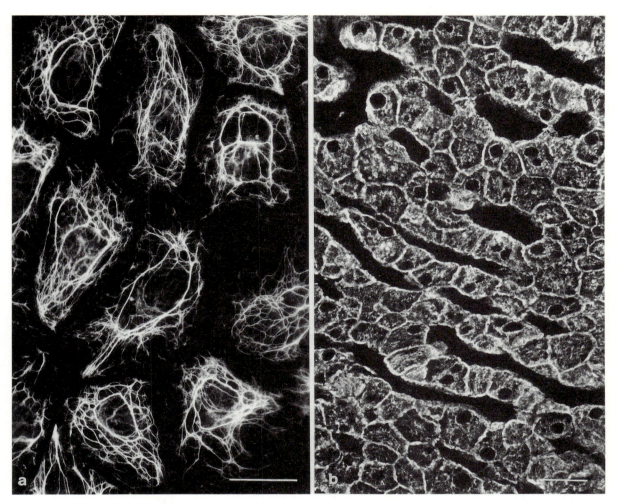

Figure 2. Immunofluorescence microscopy with anti-cytokeratin antibodies, showing the typical distribution of cytokeratin IF bundles in a monolayer of cultured human hepatocellular carcinomas of line PLC (a) and in human liver tissue (b). Bar 0.5 µm.

■ REFERENCES

1. Franke, W.W., Schmid, E., Osborn, M. and Weber, K. (1978) Proc. Natl. Acad. Sci. (USA) 75, 5034-5038.
2. Franke, W.W., Weber, K., Osborn, M., Schmid, E. and Freudenstein, C. (1978) Exp. Cell Res. 116, 429-445.
3. Franke, W.W., Schiller, D.L., Moll, R., Winter, S., Schmid, E., Engelbrecht, I., Denk, H., Krepler, R. and Platzer, B. (1981) J. Mol. Biol. 153, 933-959.
4. Franke, W.W., Schmid, E., Schiller, D.L., Winter, S., Jarasch, E.-D., Moll, R., Denk, H., Jackson, B.W. and Illmensee, K. (1982) Cold Spring Harb. Symp. Quant. Biol. 46, 431-453.
5. Fuchs, E., Tyner, A.G., Guidice, G.J., Marchuk, D., RayChaudhury, A. and Rosenberg, M. (1987) Curr. Topics Dev. Biol. 22, 5-34.
6. Heid, H.W., Moll, I. and Franke, W.W. (1988) Differentiation 37, 137-157.
7. Moll, R., Franke, W.W., Schiller, D.L., Geiger, B. and Krepler, R. (1982) Cell 31, 11-24.
8. Moll, R., Schiller, D.L. and Franke, W.W. (1990) J. Cell Biol. 111, 567-580.
9. Steinert, P.M. and Roop, D.R. (1988) Annu. Rev. Biochem. 57, 593-625.
10. Sun, T.-T., Shih, C.H. and Green, H. (1979) Proc. Natl. Acad. Sci. (USA) 76, 2813-2817.
11. Cooper, D., Schermer, A., Sun, T.-T. (1985) Lab. Invest. 52, 243-256.
12. Markl, J., Winter, S., Franke, W.W. (1989) Eur. J. Cell Biol. 50, 1-16.
13. Quinlan, R.A., Cohlberg, J.A., Schiller, D.L., Hatzfeld, M., Franke, W.W. (1984) J. Mol. Biol. 178, 265-288.
14. Coulombe, P.A., Fuchs, E. (1990) J. Cell Biol. 111, 153-169.
15. Hatzfeld, M. Weber, K. (1990) J. Cell Biol. 110, 1199-1210.
16. Steinert, P.M. (1990) J. Biol. Chem. 265, 8766-8774.
17. Hatzfeld, M., Maier, G., Franke, W.W. (1987) J. Mol. Biol. 197, 237-255.
18. Dale, B.A., Resing, K.A., Haydock, P.V. (1990) In: Cellular and Molecular Biology of Intermediate Filaments (R.D.

Goldman, P.M. Steinert, eds.) Plenum Press, New York, pp. 393-412.

19. Jackson, B.W., Grund, C., Schmid, E., Bürki, K., Franke, W.W., Illmensee, K. (1980) Differentiation 17, 161-179.

20. Oshima, R.G., Howe, W.E., Klier, F.G., Adamson, E.D., Shevinksy (1983) Dev. Biol. 99, 447-455.

21. Franz, J.K., Gall, L., Williams, M.A., Picheral, B., Franke, W.W. (1983) Proc. Natl. Acad. Sci. (USA) 80, 6254-6258.

22. Sargent, T.D., Jonas, E., Jamrich, M., Michaels, G.S., Miyatani, S., Winkles, J.A., Dawid, I.B. (1990) In: Cellular and Molecular Biology of Intermediate Filaments (R.D. Goldman, P.M. Steinert, eds.) Plenum Press, New York, pp. 335-344.

23. Quinlan, R.A., Schiller, D.L, Hatzfeld, M., Achtstätter, T., Moll, R., Jorcano, J.L., Magin, T.M., Franke, W.W. (1985) Ann. N.Y. Acad. Sci. 455, 282-306.

24. Tyner, A.L., Fuchs, E. (1986) J. Cell Biol. 103, 1945-1955.

25. Woodcock-Mitchell, J., Eichner, R., Nelson, W.G., Sun, T.-T. (1981) J. Cell Biol. 95, 580-588.

26. Osborn, M., Weber, K. (1983) Lab. Invest. 48, 372-394.

27. Moll, R., Cowin, P., Kapprell, H.-P., Franke, W.W. (1986) Lab. Invest. 54, 4-25.

28. Markl, J., Franke, W.W. (1988) Differentiation 39, 97-122.

29. Rungger-Brändle, E., Achtstaetter, T., Franke, W.W. (1989) J. Cell Biol. 109, 705-716.

30. Franke, W.W., Jahn, L., Knapp, A.C. (1989) In: Cytoskeletal Proteins in Tumour Diagnosis (M. Osborn, K. Weber, eds.) Cold Spring Harbor Laboratories, Cold Spring Harbor. pp. 151-172.

31. Knapp, A.C., Franke, W.W. (1989) Cell 59, 67-79.

32. Achtstätter, T., Hatzfeld, M., Quinlan, R.A., Parmelee, D.C., Franke, W.W. (1986) Meth. Enzymol. 134, 355-371.

33. Eichner, R., Sun, T.-T., Aebi, U. (1986) J. Cell Biol. 102, 1767-1777.

34. Magin, T.M., Hatzfeld, M., Franke, W.W. (1987) EMBO J. 6, 2607-2615.

■ Werner, W. Franke and Jürgen Kartenbeck:
Institute of Cell and Tumour Biology,
German Cancer Research Centre,
D-6900 Heidelberg,
Germany

Desmin

Desmin[1,2] is a type III intermediate filament protein characteristic of myogenic cells[3]. It is most abundant in smooth muscle, from which it was first purified[2]. In muscle, desmin filaments appear to form a structural lattice that provides support for the contractile machinery (Figure 1).

The amino acid sequence of desmin, the first obtained from an intermediate filament protein[4], revealed the basic structural features of this protein family. The native molecule is a parallel homodimer consisting of a central, mainly α-helical coiled-coil rod domain covering around 310 residues, flanked by variable nonhelical tail and head regions (Figure 2); the calculated monomer molecular weight is 53,000[4]. Results obtained using truncated molecules, produced by protein chemical and molecular genetic methods, show that the N-terminal head part is essential for filament assembly both *in vitro*[5] and *in vivo*[6], whereas the C-terminal tail, at least the latter half of it, is not required for this function[5]. The head region displays at least two subdomains, one endowing competence for filament formation (residues 7-17)[6] and a second, phosphorylatable domain (encompassing serine residues 29, 35 and 50) implicated in the regulation of the assembly-disassembly reaction[7].

At the primary sequence level mammalian, avian and amphibian desmins show greater than 90 % homology within their rod and C-terminal tail domains[4,8-10]. In contrast, the N-terminal head domain is less well conserved, but does maintain a characteristic abundance of basic residues that is common among type III intermediate filament proteins as well as a conserved nonapeptide encompassing residues 12-20 and the phosphorylation sites already mentioned. Data from chemical crosslinking molecular weight determinations and electron microscopy[11,12] indicate that an antiparallel tetramer, formed by the lateral aggregation of two dimers shifted axially by around 15 nm is the building unit of the filament (Figure 2). The tail domains appear to project, at least in part[13], from the filament surface and are presumed to be involved in peripheral filament interactions[5].

During myogenesis and in certain situations in the adult, desmin may be coexpressed with **vimentin**[3], and even **cytokeratins**[10]. Specific myopathies[14] as well as hypertrophy in smooth muscle[15] are characterized by a significant increase in desmin and the number of intermediate filaments.

Two high molecular weight proteins, **synemin** and **paranemin** have been copurified and colocalized with desmin in muscle cells[3,16]. **Plectin**[16], another high molecular weight nontissue specific protein, copurifies with intermediate filaments and colocalizes with desmin in skeletal and cardiac muscle but not in smooth muscle. The **actin** binding protein **filamin** also shows partial codistribution with desmin in smooth muscle cells[17]. In common with vimentin, the desmin head and tail domains have binding affinities for ankyrin and **lamin B**, respectively[18].

■ PURIFICATION

Desmin purification is based on the insolubility of this protein in concentrated salt solution: smooth muscle (commonly chicken gizzard or hog stomach) has been the source tissue of choice[2]. Briefly, a smooth muscle

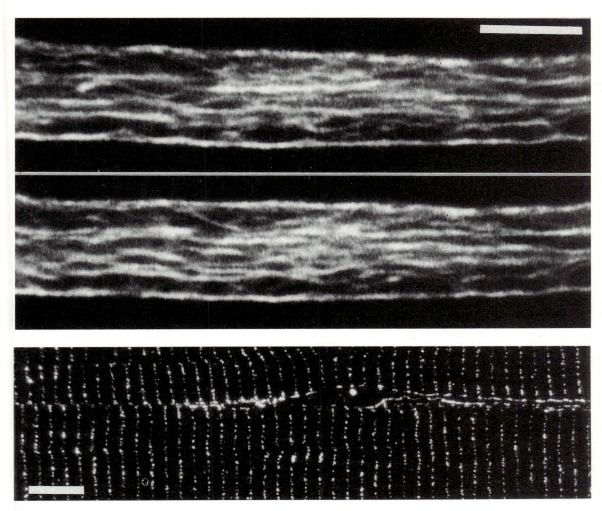

Figure 1. Localization of desmin (A,B) by consecutive confocal fluorescence images in a chicken gizzard smooth muscle cell and (B) in an ultrathin longitudinal section in rabbit psoas skeletal muscle. Desmin is concentrated at the periphery of the Z-disc. Bar 10 μm. (Courtesy of A. Draeger and E.H.K. Stelzer).

homogenate is exhaustively extracted, first with Triton X-100, then with 0.6 M KCl followed by 0.6 M KI to remove as much of the contractile and other cytoskeletal roteins as possible. Desmin is then extracted with urea[19,20] or acetic acid[2] and subsequently purified by chromatography under denaturing conditions. Desmin is also weakly soluble at low ionic strength and alkaline pH.

■ ACTIVITIES

Desmin polymerizes *in vitro* by dialysis from urea or low salt into salt containing buffers and this polymerization can be monitored by electron microscopy, turbidity[21] or quenching of fluorescence[22].

■ ANTIBODIES

Monoclonal antibodies are commercially available from various suppliers; the one developed by Debus et al.[23] is probably the best.

■ GENES

Like other proteins of this family, desmin derives from a single copy gene, fully characterized in hamster[8] and man[9]. Exon number, size and position are conserved between these two species as are the splicing signals: the introns, however, show divergence in size (in total 1 kB more in human) and sequence (less than 50 % homology). EMBL filename for the human gene: NUC: 21-8-89. NAR[9].

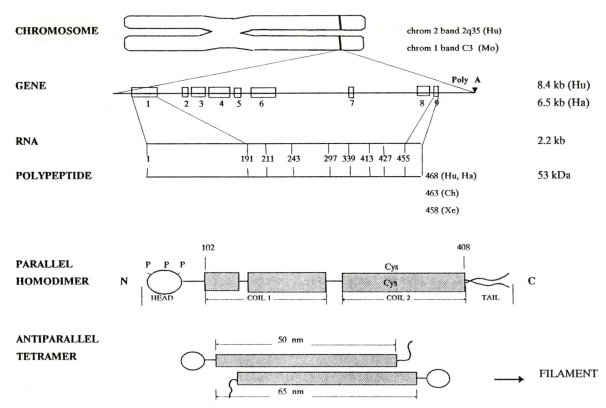

Figure 2. Molecular details of the desmin molecule. Open boxes along gene indicate exons. The corresponding splice sites are given as amino acid position[8] at the RNA and polypeptide level. The native homodimer displays a mainly α-helical coiled-coil rod domain (shaded) plus a head and tail piece. The filament building unit is a staggered antiparallel tetramer. Abbreviations: Chrom chromosome; Hu human; Ha hamster; Xe xenopus; P serine phosphorylation sites. Gene locus assignments taken from: Viegas-Péquignot, E., Li, Z., Dutrillaux, B., Apion, F. and Paulin, D. (1989). Hum. Genet. 83, 33-36; Li, Z., Mattei, M.-G., Mattei, J.-F. and Paulin, D. (1990). Genet Res. Cam. 55, 101-105.

■ REFERENCES

1. Lazarides, E. and Hubbard, B.D. (1976) Proc. Natl. Acad. Sci. 73, 4344-4348.
2. Small, J.V. and Sobieszek, A. (1977) J. Cell Sci. 23, 243-268.
3. Lazarides, D. (1982) Ann. Rev. Biochem. 51, 219-250.
4. Geisler, N. and Weber, K. (1982) EMBO J. 1, 1649-1656.
5. Kaufmann, E., Weber, K. and Geisler, N. (1985) J. Mol. Biol. 185,733-742.
6. Raats, J.M.H., Pieper, F.R., Vree Egberts, W.T.M., Verrijp, K.N., Ramaekers, F.C.S. and Bloemendal, H. (1990) J. Cell Biol. 111, 1971-1985.
7. Weber, K. and Geisler, N. (1988) In "Structure and Function of the Cytoskeleton". Ed. B.A.F. Rousset. Coloque INSERM, Vol. 171, pp 3-7, John Libbey Eurotext Ltd.
8. Quax, W., Van den Broek, L., Vree Egberts, W., Ramaekers, F. and Bloemendal, H. (1985) Cell 43, 327-338.
9. Li, Z., Lilienbaum, A., Butler-Browne, G. and Paulin, D. (1989) Gene 78, 243-254.
10. Hermann, H., Fouquet, B. and Franke, W.W. (1989) Development 105, 299-307.
11. Geisler, N., Kaufmann, E. and Weber, K. (1985) J. Mol. Biol. 182, 173-177.
12. Potschka, M., Nave, R., Weber, K. and Geisler, N. (1990) Eur. J. Biochem. 190, 503-508.
13. Birkenberger, L. and Ip, W. (1990) J. Cell Biol. 111, 2063-2075.
14. Thornell, L.-E., Eriksson, A. and Edström, L. (1983) In "Cell and Muscle Motility" Ed. R.M. Dowben and J.W. Shay, Vol. 4 pp 85-136, Plenum.
15. Malmqvist, U. and Arner, A. (1990) Circulaton Res. 66, 832-845.
16. Wiche, G. (1989) CRC Crit. Rev. Biochem. and Molec. Biol. 24, 41-67.
17. Small, J.V., Fürst, D.O. and De Mey, J. (1986) J. Cell Biol. 102, 210-220.
18. Georgatos, S.D., Weber, K., Geisler, N. and Blobel, G. (1987) Proc. Natl. Acad. Sci. (USA) 84, 6780-6784.
19. Geisler, N. and Weber, K. (1980) Eur. J. Biochem. 111, 425-433.
20. Huiatt, T.W., Robson, R.M., Arakawa, N. and Stromer, M.H. (1980) J. Biol. Chem. 255, 6981-6989.
21. Stromer, M.H., Ritter, M.A., Pang, Y.-Y.S. and Robson, R.M. (1987) Biochem. J. 246, 75-81.
22. Ip, W. and Fellows, M.E. (1990) Anal. Biochem. 185, 10-16.
23. Debus, E., Weber, K. and Osborn, M. (1983) EMBO J. 2, 2305-2312.

■ J.V. Small:
Institute of Molecular Biology,
Austrian Academy of Sciences,
Salzburg, Austria

Epinemin

Epinemin is an intermediate filament associated protein found only on vimentin filaments in nonneural cells. A biological role for this molecule has not yet been defined.

Epinemin is a monomer (or noncovalently linked dimer) of M_r 44,500 and pI 5.4; its 2-D tryptic peptide fingerprint pattern is distinct from **vimentin**. Epinemin is resistant to nonionic detergent extraction, is soluble in high salt and can thus be removed from vimentin filaments but copurifies with vimentin in low salt of anionic detergents[1]. Epinemin is present in all nonneural cells that contain vimentin filaments but is absent in neural vimentin containing cells such as white matter astrocytes and Bergmann glia[1]. By light microscopy, epinemin is uniformly distributed on vimentin filaments. However, quick-freezing deep-etching colloidal-gold-immunocytochemistry shows that epinemin is intermittently spaced, foci separated by a minimum distance of 80-100 nm (and with no obvious periodicity). The foci of epinemin molecules are distributed asymmetrically along the sides of the vimentin core polymer (Figure) and each focus binds one to three anti-epinemin antibody molecules[2]. Epinemin does not appear to crosslink either adjacent vimentin filaments to one another or vimentin filaments to other cytoskeletal networks such as 2-3 nm filaments, a class of small filaments found in fibroblasts[2].

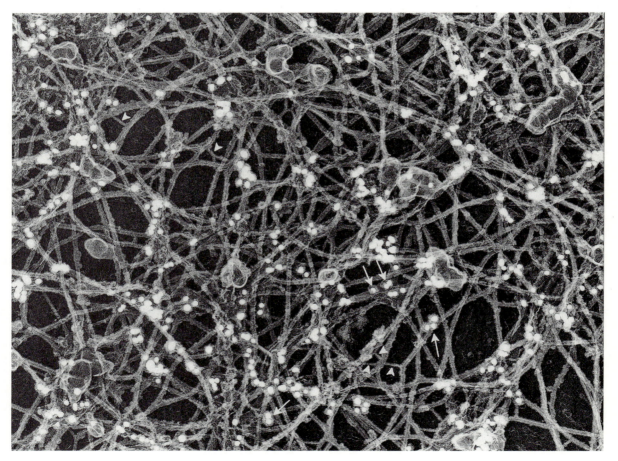

Figure. Quick-frozen deep-etched replica of a detergent extracted fibroblast cytoskeleton showing a tangle of vimentin filaments labelled with anti-epinemin antibodies followed by a second antibody coupled to colloidal gold. Asymetrically distributed foci of epinemin are easily seen (arrows). Note also the presence of small 2-3 nm filaments crossbridging between vimentin filaments (arrowheads). Mag X 87,000.

PURIFICATION/ACTIVITIES

Epinemin has not been purified and its function has not yet been determined.

ANTIBODIES

Epinemin was defined by an IgM monoclonal antibody[1].

GENES

The gene coding for epinemin has not yet been isolated.

REFERENCES

1. Lawson, D. (1983) J. Cell Biol. 97, 1891-1905.
2. Lawson, D. (1984) J. Cell Biol. 99, 1451-1460.

■ *Durward Lawson:*
  Biology Department, Medawand Building,
  University College London,
  Gower Street,
  London WCE 6BT, UK

Filaggrins

Filaggrins are a family of intermediate filament associated, cationic proteins expressed primarily in terminally differentiating mammalian epidermis. The filaggrins have the capacity to aggregate specifically keratin filaments in vitro to form tight bundles and presumably also in vivo[1,2].

Filaggrins of different mammalian species vary in size, from about 16 kDa (bovine), to 26 kDa (mouse), 35 kDa (human) and 45 kDa (rat), yet have similar amino acid compositions and are enriched in histidine, arginine, glycine and serine[1,2]. Amino acid sequences are known for rat[3], mouse[4], and human[5] and display little interspecies homology. The exact mechanism(s) by which filaggrins aggregate the **keratin** filaments is not yet known[6]. Aggregation of keratin filaments may be facilitated by metal ions, especially Zn^{2+}, which promote the rapid formation of interchain disulphide bonds in the keratin filaments[6].

Filaggrins are initially synthesized as large polyprotein precursors (profilaggrins) and temporarily stored as large irregularly shaped F-keratohyalin granules in the epidermis[7,8]. Mouse profilaggrin has a half-life of about 6 h[9]. During translation the protein is heavily phosphorylated[8-10], apparently by specific protein kinases. In all species so far studied, the profilaggrin consists of multiple filaggrin repeats that are simply arranged in tandem and are separated by short hydrophobic linker sequences. This phospho-profilaggrin molecule can be dephosphorylated and proteolytically processed into several functional filaggrin molecules by excision of the linker by specific proteases[8-10]. The released filaggrin rapidly aggregates the existing network of keratin filaments in the epidermal cells, causing a dramatic collapse of the keratin cytoskeleton and a rapid flattening of the epidermal cells, the characteristic phenotype of terminally differentiated cornified epidermis[11]. Immunogold analysis with specific antibodies shows that the filaggrin persists in the first three to five cornified cell layers but then disappears[12]. Filaggrin has a half-life of about 24 h[9], and is quickly degraded to free amino acids[7,8], which may maintain the osmolarity and thus high water content of the epidermis[7,8,13]. The released histidine is converted to urocanic acid that may have some function in the absorption of UV light in the skin, or may act as a free radical scavenger[13].

The various repeats are arranged in tandem in the profilaggrin gene (Figure), as they are in the profilaggrin protein. There are no introns in the coding portions of the rat, mouse or human genes[3-5,14], but there is a short intron in the 5'-UTR of the human gene[14]. There is a single gene per haploid genome[14], and it contains 10, 11 or 12 individual repeats in different individuals of the human population due to simple allelic variants, that segregate in a normal Mendelian manner[14]. Interestingly, neighbouring repeats in the same clones from the same individual show high degrees of sequence variation, that do not, however, change the predicted structure of the filaggrin molecules.

There is considerable morphological and histochemical evidence that expression of the profilaggrin gene is markedly altered in a number of keratinizing disorders of human skin[15,16].

PURIFICATION

Filaggrins can be prepared from cornified epidermal tissue by ion exchange chromatography[2]. The precursor profilaggrins may also be isolated by exploiting their insolubility properties[8,9].

ANTIBODIES

Antibodies to mouse[4,9,11], rat[3,9] and human[11,17-19] filaggrin/profilaggrin are available. These antibodies usually do not crossreact.

Figure. Model of the human profilaggrin gene containing 10 repeats. F=FLY... hydrophobic linker sequence.

GENES

GenBank numbers for filaggrin: J02929 (human), J05198 (mouse). The human filaggrin gene maps to chromosome 1q21[5].

REFERENCES

1. Dale, B.A., Holbrook, K.A. and Steinert, P.M. (1978) Nature 276, 728-731.
2. Steinert, P.M., Cantieri, J.S., Teller, D.C., Lonsdale-Eccles, J.D. and Steinert, P.M. (1981) Proc. Natl. Acad. Sci. 78, 4097-4101.
3. Haydock, P.V. and Dale, B.A. (1986) J. Biol. Chem. 261, 12520-12525.
4. Rothnagel, J.A. and Steinert, P.M. (1990) J. Biol. Chem. 265, 1862-1865.
5. McKinley-Grant, L., Idler, W.W., Bernstein, I.A., Parry, D.A.D., Cannizzaro, L., Croce, C.M., Huebner, K., Lessin, S.R. and Steinert, P.M. (1989) Proc. Natl. Acad. Sci. 86, 4848-4852.
6. Steinert, P.M. (1983) In The Stratum Corneum, ed. by Marks, R. and Plewig, G., Springer-Verlag, Heidelberg, pp. 25-38.
7. Scott, I.R. and Harding, C.R. (1981) Biochim. Biophys. Acta 669, 65-78.
8. Harding, C.R. and Scott, I.R. (1983) J. Mol. Biol. 170, 651-673.
9. Resing, K.A., Walsh, K.A. and Dale, B.A. (1984) J. Cell Biol. 99, 1372-1378.
10. Lonsdale-Eccles, J.D. and Dale, B.A. (1980) J. Biol. Chem. 255, 2235-2238.
11. Steinert, P.M. unpublished data.
12. Steven, A.C., Bisher, M.E., Roop, D.R. and Steinert, P.M. (1990) J. Struct. Biol. 104, 150-162.
13. Scott, I.R. and Harding, C.R. (1986) Devel. Biol. 115, 84-92.
14. Gan, S.-Q., Idler, W.W., McBride, O.W., Markova, N. and Steinert, P.M. (1990) Biochemistry 29, 9432-9440.
15. Baden, H.P., Roth, S.I., Goldsmith, L.A., Baden, S.B. and Lee, L.D. (1974) J. Invest. Dermatol. 62, 411-414.
16. Fleckman, P., Holbrook, K.A., Dale, B.A. and Sybert, V.P. (1987) J. Invest. Dermatol. 88, 640-645.
17. Kubilus, J., Scott, I.R., Harding, C.R., Yendle, J., Kvedar, J. and Baden, H.P. (1985) J. Invest. Dermatol. 85, 513-517.
18. Lynley, A.M. and Dale, B.A. (1983) Biochim. Biophys. Acta 744, 28-35.
19. Dale, B.A., Gown, A.M., Fleckman, P., Kimball, J.R. and Resing, K.A. (1987) J. Invest. Dermatol. 88, 306-313.

■ *Song-Qing Gan and Peter M. Steinert:*
Laboratory of Skin Biology,
National Institute of Arthritis and Musculoskeletal and Skin
Diseases, NIH,
Bethesda, MD 20892, USA

Filensin

Filensin (100 kDa) is a newly identified vimentin binding protein of the lens fibre cell membrane. It is derived from a larger 110 kDa precursor polypeptide which is exclusively expressed in lens tissue. Isolated filensin forms short fibrillar structures which bear a similarity to filaments assembled from neurofilament protein (NF-M). The in vivo function of filensin is presently unknown.

Filensin constitutes a major component of the lens plasma membrane. In SDS-PAGE, it migrates as a band with a molecular weight of ~100,000. This protein resists salt or Triton X-100 extraction, but can be efficiently solubilized by 0.1 M NaOH, or 7-8 M urea treatment of lens membranes. By immunoelectron microscopy and indirect immunofluorescence microscopy filensin can be localized exclusively at the periphery of the lens fibre cells[1].

Upon removal of urea and reconstitution into isotonic media at neutral pH, the isolated protein forms short, 10 nm fibrils (Figure) which resemble the polymers assembled from purified **neurofilament protein (NF-M)** subunits. The polymerization of filensin, as assessed by negative staining and rotary shadowing, is salt and pH dependent[1].

Purified filensin binds *in vitro* to the C-terminal domain of the intermediate filament protein **vimentin** and another 47 kDa urea soluble protein of the lens membrane[1]. The binding of filensin to the C-terminal domain of vimentin can be competed with purified rat liver **lamin B**[1], which also interacts with the same domain of type III intermediate filament proteins[2-4].

As indicated by northern blotting and immunochemical data, filensin is derived from a 110 kDa precursor molecule which is specifically expressed in lens tissue (Merdes, Gounari and Georgatos, unpublished).

Proteins with a molecular weight similar to that of filensin have been recently identified in mammalian and avian lens. One of them, with a molecular mass of 110 kDa has been reported to possess a sequence motif highly conserved among intermediate filament proteins[5]. A 95 kDa and a 110 kDa protein of the chicken and bovine lens, respectively, have been characterized as components of the beaded filaments, a structure unique to lens cells[6,7].

■ PURIFICATION

Filensin can be purified from crude urea extracts initially by a DEAE-cellulose column chromatography. Fractions enriched in filensin can be further purified through a hydroxylapatite column.

■ ACTIVITIES

Radiolabelled filensin binds specifically to lens vimentin under isotonic conditions as demonstrated by affinity

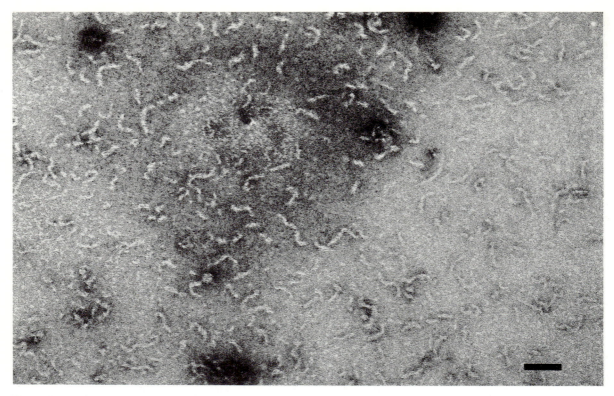

Figure. Negative staining of filensin fibrils, reconstituted by dialysis of the urea soluble protein against isotonic buffer at pH 7.4. Bar 100 nm.

chromatography and ligand-blotting assays. By the latter approach filensin reacts with a 47 kDa peripheral membrane protein of the lens cells. Purified filensin binds to PI, a synthetic peptide modelled after a segment of the C-terminal domain of peripherin. The filensin-PI binding is inhibited by purified lamin which is known to interact with PI *in vitro*.

■ ANTIBODIES

Rabbit polyclonal antibodies recognizing either the porcine[1], or the porcine and the bovine form of filensin are available. These reagents can be used for immunoprecipitation, immunoblotting, indirect immunofluorescence and immunoelectron microscopy.

■ GENES

Partial cDNA clones of filensin have been identified and preliminary sequencing data indicate the existence of nearly canonical repeats of the form *D G V(L) K E E G G P*

P E G K G in the C-terminal domain of filensin (Merdes and Gounari, unpublished).

■ REFERENCES

1. Merdes, A., Brunkener, M., Horstmann, H. and Georgatos, S.D. (1991) J. Cell Biol. 115, 397-410.
2. Georgatos, S.D. and Blobel, G. (1987) J. Cell Biol.105, 117-125.
3. Djabali, K., Portier, M.-M., Gros, F., Blobel, G. and Georgatos, S.D. (1991) Cell 64, 109-121.
4. Papamarcaki, T., Kouklis, P.D., Kreis, T.E. and Georgatos, S.D. (1991) J. Biol. Chem. 266, 21247-21251.
5. Remington, S. (1990) J. Cell Biol. 111, 44a.
6. Atreya, P.L. and Maisel, H. (1989) Biochem. Biophys. Res. Comm. 163, 589-598.
7. FitzGerald, P.G. and Gottlieb, W. (1989) Curr. Eye Res. 8, 801-811.

■ *Spyros Georgatos:*
European Molecular Biology Laboratory,
D-6900 Heidelberg
Germany

Glial Fibrillary Acidic Protein (GFAP)

GFAP, glial fibrillary acidic protein, is a member of the intermediate filament protein superfamily. It is expressed almost exclusively in astrocytes and cells of astro-glial origin and is therefore extensively used in histology and pathology as a cell type marker. GFAP is an α-helical protein which can spontaneously self-associate, via coiled-coil interactions, to form 10 nm filaments. In cells, GFAP filaments are distributed as either individual or bundles of filaments throughout the cytoplasm, extending from around the nucleus out to the cell periphery. The precise function of this protein remains obscure, although a consensus view links GFAP to maintenance of cell shape in tissues.

GFAP is a type III intermediate filament protein, along with **vimentin, desmin** and **peripherin**[1]. The calculated molecular weight for human GFAP[2] is 49,891. Like all the other intermediate filament proteins, the predicted protein structure divides GFAP into three major domains, namely a central, mainly α-helical, rod domain which is flanked at the N- and C-termini by two smaller, non-α-helical domains[3]. The rod is the most conserved domain between type III proteins, showing some 60-70% homology over the 310 residues. As suggested by this homology, GFAP can form hybrid filaments with vimentin when expressed in the same cell[4].

GFAP is purified using urea containing buffers and filament assembly *in vitro* can be initiated by removal of the urea[5] (Figure 1). Filament assembly is spontaneous, requiring no other protein factors. It occurs under defined buffer conditions over a narrow pH range from 6.75 to 7.00 and cations are required with a preference for divalent over monovalent cations[6]. The critical concentration for GFAP assembly *in vitro* is 1.6 µM, in close agreement with other type III proteins. Assembly begins by forming two chain molecules, which then self-associate to give molecular dimers, often referred to as tetramers[4,7]. In the molecule, the two rod domains are parallel and in register forming a coiled-coil. The molecule is 48 nm long. In the molecular dimer, the molecules overlap in an antiparallel manner, giving two preferred lengths, one 48 nm and the other 64 nm[8]. Filament assembly may be arrested at this point in aqueous buffers of alkaline pH and low ionic strength e.g. 2 mM Tris-HCl, pH 7.8[5].

Filament assembly can be inhibited by removing the N- or C-terminal non-α-helical domains in recombinantly produced GFAP, although the truncated GFAP constructs can still form molecular dimers[8]. A 2-3 nm region at the end of the rod, next to the C-terminal α-helical domain, contains a sequence motif conserved in all intermediate filament proteins which is important in the elongation phase of assembly, as deduced from paracrystals of the recombinantly made rod portion of GFAP[9]. The N-terminal domain is also very important in filament assembly. It contains several phosphorylation sites which, when phosphorylated *in vitro*, cause filament disassembly[10]. This is the mechanism proposed to control filament disassembly *in vivo*.

The apparent molecular weight of GFAP by SDS-PAGE is 50,000, but this value varies with different species[11,12]; for instance, murine GFAP appears ~5 kDa smaller than other mammalian GFAP. This does not reflect large sequence differences as a comparison of murine and human GFAP shows 99% homology[2]. The observed pI for GFAP is 5.7.

There is a single gene for GFAP[13,14] which is located on chromosome 11 in mouse, immediately proximal to the *Dlb-1* locus[15]. The number of exons and the position of exon-intron boundaries are similar to those of other type III genes. A single mRNA is transcribed by RNA polymerase II from the gene, although the promoter region contains two initiators, one overlapping the protein encoding sequence[16]. The position of the initiating methionine is conserved between mouse and man. Both mRNAs also have unusually long 3′ untranslated sequences of 1.3 and 1.7 kB respectively. This feature may explain the observed colocalisation of GFAP mRNA and GFAP filaments in Müller glia[17]. Expression of GFAP can be regulated by cAMP, phorbolesters, glia maturing factors and serotonin[18].

GFAP is expressed in astrocytes of the central nervous system. It also is expressed in cells of glial origin such as Müller glia of the retina and glia of the peripheral nervous system such as some Schwann cells, satellite cells and the enteric glia. Reactive astrocytes, which appear after trauma of the central nervous system are very rich in GFAP and have been used as a source for protein purification.

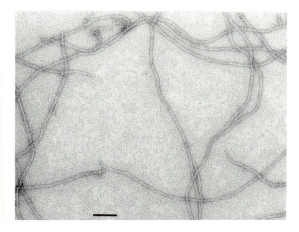

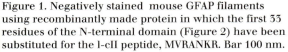

Figure 1. Negatively stained mouse GFAP filaments using recombinantly made protein in which the first 33 residues of the N-terminal domain (Figure 2) have been substituted for the l-cII peptide, MVRANKR. Bar 100 nm.

```
PORCINE:  MERRRVTSAARRSYVSSLVT VGGG   RRLGPGPRLSLARMPP
HUMAN:    MERRRITSAARRSYVSSGEMMVGGLAPGRRLGPGTRLSLARMPP
MOUSE:    MERRRITSA RRSTASATVVR GLGMPSRQLGTMPRFSLSRMPP
```

Figure 2. Sequence comparison of the non-α-helical rod domain of GFAP. The mouse sequence has been adjusted to take into account an extra T at position 326 in the published DNA sequence[13], found upon resequencing (Quinlan, R.A. unpublished data). Only those residues identical in all three sequences are in bold type.

These astrocytes are a central component of the central nervous system scars resulting from injury or disease, for example Alzheimer's disease, multiple sclerosis and adrenoleukodystrophy. In Alexander's disease, GFAP filaments become associated with the diagnostic Rosenthal fibres typical of this neurological disorder[19]. GFAP is also reported to specifically bind the scrapie prion protein[20]. Gliomas are identified on the basis of GFAP expression[21].

■ PURIFICATION

Spinal cord or brain cortex are good sources of GFAP. The protein can be purified from these tissues by conventional ion exchange chromatography techniques, maintaining the sample in denaturing concentrations of urea[10,11].

■ ACTIVITIES

GFAP filaments appear to perform an active structural role in process formation of cultured astrocytes in response to neurons as shown by the stable expression of antisense GFAP mRNA in U251 cells[22]. The discovery of a unique GFAP filament associated protein[23], increases the possibility that other important functions for GFAP filaments have yet to be discovered. Indeed, the apparent association of proenkephalin with a subpopulation of GFAP filaments within astrocytes[24] implies subtle differences in the function of GFAP filaments within the same cell.

■ ANTIBODIES

The monoclonal antibody clone G-A-5 has a broad cross-species reactivity and is extensively used on frozen as well as fixed and embedded sections.This antibody can be bought from many suppliers (e.g. Sigma, prod. no. G3893). Rabbit polyclonal antibodies are available from Dako Ltd., prod. no. Z334, and these too have a broad species crossreactivity.

■ GENES

Mouse (Acc. No. K01347)[25] and human cDNA sequences (Acc.No.J04569)[2] are published. The partial amino acid sequence for pig GFAP is also available for comparison[26]. Mouse (Acc. No. X02801)[13] and human gene sequences[14] are also published.

■ REFERENCES

1. Parysek, L.M., Chisholm, R.L., Ley, C.A. and Goldman, R.D. (1988) Neuron 1, 395-401.
2. Reeves, S.A., Helman, L.J., Allison, A. and Israel, M.A. (1989) Proc. Natl. Acad. Sci. 86, 5178-5182.
3. Quinlan, R.A. and Stewart, M. (1991) Bioessays 13, 597-600.
4. Quinlan, R.A. and Franke, W.W. (1983) Eur. J. Biochem. 132, 477-484.
5. Rueger, D.C., Huston, J.S., Dahl, D. and Bignami, A. (1979) J. Mol. Biol. 135, 53-68.
6. Yang, Z.W. and Babitch, J.A. (1988) Biochemistry 27, 7038-7045.
7. Quinlan, R.A., Hatzfeld, M., Franke, W.W., Lustig, A., Schulthess, T. and Engel, J. (1986) J. Mol. Biol. 192, 337-349.
8. Quinlan, R.A., Moir, R.D. and Stewart, S. (1989) J. Cell Sci. 93, 71-83.
9. Stewart, M., Quinlan, R.A. and Moir, R.D. (1989) J. Cell Biol. 109, 225-234.
10. Inagaki, M., Gonda, Y., Nishizawa, K., Kitamura, S., Sato, C., Ando, S., Tanabe, K., Kikuchi, S. and Nishi, Y. (1990) J. Biol. Chem. 265, 4722-4729.
11. Dahl, D. and Bignami, A. (1983) In Handbook of Neurochemistry 5, 127-150. Ed. A. Lajtha, Plenum Publ.
12. Yokoyama, K., Mori, H. and Kurokawa, M. (1981) FEBS Lett. 135, 25-29.
13. Balcarek, J. and Cowan, N.J. (1985) Nucleic Acids Res. 13, 5527-5543.
14. Brenner, M., Lampel, K., Nakatani, Y., Mill, J., Banner, C., Mearow, K., Dohadwala, M., Lipsky, R. and Freese, E. (1990) Mol. Brain Res. 7, 277-286.
15. Bernier, L., Colman, D.R. and D'Estachio, P. (1988) J. Neurosci. Res. 20, 497-504.
16. Nakatani, Y., Brenner, M. and Freese, E. (1990) Proc. Natl. Acad. Sci. 87, 4289-4293.
17. Sarthy, P.V., Fu, M. and Huang, J. (1989) Mol. Cell. Biol. 9, 4556-4559.
18. Le Prince, G., Copin, M.-C., Hardin, H., Belin, M.-F., Bouilloux, J.-P. and Tardy, M. (1990) Dev. Brain Res. 51, 295-298.
19. Iwaki, T., Kume-Iwaki, A., Liem, R.K.H. and Goldman, J.E. (1989) Cell 57, 71-78.
20. Oesch, B., Teplow, D.B., Stahl, N., Serban, D., Hood, L.E. and Pruisner, S.B. (1990) Biochemistry 29, 5848-5855.
21. Osborn, M., Altmannsberger, M., Debus, E. and Weber, K. (1985) Ann. N.Y. Acad. Sci. 455, 649-668.
22. Weinstein, D.E., Shelanski, M. and Liem, R.K.H. (1991) J. Cell Biol. 112, 1205-1213.
23. Abd-El-Basset, E.M., Kalnins, V.I., Ahmed, I. and Federoff, S. (1989) J. Neuropathol. Exp. Neur. 48, 245-254.
24. Spruce, B., Curtis, R., Wilkin, G.P. and Glover, D. (1990) EMBO J. 9, 1787-1795.
25. Lewis, S.A., Balcarek, J.M., Krek, V., Shelanski, M. and Cowan, N. (1984) Proc. Natl. Acad. Sci. 81, 2743-2746.
26. Geisler, N. and Weber, K. (1982) EMBO J. 2, 2059-2063.

■ Roy A. Quinlan:
Department of Biochemistry,
The University,
Dundee, DD1 4HN, UK

α-Internexin

α-Internexin is a novel neuronal intermediate filament protein. Originally identified as an intermediate filament associated protein, recent purification and assembly studies, as well as cDNA cloning and sequencing have shown that it is the subunit of type IV intermediate filaments found predominantly in the central nervous system.

α-Internexin was originally identified in intermediate filament protein preparations from rat spinal cord and optic nerve as a protein that migrated slightly ahead of the low molecular weight neurofilament protein (**NF-L**) on SDS-PAGE[1]. Its apparent molecular weight was ~68,000, and it was therefore sometimes referred to as a 68 kDa neurofilament subunit[2]. The protein was purified to homogeneity and shown to be different from NF-L by peptide mapping and immunological studies. The original description of the protein also showed that it was axonally transported, recognized by an antibody which reacts with all intermediate filament proteins[3] and that it was able to associate with various other intermediate filament proteins[1]. Initial failure to assemble the protein into filaments led to the classification of α-internexin as an intermediate filament associated protein[1].

Recent evidence has shown that α-internexin is actually an intermediate filament protein that can self-assemble *in vitro*[2,4]. The protein sequence as determined from a full length cDNA confirmed the presence of the highly conserved α-helical rod with a typical heptad repeat as described for all other intermediate filament proteins[5]. The highest homology was with **NF-M**, the middle molec-

ular weight neurofilament protein. A particularly interesting homology is in the C-terminal tail, involving a sequence which is conserved in NF-M from a wide variety of species, and is thought to enable NF-M to interact with other cellular elements[6]. The calculated molecular weight of the protein is 55 kDa, indicating that like the **neurofilament triplet proteins**, α-internexin migrates anomalously on SDS-PAGE. The sequence also reveals a charged C-terminal tail which is particularly rich in glutamic acid, which is again similar to NF-M. The characterization of a genomic clone for α-internexin has shown that it is clearly a type IV intermediate filament protein.

Immunolocalization studies have shown that in the adult rat, α-internexin is present primarily in the central nervous system, although low levels of the protein are observed in the peripheral nervous system[2,4]. α-Internexin is also the only intermediate filament present in the parallel fibres of the granule neurons in the cerebellum. In large axons α-internexin appears to be less abundant than the neurofilament triplet proteins. During development, α-internexin mRNA reaches a peak at embryonic day 16 in the rat brain, a point when NF-L mRNA is still relatively low; α-internexin mRNA decreases after this stage,

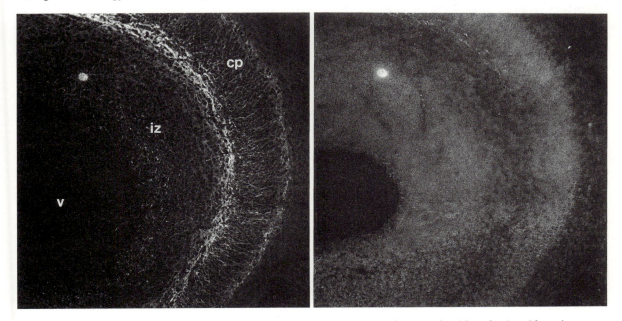

Figure. Double immunofluorescent staining of the cerebral cortex of an embryonic day 16 rat brain with anti-α-internexin antibodies during development. cp = cortical plate; iz = intermediate zone; v = ventricle. (Picture provided by M.P. Kaplan).

whereas NF-L mRNA increases into adulthood[5]. Immunocytochemical studies showed that α-internexin can be detected before NF-L in a number of areas in the central nervous system, including the cortex, the retina and the olfactory bulb[4]. However, in other regions, NF-L, NF-M and α-internexin appear to be expressed at the same time.

■ PURIFICATION

α-Internexin can be purified from rat brain, optic nerve or spinal cord by ion exchange chromatography. The original purification started with a preparation of Triton-X 100 insoluble proteins from rat spinal cord dissolved in 8 M urea. A two step purification using hyroxylapatite chromatography and DEAE chromatography yields a single band on SDS-PAGE[1,4]. A variation on this purification is to use FPLC with similar columns and gradients[2].

■ ACTIVITIES

Purified α-internexin can be reassembled by replacing the urea by dialysis against phosphate buffered saline to yield intermediate sized filaments[2,4]. It can bind to a number of other intermediate filament proteins[1] in vitro, presumably through interactions involving the α-helical rod region.

■ ANTIBODIES

Two monoclonal antibodies against α-internexin have been described[4]. These antibodies crossreact against several other mammalian species, but showed no reaction against the avian protein. A polyclonal antibody has also been described[2].

■ GENES

α-Internexin has been cloned from an adult rat brain library and the entire coding region has been published[5,7]. The GenBank/EMBL Data bank accession number is M73049.

■ REFERENCES

1. Pachter, J.S. and Liem, R.K.H. (1985) J. Cell Biol. 101, 1316-1322.
2. Chiu, F.C., Barnes, E.A., Das, K., Haley, J., Socolow, P., Macaluso, F.P. and Fant, J. (1990) Neuron 2, 1435-1445.
3. Pruss, R.M., Mirsky, R., Raff, M.C., Thorpe, R., Dowding, A.J. and Anderton, B.H. (1981) Cell 27, 419-428.
4. Kaplan, M.P., Chin, S.S.M., Fliegner, K.H. and Liem, R.K.H. (1990) J. Neurosci. 10, 2735-2748.
5. Fliegner, K.H., Ching, G.Y. and Liem, R.K.H. (1990) EMBO J. 9, 749-755.
6. Shaw, G. (1989) Biochem. Biophys. Res. Comm. 162, 294-299.
7. Ching, G.Y. and Liem, R.K.H. (1991) J. Biol. Chem. 266, 19459-19468.

■ *Ronald K.H. Liem:*
Departments of Pathology and Anatomy & Cell Biology,
Columbia University,
New York, NY, USA

Lamins

Lamins comprise a distinct class of intermediate filament proteins that are assembled into a filamentous meshwork lining the inner nuclear membrane called the nuclear lamina[1]. The lamina is thought to provide a framework for organizing the structure of the nuclear envelope and an anchoring site for chromosomes at the nuclear periphery. Thus, lamins may play a major role in defining higher level nuclear architecture.

Lamins have been characterized in a wide variety of higher eukaryotes, and range in molecular mass between 60 and 75 kDa[1-3]. Multiple lamin isotypes are found in vertebrates, where the lamina of a particular cell type may contain up to four or more distinct lamins[1,4]. Different members of the lamin family are expressed in a developmentally regulated fashion, a feature that may be important for changes in nuclear architecture occuring during development[1-3]. Vertebrate lamins have been classified into "A" and "B" subgroups[2,4]. α-type lamins which include lamins A and C are more basic (with pIs of about 7) and become detached from membranes following nuclear envelope disassembly in M-phase, while β-type lamins which include lamin B1 and lamin B2[5,6] are more acidic (with pIs around 6) and remain membrane bound during M-phase.

Similar to other intermediate filament proteins, lamins contain an internal α-helical rod domain composed mostly of heptad repeats flanked by N- and C-terminal "head" and "tail" domains[2]. The internal α-helical domain of lamins is about 360 residues long, and the tail domain contains most of the remaining lamin mass. Lamins share sequence homology with other classes of intermediate filament proteins primarily in the α-helical domain, the region that is thought to form the intermediate filament backbone. The basic lamin protomer is a parallel unstaggered dimer formed by coiled-coil interactions in the α-helical region which give rise to a 50 nm long rod domain[7,8] (Figure 1). Lamin protomers are probably packed in a half-staggered fashion to form lamin filaments. The structure of the lamina has been most clearly visualized in *Xenopus* oocytes, where the lamina forms a quasi tetragonal lattice of 10 nm filaments with a crossover spacing of 50 nm[7].

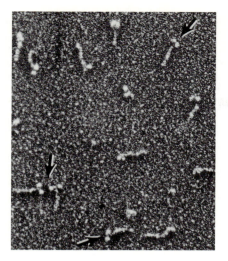

Figure 1. Electron micrograph of rotary shadowed dimers of rat liver lamins A and C. The C-terminal tail domains flanking the 50 nm rod domain are indicated by arrows.

Different members of the lamin family can be modified posttranslationally by phosphorylation[5], carboxyl-methylation[9], proteolysis[10] and C-terminal isoprenylation[3,11]. While lamins are almost entirely assembled at the nuclear envelope during interphase (Figure 2), they are reversibly depolymerized to protomers during M-phase concomittant with disassembly and reformation of the nuclear envelope[1]. Lamina disassembly during this period is mediated by hyperphosphorylation of lamins by the p34[cdc2] kinase[5,12]. M-phase specific demethylation of a lamin B isotype is correlated with lamina disassembly in mammalian cells[9]. Interphase lamins also are phosphorylated, in part in response to extracellular signalling[13,14].

■ PURIFICATION

The simplest procedures for isolation of lamins involve preparation of lamina enriched fractions from nuclear envelopes or nuclei based on the intrinsic insolubility of lamins in solutions containing low or high concentrations of monovalent salt at neutral pH[1]. Subsequently, lamins can be solubilized from this material in 6-8 M urea and purified by conventional column chromatography[7,15,16]. Purified lamins are soluble in the absence of urea at alkaline pH in high concentrations of monovalent salt[7].

■ ACTIVITIES

No enzymatic activity but isolated lamins form 10 nm filaments and paracrystalline filament aggregates after dialysis into physiological buffers[7,8]. In addition, purified lamins A and C interact with specific binding sites on mitotic chromosomes to assemble into a supramolecular structure[16]. A number of distinct integral membrane proteins of the nuclear envelope have been described that are potential membrane receptors for lamins[6,17].

■ ANTIBODIES

Polyclonal and monoclonal antibodies have been described that react with mammalian, amphibian, chicken, clam and *Drosophila* lamins[1-3].

■ GENES

cDNA clones have been isolated and sequenced for four mammalian lamin isotypes[4], three chicken isotypes[18], four *Xenopus* isotypes[19] and one *Drosophila* isotype[20].

A

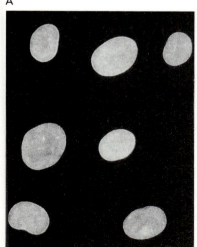

B

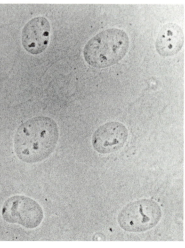

Figure 2. Immunofluorescence (A) and phase-contrast (B) micrographs of NRK cells stained with a monoclonal antibody recognizing nuclear lamins.

■ REFERENCES

1. Gerace, L. and Burke, B. (1988) Ann. Rev. Cell Biol. 4, 335-374.
2. Nigg, E. (1989) Curr. Opin. Cell Biol. 1, 435-440.
3. Burke, B. (1990) Curr. Opin. Cell Biol. 2, 514-520.
4. Hoger T., Zatlouka, K., Waizenegger, I. and Krohne, G. (1990) Chromosoma 99, 379-390.
5. Ottaviano, Y. and Gerace, L. (1985) J. Biol. Chem. 260, 624-632.
6. Senior, A. and Gerace, L. (1988) J. Cell Biol. 107, 2029-2036.
7. Aebi, U., Cohn, J., Buhle, L. and Gerace, L. (1986) Nature 323, 560-564.
8. Heitlinger, E., Peter, M., Haner, M., Lustig, A., Aebi, U. and Nigg, E. (1991) J. Cell Biol. 113, 485-495.
9. Chelsky, D., Olson, J. and Koshland, D. (1987) J. Biol. Chem. 262, 4303-4309.
10. Weber, K., Plessmann, U. and Traub, P. (1989) FEBS Lett. 257, 411-414.
11. Beck, L., Hosick, T. and Sinensky, M. (1988) J. Cell Biol. 107, 1307-1316.
12. Peter, M., Nakagawa, J., Doree, M., Labbe, J. and Nigg, E. (1990) Cell 61, 591-602.
13. Hornbeck, P., Huang, K. and Paul, W. (1988) Proc. Nat. Acad. Sci. (USA) 85, 2279-2283.
14. Friedman, D. and Ken, R. (1988) J. Biol. Chem. 263, 1103-1106.
15. Georgatos, S. and Blobel, G. (1987) J. Cell Biol. 105, 117-125.
16. Glass, J. and Gerace, L. (1990) J. Cell Biol. 111, 1047-1057.
17. Worman H., Yuan, J., Blobel, G. and Georgatos, S. (1988) Proc. Nat. Acad. Sci. (USA) 85, 8531-8534.
18. Vorburger, K., Lehner, C., Kitten, G., Eppenberger, H. and Nigg, E. (1989) J. Mol. Biol. 208, 405-415.
19. Krohne G., Wolin, S., McKeon, F., Franke, W. and Kirschner, M. (1987) EMBO J. 6, 3801-3807.
20. Gruenbaum, Y., Landesman, Y., Drees, B., Bare, J., Saumweber, H., Paddy, M., Sedat, J., Smith, D., Benton, B. and Fisher, P. (1988) J. Cell Biol. 106, 585-596.

■ *Larry Gerace:*
The Scripps Research Institute,
La Jolla, CA 92037, USA

Nestin

Mammalian neural precursor cells and myoblasts transiently express a large (~200 kDa) intermediate filament protein called nestin (neural stem cell protein). Like the neurofilament heavy chain protein, nestin has a large, acidic C-terminal tail bearing a repeated motif; unlike all other known intermediate filaments it has virtually no N-terminal "head" domain.

Nestin (calculated MW 200 kDa) was originally detected in the developing rat central nervous system as an epitope expressed by somitic myoblasts and the immediate mitotic precursors to neurons but not by mature neurons, glia or muscle[1-3].

The cDNA sequence of nestin shows that it is not closely related to other presently characterized intermediate filaments; while the coiled-coil "core" domain contains the typical intermediate filament pattern of conserved, heptad bearing regions and nonheptad, less conserved "spacers", it is only between 16-29% identical to that of other intermediate filament core sequences, and the noncore domains of nestin do not bear significant sequence homology to other intermediate filaments, nor indeed to anything else in the database[3]. The very large (>1400 amino acids) C-terminal domain contains about 35 tandem copies of an 11 amino acid motif S/P-L-$\underline{E}$-E/K-$\underline{E}$-X-$\underline{Q}$-E-S/L-L-R (underlined amino acids are strongly conserved) in a 500 amino acid region. The virtual absence of an N-terminal domain may be significant in light of evidence suggesting that this domain plays a role in the self-assembly of other intermediate filament homopolymers. Nestin, which colocalizes with **vimentin**, may be obliged to form a vimentin copolymer in much the same manner that the assembly *in vitro*[4] of neurofilment high (**NF-H**) and middle (**NF-M**) chains depends on neurofilament low (**NF-L**) chain.

■ PURIFICATION

The nestin protein has not been purified.

■ ACTIVITIES

None have been determined.

■ ANTIBODIES

Monoclonal mouse anti-rat nestin antibody recognizes rat nestin by immunoblotting and immunocytochemically[1]; polyclonal rabbit anti-rat fusion protein antibodies and anti-peptide antisera recognize rat, mouse, and human nestin (Marvin and McKay, unpublished).

■ GENES

Full length rat cDNA (GenBank M34384) has been published[3].

■ REFERENCES

1. Hockfield, S. and McKay, R.D.G. (1985) J. Neurosci. 5 (12), 3310-3328.
2. Frederiksen, K. and McKay, R.D.G. (1988) J. Neurosci. 8 (4) 1144-1151

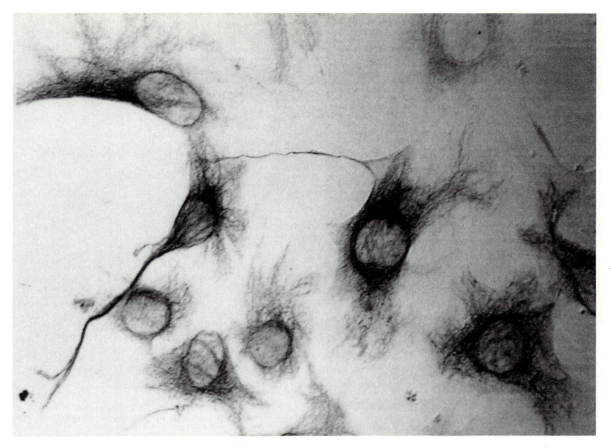

Figure. Rat hippocampal neural precursor cell line stained with anti-nestin monoclonal antibody[5].

3. Lendahl, U., Zimmerman, L. and McKay, R. (1990) Cell 60, 585-595.
4. Traub, N. (1985) Intermediate Filaments: A Review (Berlin: Springer-Verlag) pp. 113-116.
5. Renfranz, P., Cunningham, M. and McKay, R. (1991) Cell 66, 713-729.

■ Lyle B. Zimmerman and Ronald D.G. McKay:
Department of Biology,
Massachusetts Institute of Technology
Cambridge, MA, USA

Neurofilament Triplet Proteins (NF-L, NF-M, NF-H)

The neurofilament triplet proteins, originally defined as polypeptides comigrating in slow axonal transport[1], are now recognized as the major components of the neurofilaments (NF), the intermediate filaments (IF) typical for most mature neurones. Neurofilaments are thought to be a major determinant of axonal caliber[2].

Neurofilaments purified from mammalian or avian brain and spinal cord consist of the triplet proteins[3,4] with molecular weights on SDS-PAGE of approximately 68 K (NF-L), 160 K (NF-M) and 200 K (NF-H); L, M and H indicate low, middle and high apparent molecular weights. The molecular weight values for NF-M and NF-H are gross overestimates versus the true chemical values of 102 kDa and 112 kDa respectively[5-7]. The discrepancies arise from several unusual properties of the C-terminal tail domains, i.e. hyperphosphorylation, extended regions very rich in glutamic acid residues and repetitive sequence segments.

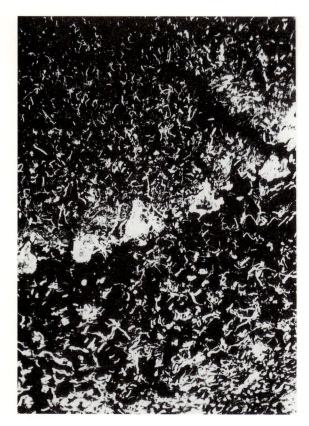

Figure 1. Frozen section of rat cerebellum in indirect immunofluorescence microscopy using monoclonal antibody to NF-H (N52). (Courtesy Dr. Mary Osborn.)

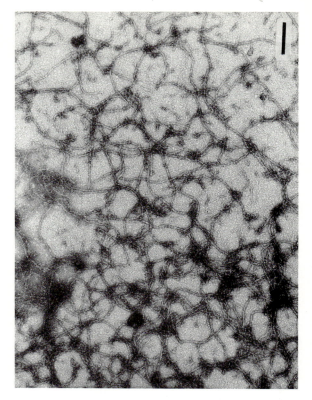

Figure 2. Filaments formed by purified porcine NF-L *in vitro* are negatively stained with uranylacetate[5]. Bar 0.2 μm.

Neurofilament triplet proteins have the canonical features of IF proteins. A highly α-helical domain able to form double stranded coiled-coils (the rod with 310 residues) is flanked by nonhelical head and tail domains[8]. The NF tail domains are unique among IF proteins. Mammalian NF-M and particularly NF-H harbour sequence repeats in which the serine residues present as KSP-motifs are phosphorylated[6-10]. With some 40 repeats and two motifs per repeat, axonal NF-H may well contain 40 to 80 serine phosphates and therefore be the major phosphoprotein of the axonal cytoskeleton[11,12]. Phosphorylation of NF-H seems lower and/or different in dendritic NF[13]. For possible NF kinases see for instance[14]. NF-L, the major triplet protein, forms IF *in vitro* either on its own or together with NF-M and NF-H[4]. Whether either NF-M or NF-H can form true IF on their own is undecided. The triplet proteins are coassembled *in vivo*. Upon transfection NF-L and NF-M incorporate into the endogenous **vimentin** IF system of fibroblasts[15] in line with the pronounced homology of the rod sequences[16]. Transgenic mice over-expressing NF-L do not show an increase in axonal diameter[15]. A more detailed review of NF is given elsewhere[17].

There is a striking delay in the expression of NF-H (Figure 1) versus in the other two triplet proteins during embryogenesis[18,19]. The genes of the NF triplet proteins (IF type IV genes) have an intron pattern quite distinct from that of the nonneuronal IF genes (types I, II and III)[6,7,20]. Interestingly, the type IV intron pattern holds also for **nestin**. This IF protein of neuroepithelial cells and of the stem cells of the central nervous system has a giant tail domain[21]. α-**Internexin** seems to be an early embryonal neurofilament protein, which is later substituted by NF-L[22]. **Peripherin**, a marker of the peripheral nervous system, is encoded by a gene with the intron pattern of type III genes such as vimentin[23].

■ PURIFICATION

Adult bovine or porcine spinal cord is an excellent source. The soluble extract is used to form NF at 37°C with 20% glycerol[3]. The filaments are harvested and dissolved in 6-8 M urea. The triplet proteins are separated by chromatography on DEAE-cellulose in the presence of urea using a salt gradient[4].

ACTIVITIES

No enymatic activity. Filament formation occurs *in vitro* (Figure 2) after removal of urea (see above).

ANTIBODIES

Numerous good monoclonal antibodies for each protein have been reported, and many are commercially available. Note that many antibodies to NF-M and NF-H detect only phosphorylated epitopes. Others, however, are independent of the phosphorylation state. Many antibodies show broad crossspecies reactivity.

GENES

The Gene EMBL bank contains the genes for human NF-H (X15306 to X15309), human NF-M (Y00067), human NF-L (X05608), murine NF-H (M23349), M24494 to M24496), murine NF-M (X05640), murine NF-L (M13016) and chicken NF-M (X17102). Additional cDNA sequences cover rat NF-H (J04517), rat NF-M (M18628) and rat NF-L (M25638). The NBRF bank contains the protein chemically obtained sequence of porcine NF-L (Protein: Qfpgl) and a partial sequence of porcine NF-M (Protein: Qfpgm). All data are accessible by searching for "neurofilament". Additional partial sequences on porcine NF-H are known[9,24] and so is a nearly complete cDNA clone for a NF-M-like protein from Torpedo[25]. Indicative of a recent gene duplication human NF-L and NF-M map on chromosome 8 band p21[26].

PATHOLOGY

Antibodies to triplet proteins have been used to distinguish neuroblastomas[27] and seem to stain approximately one third of primitive neuroectodermal tumours (PNETS)[28].

REFERENCES

1. Hoffman, P.N. and Lasek, R.J. (1975) J. Cell Biol. 66, 351-366.
2. Hoffman, P.N., Cleveland, D.W., Griffin, J.W., Landes, P.W., Cowan, N.J. and Price, D.L. (1987) Proc. Natl. Acad. Sci. (USA) 84, 3472-3476.
3. Delacourte, A., Filliatreau, C., Boutteau, F., Biserte, G. and Schrevel, J. (1980) Biochem. J. 191, 543-546.
4. Geisler, N. and Weber, K. (1981) J. Mol. Biol. 151, 565-571.
5. Kaufmann, E., Geisler, N. and Weber, K. (1984) FEBS Lett. 170, 81-84.
6. Myers, M.W., Lazzarini, R.A., Lee, V.M.-Y.L., Schlaepfer, W.W. and Nelson, D.L. (1987) EMBO J. 6, 1617-1626.
7. Lees, J.F., Schneidmann, P.S., Skuntz, S.F., Carden, M.J. and Lazzarini, R.A. (1988) EMBO J. 7, 1947-1955.
8. Geisler, N., Kaufmann, E., Fischer, S., Plessmann, U. and Weber, K. (1983) EMBO J. 2, 1295-1302.
9. Geisler, N., Vandekerckhove, J. and Weber, K. (1987) FEBS Lett. 221, 403-407.
10. Lee, V.M.-Y., Otvos, L., Carden, M.J., Hollosi, M., Dietzschold, B. and Lazzarini, R.A. (1988) Proc. Natl. Acad. Sci. (USA) 85, 1998-2002.
11. Julien, J.-P. and Mushynski, W.E. (1982) J. Biol. Chem. 257, 10467-10470.
12. Jones, S.M. and Williams Jr., R.C. (1982) J. Biol. Chem. 257, 9902-9905.
13. Sternberger, L. and Sternberger, N. (1983) Proc. Natl. Acad. Sci. (USA) 80, 6126-6130.
14. Wible, B.A., Smith, K.E. and Angelides, K.J. (1989) Proc. Natl. Acad. Sci. (USA) 86, 720-724.
15. Monteiro, M.J., Hoffman, P.A., Gearhart, J.D. and Cleveland, D.W. (1990) J. Cell Biol. 111, 1543-1557.
16. Geisler, N., Plessman, U. and Weber, K. (1983) FEBS Lett. 163, 22-24.
17. Liem, R.K.H. (1990) Curr. Opin. Cell Biol. 2, 86-90.
18. Shaw, G. and Weber, K. (1982) Nature 298, 277-279.
19. Pachter, J.S. and Liem, R.K.H. (1984) Dev. Biol. 103, 200-210.
20. Lewis, S.A. and Cowan, N.J. (1986) Mol. Cell Biol. 6, 1529-1534.
21. Lendahl, U., Zimmerman, L.B. and McKay, R.D.G. (1990) Cell 60, 585-595.
22. Fliegner, K.H., Ching, G.Y. and Liem, R.K.H. (1990) EMBO J. 9, 749-755.
23. Thompson, M.A. and Ziff, E.B. (1989) Neuron 2, 1043-1053.
24. Geisler, N., Fischer, S., Vanderkerckhove, J., Van Damme, J., Plessmann, U. and Weber, K. (1985) EMBO J. 4, 57-63.
25. Linial, M. and Scheller, R.H. (1990) J. Neurochem. 54, 762-770.
26. Hurst, J., Flavell, D., Julien, J.P., Meijer, P., Mushinski, W. and Grosveld, F. (1987) Cytogenet. Cell Genet. 45-30-32.
27. Osborn, M., Dirk, T., Käser, H., Weber, K. and Altmannsberger, M. (1986) Am. J. Pathol. 122, 433-442.
28. Molenaar, W.M., Jansson, D.S., Gould, V.E., Rorke, L.B., Franke, W.W., Lee, V.M.-Y., Packer, R.J. and Trojanowski, J.Q. (1989) Lab. Invest. 61, 635-643.

■ *Klaus Weber:*
Department of Biochemistry,
Max Planck Institute for Biophysical Chemistry,
D-3400 Göttingen,
Germany

Paranemin

Paranemin is a developmentally regulated 280 kDa polypeptide that is tightly associated with desmin or vimentin containing intermediate filaments. It is distinct from synemin in terms of abundance, size, charge, peptide map, antigenicity, and cell type distribution. It is similar to synemin in that it exhibits developmental changes, and is found only in association with filaments that contain desmin and/or vimentin. The molar ratio of paranemin to core monomers is approximately 4:100.

Paranemin (from the Greek para [with, beside] and nema [filament]) was originally characterized as a component of suspensions of intermediate filaments that were purified from homogenates of embryonic chicken skeletal muscle[1]. Paranemin has a pI of ~4.5 (in urea), and often appears as a closely spaced doublet of ~280 kDa by SDS-PAGE[1]. The solubility properties of paranemin have not been distinguished from the **desmin** and/or **vimentin** filaments with which paranemin is always associated; all are insoluble in physiological and high salt solutions, as well as in nonionic detergents[1,2]. Double immunofluorescence with specific antibodies always shows the same intracellular distribution for paranemin as for the associated desmin and/or vimentin[1,2].

Paranemin undergoes a number of developmental changes in chickens[1,2]. All developing muscle cells express paranemin simultaneously with desmin, vimentin and **synemin**. In contrast to synemin, however, paranemin is not found in adult skeletal muscle or visceral smooth muscle, but does remain in adult cardiac muscle. Paranemin is found in the smooth muscle cells of elastic arteries, but not in the smooth muscle cells of adult muscular arteries; conversely, it is found in the endothelial cells of muscular arteries and capillaries, but not in the endothelial cells of elastic arteries. Paranemin is found in cardiac conducting fibres and Schwann cells. Synemin is not found in endothelial cells or Schwann cells, but is found in conducting fibres and all vascular smooth muscle cells; thus, paranemin and synemin are not mutually exclusive or mutually expressed, but have partially overlapping cell type distributions, even in adult tissues. Furthermore, expression of desmin or vimentin seems to be necessary but not sufficient for the expression of paranemin and synemin. The expression of paranemin does not parallel the expression of any other known intermediate filament protein.

■ PURIFICATION

Paranemin is enriched in gel-filtered suspensions of intermediate filaments that are generated by gentle homogenization of embryonic chicken skeletal muscle[1]. Other methods that enrich for intermediate filament proteins should also enrich for paranemin. Purification by preparative SDS-PAGE has been used to generate antigen, but paranemin has not been purified under nondenaturing conditions.

■ ACTIVITIES

No enzymatic activity is known. Paranemin is routinely identified by its solubility properties, mobility during SDS-PAGE and immunoreactivity.

■ ANTIBODIES

A rabbit anti-chicken paranemin antiserum recognizes only paranemin in chickens[1]; its crossreactivity with presumptive paranemins from other species or classes of organisms has not been investigated.

■ GENES

Uncharacterized.

■ REFERENCES

1. Breckler, J. and Lazarides, E. (1982) J. Cell Biol. 92, 795-806.
2. Price, M.G. and Lazarides, E. (1983) J. Cell Biol. 97, 1860-1874.

■ *Bruce L. Granger:*
Veterinary Molecular Biology,
Montana State University,
Bozeman, MT 59717, USA

Peripherin

Peripherin is a type III intermediate filament protein[1]. It is detected in well defined sets of neurons where it is coexpressed with neurofilament proteins to the exception of the olfactory neurons. It is also expressed in a few cell lines which are utilized as neuronal models.

The apparent molecular mass of peripherin as determined on 2-D SDS-PAGE is 57-58 kDa[2-4]. A few discrete isoforms of similar molecular mass but slightly different isoelectric points are observed in cellular extracts from diverse neural tissues or species[2,5]. The cDNA cloning of mouse neuroblastoma poly(A+)-RNA showed that three isoforms of different molecular masses (a major species of 57-58 kDa, and two minor ones of 56 and 61 kDa respectively) arise from alternative splicing of a unique peripherin gene[6].

Treatment of rat pheochromocytoma PC12 cells with NGF regulates synthesis of peripherin[2,7,8] and induces a large increase in its phosphorylated form[9]. Other studies on the mouse neuroblastoma N1E 115 cell line and on primary cultures of rat sympathetic neurons demonstrate that peripherin is phosphorylated on its N-terminal domain[10].

In well defined neuronal populations peripherin is coexpressed with the **neurofilament triplet proteins**[8,11-14]; peripherin is detected in neurons originating from the neural tube (motoneurons, ganglionic cells of the retina and preganglionic neurons in the mediolateral columns of the spinal cord), from the neural crest (sensory neurons from the dorsal root ganglia and every sympathetic neuron) and from placodes which are thickenings of the cephalic ectoderm (olfactory neurons and neurons from the acoustic ganglia). A common feature of all these neurons is that their axons lie mostly at the outside of the central nervous system. This means that, during development, they have to find their way through a nonneural environment in response to specific signals from their respective targets.

Peripherin shares with **vimentin** and **desmin**, two other type III intermediate filament proteins, an affinity for **lamin B**[15].

Peripherin has also been called 57 kDa neural intermediate filament protein[16].

■ PURIFICATION

Peripherin has been purified from cultures of the mouse neuroblastoma N1E 115 cell line employing established chromatographic methods[15]. Since this cell line also expresses high amounts of vimentin whose molecular mass and isoelectric point are close to that of peripherin, purity of the preparation has to be controlled very carefully.

■ ACTIVITIES

Although the function of peripherin is not yet known with certitude, the expression of this protein in neurons with peripheral axons and its attachment to the nuclear membrane[15] suggest that it may have a role in signal transduction. It has also been proposed that peripherin phosphorylation may provide the means to modulate particular aspects of neurite outgrowth, maintenance and repair[9]. Moreover, axotomy induces an up-regulated expression of peripherin while the neurofilament triplet proteins are down-regulated[17]; this suggests that peripherin may be selectively recruited in regenerating sprouts.

■ ANTIBODIES

Polyclonal antisera against mouse neuroblastoma N1E 115[13] and pheochromocytoma PC 12[3] total peripherin have been prepared, as well as against synthetic peptides corresponding to the unique C-terminus of rat peripherin[4,15].

■ GENES

The peripherin gene has been cloned and sequenced from rat pheochromocytoma PC 12[18] (GenBank M26232) and from mouse neuroblastoma N1E 115 (EMBL Data Base X59840). A partial cDNA sequence exists for mouse neuroblastoma N1E 115 peripherin[6] (EMBL Data Base X15475).

■ REFERENCES

1. Greene, L.A. (1989) Trends Neurosci. 12, 228-230.
2. Portier, M.-M., Brachet, P., Croizat, B. and Gros, F. (1984) Dev. Neurosci. 6, 215-226.
3. Parysek, L.M. and Goldman, R.D. (1987) J. Neurosci. 7, 781-791.
4. Aletta, J.M., Angeletti, R., Liem, R.K.H., Purcell, C., Shelanski, M.L. and Greene, L.A. (1988) J. Neurochem. 51, 1317-1320.
5. Portier, M.-M., de Néchaud B. and Gros, F. (1984) Dev. Neurosci. 6, 335-344.
6. Landon, F., Lemonnier, M., Benarous, R., Huc, C., Fiszman, M., Gros, F. and Portier, M.-M. (1989) EMBO J. 8, 1719-1726.
7. Leonard, D.G.B., Ziff, E.B. and Greene, L.A. (1987) Mol. Cell. Biol. 7, 3156-3167.
8. Leonard, D.G.B., Gorham, J.D., Cole, P., Greene, L.A. and Ziff, E.B. (1988) J. Cell Biol. 106, 181-193.
9. Aletta, J.M., Shelanski, M.L. and Greene, L.A. (1989) J. Biol. Chem. 264, 4619-4627.
10. Huc, C., Escurat, M., Djabali, K., Derer, M., Landon, F., Gros, F. and Portier, M.-M. (1989) Biochem. Biophys. Res. Com. 160, 772-779.
11. Parysek, L.M. and Goldman, R.D. (1988) J. Neurosci. 8, 555-563.
12. Escurat, M., Gumpel, M., Lachapelle, F., Gros, F. and Portier, M.-M. (1988) C.R. Acad. Sci. Paris 306 (série III) 447-456.

13. Escurat, M., Djabali, K., Gumpel, M., Gros, F. and Portier, M.-M. (1990) J. Neurosci. 10, 764-784.
14. Troy, C.M., Brown, K., Greene, L.A. and Shelanski, M.L. (1990) Neuroscience 36, 217-237.
15. Djabali, K., Portier, M.-M., Gros, F., Blobel, G. and Georgatos, S.D. (1991) Cell 64, 109-121.
16. Brody, B.A., Ley, C.A. and Parysek, L.M. (1989) J. Neurosci. 9, 2391-2401.
17. Oblinger, M.M., Wong, J. and Parysek, L.M. (1989) J. Neurosci. 9, 3766-3775.
18. Thompson, M.A. and Ziff, E.B. (1989) Neuron 2, 1043-1053.

■ *Marie-Madeleine Portier and Françoise Landon:*
Collège de France,
Biochimie Cellulaire,
Paris, France

Plectin

Plectin[1,2] is a ubiquitous and abundant cytomatrix protein of high molecular weight. Its proposed function is that of a versatile crosslinking element of intermediate filaments and of the cytoplasm in general.

Plectin has been identified over a wide range of different tissues and cell types[2]. Depending on the cell type, plectin has been localized either throughout the cytoplasm, where it partially codistributes with intermediate filaments (IF), or in peripheral regions at cellular junctions or membrane attachment sites of IF and microfilaments, or at both locations[3,4].

Plectin isolated from various sources migrates as a single band of apparent molecular weight of 300,000 in SDS-PAGE. However, the actual mass of the polypeptide chain is 466 kDa, as deduced from cDNA sequencing. In ultrastructure, plectin molecules resemble dumb bells, consisting of an ~190 nm long rod section flanked by two globes of ~9 nm diameter[5] (Figure 1). On the molecular level a variety of proteins has been shown to interact with plectin: various IF proteins (**vimentin**; **desmin**; the **neurofilament triplet proteins NF-L, NF-M,** and **NF-H**;

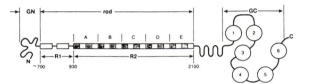

Figure 2. Macromolecular structure of plectin proposed on the basis of its predicted amino acid sequence. GN and GC, NH_2- and COOH-terminal domains, respectively. Rod, central fibrous domain with α-helical coiled-coil conformation. R1 and R2, subdomains of the rod. A-D, tandem repeats of ~200 amino acid residues. 1-6, Six repeats of ~300 amino acid residues, each containing 10 tandem repeats of a 19-residue motif. Plectin molecules probably are formed by the parallel and in register arrangement of two chains; antiparallel lateral association of two such molecules is likely to yield dumb bell structures as shown in Figure 1.

GFAP; skin **cytokeratins 10** and **11**; **lamin B**), **MAP1** and **MAP2**, and α-**spectrin** and its **fodrin** analogue[6-8]. In addition, plectin-plectin self-interaction gives rise to complex network-type arrays[5]. In domain analyses using monoclonal antibodies, the vimentin and lamin B binding sites were mapped to the central rod domain of plectin[9], self-association seems to involve more peripheral rod sections and the globular end domains (Figure 2). Plectin serves as target for several protein kinases *in vivo* and *in vitro*[8,10]. Phosphorylation by kinases A and C has differential effects on plectin's *in vitro* interactions with vimentin and lamin B[8]. DNA sequence data[11] revealed a significant homology between the C-termini (but no other regions) of plectin and human **desmoplakins**[12], including tandem repeats of a 19 amino acid motif, found also in bullous **pemphigoid antigen**[13].

Based on the variety of identified ligands and its subcellular localization, the following functions have been proposed for plectin[2,14]: (1) interlinking of IF, (2) anchorage of IF at the plasma membrane and the nuclear envelope, (3) crosslinking of IF with microtubules (and/or

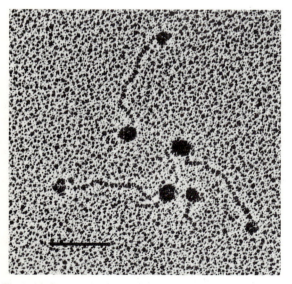

Figure 1. Rotary shadow electron micrograph of rat glioma C_6 cell plectin. Bar 100 nm.

microfilaments), and (4) cytoplasmic network formation through self-association.

In baby hamster kidney cells plectin has also been referred to as IFAP-300K[15].

■ PURIFICATION

Plectin is purified routinely from Triton X-100/0.6M KCl-insoluble residues (crude IF preparations) of cultured cells or bovine eye lens tissue[2]. Solubilization of such residues in 8M urea or 1% sodium lauroylsarcosinate[16] followed by gel permeation chromatography yields plectin preparations of over 90% purity.

■ ACTIVITIES

Plectin has no known enzymatic activities. In solid phase binding assays plectin shows specific affinities to certain IF proteins, spectrin-type proteins, MAP1 and MAP2, and to itself[6-8].

Codistribution with vimentin occurs over repeated cycles of assembly and disassembly[7]. Self association of dumb bell shaped plectin molecules to structures of higher order in dependence of salt concentration has been monitored by rotary shadowing electron microscopy[5].

■ ANTIBODIES

Rabbit antibodies to rat plectin recognize homologues in a variety of cultured cell lines including man, monkey, rat, mouse, hamster, cow and rat kangaroo[2,4,17], as well as in various tissues from man, rat and cow[2,4]. A series of monoclonal antibodies shows characteristics largely similar to those of the antisera (9; Wiche, unpublished data).

■ GENES

Rat plectin has been cloned and sequenced (EMBL sequence data library accessory number: X59601)[11]. The predicted amino acid sequence is 4,140 residues long. Human cDNA clones have also been isolated. Northern blot analysis revealed a ~15 kB mRNA.

■ REFERENCES

1. Wiche, G., Herrmann, H., Leichtfried, F. and Pytela, R. (1982) CSH Symp. Quant. Biol. 46, 475-482.
2. Wiche, G. (1989) CRC Crit. Rev. Biochem. 24, 41-67.
3. Wiche, G., Krepler, R., Artlieb, U., Pytela, R. and Denk, H. (1983) J. Cell Biol. 97, 887-901.
4. Wiche, G., Krepler, R., Artlieb, U., Pytela, R. and Aberer, W. (1984) Exp. Cell Res. 155, 43-49.
5. Foisner, R. and Wiche, G. (1987) J. Mol. Biol. 198, 515-531.
6. Herrmann, H. and Wiche, G. (1987) J. Biol. Chem. 262, 1320-1325.
7. Foisner, R., Leichtfried, F.M., Herrmann, H., Small, J.V., Lawson, D. and Wiche, G. (1988) J. Cell Biol. 106, 723-733.
8. Foisner, R., Traub, P. and Wiche, G. (1991) Proc. Natl. Acad. Sci. (USA) 88, 3812-3816.
9. Foisner, R., Feldman, B., Sander, L. and Wiche, G. (1991) J. Cell Biol. 112, 397-405.
10. Herrmann, H. and Wiche, G. (1983) J. Biol. Chem. 258, 14610-14618.
11. Wiche, G., Becker, B., Luber, K., Weitzer, G., Castañon, M.J., Hauptmann, R., Stratowa, C. and Stewart, M. (1991) J. Cell Biol. 114, 83-99.
12. Green, K.J., Parry, D.A.D., Steinert, P.M., Virata, M.L.A., Wagner, R.M., Angst, B.D. and Nilles, L.A. (1990) J. Biol. Chem. 265, 2603-2612.
13. Stanley, J.R., Tanaka, T., Mueller, S., Klaus-Kovtun, V. and Roop, D. (1988) J. Clin. Invest. 82, 1864-1870.
14. Foisner, R. and Wiche, G. (1991) Curr. Opin. Cell Biol. 3, 75-81.
15. Yang, H.-Y., Lieska, N., Goldman, A.E. and Goldman, R.D. (1985) J. Cell Biol. 100, 620-631.
16. Weitzer, G. and Wiche, G. (1987) Eur. J. Biochem. 169, 41-523.
17. Wiche, G. and Baker, M. A. (1982) Exp. Cell Res. 138, 15-21.

■ *Gerhard Wiche:*
Institute of Biochemistry, University of Vienna,
1090 Vienna,
Austria

Synemin

Synemin is a 230 kDa polypeptide that is associated with desmin or vimentin containing intermediate filaments in a small number of cell types. In chicken, synemin is found in smooth, skeletal and embryonic cardiac muscle, as well as in erythrocytes, lens cells and ependymal tanycytes. The molar ratio of synemin to core monomers is 1-2:100. Synemin may have a role in crosslinking of intermediate filaments.

Synemin (from the Greek syn [with] and nema [filament]) has solubility properties that are similar to those of **desmin** and **vimentin**. All can be solubilized to some extent by buffers of very low ionic strength, but solubilization otherwise requires denaturing conditions[1-4]. After purification in 6 M urea, dialysis of synemin against a low salt buffer generates a soluble, globular, tetrameric form that can bind to desmin *in vitro*[5]. Synemin is rich in glutamic acid and serine, and has a pI of 5.3 in urea[5]. Synemin is a major phosphate acceptor in smooth muscle[5,6], with 95% of the phosphate being attached to serine and 5% to threonine residues[5]. Synemin is distinct from **paranemin**.

Synemin is not an obvious appendage of intermediate filaments when native filaments are examined by electron microscopy[3,4]. Decoration of such filaments with anti-synemin immunoglobulins reveals a regular 180 nm periodicity along the core vimentin filaments of chicken erythrocytes[3] and lens cells[4]. The latter technique has also led to the suggestion that synemin may crosslink intermediate filaments through self-interaction[3], but this remains to be confirmed by nonimmunological techniques[4]. The shape, topology and monomeric/oligomeric state of synemin molecules that are associated with intermediate filaments have not yet been determined. The calculated molar ratio of synemin molecules to core monomers (0.01-0.02) assumes equivalent Coomassie blue binding capabilities and typical mobility characteristics during SDS-PAGE[1,2].

Synemin exhibits several interesting, but as yet unexplained, developmental changes. In chicken striated muscle, synemin synthesis is induced along with desmin in postmitotic myoblasts[1]. Synemin, desmin and vimentin thereafter have indistinguishable distributions, as judged by immunofluorescence microscopy. They coexist in wavy filaments throughout the cytoplasm of early myotubes, and then coalesce around the Z-disks of the developing myofibrils[1]. Whereas all three of these proteins persist in adult skeletal muscle fibres as interlinked collars around the registered Z-disks[1], synemin and vimentin disappear from adult cardiac muscle[7]. Synemin remains in adult cardiac conducting fibres, however[7].

During chicken erythropoiesis, the periodicity of synemin along the core vimentin filaments decreases from 230 nm in embryonic erythroid cells to 180 nm in adult erythrocytes[3]. The rate of assembly of newly synthesized synemin into these filaments seems to be limited by the rate of vimentin assembly and core filament elongation[8]. Whether synemin associates peripherally with the core filament or has a copolymerizing domain has not yet been determined. Synemin coexists with the major neurofilament subunit in erythrocytes from young chickens[9], but is not found with neurofilaments in neurons[7].

Synemin coexists with vimentin in ependymal tanycytes, especially in the cell processes that extend radially from the central canal to the dorsal and ventral fissures of the chicken spinal cord[4]. Synemin appears here very early in embryogenesis, in a small group of cells on the ventral side of the neural tube[4]; a similar distribution has been noted for **annexin**, an EGF receptor kinase substrate[10].

■ PURIFICATION

Synemin was originally identified and enriched as a component of desmin containing intermediate filaments from chicken smooth (gizzard) muscle, where it was found to persist through cycles of filament depolymerization and repolymerization (using 1 M acetic acid)[1]. Purification of synemin (independently of SDS-PAGE) was accomplished by solubilizing intermediate filaments from gizzard cytoskeletons with 6 M urea, and then subjecting the extract to hydroxylapatite, anion exchange, and phosphocellulose chromatography (all in 6 M urea)[5]. Synemin has not been separated from desmin or vimentin under nondenaturing conditions, although all three can be solubilized from gizzard, lens and erythrocyte cytoskeletons by low ionic strength, divalent cation-free solutions[1,2,4].

■ ACTIVITIES

No enzymatic activity is known. Synemin may crosslink intermediate filaments[3,4]. Synemin is routinely identified by its solubility properties, mobility during SDA-PAGE and immunoreactivity.

■ ANTIBODIES

A rabbit anti-chicken gizzard synemin antiserum recognizes at least avian[1] and amphibian[11] synemins, but not mammalian synemins[1,4]. Presumptive bovine synemin has been identified[6], but antibodies that recognize mammalian synemins have not been reported.

■ GENES

Uncharacterized.

■ REFERENCES

1. Granger, B.L. and Lazarides, E. (1980) Cell 22, 727-738.
2. Granger, B.L., Repasky, E.A. and Lazarides, E. (1982) J. Cell Biol. 92, 299-312.
3. Granger, B.L. and Lazarides, E. (1982) Cell 30, 263-275.
4. Granger, B.L. and Lazarides, E. (1984) Mol. Cell. Biol. 4, 1943-1950.
5. Sandoval, I.V., Colaco, A.L.S. and Lazarides, E. (1983) J. Biol. Chem. 258, 2568-2576.
6. Park, S. and Rasmussen, H. (1986) J. Biol. Chem. 261, 15734-15739.
7. Price, M.G. and Lazarides, E. (1983) J. Cell Biol. 97, 1860-1874.
8. Moon, R.T. and Lazarides, E. (1983) Proc. Natl. Acad. Sci. (USA) 80, 5495-5499.
9. Granger, B.L. and Lazarides, E. (1983) Science 221, 553-556.
10. McKanna, J.A. and Cohen, S. (1989) Science 243, 1477-1479.
11. Centonze, V.E., Ruben, G.C. and Sloboda, R.D. (1986) Cell Motil. Cytoskel. 6, 376-388.

■ *Bruce L. Granger:*
Veterinary Molecular Biology,
Montana State University,
Bozeman, MT 59717, USA

Vimentin

Vimentin is the type III intermediate filament protein characteristic of - but not restricted to - fibroblasts and other mesenchymally derived cell types in situ[1]. Vimentin is characteristically expressed in a variety of cell lines in vitro. It is growth regulated[2].

Vimentin (Mr 54 kDa, pI 5.3) usually appears in several isoforms thought to reflect different phosphorylated states. Vimentin shares the common features associated with intermediate filament (IF) proteins, i.e. a highly α-helical rod domain (~310 residues) flanked by nonhelical head and tail domains. Purified vimentin assembles into homopolymeric 10 nm filaments[3] (Figure 1). Vimentin is a type III IF protein[4,5] and therefore more closely related to **desmin**, **GFAP** and **peripherin**, than to other IF proteins (see Introduction). Pig[4], hamster[5], human[2], chicken, mouse[6] and *Xenopus*[7] vimentins have a chain length of 458-466 residues and the sequences are highly homologous (e.g. identity mouse to hamster 98.7%, to human 96%, to chicken 88%)[6]. Vimentin is a substrate for certain protein kinases with phosphorylation restricted to the head domain[8-11].

In cultured cells in interphase, vimentin IF run from the nucleus, where they seem to abut the nuclear membrane, to - or close to - the plasma membrane (Figure 2). After treatment with colchicine to depolymerize microtubules, vimentin IF form coils near the nucleus. Thus *in vivo* IF and microtubules may interact. Although high binding affinity of vimentin to nucleic acids has been noted *in vitro*[12], there is no evidence for nuclear vimentin. Vimentin can be found alone (e.g. 3T3, HS27, teratocarcinoma EC cells), or together with desmin (e.g. BHK or RD), GFAP (e.g. U333CG/343MG), **NF-L** (e.g. SK-LC-17), or with **cytokeratins** (e.g. HeLa). Microinjection experiments show that, with the exception of the keratins, these other IF proteins form copolymers with vimentin. In mitotic cells, vimentin phosphorylation is increased[8,11]. This may change the interaction of vimentin IF with other structures without drastic rearrangement, or could cause the filament to ball transition seen during mitosis in the occasional cell type.

Vimentin is the IF protein typical of fibroblasts, osteocytes, chondrocytes, melanocytes, Langerhans cells of the skin and endothelial cells. Most but not all lymphatic cells are vimentin positive. Usually it is not expressed in normal adult epithelial, muscle or neuronal cells[1,13,14]. During murine embryogenesis, vimentin is found in parietal endoderm, where it is coexpressed with keratins[15,] and in primary mesenchymal cells at the primitive streak stage[16]. In day 15 embryos high expression of vimentin is noted using cDNA probes in brain, heart, lung, skeletal muscle and eye lens. In brain and heart, vimentin expression is strongly reduced at birth, but in lens and lung high levels persist even in the adult animal[6]. These results confirm earlier fluorescence studies of animal and human tissues[1,13,14].

Vimentin positive tumours[13,17,18] include nonmuscle soft

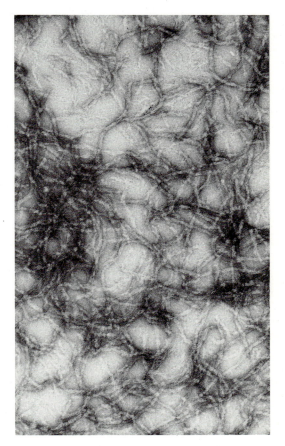

Figure 1. Electron micrograph of 10 nm filaments reassembled from purified vimentin.(Courtesy Norbert Geisler)

tissue tumours[13] (n.b.: synovial carcinomas and epitheloid sarcomas coexpress vimentin and keratin), bone tumours, most but not all lymphomas[18] and leukemias. Most muscle sarcomas coexpress desmin and vimentin. Melanomas are vimentin positive but some seem to coexpress keratin. Carcinomas of some sites, e.g. kidney and thyroid, coexpress keratin and vimentin in most cases, while carcinomas of other sites, e.g. breast, coexpress vimentin in some (~20%) cases, and still other carcinomas, e.g. gastrointestinal tract, express only keratin. Vimentin coexpression in node negative breast carcinomas may be associated with a poorer prognosis[19].

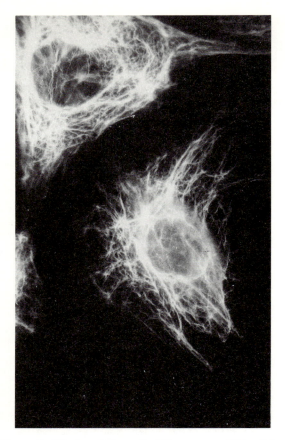

Figure 2.: Display of vimentin IF in a mouse 3T3 cell in culture.

Vimentin seems growth regulated[2]. Levels are increased by rapid growth, by serum or PDGF, and in G_0 cells by mitogens. Keratin and vimentin coexpression in epithelial cell lines is common[20]. NBT II cells, (normally vimentin negative keratin positive) switch on vimentin and down regulate keratin when grown in Ultraser G[21]. Vimentin can be hormonally regulated in a few epithelial cell lines and vimentin coexpression is a feature of hormone independent breast carcinoma cell lines[22]. Vimentin expression is induced during regeneration of kidney tubular epithelium in rat and man[23]. Overexpression of vimentin in transgenic mice results in cataracts followed by pronounced lens degeneration[24].

■ PURIFICATION

Eye lens is a useful source. After homogenization and centrifugation the insoluble material is extracted by 8 M urea. The supernatant is processed by ion exchange chromatography in the presence of urea[3].

■ ACTIVITIES

No enzymatic activity but the purified protein forms IF in vitro. Antibody injection experiments, as well as the existence of cell lines lacking IF, suggest that vimentin or other IF proteins are not necessary for cell division, movement on a substratum, etc. Functions should probably be sought in tissues rather than in individual cells.

■ ANTIBODIES

Several good monoclonal antibodies are commercially available, e.g. V9[14] (epitope between the single cysteine residue and the C-terminus) or MVI. V9 reacts with methanol or acetone fixed cells and reactivity has also been reported on formaldehyde fixed paraffin embedded tissues[19]. Not all vimentin antibodies give equivalent staining patterns; a few, e.g. PKVI stain only mitotic cells. ~90% of patients with chronic hepatitis or primary biliary cirrhosis have high levels of autoantibodies to vimentin.

■ GENES

The Gene EMBL bank contains the vimentin genes from hamster[5] (K00921 - 27)(M16718) human[2] (M200027 - 28) (M18888-95, M25246, M17888, X51907) and chicken (M15850-52, X00185). For the mouse cDNA sequence see[6], and for Xenopus see[7]. The NBRF bank contains residues 1-275 of the pig sequence determined by amino acid sequencing[4] (protein: Vepg). The vimentin gene occurs once in the haploid genome. Two vimentin mRNAs differing in the 3' end are observed in chicken. The gene structure is typical of class-III IF, with the intron positions being shared with desmin, GFAP, and peripherin. Vimentin gene maps to chromosome 10p13[25] in man and to chromosome 2A[26] in mouse.

■ REFERENCES

1. Franke W.W., Schmid, E., Osborn, M. and Weber, K. (1978) Proc. Natl. Acad. Sci. (USA) 75, 5034-5038.
2. Ferrari S., Battini, R., Kaczmarek, L., Rittling, S., Calabretta, B., De Reil, J.K., Philiponis, V., Wei-Fang, J. and Baserga, R. (1986) Mol. Cell Biol. 6, 3614-3620.
3. Geisler N. and Weber K. (1981) FEBS Lett. 125, 253-256.
4. Geisler N. and Weber K. (1981) Proc. Natl. Acad. Sci. (USA) 78, 4120-4123.
5. Quax W., Egberts, W.V., Hendriks, W., Quax-Jeuken, Y. and Bloemendal, H. (1983) Cell 35, 215-223.
6. Capetanaki Y., Kuisk, I., Rothblum, K. and Starnes, S. (1990) Oncogene 5, 645-655.
7. Hermann H., Fouquet, B. and Franke, W.W. (1989) Development 105, 279-298.
8. Evans R.M. (1988) Eur. J. Cell Biol. 46, 152-160.
9. Inagaki M., Gonda, Y., Matsuyama, M., Nishizawa, K., Nishi, Y. and Sato, C. (1988) J. Biol. Chem. 263, 5970-5978.
10. Geisler N., Hatzfeld, M. and Weber, K. (1989) Eur. J. Biochem. 183, 441-447.
11. Chou Y.-H., Rosevear, E. and Goldman, R.D. (1989) Proc. Natl. Acad. Sci (USA) 86, 1885-1889.
12. Traub P. Intermediate Filaments Springer Verlag (1985) .
13. Osborn M. and Weber K. (1983) Lab. Invest. 48, 372-394.
14. Osborn M., Debus, E. and Weber, K. (1987) Eur. J. Cell Biol. 34, 137-143.
15. Lane E.B., Hogan, B.L.M., Kurkinen, M. and Garrels, J.I. (1983) Nature 303, 701-704.

16. Franke W.W., Grund, C., Kuhn, C., Jackson, B.W. and Illmensee, K. (1982) Differentiation 23, 43-59.
17. Miettinen M. (1990) Path. Ann. 25, part 1, 1-36.
18. Möller P., Momburg, F., Hofmann, W.J. and Mattthei-Maurer, D.U. (1988) Blood 71, 1033-1038.
19. Domagala W., Lasota, J., Dukowicz, A., Markiewski, M., Striker, G., Weber, K. and Osborn, M. (1990) Am. J. Pathol. 137, 1299-1304.
20. Franke W.W., Schmid, E., Winter, S., Osborn, M. and Weber, K. (1979) Exp. Cell Res. 123, 25-46.
21. Boyer B., Tucker, G.C., Vallés, A.M., Franke, W.W. and Thiery, J.P. (1989) J. Cell Biol. 109, 1495-1509.
22. Sommers C.L., Walker-Jones, D., Heckford, S.E., Worland, P., Valverius, E., Clark, R., McCormick, F., Stampfer, M., Abularach, S. and Gelmann, E.P. (1989) Cancer Res. 49, 4258-4263.
23. Gröne H.J., Weber, K., Gröne, E., Helmchen, U. and Osborn, M. (1987) Am. J. Pathol. 129, 1-8.
24. Capetanaki Y., Smith, S. and Heath, J.P. (1989) J. Cell Biol. 109, 1653-1664.
25. Mattei M.G., Lilienbaum, A., Lin, L.Z., Mattei, J.-F. and Paulin, D. (1989) Genetical Research 53, 183-185.
26. Ferrari S., Cannizzaro, L.A., Battini, R., Huebner, K. and Baserga, R. (1987) Human Genetics 41, 616-626.

■ Mary Osborn:
Max Planck Institute for Biophysical Chemistry,
D-3400 Goettingen,
Germany

4

Motor Proteins

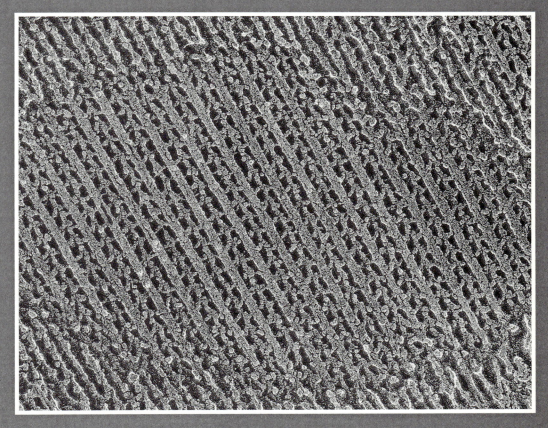

Transmission electron microscope image of freeze-etched, platinum replicated flight muscle from fly thorax (extracted with glycerol to remove cytoplasmic proteins and ATP, thus creating the stiff 'rigor' state) in which thick myosin filaments and intervening thin actin filaments are interconnected by myriads of tilted myosin cross-bridges. This interior view of the myofilament lattice was generated by freezing the glycerinated muscle, cracking it open with a knife, and momentarily freeze-drying or 'etching' the exposed surface to remove enough ice to expose the filaments.

(Courtesy of Dr John Heuser, Washington University, St Louis).

Motor Proteins

Through the actions of proteins that convert chemical energy into force and motion, cells can accomplish a remarkable number of animated tasks. Some forms of motility are ubiquitous to all eukaryotes; examples include the segregation of chromosomes, cytokinesis and the transport of intracellular membranous organelles. In addition, some cells, such as muscle or ciliated cells, have developed highly specialized arrays of motor proteins that generate large forces in an orchestrated fashion.

Cells have evolved a variety of different mechanisms for the generation of motility. **Actin** filaments and microtubules hydrolyze nucleotide and can perform mechanical work through their assembly and disassembly[1-3]. Another polymer, **caltractrin**, also performs work, in this case through reversible contractions and extensions coupled to changes in calcium concentration. The bacterial flagellar motor, ribosomes, and DNA/RNA helicases are other examples of protein complexes that undergo directional motion.

The "motors" that we will consider here are eukaryotic proteins that hydrolyze nucleotide and harness the derived energy to move unidirectionally along cytoskeletal polymers. Motor proteins are most broadly categorized according to the type of cytoskeletal polymer with which they interact. **Myosins** refer collectively to the large family of related motors that move along actin filaments. Two rather different types of force generating motors, **dyneins** and **kinesins**, interact with microtubules. **Dynamin** may represent a third type microtubule motor, although further work is required to establish whether it has force-generating capability. So far, motors that utilize intermediate filaments as tracks have not been described. Here the properties and functions of actin and microtubule based motors are summarized. A number of more detailed reviews are also available[4-9].

■ THE IMPORTANCE OF BEING POLAR

Motors recognize the asymmetry of microtubules and actin filaments and move along these tracks with a defined polarity. All myosins discovered to date move towards the plus-end (barbed end) of actin. In contrast, microtubules interact with both plus-end directed (generally kinesins) and minus-end directed (generally dyneins) motors. For cells to use polarity specific motors constructively, filament assembly within the cytoplasm is nucleated from specific sites and with a defined polarity. Microtubules in most cells are nucleated with their minus ends anchored at centrosomes or basal bodies. The cortical actin filaments, on the other hand, have their plus (barbed) ends attached at unidentified sites at the plasma membrane.

The regular filament polarity in muscle sarcomeres, mitotic spindles and cilia enables the force generating activities of motor arrays in these specialized motile apparati to be coordinated and directed. The ordered polarity of filaments also provides a navigational system for transport. Since microtubule polarity is generally similar throughout the cytoplasm with respect to the centre of most cells, an organelle can be directed towards or away from the cell surface by interacting with either a plus-end or a minus-end directed motor respectively[7]. Similarly in axons, which have the plus-ends of microtubules uniformly oriented towards the nerve terminal, vesicles can be delivered selectively by polarity specific motors to either the cell body or the nerve ending. The general polarity of cortical actin also allows myosins to transport membrane glycoproteins towards the leading edge of migrating cells[10] or (if anchored) to generate tension in the opposite direction from the membrane[11]. The actin cables in *Nitella*, which are used for transporting membranes, are another example of polarized actin filament arrays[12].

In some differentiated cells, filament organization is altered in order to conduct specialized transport tasks. Microtubules in MDCK kidney epithelial cells are nucleated at their minus-ends near the apical membrane and drape the cytoplasm from the apical to the basal ends, presumably in order to transport components between these domains[13]. Even more curious is the rearrangement of microtubules that occurs in neurons when they establish axonal and dendritic processes. At this time, microtubule polarity in dendrites (but not axons) becomes mixed (plus-ends point towards and away from the nucleus in equal numbers), which may allow distinct components to be transported to axonal and dendritic branches[14].

Neurons and epithelial cells clearly can reorganize the polarity of their filaments, but can cells also control the direction that motors travel along these filaments? Until recently, such an idea would have sounded preposterous, as motors of the same general class (i.e. kinesins or dyneins) were all believed to move in the same direction. Two recent and surprising results have shattered such dogma. First, the ***ncd*** (see **kinesin related proteins**) motor from *Drosophila*, which is similar to kinesin by sequence homology, was found to move towards the microtubule minus-end, the direction characteristic of dynein but not of kinesin[15,16]. Second, evidence has been presented that a dynein-like molecule from *Reticulomyxa* can switch its direction of movement along microtubules; the choice of direction may be determined by a posttranslational modification of the motor, possibly phosphorylation[17,18]. Thus, subtle structural features appear to define a motor's direction of movement[19].

MOTOR PROTEINS ARE COMPLEX MACHINES

Motors are large proteins with intriguing morphologies. By electron microscopy, dynein, kinesin and the muscle myosins appear as elongate molecules (40-100 nm in length) that have two or more oblate shaped heads (the ATP hydrolyzing domains) connected to long rod domains (Figure 1). In the case of kinesin and dynein, a different shaped globular "tail" domain is found at the opposite end of the rod. Several polypeptide chains can contribute to this overall structure. Ciliary outer arm dynein (1,900 kDa), the most complex cytoskeletal motor, has as many as ten polypeptides, three of which are 450-500 kDa.

What are the functions of these various structural domains and subunits? The answer emerging from a variety of studies is that the different domains are specialized for tasks related to chemomechanical transduction, regulation and binding interactions. The force generating domains of motors have been partially delineated through ATPase and in vitro motility measurements of proteolytic motor fragments and of truncated motors expressed in bacteria. In the case of myosin, the globular S1 domain (900 amino acids) has actin-activated ATPase activity, moves actin in vitro and generates forces comparable to intact **sarcomeric myosin**[20,21]. Similarly, an approximately 350 amino acid globular domain of the kinesin heavy chain, which contains an ATP and a micro-

tubule binding site, has motor activity[22] (Goldstein; personal communication). The motor domain is highly conserved in related family members of a particular class of motor proteins (see below). Kinesins, myosins, and dynamins, however, show no sequence homology to one another, suggesting that they either arose independently or very distantly in evolution. The sequence of the 450 kDa ATP binding polypeptide of **axonemal dynein** also shows no homology to the other motor classes[23,24]. This heavy chain is also unusual in having four consensus sites for ATP binding.

Motors must be tethered to an object through an "attachment" domain to generate force or tension. Various forms of motility are achieved using three basic arrangements of motors and cytoplasmic filaments. Directional transport is achieved by attaching motors onto a "cargo" which is moved through the cytoplasm along stationary filaments. The types of cargo transported within cells are numerous and include membranous organelles, chromosomes, protein complexes, and even possibly RNA. The interaction between motors and these various cargoes must be specific, or else motors would move and pull on structures at random, thereby creating havoc in the cytoplasm. A second type of organization is achieved by motors which crossbridge and generate a sliding force between parallel arrays of cytoskeletal filaments. Axonemal beating and anaphase B chromosome movements[25] are elicited by motors arranged in such a manner. A third type of force generating organization is created through the self-assembly of motors into a filament, as occurs with muscle myosin and **cytoplasmic myosin II**.

In contrast to the chemomechanical domain, the "attachment" or "tail" domains of motors vary considerably in structure and primary sequence, most likely reflecting the diversity of molecular interactions and arrangements of motors within cells[26]. Sarcomeric myosin, for example, has an α-helical coiled-coil domain, that induces the self-assembly of myosin into filaments; some **myosin I** motors, in contrast, contain "tail" domains that can attach to actin filaments or to phospholipid head groups[5]. Evolution has clearly tinkered with the binding properties of motors in order to select for a variety of force generating activities, as will be explored below.

SUPERFAMILIES OF MOTOR PROTEINS

A remarkable number of new motor proteins have been uncovered in the last couple of years. Prior to this recent turn of events, new motor proteins were discovered every few human generations rather than every few months. Myosin was the first named by Kuhne in 1864[27], and it remained the sole motor under investigation, until dynein was isolated from Tetrahymena cilia by Ian Gibbons[28] a century later. Muscle myosin and axonemal dynein became the paradigm for actin and microtubule based motility, and it was assumed that various forms of cytoplasmic based movements would be driven by these or like motors. Indeed, the discovery of a small, single headed myosin-like molecule by Pollard and Korn[29] in

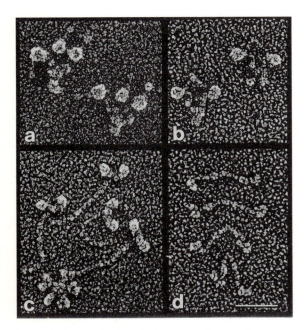

Figure 1. Rotary shadowed electron microscopic images of outer arm dynein (a), cytoplasmic dynein (b), sarcomeric myosin (with an IgM attached to the myosin tail) (c) and conventional kinesin (d). Bar 50nm. Photographs kindly supplied by Dr. John Heuser, Washington University, St. Louis, MO.

Acanthamoeba in the early 1970s was met with the skepticism that this small myosin may be a proteolytic fragment of a two-headed, muscle-like myosin. However, we now know that such small myosins are common place, and that cytoplasmic motions generally employ a host of motors that are quite different from the conventional motors operating in muscle and flagella.

The population explosion of motor proteins can be attributed to two different experimental approaches. First, the development of in vitro motility assays (discussed later) allowed the activities of motors to be detected even in crude cell homogenates. In combination with biochemical fractionation, such assays led to the identification of kinesin, cytoplasmic dynein and possibly dynamin as force generating enzymes. A second approach for identifying new motors has relied upon genetics and molecular biology. Many motors were discovered serendipitously through genetics. Mutations with phenotypic defects in processes as disparate as yeast mating, fungal mitosis, worm movements, Drosophila vision, and determination of mouse coat color have turned out to lie in genes that encode proteins with a "motor" domain[30-32]. With this flood of new sequence information, regions of high conservation within the kinesin and myosin motor domains became apparent, and this information stimulated motor "hunts" using polymerase chain reaction.

From these molecular genetic searches, it now appears as though all eukaryotes have large superfamilies of motor proteins. When this article was written, the number of kinesin related genes in Drosophila was at least twelve (Goldstein, personal communication) and may be as high as thirty[33]. The number of myosins in Dictyostelium has reached six (see **protozoan myosin I**) and may be as high as twelve or more. Genes for many dynamin-like proteins have also been uncovered[34]. The extent of the dynein family has not yet been explored through molecular biology, but genetic studies in Chlamydamonas as well as biochemical investigations have revealed that several distinct dynein species comprise the outer and inner arms of axonemal outer doublets and that yet another dynein exists in the cytoplasm.

All members of a motor superfamily share a similar sized, highly homologous force transducing domain (see Figures 2 and 3). The degree of conservation within the motor domain is generally 30-40% between the various kinesin and myosin family members; certain elements of the motor domain involved in ATP and filament binding, however, are even more highly conserved. Sequence comparison may ultimately provide some insight into structural motifs that are important for the force transducing process. In most cases, the motor domains are located at the N-terminus, although KAR3 (yeast; see kinesin related proteins), ncd (a Drosophila myosin-like protein) and nina C (Drosophila ; see myosin related proteins) all have their motor domain in the C-terminal portion of the protein.

Outside of the motor domain, there is relatively little amino acid homology between the various members within the kinesin and myosin motor superfamilies, and even their predicted secondary structures appear quite different. The functions of these nonmotor regions are not definitely known, although they are thought to target motors to different macromolecular structures, as discussed earlier. The nonmotor domains can also be important for regulation. For instance, phosphorylation of the tail of **smooth muscle myosin** causes it to bend, which in turn inhibits the ATPase activity of the motor head. Even more unusual is the ninaC gene product, which has an N-terminal putative protein kinase domain spliced onto a myosin motor domain[35]. Perhaps this kinase regulates the motor by autophosphorylation or, alternatively, the kinase may be delivered to its substrate targets by an active transport process.

It should be mentioned that many of the new motors are classified as motors solely because of their sequence homology with either myosin or kinesin. Other than conventional dynein, kinesin and myosin, only myosin I from Acanthamoeba, **brush border myosin** from intestinal epithelia, p190/myo2/dilute, ncd (kinesin-like), Eg5 (kinesin-like) and MKLP-1 (kinesin-like) have actually been demonstrated to generate force in vitro. In addition, the native molecular weight and subunit composition of many motors have not been determined. The reward and necessity of biochemically characterizing motors uncovered by genetic or molecular biological screens is demonstrated in the work with the kinesin-like motor ncd, which surprisingly was shown to move towards the microtubule minus-end. This result illustrates that one cannot, at present, assign important functional attributes to motors simply based upon amino acid comparisons.

In addition to biochemical characterization, the daunting task of deciphering the biological functions of these various newly discovered motors is underway. Clearly, the phenotypes of cells with deletions in motor genes can potentially reveal a great deal regarding the functions of these proteins. Determining the cellular location of these proteins will also be important for deciphering their biological role(s).

■ WHY SO MANY MOTORS?

Why do cells have so many actin and microtubule based motors and what are their functions? Very likely, cells dedicate different motors to different force generating tasks. Certain motile processes may even be finely subdivided. Recent genetic experiments with unc-104 in C. elegans (see kinesin related proteins), for instance, suggest that different types of vesicles are transported to nerve terminals via different kinesin motors[36]. Differentiated cells also appear to have evolved their own specific motors (such as brush border myosin in intestinal epithelia) to carry out functions unique to their specialized structure and function.

Other findings, however, indicate that the notion of "one task-one motor" cannot be strictly correct. In many instances, knock-outs of myosin or kinesin genes in yeast, Aspergillus and Dictyostelium produce little or no adverse affect upon the cell, suggesting that there must be considerable overlap and redundancy in the function of various motors. This is apparent in S. cerevisiae, where multiple kinesin deletions are required to cripple the cells' ability to segregate its chromosomes (see kinesin related proteins; *KAR3*), and in *Dictyostelium*, where knock-outs of individual myosin I genes do not hamper cell migration (see Protozoan Myosins). One could contend that several different back-up mechanisms for mitosis and cell movement may confer a survival advantage, since these cellular processes are such critical ones. Yet by such an argument, it is surprising that cells rely entirely upon a single motor, myosin II, to carry out cytokinesis (see cytoplasmic myosin II).

Although motor domains and filament subunits (**tubulin** and **actin**) are very highly conserved, the functions carried out by these motors and polymers are not necessarily the same in all organisms. As an example, transport of membranous organelles occurs along microtubules in vertebrate cells, yet such processes are actomyosin driven in certain plants and lower eukaryotes such as *S. cerevisiae*[37]. Thus, although virtually all eukaryotes require a force generating process to arrange and transport membranous organelles, some organisms preferentially exploit actin filaments[38] while others use microtubules[39] for such tasks. Melanophores[40] and *Acanthamoeba*[41,42] (and perhaps all cells) appear to use both microfilaments and microtubules for membrane transport.

Migration of cells as distantly related as *Dictyostelium* and human fibroblasts depends upon a cortical actin array (and perhaps myosin I). Nematode sperm, however, quite efficiently crawl along surfaces, yet possess few, if any, actin filaments or microtubules! Instead of utilizing these ubiquitous filaments, these cells are packed with a polymer composed of a protein called **major sperm protein**, which appears to be unique to nematodes. Have nematodes abandoned convention and evolved their own unique motile machinery?

Different organisms can also use the same types of components in a motile process, but use them in various ways. Although all cells utilize microtubules and microtubule based motors for meiosis and mitosis, the precise mechanism for separating of chromosomes and chromatids can vary considerably in different cell types[43]. For instance, in mitotic chromosome segregation, mammalian cells principally employ an anaphase A mechanism (movement of chromosomes down a kinetochore fibre towards the pole), while yeast rely on anaphase B separation (elongation of the spindle). Even the mechanisms of anaphase B can be quite different in different species. Diatoms use a motor in their spindle overlap zone to push the spindle apart[25], while certain fungi have motors that act upon the astral microtubules to pull the spindle poles apart. Thus, cells can use different mechanical strategies to move similar macromolecular structures in the same direction; one must therefore be cautious about generalizing motile mechanisms to all cell types.

■ SOLVING THE FORCE GENERATING MECHANISM

Contrary to what is presented in most college biology textbooks, the mechanism of force generation is far from solved. The classical model for muscle contraction is that the myosin head undergoes a large conformational change during its ATPase cycle that changes its angle relative to actin, thereby generating a relative sliding motion[44]. Dynein and kinesin are thought to undergo similar mechanical actions. Although this hypothesis is conceptually attractive and indeed may be correct, a rotation of the myosin head as such has never been demonstrated, despite intensive investigations. Even the distance that motors move along filaments during one cycle of ATP hydrolysis is in considerable debate; the current estimates vary from 10 nm[45] to as great as 200 nm[46] under zero load. Clearly the latter results, if true, would be incompatible with the conventional notion of a conformational change in myosin occurring once per ATP hydrolysis cycle. Furthermore, motors are now found to have a number of remarkable properties which are difficult to explain by conventional models, including the abilities to induce torque[47], move bidirectionally[18], hold onto filaments for 99.9% of their ATPase cycle[48,49], and move as slow as 0.01 μm/sec[50] or as fast as 80 μm/sec (*Nitella* membrane myosin). Alternatives to the "rotating crossbridge" model have been proposed[51], but they stand on equally tenuous ground. Alas, the force generating mechanism seems more elusive now than it did a decade ago.

Why has the force generating mechanism been so difficult to elucidate? Clearly the problem is complex and multifaceted. A detailed mechanistic understanding will require high resolution structural information on the motors and filaments, knowledge of the chemical intermediates in the ATPase cycle and an understanding of the mechanical properties and interactions between motors and filaments at different stages of the ATPase cycle. Some aspects of the problems are far advanced; for example, the kinetics of the myosin and dynein ATPase cycles have been dissected in great detail[4,52]. On the other hand, the structure of a motor protein at atomic resolution has not yet been reported, although the structure of the myosin motor domain will be revealed soon.

New assays and technology have brought renewed optimism for understanding the force generating problem. In previous decades, motor activity was studied almost exclusively by examining the mechanical and structural properties of contracting muscle. While this system has numerous advantages, such as the great degree of order of the actin and myosin molecules, it is also important to study motile properties of purified motors in an *in vitro* setting. The *in vitro* motility assay most commonly employed involves first coating a surface with purified motors and then examining the ability of these attached motors to translocate their complementary filament across the surface in the presence of ATP. Motors that perform considerably different biological functions, such as muscle myosin, ciliary dynein, and a mitotic kinesin, can now be studied and compared using this type of motility assay. An alternative approach is to coat the surface of a submicron sphere with a motor and then observe the translocation of the sphere along a stationary filament. The above assays enable one to perform experiments on motors that would not be possible in their normal biological habitat. For example, mitotic microtubule motors, which are present in vanishingly small quantities in cells, can be expressed in bacteria and their motility studied *in vitro* [15,16].

Remarkable advances are also being made in the ways that motors can be manipulated *in vitro*. Using low dilution assays, it is now possible to examine movements generated by relatively few motors; in the case of kinesin[49] and myosin[53], microtubule or actin translocation induced by single molecules can be observed. Previously, ion channels were the only proteins that could be assayed (through patch clamp technology) at the level of individual molecules. Optical tweezers can also be used to induce motility of a bead containing a single motor on a microtubule[48], an interaction that would be rarely achieved if the two components had to diffuse to come in contact. Other clever assays are being developed to measure motor forces. In one approach, an actin filament attached firmly onto a flexible glass needle is "cast" over a lawn of surface-bound myosin heads[20]. The motors move the actin filament which causes the needle to bend until the forces are balanced; the degree of bending of this calibrated needle reveals the amount of force exerted by the motors. Optical traps, which can halt motor induced transport of beads, organelles[54] and chromosomes[55], are also being used to measure motor forces.

Measurements of motor induced forces and displacements can also be made with remarkable precision. Nanometer sized displacements of objects can be computed from video records[56], and similar spatial precision with millisecond temporal resolution can be achieved with photodiode detectors[57]. Such high spatial and temporal resolution measurements can potentially reveal the quantal events in force production. Such information should provide a more detailed understanding of the mechanochemical cycle.

■ REFERENCES

1. Coue, M., Lombillo, V.A. and McIntosh, J.R. (1991) J. Cell Biol. 112, 1165-1175.
2. Koshland, D.E., Mitchison, T.J. and Kirschner, M.W. (1988) Nature 331, 499-504.
3. Tilney, L.G. and Inoue, S. (1982) J. Cell Biol. 93, 820-827.
4. Cooke, R. (1986) CRC Crit. Rev. Biochem. 21, 53-118.
5. Pollard, T.D., Stephen, K., Dobberstein, S.K. and Zot, H.G. (1991) Annu. Rev. Physiol. 53, 653-681.
6. Porter, M.E. and Johnson, K.A. (1989) Annu. Rev. Cell Biol. 5, 119-151.
7. Vale, R.D. (1987) Annu. Rev. Cell Biol. 3, 347-378.
8. Vallee, R.B. and Shpetner, H.S. (1990) Annu. Rev. Biochem. 59, 909-932.
9. Warrick, H.W. and Spudich, J.A. (1987) Annu. Rev. Cell Biol. 3, 379-421.
10. Sheetz, M.P., Turney, S., Qian, H. and Elson, E.L. (1989) Nature 340, 284-288.
11. Smith, S.J. (1988) Science 242, 708-715.
12. Kersey, Y.M. and Wessells, N.K. (1976) J. Cell Biol. 68, 264-275.
13. Bacallao, R.C., Antony, C., Dotti, C., Karsenti, E., Stelzer, E.H.K. and Simons, K. (1989) J. Cell Biol. 109, 2817-2832.
14. Baas, P.W., Black, M.M. and Banker, G.A. (1989) J. Cell Biol. 109, 3085-3094.
15. McDonald, H.B., Stewart, R.J. and Goldstein, L.S. (1990) Cell 63, 1159-1165.
16. Walker, R.A., Salmon, E.D. and Endow, S.A. (1990) Nature 347, 780-782.
17. Euteneuer, U., Koonce, M.P., Pfister, K.K. and Schliwa, M. (1988) Nature 332, 176-178.
18. Schliwa, M., Shimizu, T., Vale, R.D. and Euteneur, U. (1991) J. Cell Biol. 112, 1199-1203.
19. Malik, F. and Vale, R.D. (1990) Nature 347, 713-714.
20. Kishino, A. and Yanagida, T. (1988) Nature 344, 74-76.
21. Toyoshima, Y.Y., Kron, S.J., McNallay, E.M., Niebling, K.R., Toyoshima, C. and Spudich, J.A. (1987) Nature 328, 536-539.
22. Yang, J.T., Saxton, W.M., Stewart, R.J., Raff, E.C. and Goldstein, L.S. (1990) Science 249, 42-47.
23. Gibbons, I.R., Gibbons, B.H., Mocz, G. and Asai, D.J. (1991) Nature 352, 640-643.
24. Ogawa, K. (1991) Nature 352, 643-645.
25. Hogan, C.J. and Cande, W.Z. (1989) Cytoskeleton 16, 99-103.
26. Schroer, T.A. (1991) Curr. Opin. Cell Biol. 3, 133-137.
27. Kuhne, W. (1864) Untersuchungen über das Protoplasma und die Kontraktilität. Leipzig, Engelmann.
28. Gibbons, I.R. (1963) Proc. Natl. Acad. Sci. (USA) 50, 1002-1010.
29. Pollard, T.D. and Korn, E.D. (1973) J. Biol. Chem. 248, 4682-4690.
30. Endow, S.A. (1991) Trends Bio. Sci. 16, 221-225.
31. Kiehart, D.P. (1990) Cell 60, 347-350.
32. Vale, R.D. and Goldstein, L.S.B. (1990) Cell 60, 883-885.
33. Endow, S.A. and Hatsumi, M. (1991) Proc. Natl. Acad. Sci. (USA) 88, 4424-4427.
34. Hollenbeck, P.J. (1990) Nature 347, 229.
35. Montell, C. and Rubin, G.M. (1988) Cell 52, 757-772.
36. Hall, D.W. and Hedgecock, E.M. (1991) Cell 65, 837-847.
37. Johnston, G.C., Prendergast, J.A. and Singer, R.A. (1991) J. Cell Biol. 113, 539-551.
38. Kachar, B. and Reese, T.S. (1988) J. Cell Biol. 106, 1545-1552.
39. Dabora, S.L. and Sheetz, M.P. (1988) Cell 54, 27-35.
40. Mercer, J.A., Seperack, P.K., Strobel, M.C., Copeland, N.G. and Jenkins, N.A. (1991) Nature 349, 709-713.
41. Adams, R.J. and Pollard, T.D. (1986) Nature 322, 754-756.
42. Kachar, B., Albanesi, J.P., Fujisaki, H. and Korn, E.D. (1987) J. Biol. Chem. 262, 16180-16185.
43. Carpenter, A.T.C. (1991) Cell 64, 885-890.
44. Huxley, H.E. (1969) Science 164, 1356-1364.
45. Toyoshima, Y.Y., Kron, S.J. and Spudich, J.A. (1990) Proc. Natl. Acad. Sci. (USA) 87, 7130-7134.
46. Harada, Y., Sakurada, K., Aoki, T., Thomas, D.D. and Yanagida, T. (1990) J. Mol. Biol. 216, 49-68.
47. Vale, R.D. and Toyoshima, Y.Y. (1988) Cell 52, 459-469.
48. Block, S.M., Goldstein, L.S. and Schnapp, B.J. (1990) Nature 348, 348-352.
49. Howard, J., Hudspeth, A.J. and Vale, R.D. (1989) Nature 342, 154-158.
50. Mooseker, M.S. and Coleman, T.R. (1989) J. Cell Biol. 108, 2395-2400.
51. Vale, R.D. and Oosawa, F. (1990) Adv. Biophys. 26, 97-134.
52. Johnson, K.A. (1985) Annu. Rev. Biophys. Biophys. Chem. 14, 161-188.
53. Uyeda, T.Q.P., Warrick, H.M., Kron, S.J. and Spudich, J.A. (1991) Nature 352, 307-311.
54. Ashkin, A., Schulze, K., Dziedzic, J.M., Euteneur, U. and Schliwa, M. (1990) Nature 348, 346-348.
55. Berns, M.W., Wright, W.H., Tromberg, B.J., Profeta, G.A., Andrews, J.J. and Walter, R.J. (1989) Proc. Natl. Acad. Sci. (USA) 86, 4539-4543.
56. Gelles, J., Schnapp, B.J. and Sheetz, M.P. (1988) Nature 331, 450-453.
57. Kamimura, S. and Kamiya, R. (1989) Nature 340, 476-478.
58. Roof, D.M., Meluh, P.M. and Rose, M.D. (1992) J. Cell Biol. 118, 95-108.

59. Stewart, R.J., Pesavento, P.A., Woerpel, D.N. and Goldstein, L.S.B. (1991) Proc. Natl. Acad. Sci. (USA) 88, 8470-8474.
60. Sawin, K.E., Mitchison, T.J. and Wordeman, L.G. (1992) J. Cell Sci. 101, 303-313.
61. Cole, D.G., Cande, W.Z., Baskin, R.J., Skoufias, D.A., Hogan, C.J., and Scholey, J.M. (1992) J. Cell Sci. 101, 291-301.

■ Ronald D. Vale:
Department of Pharmacology,
University of California,
San Francisco, CA 94143, USA

```
          0                                                     50
Unc104    .......... ....MSSVKV AVRVRPFNQR EISNTSKCVL QVNGNTTTIN GHSINKENFS FNFD...HSY WSFARNDPHF
Krg2      .......... .......... .......... .......... .......... ......EIGE FKFDHVFASH
Krg1      TLTPPTCNNG AATSDSNIHV YVRCRSRNKR EIEEKSSVVI STLGP...QG KEIILSNGSH QSYSS.SKKT YQFDQVFGAE
Cut7      ....HALHDE NE.TNINVVV RVRGRTDQEV RD..NSSLAV STSGA...MG AELAIQSDPS SMLVT...KT YAFDKVFGPE
Bimc      ......EREI NEDTSIHVVV RCRGRNEREV KE..NSGVVL QTEGV...KG KTVELSMGPN AVSN....KT YTFDKVFSAA
Flyncd    ...RKELHNT VMDLRGNIRV FCRIRPPLES EE..NRMCCT WTYHD....E STVELQSIDA QAKSKMGQQI FSFDQVFHPL
Kar3      ...RRTLHNE LQELRGNIRV YCRIRPALKN LENSDTSLIN VNEFDDNSGV QSMEVTKIQN TAQV....HE FKFDKIFDQQ
Nod       ........M EGAKLSAVRI AVREAPYRQF LG..RREPSV VQFPP.WSDG KSLIVDQ... ........NE FHFDHAFPAT
Squidkin  ........M DVASECNIKV ICRVRPLNEA EE.RAGSKFI LKFP....TD DSISIAG... ........KV FVFDKVLKPN
Flykin    ....MSAER EIPAEDSIKV VCRFRPLNDS EE.KAGSKFV VKFP.NNVEE NCISIAG... ........KV YLFDKVFKPN
Humkin    ........M ADLAECNIKV MCRFRPLNES EV.NRGDKYI AKFQ...GED TVVIAS.... ........KP YAFDRVFQSS
Kinfam    .......m ....esnikv .cr.rplnes ee..nsskv. vtfg.....g ks.i.s.... ........k. .Fdkvf.p.

          80                                                    130
Unc104    ITQKQVYEEL GVEMLEHAFE GYNVCIFAYG QTGSGKSYTM MGKANDPDE. .......... .MGIIPRLCN DLFARIDNNN
Krg2      CTNLEVYERT SRPMIBKLLM SFNATIFAYG MTGSGKTFTM SGNEQE.... .......... .LGLIFLSVS YLFTNIMEQS
Krg1      SDQETVFNAT AKNYIKEMLH GYNCTIFAYG QTGTGKTYTM SGDINILGDV QSTDNLLLGE HAGIIPRVLV DLFKELSSLN
Cut7      ADQLMLFENS VAPMLEQVLN GYNCTIFAYG QTGTGKTYTM SGDLSDSDGI LSEG...... .AGLIPRALY QLFSSLDNSN
Bimc      ADQITVYEDV VLPIVTEMLA GYNCTIFAYG QTGTGKTYTM SGDMTDTLGI LSDN...... .AGIIPRVLY SLFAKLADTE
Flyncd    SSQSDIFEMV S.PLIQSALD GYNICIFAYG QTGSGKTYTM DGVPES.... .......... .VGVIPRTVD LLFDSIRGYR
Kar3      DTNVDVFKEV G.QLVQSSLD GYNVCIFAYG QTGSGKTFTM LNPG...... .......... .DGIIPSTIS HIFNWINKLK
Nod       ISQDEMYQAL IILPLVDKLLE GFQCTALAYG QTGTGKSYSM GMTPPGEILP EH........ .LGILPRALG DIFERVTARQ
Squidkin  VSQE.VYNVG AKPIVADVLS GCNGTIFAYG QTSSGKTHTM EGVLDKPSM. .......... .HGIIPRIVQ DIFNYIYGMD
Flykin    ASQEKVYNEA AKSIVTDVLA GYNGTIFAYG QTSSGKTHTM EGVIGDSVK. .......... .QGIIPRIVN DIFNHIYAME
Humkin    TSQEQVYNDC AKKIVKDVLE GYNGTIFAYG QTSSGKTHTM EGKLHDPEG. .......... .MGIIPRIVQ DIFNYIYSMD
Kinfam    asqe.vyeev akpiv.dvle gynctifAYG qTgsGKtytM sg.lnd.... .......... .aGiipriv. dlFn.iy.mn

          160                                                   210
Unc104    DKDVQ.YSVE VSYMEIYCER VKDLLNPNSG .....GNLRV REHPLLGP.. .......... .....YVDD LTKMAVCSYH
Krg2      MNGDKKFDVI ISYLEIYNER IYDLLESGLE ESGSRISTPS RLYMSKSNSN GLGVELKIRD DSQYGVKVIG LTERRCESSE
Krg1      KE....YSVK ISFLELYNEN LKDLLSDSED DDPAVNDPKR QIRIFDNNNN NSSI...... ....MVKG MQEIFINSAH
Cut7      QE....YAVK CSYYELYNEE IRDLLVSEE. .......LRK PARVFEDTSR RGNV...... ......VITG IEESYIKNAG
Bimc      ......STVK CSFIELYNEE LRDLLSAEE. ........NP KLKIYDNEQK KGHMST.... ....LVQG MEETYIDSAT
Flyncd    NLGWE.YEIK ATFLEIYNEV LYDLLSNEQ. ......KDM EIRMAKNNKN DI........ ....YVSN ITEETVLDPN
Kar3      TKGWD.YKVN CEFIEIYNEN IVDLLRSDNN N.....KEDT SIGLKHEIRH DQETKTT... ....TITN VTSCKLESEE
Nod       ENNKDAIQVY ASFIEIYNEK PFDLLGST.. ......PHMP MVAARCQR.. .......... .......... CTCLPLHSQA
Squidkin  ENPEF.HIHK ISYYEIYDLK IRDLLDVT.. ......KTNL AVHEDKNRVP .......... .....FVKG ATERFVSSPE
Flykin    VNLEF.HI.K VSYYEIYMDK IRDLLDVS.. ......KVNL SVHEDKNRVP .......... .....YVKG ATERFVSSPE
Humkin    ENLEF.HI.K VSYFEIYLDK IRDLLDVS.. ......KTNL SVHEDKNRVP .......... .....YVKG CTERFVCSPD
Kinfam    en....yivk .sy.EiYnek irDLL.s... ......ktnl svhedknrv. .......... .....yvkg .terfv.spe
```

Figure 2. Sequences of the motor domain from members of the kinesin superfamily. Kinfam represents a consensus sequence for unc104 (*C. elegans*), KRG1(KIP1) (*S. cerevisiae*), KRG2(KIP2) (*S. cerevisiae*), cut 7 (*S. pombe*), bimC (*A. nidulans*), flyncd (ncd) (*Drosophila*), KAR3 (*S. cerevisiae*), squid kinesin (*L. paeli*), fly kinesin (*Drosophila*), and human kinesin (*H. sapiens*). Characteristic "signatures" of the kinesin motor domain are revealed the regions of high homology. Degenerate PCR primers designed from the regions Kinfam 106–112, 175–182, 306–314, 383–388 have

```
         240                                                              290
Unc104   DICNLMDEGN KARTVAATNM NSTSSRSHAV FTIVLTQKRH CADSNLDTEK H......... ..SKISLVDL AGSERANSTG
  Krg2   ELLRWIAVGD KSRKIGETDY NARSSRSHAI VLIRLTSTNV KNGTSRS... .......... ..STLSLCDL AGSERA...T
  Krg1   EGLNLLMQGS LKRKVAATKC NDLSSRSHTV FTITTNIVEQ DSKDHGQNKN FVKI...... ..GKLNLVDL ACSENIMREC
  Cut7   DGLRLLREGS HRRQVAATKC NDLSSRSHSI FTITLHRKVS SGMTDETNSL TINNNSDDLL RASKLHMVDL AGSENIGRSG
  Bimc   AGIKLLQQGS HKRQVAATKC NDLSSRSHTV FTITVNIKRT TESGEEYVCP .......... .GKLNLVDL AGSENIGRSG
Flyncd   HLRHLMHTAK MNRATASTAG NERSSRSHAV TKLELIGRHA EKQEISV... .......... ..GSINLVDL AGSE....SP
  Kar3   MVEIILKKAN KLRSTASTAS NEHSSRSHSI FIIHLSGSNA KTGAHSY... .......... ..GTLNLVDL AGSERINVSQ
   Nod   DLHHILELGT RNRRVRPTNM NSNSSRSHAI VTIHVKSKTH H......... .......... ..SRMNIVDL AGSEGVRRTG
Squidkin EVMEVIDEGK NNRHVAVTNM NEHSSRSHSV FLINVKQENV ETQKKLS... .......... ..GKLYLVDL AGSEKVSKTG
Flykin   DVFEVIEEGK SNRHIAVTNM NEHSSRSHSV FLINVKQENL ENQKKLS... .......... ..GKLYLVDL AGSEKVSKTG
Humkin   EVMDTIDEGK SNRHVAVTNM NEHSSRSHSI FLINVKQENT QTEQKLS... .......... ..GKLYLVDL AGSEKVSKTG
Kinfam   .vl.lldegk knRhvaaTnm NehSSRSHsv fti..kqkn. etq.kls... .......... ..gklnlvDL AgSE..srtg

         320                                                              370
Unc104   AEGQRLKEGA NINKSLTTLG LVISKLAEES TKKKK..... .......... ........SN KGVIPYRDSV LTWLLRENLG
  Krg2   GQQERRKEGS FINKSLLALG TVISKLSADK MNSVGSNIPS PSASGSSSSS GNATNNGTSP SNHIPYRDSK LTRLLQPALS
  Krg1   AENKRAQEAG LINKSLLTLG RVINALVD.. .......... .......... ......N GNNTPYRESK LTRLLQDSLS
  Cut7   AENKRARETG MINQSLLTLG RVINALVE.. .......... .......... ......K AHHIPYRESK LTRLLQDSLG
  Bimc   AENKRATEAG LINKSLLTLG RVINALVD.. .......... .......... ......K SQHIPYRESK LTRLLQDSLG
Flyncd   KTSTRMTETK NINRSLSELT NVILALLQ.. .......... .......... ......K QDHIPYRNSK LTHLLMPSLG
  Kar3   VVGDRLRETQ NINKSLSCLG DVIHALGQPD S......... .......... .....T KRHIPFRNSK LTYLLQYSLT
   Nod   HEGVARQEGV NINLGLLSIN KVVMSMAAG. .......... .......... HTVIPYRDSV LTTVLQASLT
Squidkin AEGAVLDEAK NINKSLSALG NVISALADG. .......... .......... ......N KSHVPYRDSK LTRILQESLG
Flykin   AEGTVLDEAK NINKSLSALG NVISALADG. .......... .......... ......N KTHIPYRDSK LTRILQESLG
Humkin   AEGAVLDEAK NINKSLSALG NVISALAEG. .......... .......... STYVPYRDSK MTRILQDSLG
Kinfam   aegkrldEak nINksL..lg nVisalad.. .......... .......... n kthiPyRdSk lTrlLqdsLg

         400                                                              450
Unc104   GNSKTAMLAA LSPADINFDE TLSTLRYADR AKQIVCQAVV N......... .......... ....
  Krg2   GDSIVTTICT VDTRNDAAAE TMNTLRFASR AKNVALHVSK KSIISNGNND GDKDRTIELL RRQLEEQRRM I...
  Krg1   SMTNTGTTAT TSFAKISMEE TASTLEYATR AKSIKNTPQV NQSLSKDTCL KDYIQEIEKL RNDLKNSRNK QGIF
  Cut7   GKTKTSMIVT VSSTNTNLEE TISTLEYAAR AKSIRNKPQN .......... .......... ....
  Bimc   GRTKTCIIAT MSPARSNLEE TISTLDYAFR AKNIRNKPQI .......... .......... ....
Flyncd   GNSKTLMFIN VSPFQDCFQE SVKSLRFAAS VNSCKMTKAK .......... .......... ....
  Kar3   GDSKTLMFVN ISPSSSHINE TLNSLRFASK VNSTRLVSRK .......... .......... ....
   Nod   AQSYLTFLAC ISPHQCDLSE TLSTLRFGTS AKKLRLNPMQ .......... .......... ....
Squidkin GNARTTMVIC CSPASYNESE TKSTLLFGQR AKTIKNVVSV NEELTADEWK RRYEKEKERV TKLKATMAKL EAEL
Flykin   GNARTTIVIC CSPASFNESE TKSTLDFGRR AKTVKNVVCV NEELTAEEWK RRYEKEKEKN ARLKGKVEKL EIEL
Humkin   GNCRTTIVIC CSPSSYNESE TKSTLLFGQR AKTIKNTVCV NVELTAEQWK KKYEKEKEKN KILRNTIQWL ENEL
Kinfam   gnskttm... .spas.n.sE t.stLrfa.r aksikn..qv .......... .......... ....
```

been used successfully to identify several new kinesin genes in *S. cerevisiae*[58], *Drosophila*[33,59] and Dictyostelium (McCaffrey and Vale, unpublished). Anti-peptide antibodies[60,61] have also been generated against two conserved regions of the motor domain (LVDLAGSE, kinfam aa 306; SSRSHSVF, kinfam aa 264; HIPYRNSKLT, kinfam aa 384); they react with a variety of kinesin-related motor proteins and have proven to be useful reagents. For further information, see contributions on Kinesin and on Kinesin-Related Proteins.

```
                                                                                                          120
                                MG KAAVEQRGVD DLVLMPKITE QDICANLEKR YFNDLIYTNI
                                M  TLLEGSVGVE DLVLLEPLEQ ESLIRNLQLR YEKKEIYTYI

  1
Acmy1l                MA ASELYTKFAR VVIPDPEEVW KSAELLKDYK PGDKVLLLHL EEGKDLEYRL DPKTGLPHL
Bbm1hc          MTGGQSCS SNMIVWIFDE KEVFVKGELM STDINKNKFT GQEEQIGTVH PLDSTEVSNL SQVRISDVFP
Moudil MNPIHDRTSD YHKYLKVKQG DSDLVKLIVS DKRYIWYNPD PKERDSYECG EIVSETSDSF TFKTVDGQDR QVKKDDANQR  RNPDLLVGEN DTALSYLHE. PAVLHNLRFI DSK.LIVTYC
Scmyol MEHEKDPGWQ YLRRTREQVL EDQSKPYDSK KNVWIPDPEE GYLAGEITAT KGDQVIVTA. REMSVIQ..V TLKKELVQE.  VNPSTFDKVE NMSELTHLNE PSVLYNLEKR YDCDLIYTYS
Ddmyo2 MPKPVANQED EDPTYLFVS  LEQRRIDQSK PYDSKKSCWI PDEKEGYLLG EIKATKGDIV SVGLQ.GGET RDLKKDLLQQ  .NPIKFDGVE DMSELSYLNE PAVFHNLRVR YNQLIYTYS
Nemunc MSQKPLSDDE KFLFVDKNFV NNPLAQADWS AKKLVWPSE  KHGFEAASIK EEKGDEVTVE LQEN..GKKV TLSKDDIQK.  MNPPKFEKTE DMSNLSFLND ASVLHNLRSR YAAMLIYTYS
Droifm MGDSEMAVFG AAAPYLRKSE KERLEAQTRP FDLKKDVFVP DDKQEFVKAK IVSREGGKVT AETEY.GKTV TVKEDQVMQQ  MNPPKFSKVE DMSNLTYLND ASVLHNLRQR YYNRLIYTYS
Chksmo                                                                                      MNPPKFSKVE DMAELTCLNE ASVLHNLRER YFSGLIYTYS
Humcar                                                                                      .NPPKFDKIE DMAMLTFLHE PAVLYNLKDR YGSWMIYTYS
Consen m.......... ........l.. ...l...... ........e.. ....e....k .....e.g.v. .......e.v  t.kkdd..q.  .nppkfdkve dmseltylne psvlhNLkr  yf.dlIYtYs

 121                                                                                                        240
Acmy1l GPVLISVNPF RRIDALLTDE CLHCYRGRYQ HEQ.......  ....PPHYA LAEAAYRGVK SENINQCVII SCESGAGKTE AVSGNSGGVD FVKH.......
Bbm1hc GNVLVSVNPY QQLPIYDLEF V.AKYRDYTF YELK......  ....PHIYA LANMAYQSLR DRDRDQCILI TCESGAGKTE AVCGKGEQVN SV........  .KEQLL
Moudil GIVLVAINPY EQLPIYGEDI INAYSGQNMG DMD.......  ....PHIFA VAEEAYKQMA RDERNQSIIV SGESGAGKTV TVSGSASEAN VE........  ....EKVL
Scmyol GLFLVAINPY HNLNLYSEDH INLYHNKENR LSKSRLDENS HEKLPPHIFA IAEEAYENLL SEGKDQSILV NTKKILQYLA SITSGSPSNI APVSGSSIVE SFEM....KIL
Ddmyo2 GLFLVAVNPF KRIRIYTQEM VDIFKGRRRN EVA.......  ....PHIFA ISDVAYRSML DDRQNQSLLI TCESGAGKTE SVAGRNQANG SGV.......  ...LEQIL
Nemunc GLFCVAINPY KRPLYTDSC  ARMFMGKRRN EV........  ....PPHIFA VSDEAYRNML QHENQSMLI  TCESGAGKTE AVGASQQEGG AEVDPNKKKV T...LEDQIV
Droifm GLFCVAINPY KRYPVYTNRC AKMYRGKRRN EV........  ....PPHIYA ISDGAYVDML TNHVNQSMLI TCESGAGKTE NTKKVIAYFA TVGASKTDE  AAKSKGS...  ...LEDQVV
Chksmo GLFCVINPY  KQLPIYSEKI IDMYKGKRH  EM........  ....PPHIYA IADTAYRSML QDREDQSILC TCESGAGKTE NTKKVIQYLA VVASSHKGR  TPASLKVHLF PYGELEKQLL
Humcar GLFCVTVNPY KWLPVYTPEV VAAYRGKKRS EA........  ....PPHIFS ISDNAYQYML TDRENQSILI TCESGAGKTV NTKRVIQYFA VIAAIGDRSK KDQSPGK...  ..GTLEDQII
Consen Glflvainpy krlplyte.. ...y.gkrrn em........  ....pPHifa iadeAYrsml .drenQsili tCESGAGKTe ntKkviqYfA avags.....  .a.vs.....  ....ledqil

 241                                                                                                        360
Acmy1l .SNPLLEAFG NAKTLRNNNS SRFGKYFEIH FNRLGEPCGG RITNYLLEKS RVTFQTRGER SFHIFYQLLA GASDAEAQEM QLYAPENFNY LNQSACYTVD GIDDIKEFAD TRNAINVM.G
Bbm1hc QSNPVLEAFG NAKTIRNNNS SRFGKYMDIE FDFKGFPLGG VITNYLLEKS RVVKQLEGER NFHIFYQLLA GADAQLLKAL KLERDTGGYA YLNPDTSRVD GMDDDANFKV LQSAMTVI.G
Moudil ASNPIMESIG NAKTTRNDNS SRFGKYIEIG FDKRYRIIGA NMRTYLLEKS RVVQFAEEER NYHIFYQLCA SAKLPEFKML RLGNADSFHY TKQGGSPMIE GVDDAKEMAH TRQACTLL.G
Scmyol QSNPILESFG NAQTVRNNNS SRFGKFIKIE FNEHGMINGA HIEWYLLEKS RIVHQNSKER NYHIFYQLLS GLDDSELKNL RLKSRNVRDY KILSNSNQDI IPGIDVENFK ELLSALSIIG
Ddmyo2 QANPILEAFG NAKTTRNNNS SRFGKFIEIQ FNNAGFISGA SIQSYLLEKS RVVFQSETER NYHIFYQLLS GATAEEKKAL HLAGPESFNY LNQSGCV.DI KGVSDGEEFK ITRQAMDIVG
Nemunc QTNPVLEAFG NAKTVRNNNS SRFGKFIRIH FNKHGRLASC DIEHYLLEKS RVIRQAPGER CYHIFYQIYS DRPPELKKEL LLDL.PIKDY WFVAQAELII DGIDDVEEFQ LTDEAFDILN
Droifm QTNPVLEAFG NAKTVRNDNS SRFGKFIRIH FGPTGKLAGA DIETYLLEKA RVISQQSLER CYHIFYQIMS GSVPGVKEYC LLSNNI.YDY RIVSQGKTTI PSVNDGEEWV AVDQAFDILG
Chksmo QANPILEAFG NAKTVKNDNS SRFGKFIRIN FDVTGYIVGA NIETYLLEKS RAIRQAKDER TFHIFYYLIA GASEQMRNDL LLEGFN.NYT FLSNGHYPIP AQQDDEMFQE TLE.AMTIMG
Humcar QANPALEAFG NAKTVRNDNS SRFGKFIRIH FGATGKLASA DIETYLLEKS RVIFQLKAER DYHIFYQLLS NKKPELLDML LITNNPYDYA FISQGETTVA SIDDAEELMA TDN.AFDVLG
Consen qsNPlLEafG NAkTvrNnNs SRFGKfirIh Fnktg.laga dietYLLEKs Rv.fQaegER nyHIFyqlla ga.pee.k.l llen...f.y .iqsg.vt.i g.ddd.e.f. t..aafd.ilg

 361                                                                                                        480
Acmy1l MTAEEQRQVF HLVAGILHLG NVAFHDGKG  TAAVHDRTPF ALKNALLFRV LNTGGAGAKK MSTYNPQNV  EQAASARDAL AKTIYSRMFD WIVSKVNEAL QKQGSGDHN  NNM.......
Bbm1hc FSDEEIRQVL EVAALVLKLG NVELINEFQA NGVPASGIRD GRGVQEIGEL VGLNSVELER ALCSRTMETA KEKVTTLNV  IQAQYARDAL AKNIYSRLFN WLVRINESI  KVGTGEKRKV
Moudil ISESYQMGIF RILAGILHLG NVGFASRDSD SCTIPPKHEP LTIFCDLMGV IMKRCVTALP PKAATATEIY IKPISKLQAT NARDALAKYI YAKLFNWIVD HVNQALHSV  KQHS......F
Scmyol FSKDQIRWIF QVVAIILLIG NIEFVSDRAE QASFKNDVSA ICSNLGVDEK DFQTAILRPR SKAGKEWVSQ SKNSQQAKFI LNALSRNLYE RLFGIYVDMI IPGIDVENFK LN.......Y
Ddmyo2 FSQEEQMSIF KIIAGILHLG NIKFEKGAGE GAVLKDKTAL NAASTVLVSI HQSLKALMNT GCEEFLKALT KPRVKVGTEW NWAVGAMAKG LYSRVFNWLV KKCNLTLDQ  GID...RDYF
Nemunc FSAVEKQDCY RLMSAHHMG  NMKFKQRPRE EQAEPDGTVE AEKASNMYGI GCEEFLKALI KPRVKVGNEF VSKGQNCEQV NWAVGAMAKG LYSRVFNWLV KKCNLTLDQ  GID...RDYF
Droifm FTKQEKEDVY RITAAVMEHG GMKFKQRGRE EQAEQDGEEE GGRVSKLFGC DTAELYKNLL KPRVKVGNEF VTQGRNVQGV TNSIGALCKG VFDRLFKWLV KKCNETLDTQ QK....RQHF
Chksmo FTEEEQTSIL RVVSSVLQLG NIVFKKERNT DQASMPDNTA AQKVCHLMGI NVTDFTRSIL TPRIKVGRDV VKAQTKEQA  DFAIEALAKA KFERLFRWIL TRVNKALDKT KRQGAS...F
Humcar FTSEEKNSMY KLTGAIMHFG NMKFKLKQRE EQAEPDGTEE ADKSAYLMGL NSADLLKGLC HPRVKVGNEY VTKGQNVQQV IYATGALAKA VYERMFNWMV TRINATLETK QP....RQYF
Consen fs.eeqrsif r.va.ilhlG n.kfk.r.re eqaepdgtee a.kv.lmgi  .....kal.  pr.kvg.e.  v.kgnvqqv  .naigalak. v.erlfnwlv .k.n.tld.. kn........f
```

Figure 3. Sequences of the motor domain from members of the myosin superfamily. The consensus sequence (consen) is shown below Acanthamoeba myosin 1L (acmy1L), brush border myosin I heavy chain (Bbm1hc), mouse dilute gene (Moudil), Dictyostelium discoideum myosin II (Ddmyo2), *C. elegans* skeletal myosin heavy chain (unc-54)(Nemunc), *Drosophila* intermediate flight muscle myosin (Droifm), chicken smooth muscle myosin (Chksmo), human β-cardiac myosin (Humcar). PCR degenerate oligonucleotides around aa FGNAKT (consen. aa 249),

```
                                                                                                              500
481
Acmyl1  IGVLDIFGFE IFEQNGPEQF CINYVNEKLQ QYFIELTLKA EQEEYVNEGI QWTPIKYFNR KVVCE.LIEG KR.PPGIFSL LDDICFTMHA QSDGMDGKFL LHFRGMNN..
Bbm1hc  MGVLDIYGFE ILEDNSPEQF VINYCNEKLQ QVFIEMTLKE EQEEYKREGI PWVKVEYFDN GIICN.LIEH NQR..GILAM LDEECLRPGV VSDSTFLAKL NQLFSKHSHY ESKVTQNAQR
Moudil  IGVLDIYGFE TFEINSPEQF CINYANEKLQ QQPNMHVFKL EQEEYMKEQI PWTLIDFYDN QPCIN.LIES KL...GIIDL LDEECKMPKG TDDTWAQ.KL YNTHLNKCAL FEKPRMSNK.
Scmyo1  IGLLDIAGFE IFENNSFEQL QFFNHHMFVL EQSEYLKENI NWTFIDFGLD SQAIIDLIDG RHT..GFSS. LDEHL.FPNA TDNTLIT.KL ISTWDQNSSK FKRSRLKNG.
Ddmyo2  SSEVDISGFE IFKVNSFEQL QFFNHHMFKL EQEEYLKEKI NWTFIDFGLD LQACIELIEK PL...GIISM LDEECIVPKA TDLTLAS.KL VDQHLGKHPN FEKPKPPKGK
Nemunc  IGVLDIAGFE IPDFNSFEQL WINFVNEKLQ QFFNHHMFVM EQEEYAREGI QWVFIDFGDT LQACIELIEK PL...GIISM LDEECIVPKA TDQTFSE.KL TNTHLGKSAP FQKPKPPKG
Dro1fm  IGVLDIAGFE IFEYNGFEQL CINFTNEKLQ QFFNHIMFVL EQEEYKKEGI NWDFIDFGMD LLACIDLIEK PM...GILSI LEEESMFPKA TDQTFSE.KL TNTHLGKSAP FQKPKPPKG
Chksmo  LGILDIAGFE IFEINSFEQL CINYTNEKLQ QLFNHTMFIL EQEEYQREGI EWNFIDFGLD LQPCIELIER PTNPPGVLAL LDEECWFPKA TDTSFVE.KL IQEQGNHA.K FQKSKQLKDK
Humcar  IGVLDIAGFE IFDFNSFEQL CINFTNEKLQ QFFNHHMFVL EQEEYKKEGI EWTFIDFGWD LQACIDLIEK PM...GIMSI LEEECMFPKA TDMTFKA.KL FDNHLGKSAN FQKPRNIGK
Consen  igvlDIaGFE ife.NsFEQl cINytNEKLQ QfFnhhmf.l EQeEYkkEgI qWtfidfgld lqacidLiek p....gilsl Ldeecmfpka tddtfa..kL i.thlgksak f.kpr..kg.

                                                                                                              720
601
Acmyl1  .........A FSIKHYAGEV TYEAEGFCEK NKDTLFDDLI AVIQESENR. LLVSWFPE.D TKQLQKKRPT TAG.......  .........  .........  .........  FKLKTSCDAL MEALSRCSPH
Bbm1hc  QYDHSMGCLSC FRICHYAGKV TYNVNSFIDK NNDLLFRDLS QAMWKARHP. LLRSLFPEGD PKQASLKRPP TAGAQ..... .........  .........  .........  ..FKSSVTTL MKNLYSKNPN
Moudil  .........A FIIKHFADKV EYQCEGFLEK NKDTVFEEQI KVLKSSKFRM .LPELFQDDE KAISPTSATS SGRTPLIRVP VKPTKGRPGQ TAKEHKKTVG HQFRNSLHLL METLNATTPH
Scmyo1  .......... FLLKHYAGDV EVTVEGWLSK N.DPLNDNLL SLLSSQ.ND IISKLFQPEG GKNLLVCGVE ANISNQEVKK SARTSTFKTI SSRH......  .........  ...REQITL LNQLASTHPH
Ddmyo2  .......... FGVTHYAGQV MYEIQDWLEK NKDPLQQRSE LCFKDSSDNV VT.KLFNDFN IASRAKKGAN FITVAAQ...  .........  .........  .........  ...YKEQLASL MATLETTNPH
Nemunc  QGE.....AH FAMRHYAGTV RYNCLNWLEK NKDPLNDTVV SAMKQSKGND LLVEIWQDYT TQEEAAAKAK EGGGGKKKG KSGSFMTVSM .........  .........  ...LYRESLNNL MTMLNKTHPH
Dro1fm  QQA.....AH FAIAHYAGCV SYNITGWLEK NKDPLNDTVV DQFKKSQ.NK LLIEIFADHA GQSGGGEQAK GCGRGKKGGGF ATVSSA....  .........  .........  ...YKEQLNSL MTTLRSTQPH
Chksmo  TE......AH FCILHYAGKV TYNASAWLTK NMDPLNDNVT SLLNQSSD.K FVADLWKDVD RIVGLDQMAK MTESSLPSAS KTKKGMFRTV GQ........  .........  ...LYKEQLTKL MTTLRNTNPN
Humcar  PE......AH FSLIHYAGIV DYNIIGHLQK NKDPLNETVV GLYQKSSLKL LSTLFANYAG ADAPIEKGKG KAKKGSSFQT VSALH.....  .........  ...RENLNKL MTNLRSTHPH
Consen  ........F .ikHYAgkV tYniegvleK NkDplndtvv slikkss.n. llvelfqd.d .k..kkgak t.g.......  .........  .........  ...yke.ln.L mttLrst.Ph

                                                                                                              840
721
Acmyl1  YIRCIKPNDN KAYHDWDATR TKHQVQYLGL LENRVRRAG FAYRAEFDRF LRRYKKLSPK TWGIWGEWSG APKDGCQTLL NDLGLDTSQW QLGKSKVFIR YPETLFHLEE CLDRKDYDCT
Bbm1hc  YIRCIKPNEH QQRGHFSFEL VSVQAQYLGL LENRVRRAG YAYRQAYGSF LERYRLLSRS TWPRWNGDQ EGVEKVLGEL SMSSEELAFG KT...KIFIR SPKTLFYLEE QRRLRLQQLA
Moudil  YVRCIKPNDF KPFFTFDEKR AVVQLRACGV LETIRISARG FPSRWTYQEF FSRYRVLMKQ KDVLGDR...  ..KQTCKNVL EKLILDKDKY QFGKTKIFFR AGQVAYLEKL RADKLRACIR
Scmyo1  FVRCIIPNNV KKVKTFNRSL ILDQLRCNGV VLDQLRCNGV YPNRIAFQEF FQRYRILYPR KFNHHDFSSK L.KASTKQNC RFLLTLSQLD TKVYKIGILT VFQKLEYWSD LEKQKDVNVN
Ddmyo2  FVRCIIPNNK QLPAKLEDKV VLNQLTCNGV LEGIRITRKG FPNRIIYADF VKRYYLLAPN VPRDAEDSQ.  ..KATDAVLK HLNIDPEQY. RFGITKIFFR AGQLARIEEA REQRISEIIK
Nemunc  FIRCIIPNEK KQSGMIDAAL VLNQLTCNGV LEGIRICRKG FPNRTLHPDF VQRYAILAAK EAKSDDDKKK CAEAIMSKLV NDGSLSEEMF RIGLTKVFFK AGVLAHLEDI RDEKLATILI
Dro1fm  FVRCIIPNEM KQPGVVDAHL VMHQLTCNGV LEGIRICRKG FPNRMYPDF KMRYMLAPI MAAEKVA...  .KNAAGKCL EAVGLDPDMY RIGRTKVFFR AGVLGQMEEF RDERLGKIMS
Chksmo  FVRCIIPNHE KRAGKLDAHL VLEQLRCNGV LEGIRICROG FPNRIVFQEF RQRYEILAAN AIPKG...FM DGKQACILMI KALELDPNLY RIQGSKIFFR TGVLAHLEEE RDERLSRIIT
Humcar  FVRCIIPNET KSPGVMDNPL VMHQLRCNGV LEGIRICRKG FPNRILYGDF RQRYRILNPA AIPEQ..FI DSRRGAEKLL SSLDIDHNQY KFGHTKVFFK AGLLGLLEEM RDERLSRIIT
Consen  fvRCIIPNek kqpg.fda.l vlhQlrcnGv LEgiRicrkG fpnRl.yqdF .qRYrilap. ..pegd....  ..ka.ckkll ....l.ld..y r.g.tk.ffr ag.la.leee rderl..iit

                                                                                                              960
841
Acmyl1  LRIQKAWRHW KSRKHQLEQR KMAADLLGK KERQRHSVNR KYEFDYINYD ANYPLQDCVR SSGRDKEATA FTDQVLVLNR RGKPERRDLI VTNEAVVFAM RKKKSQQVVY NLKRRIPLGE
Bbm1hc  TLIQKTYRGW RCRTHYQLMR KSQIVISSWF RGNMQKKHYR KMKASALLIQ AFVRGWKARK NYRKYFRSGA ALILSNFIYK SMVQKFLLGL KNDLPSPSIL DKKWPSAPYK YPNTANHELQ
Moudil  ..IQKTIRGW LLRKRYLCMQ RAAITVQRYV RGY.QARCYA KFLRRTKAAT TIQKYWRMTV VRRRYKIRRA ATIVIQSYLR GYLTRNRYRK ILREYKAVII QKRVRGWLAR THYKRTMKAI
Scmyo1  NIMIKLTATI RGTVRKEIT YHLQKLKKTR VIGNTFRLYN RLVKEDPWFN LFIRIKPLLT SSNDMRTKKF NEQ.INKLKN DLQEMESKKK TELNSQITK INTNITETPQ
Ddmyo2  AIQAATRGWI ARKVYKQARE HTVAARIIQQ NLRAYIDFKS ......WPWK LFSKARPLLK RRNFEKEIKE KEREILELKS NLTDSTQKD SLAKEEKLRK ELEESSAKLV ESTKTQLSDA
Nemunc  GFQSQATWHL GLKDRKRRME QRAGLLIVQK NLRKYLQLRT .....WFWYK LYGKVKPMLK AGKEAEELEK INDKVKALED LEEKAKAAE LHAAEVVRK ELEALNAKLL AEKTALLDSL
Dro1fm  WMQAWARGYL SRKGFKKLQE QRVALKKVVQR NLRKYLQLRN .....WFWYK LWQKVKPLLN VSRIEDEIAR LEEKAKAEAE KDEELQRTKE VTRQEEMQA KDEELQRTKE EEKNLQKLQ QAETELWAEA
Chksmo  AFQAQCRGYL ARKAFAKRQQ QLTAMKVIQR NCAAYLKLRN .....WQWMR LFTKVKPLLQ SAEREKEMAS MKEEFTRLKE ALEKSEARRK ELEQKHQLC EEKNLQKLQ QAEEDNLADA
Humcar  RIQAQSRGVL ARMEYKKLLE RRDSLLVIQW NIRAFMGVKN ......vpw.k lfqkvkplik
Consen  aiqaktrg.l ark..kkl.e qraallviqr n.rayl.l.n ......vpw.k lfqkvkplik .srreke.a .eeei..lke .l.kee.lrk elek.akll .ek.lqlql qaek.tl.da
```

Figure 3 cont. GAGKT (consen. aa 195), and ERNHIFY (consen. aa 299) have been useful for identifying myosin genes from all classes, while oligonucleotides designed around GVLDIAGFEI (consen. aa 482) and YLGLLENV (consen. aa 747) have been useful for isolating myosin I genes (PCR information provided by H. Goodsen, Stanford University, and M. Titus, Duke University). For more information, see contributions on Sarcomeric Myosins, Cytoplasmic Myosin II, Protozoan Myosins, and Brush Border Myosin.

Axonemal Dyneins

Dynein defines a family of very high molecular weight microtubule based, force generating ATPases that are responsible for motility in a variety of cell systems[1]. The Dynein family can be divided into two subfamilies, one being the axonemal dyneins that are located in the cilia and flagella of eukaryotes and the other being the cytoplasmic dyneins recently isolated from a variety of sources. Axonemal dynein is localized in the outer and inner arms on the doublet microtubules where its function is to generate the relative sliding motion between adjacent doublet microtubules that leads to the characteristic oscillatory bending movement of cilia and eukaryotic flagella[2].

Axonemal dyneins from most sources consist of two globular heads joined by slender stems to a complex basal region[3-5] (Figure). The outer arm dyneins of *Tetrahymena* and *Chlamydomonas* are exceptions that contain three globular heads[5,6], although the inner arm dyneins from these species contain the usual two heads. Two-headed dyneins sediment at ~20 S and have a molecular mass of ~1,300 kDa[2,6], while three-headed dyneins sediment at 24-28 S and have a molecular mass of 1,640-1,900 kDa[7,8], as determined by sedimentation analysis or scanning transmission electron microscopy. A characteristic property of all known dyneins is their possession of very large heavy chain subunits (>400 kDa). In addition to these heavy chains, axonemal dyneins possess two or three intermediate size chains of 75-120 kDa, and about four light chains of 15-30 kDa[2,7]. Axonemal dyneins are attached at their basal ends to the A-tubule of the axonemal doublet microtubules while the globular heads of the dynein undergo ATP-driven crossbridge cycling with **tubulin** subunits in the B-tubule of the adjacent doublet[3,7,8]. Ciliary and flagellar axonemes possess a single species of outer arm dynein located with a periodicity of 24 nm along the A-tubule of each doublet[8]. However, the axonemes contain three structurally distinct species of inner arm that are distributed along the length of each A-tubule in unevenly spaced groups of three[8,9]. Since each of the two to three heavy chains in an outer arm and most of the heavy chains in the three species of inner arm are electrophoretically and immunologically distinct, it appears that an axoneme contains seven to nine isoforms of dynein heavy chain[9,10]. Ciliary and flagellar dyneins from a single species possess at least some different heavy chains[11,12]. Many dynein subunits are phosphorylated *in vivo*[13], but the significance of this and other possible post-translational modifications is not known.

The ATP binding domain involves a multiply folded region of ~100 kDa located in the middle region of the heavy chain[14,15] that contains the amino acid sequence motifs for four nucleotide binding sites[16,17]. Axonemal dyneins have a moderately high degree of specificity for ATP as substrate[2]. Deoxy-ATP, 2-azido-ATP, 2-chloro-ATP and $(S_p)ATP[\alpha S]$ are sufficiently good substrate analogs to support motility, whereas ITP, GTP, CTP and UTP are hydrolyzed at only about 10% of the rate of ATP and do not support motility[2,18,19]. Vanadate (V_i) is a potent inhibitor of dynein ATPase activity (K_i ~1μM)[20], probably through acting as a phosphate analog and forming a stable dynein-ADP-V_i complex that acts as a dead-end kinetic block[5]. Irradiation of the dynein-MgADP-V_i complex with near UV light (365 nm) causes photolytic cleavage at a highly specific site on each of the dynein heavy chains[21]; this photocleavage reaction can be used both for linear mapping of the heavy chain[14,15] and as a diagnostic criterion for dynein.

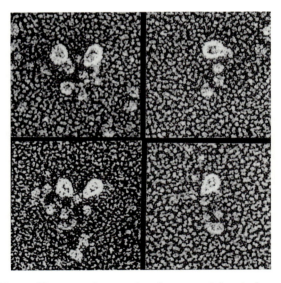

Figure. Electron micrographs of axonemal dynein from sea urchin sperm flagella, prepared by adsorption onto mica, freezing, etching and shadow casting. *Left*, molecules of intact outer arm dynein; *right*, isolated β heavy chain/intermediate chain subfraction. Magnification: 350,000. (Micrographs kindly provided by Dr. Win Sale, Emory University, Atlanta).

■ PURIFICATION

Axonemal dynein has been obtained from flagella of *Chlamydomonas*, from sperm of sea urchin, trout and bull, and from cilia of *Tetrahymena*[22-26]. The dynein is usually solubilized from the demembranated flagella or cilia by extraction with 0.6 M NaCl. It can be purified by sucrose density gradient centrifugation or by chromatography.

The ease of purification differs greatly with different species and in some cases chromatography gives a very poor yield. Solutions of dynein tend to lose physiological activity within a day or two upon storage at 4°C, but they can be stored by freezing at -80°C in the presence of 20% sucrose[22]. In outer and inner arm dyneins from some species, the individual heavy chain subunits can be separated by a combination of treatment at low ionic strength and density gradient centrifugation, or by chromatography on a MonoQ column[23,26].

■ ACTIVITIES

All dynein heavy chains possess a site for binding and hydrolysis of MgATP. The kinetics of ATP hydrolysis resemble those of **myosin**, but with the rate being stimulated up to 30-fold by tubulin[27]. The enzymatic properties of the different heavy chains in outer arm dynein have differences that suggest that the individual heavy chains have distinct functions in motility. The physiological activity of axonemal dyneins can be studied in demembranated systems reactivated with exogenous ATP. Depletion of the normal level of dynein, either by salt extraction or by mutation, results in a decreased beat frequency[2]. The function of axonemal dynein can be assayed by its ability to restore the beat frequency to a dynein depleted system[2]. Function can also be assayed by measuring the velocity with which the doublets slide apart in axonemes that have been briefly pretreated with trypsin and then exposed to ATP[2,28,29]. The isolated heavy chain subunits are not functional in either of these assays. In some cases however they are able to perform the more basic motile function of translocating microtubules over a glass surface coated with the dynein subunit in the presence of ATP[30,31]. Axonemal dyneins, like **cytoplasmic dyneins**, normally transport themselves and their attachments toward the minus-end of the microtubule that they are moving along[32]. Certain axonemal dyneins also have the unusual property of rotating microtubules as they transport them *in vitro*[32]; this rotation may be related to the propensity of the paired central tubules to rotate during normal ciliary beating in some species[33].

■ ANTIBODIES

Rabbit polyclonal antisera have been prepared against axonemal dyneins from various species, as well as against an ATPase containing fragment of outer arm from sea urchin sperm[34]. Monoclonal antibodies have been described against several of the heavy and intermediate chains of outer arm dyneins from sea urchin sperm, *Tetrahymena*, and *Chlamydomonas*[35,36]. The polyclonal and monoclonal antibodies against dynein heavy chains are usually species specific, but some antibodies against the intermediate chains crossreact against a broad range of species including vertebrates, invertebrates, and *Chlamydomonas*.

■ GENES

Genes for the three heavy chains in outer arm dynein of *Chlamydomonas* have been identified and restriction mapped[37]. Complete sequences have been published for the β heavy chain of axonemal dynein from two species of sea urchin[16,17] (EMBL D01021, and X59603) and for the 70 kDa the intermediate chain of dynein from the ODA6 mutant of *Chlamydomonas*[38] (EMBL X55382).

■ REFERENCES

1. Gibbons, I.R. and Rowe, A.J. (1965) Science 149, 424-426.
2. Gibbons, I.R. (1981) J. Cell Biol. 91, 107s-124s.
3. Gibbons, I.R. (1988) J. Biol. Chem. 263, 15837-15840.
4. Sale, W.S., Goodenough, U.W. and Heuser, J.E. (1985) J. Cell Biol. 101, 1400-1412.
5. Johnson, K.A. (1985) Annu. Rev. Biophys. Biophys. Chem. 14,161-188.
6. Goodenough, U.W. and Heuser, J.E. (1984) J. Mol. Biol. 180, 1083-1118.
7. Wells, C., Molina-Garcia, A., Harding, S.E. and Rowe, A.J. (1990) J. Muscl. Res. Cell Motil. 11, 344-350.
8. Goodenough, U.W. and Heuser, J.E. (1985) J. Cell Biol. 100, 2008-2018.
9. Piperno, G., Ramanis, Z., Smith, E.F. and Sale, W.S. (1990) J. Cell Biol. 110, 379-389.
10. Gibbons, I.R., Fronk, E., Gibbons, B.H. and Ogawa, K. (1976) In Cell Motility. Book C. (Goldman, R., Pollard, T. and Rosenbaum, J. eds.) Cold Spring Harbor Laboratory, Cold Spring Harbor, N.Y. pp 915-932.
11. Linck, R. (1973) J. Cell Sci. 12, 951-981.
12. Ogawa, K., Yokota, E., Hamada, Y., Wada, S., Okuno, M. and Nakajima, Y. (1990) Cell Motil. Cytoskel. 16, 58-67.
13. Piperno, G., Ramanis, Z., Smith, E.F. and Sale, W.S. (1981) J. Cell Biol. 88, 73-79.
14. Mocz, G., Tang, W.-J.Y. and Gibbons, I.R. (1988) J. Cell Biol. 106, 1607-1614.
15. King, S.M. and Witman, G.B. (1989) In Cell Movement: The Dynein ATPases (Warner, F.D., Satir, P. and Gibbons, I.R. eds) Alan. R. Liss, New York, Vol. 1, 61-75.
16. Gibbons, I.R., Gibbons, B.H., Mocz, G. and Asai, D.J. (1991) Nature 352, 640-643.
17. Ogawa, K. (1991) Nature 352, 643-645.
18. Omoto, C.K. and Nakamaye, K. (1989) Biochim. Biophys. Acta 999, 221-224.
19. Shimizu, T., Okuno, M., Marchese-Ragona, S.P. and Johnson, K.A. (1990) Eur. J. Biochem. 191, 543-550.
20. Gibbons, I.R., Cosson, M.P., Evans, J.A., Gibbons, B.H., Houck, B., Martinson, K.H., Sale, W.S. and Tang, W.-J.Y. (1978) Proc. Natl. Acad. Sci. (USA) 75, 2220-2224.
21. Gibbons, I.R. and Mocz, G. (1991) Methods Enzymol. 196,428-442.
22. Bell, C.W., Fraser, C., Sale, W.S., Tang, W.-J.Y. and Gibbons, I.R. (1982) Methods Enzymol. 85, 450-475.
23. King, S.M., Otter, T. and Witman, G.B. (1986) Methods Enzymol. 134, 291-306.
24. Johnson, K.A. (1986) Methods Enzymol. 134, 306-324.
25. Gagnon, C. (1986) Methods Enzymol. 134, 318-324.
26. Goodenough, U.W., Gebhart, B., Mermall, V., Mitchell, D.R. and Heuser, J.E. (1987) J. Mol. Biol. 194, 481-494.
27. Shimizu, T., Marchese-Ragona, S.P. and Johnson, K.A. (1989) Biochemistry 28, 7016-7021.

28.Yano, Y. and Miki-Noumura, T. (1981) J. Cell Sci. 48, 223-239.
29.Kamiya, R., Kurimoto, E., Sakakibara, H. and Okagaki, T. (1989) In Cell Movement: The Dynein ATPases. (Warner, F.D., Satir, P. and Gibbons, I.R. eds.) Alan. R. Liss, New York, Vol. 1, 209-218.
30.Sale, W.S. and Fox, L.A. (1988) J. Cell Biol. 107, 1793-1798.
31.Vale, R.D., Soll, D.R. and Gibbons, I.R. (1989) Cell 59, 915-925.
32.Vale, R.D. and Toyoshima, Y.Y. (1989) Cell 52, 459-469.
33.Omoto, C.K. and Kung, C. (1980) J. Cell Biol. 87, 33-46.
34.Ogawa, K. and Mohri, H. (1975) J. Biol. Chem. 250, 6476-6483.
35.Piperno, G. (1984) J. Cell Biol. 98, 1842-1850.

36.King, S.M. and Witman, G.B. (1990) J. Biol. Chem. 265, 9807-9811.
37.Mitchell, D.R. (1989) Cell Motil. Cytoskel. 14, 435-445.
38.Mitchell, D.R. and Kang, Y. (1991) J. Cell Biol. 113, 835-842.

■ Ian Gibbons:
Pacific Biomedical Research Center,
University of Hawaii,
Honolulu, HI 96822, USA

Brush Border Myosin I

Brush border (BB) myosin I[1,2] is a single-headed, membrane associated myosin expressed in vertebrate intestinal epithelial cells. It is primarily localized within the microvilli of the apical, BB surface where it comprises the bridges that laterally tether the microvillar actin core to the plasma membrane.

BB myosin I has been purified and extensively characterized from only one species, the chicken[2-9]. However, immunological studies on brush borders from other vertebrate species indicate that this protein is a ubiquitous element of the vertebrate brush border cytoskeleton. Electron microscopy of chicken BB myosin I reveals a tadpole shaped molecule ~30 nm in length[4]. BB myosin I consists of a heavy chain (hc) of ~119 kDa and multiple (3-4) calmodulin (CM) light chains. The heavy chain of BB myosin I comprises discrete structural and functional domains (Figure). Identification and characterization of these domains has included studies on proteolytic fragments of BB myosin I[7-10] as well as the determination of the deduced primary structure of both the bovine[11,12] and chicken[13] protein. Like the **protozoan myosins I**, BB myosin I has an N-terminal myosin "head" domain of ~80 kDa homologous in primary structure to that of other **myosins**. Adjacent to the head is a "neck" domain of ~15 kDa that contains multiple CM binding sites. Based on cDNA sequence analysis, a second transcript of chicken BB myosin I hc with an 87 Bp (29 amino acid) insert within the neck region of the tail has been identified[14]. This insert encodes an additional CM binding domain. This presumed alternatively spliced transcript is expressed at substantially lower levels than the smaller form; the levels of protein expression and the subcellular distribution of this higher molecular weight form of BB myosin I has not yet been determined. Next to the neck domain of BB myosin I hc is an ~20 kDa region which effects the binding of BB myosin I to acidic phospholipids[9]. This C-terminal domain, which is rich in basic amino acids, shares weak similarity to the presumed phospholipid binding domain of protozoan myosins I[9].

The functions for BB myosin I are not known. Studies on the expression of BB myosin I during enterocyte differentiation suggest that its assembly into the brush border is dependent on the maturation of the microvillar membrane[15]. Thus, one possible function for this motor is the delivery of newly synthesized membrane components to the apical domain of the cell.

■ PURIFICATION

There are several protocols[3,5,6] for the purification of BB myosin I from isolated chicken intestinal brush borders. Most of these methods involve purification of the protein from ATP extracts of brush borders by a combination of gel filtration and ion exchange chromatography (both anion and cation).

■ ACTIVITIES

A number of laboratories have contributed to the extensive characterization of the *in vitro* properties of BB myosin I[2-9,16,17]. Like other myosins, BB myosin I exhibits ATP dependent binding to **actin**. BB myosin I also crosslinks actin filaments in the absence of ATP[9,16], although there is no evidence for a second, ATP insensitive actin binding site on the tail domain, comparable to that of protozoan myosins I. The most striking feature of the actin binding properties of BB myosin I is that it will form helical bridges comparable to that *in vivo* when added to actin bundles containing the microvillar core proteins **villin** and **fimbrin**[18]. BB myosin I exhibits K+-EDTA and Ca2+-ATPase activ-

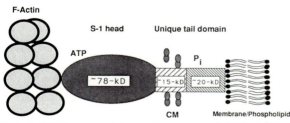
Figure. The functional domain structure of chicken BB myosin I.

ities characteristic of most myosins. Most importantly, the MgATPase of BB myosin I is activated by actin. In addition, the MgATPase of BB myosin I is activated by 5-10 μM Calcium[2,4,16,17]. This Ca^{2+}-activation occurs whether or not actin is present. Recent studies[2,16] indicate that this activation is mediated by the CM light chains, which apparently serve as Ca^{2+}-sensitive suppressors of the MgATPase of BB myosin I. In vitro motility studies[2,6,19,20] have demonstrated that BB myosin I is an active mechanoenzyme that promotes movement in the same direction as **sarcomeric myosin**[19]. However, differences in both rate and Ca^{2+}-sensitivity for motility have been reported. There is preliminary evidence for the presence of a microvillar membrane glycoprotein in both porcine[21] and chicken BBs[20] that effects the binding of BB myosin I to the membrane.

■ ANTIBODIES

Several laboratories have prepared polyclonal antibodies against the hc of BB myosin I[15,22,23]. Four monoclonal antibodies (termed CX1-4) reactive with the head domain of chicken BB myosin I have also been raised[8,9,13]. CX1-4 react with a variety of immunogens in the 90-190 kDa range in a number of cell types and tissues as well as with the head region of striated muscle myosins. None of these monoclonal antibodies are suitable for immunocytochemical use.

■ GENES

A full length cDNA encoding bovine BB myosin I heavy chain[11] has been sequenced (GenBank/EMBL J02819) as has a partial length chicken cDNA[13].

■ REFERENCES

1. Louvard, D. (1989) Current Opinions in Cell Biol. 1, 51-57.
2. Mooseker, M.S., Wolenski, J.S., Coleman, T.R., Hayden, S.M., Cheney, R.E., Espreafico, E., Heintzelman, M.B. and Peterson, M.D. (1991) In: Ordering the Membrane-Cytoskeleton Trilayer. M.S. Mooseker and J.S. Morrow, editors. Academic Press. 31-35.
3. Swanljung-Collins, H., Montibeller, J. and Collins, J.H. (1987) Meth. Enzym. 139, 137-148.
4. Conzelman, K.A. and Mooseker, M.S. (1987) J. Cell Biol. 105, 313-324.
5. Coluccio, L.M. and Bretscher, A. (1987) J. Cell Biol. 105, 325-333.
6. Collins, K., Sellers, J.R. and Matsudaira, P.T. (1990) J. Cell Biol. 110, 1137-1147.
7. Coluccio, L.M. and Bretscher, A. (1988) J. Cell Biol. 106, 367-373.
8. Carboni, J.M., Cozelman, K.A., Adams, R.A., Kaiser, D.A., Pollard, T.D. and Mooseker, M.S. (1988) J. Cell Biol. 107, 1749-1757.
9. Hayden, S.M., Wolenski, J.S. and Mooseker, M.S. (1990) J. Cell Biol. 111, 443-451.
10. Coluccio, L.M. and Bretscher, A. (1990) Biochemistry 29, 11089-11094.
11. Hoshimaru, M. and Nakanishi, S. (1987) J. Biol. Chem. 262, 455-459.
12. Hoshimaru, M., Fujio, Y., Sobue, K., Sugimoto, T. and Nakanishi, S. (1989) J. Biochem. 206, 455-459.
13. Garcia, A., Coudrier, E., Carboni, J., Anderson, J., Vandekerhove, J., Mooseker, M., Louvard, D. and Arpin, M. (1989) J. Cell Biol. 109, 2895-2903.
14. Halsall, D.J. and Hammer III, J.A. (1990) FEBS Lett. 267, 126-130.
15. Heintzelman, M.B. and Mooseker, M.S. (1991) In: The cytoskeleton and development. E. Bearer, Editor, Academic Press. 93-122.
16. Wolenski, J.S., Hayden, S.M. and Mooseker, M.S. (1990) J. Cell Biol. 111, 168a.
17. Swanljung-Collins, H. and Collins, J.H. (1991) J. Biol. Chem. 266, 1312-1319.
18. Coluccio, L.M. and Bretscher, A. (1989) J. Cell Biol. 108, 495-502.
19. Mooseker, M.S. and Coleman, T.R. (1989) J. Cell Biol. 108, 2395-2400.
20. Mooseker, M.S., Conzelman, K.A., Coleman, T.R., Heuser, J.E. and Sheetz, M.P. (1989) J. Cell Biol. 109, 1153-1161.
21. Coudrier, E., Reggio, H. and Louvard, D. (1983) EMBO J. 2, 469-474.
22. Glenney, J.R., Osborn, M. and Weber, K. (1982) Exp. Cell Res. 138, 199-205.
23. Fath, K.R., Obenauf, S.D. and Burgess, D.R. (1990) Development 109, 449-459.

■ M.S. Mooseker and J.S. Wolenski:
Dept. of Biology,
Yale University
New Haven, CT, USA

Caltractin

Caltractin is a major component of a Ca^{2+}-sensitive contractile fibre system that physically links the basal body complex to the nucleus in Chlamydomonas. It is a member of the Ca^{2+}-modulated family of proteins that includes calmodulin and troponin C. It appears to play a major role in the normal disposition and segregation of the basal bodies and, hence, the spatial and temporal distribution of microtubules in the cell.

Caltractin[1,2], a 20 kDa Ca^{2+}-binding protein, is a component of the basal body complex, which constitutes the major microtubule organizing centre in *Chlamydomonas* and the functional homologue of the centrosome in other eukaryotes. The deduced amino acid sequence of caltractin cDNA shares a significant sequence relatedness with calmodulin from *Chlamydomonas* and other organisms, rabbit skeletal muscle **troponin C**, and the deduced

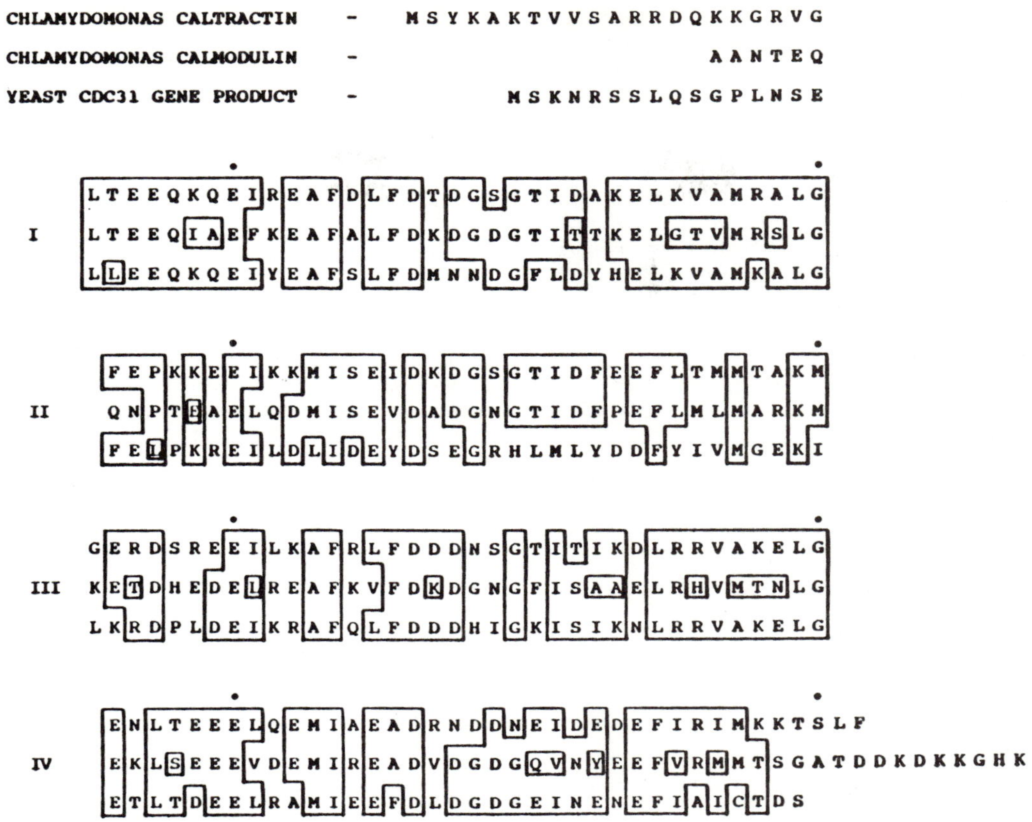

CHLAMYDOMONAS CALTRACTIN   –   M S Y K A K T V V S A R R D Q K K G R V G

CHLAMYDOMONAS CALMODULIN   –   A A N T E Q

YEAST CDC31 GENE PRODUCT   –   M S K N R S S L Q S G P L N S E

I
L T E E Q K Q E I R E A F D L F D T D G S G T I D A K E L K V A M R A L G
L T E E Q I A E F K E A F A L F D K D G D G T I T T K E L G T V M R S L G
L L E E Q K Q E I Y E A F S L F D M N N D G F L D Y H E L K V A M K A L G

II
F E P K K E E I K K M I S E I D K D G S G T I D F E E F L T M M T A K M
Q N P T E A E L Q D M I S E V D A D G N G T I D F P E P L M L M A R K M
F E I P K R E I L D L I D E Y D S E G R H L M L Y D D F Y I V M G E K I

III
G E R D S R E E I L K A F R L F D D D N S G T I T I K D L R R V A K E L G
K E T D H E D E L R E A F K V F D K D G N G F I S A A E L R H V M T N L G
L K R D P L D E I K R A F Q L F D D D H I G K I S I K N L R R V A K E L G

IV
E N L T E E L Q E M I A E A D R N D D N E I D E D E F I R I M K K T S L F
E K L S E E V D E M I R E A D V D G D G Q V N Y E E F V R M M T S G A T D D K D K K G H K
E T L T D E E L R A M I E E F D L D G D G E I N E N E F I A I C T D S

Figure 1. Comparison of the amino acid sequence of *Chlamydomonas* caltractin with those of *Chlamydomonas* calmodulin and *Saccharomyces* CDC31 gene product[9].

amino acid sequence of the CDC31 gene product of *Saccharomyces cerevisiae,* which is required for the duplication of the spindle pole body that serves as the yeast homologue of the centrosome[3]. The linear sequence identity of caltractin with calmodulin and the yeast CDC31 protein spans the four well described Ca^{2+}-binding domains of the EF hand structure[4] (Figure 1).

Immunofluorescence studies have localized caltractin in *Chlamydomonas* to the connections between the paired basal bodies, the two striated flagellar roots (SFRs) that attach the basal bodies to the underlying nucleus, and a diffuse array of filaments that envelops the nucleus (Figure 2). The SFRs, which are Ca^{2+}-sensitive organelles[5,6], appear to play a major role in several important aspects of cell activity. A mutant that has a variant number of basal bodies that localize to aberrant positions within the cell, and thus a variable flagellar number phenotype, is defective in SFRs[5]. SFRs may also have roles in nuclear movement and initiation of gene expression of flagellar precursors during flagellar regeneration[7], and in the movement of chromosomes in mitosis[6].

Chlamydomonas caltractin is homologous to a 20 kDa Ca^{2+}-binding acidic phosphoprotein that constitutes the major structural component of the SFRs in the marine alga *Tetraselmis*[8], both in antigenicity[2,8] and in amino acid sequence[9]. Antigenic determinants related to the 20 kDa *Tetraselmis* SFR protein, collectively termed "centrins"[6], have been localized to the basal bodies of other flagellated algae[10,11], the centrosome region in two cultured mammalian cell lines[12,13] and the basal bodies of human sperm cells[12]. Antigenic determinants related to *Chlamydomonas* caltractin has also been demonstrated recently in the low molecular mass components of the locomotory organelles of two ciliated protozoa: the spasmoneme of *Vorticella*[14] and the myoneme of *Stentor*[15].

■ PURIFICATION

Purification of caltractin from *Chlamydomonas* and the 20 kDa SFR protein from *Tetraselmis* is based on extraction with high salt, followed by Ca^{2+}-dependent hydrophobic interaction chromatography on a phenyl-Sepharose column[2,16].

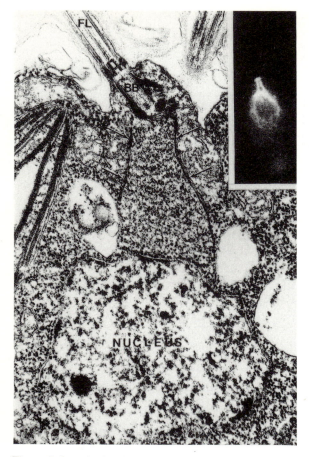

Figure 2. Longitudinal thin section electron micrograph through the apical portion of a *Chlamydomonas* cell at interphase. (Inset) Indirect immunofluorescence staining of a *Chlamydomonas* cell at interphase with a rabbit anti-caltractin immune serum[2].

■ ACTIVITIES

Caltractin most likely participates in the normal disposition and segregation of the basal body complex[5]. The association of caltractin and its homologues to microtubule organizing centres of divergent structures suggests a functionally conserved role. Caltractin based filaments appear to exemplify a novel class of cytoplasmic structures that through a Ca[2+]-mediated contractile mechanism operate distinctly from but yet in concert with microtubules, intermediate filaments, and microfilaments in organizing and integrating cell morphology and behaviour[17].

■ ANTIBODIES

Polyclonal antibodies made against caltractin from *Chlamydomonas* has been documented[2]. Polyclonal and monoclonal antibodies made against the 20 kDa *Tetraselmis* SFR protein have been described[5,6].

■ GENES

A full length clone isolated from a *Chlamydomonas* cDNA library that contains the entire coding sequence of caltractin has been reported[1] (EMBL X12634) and the sequence of the gene locus for *Chlamydomonas* caltractin has been described[18] (EMBL X57973). Genomic sequence of the CDC31 gene product from *Saccharomyces* has also been published[3].

■ REFERENCES

1. Huang, B., Mengersen, A. and Lee, V.D. (1988) J. Cell Biol. 107, 133-140.
2. Huang, B., Watterson, D.M., Lee, V.D. and Schibler, M.J. (1988) J. Cell Biol. 107, 121-131.
3. Baum, P., Furlong, C. and Byers, B.. (1986) Proc. Natl. Acad. Sci. (USA) 83, 5512-5516.
4. Kretsinger, R.H. (1980) Crit. Rev. Biochem. 8, 119-174.
5. Wright, R.L., Salisbury, J. and Jarrik, J.W. (1985) J. Cell Biol. 101, 1903-1912.
6. Salisbury, J.L., Baron, A.T. and Sanders, M.A. (1988) J. Cell Biol. 107, 635-641.
7. Salisbury, J.L., Sanders, M.A. and Harpst, L. (1987) J. Cell Biol. 105, 1799-1805.
8. Salisbury, J.L., Baron, A., Surek, B. and Melkonian, M. (1984) J. Cell Biol. 99, 962-970.
9. Lee, V.D. and Huang, B. (1990) In Calcium as an Intracellular Messenger in Eucaryotic Microbes (O'Day, D.H., ed.), 245-257, American Society for Microbiology, Washington, D.C.
10. Schulze, D., Robenek, H., McFadden, G.I. and Melkonian, M. (1987) Eur. J. Cell Biol. 45, 51-61.
11. Melkonian, M., Schulze, D., McFadden, G.I. and Robenek, H. (1988) Protoplasma 144, 56-61.
12. Salisbury, J.L., Baron, A.T., Coling, D.E., Martindale, V.E. and Sanders, M.A. (1986) Cell Motil. Cytoskel. 6, 193-197.
13. Baron, A.T. and Salisbury, J.L. (1988) J. Cell Biol. 107, 2669-2678.
14. Vacchaino, M. and Buhse, H., unpublished results.
15. Huang, B. and Maloney, M., unpublished results.
16. Salisbury, J.L., Aebig, K.W. and Coling, D.E. (1986) Meth. Enzymol. 134, 408-414.
17. Roberts, T.M. (1987) Cell Motil. Cytoskel. 8, 130-142.
18. Lee, V.D., Stapleton, M. and Huang, B. (1991) J. Mol. Biol. 220, 221, 175-191.

■ *Vincent D. Lee and Bessie Huang:*
Department of Cell Biology,
The Scripps Research Institute,
La Jolla, CA, USA

Cytoplasmic Dynein (MAP1C)

Cytoplasmic dynein[1,2] (MAP1C) is a large, two-headed ATPase which produces force along microtubules in the retrograde direction (toward the microtubule minus-end). Its physical and enzymatic properties are similar to those of two-headed ciliary and flagellar dyneins, though a number of features distinguish the cytoplasmic and axonemal forms of the enzyme. Cytoplasmic dynein is thought to be responsible for retrograde organelle movement, and possibly for some aspects of chromosome movement.

Cytoplasmic dynein consists of two globular "heads" (13.4 nm) attached through stalks to a complex basal domain, and is morphologically indistinguishable by present criteria from two-headed **axonemal dyneins**[3]. The molecule has a mass of 1,200 kDa as determined by scanning transmission electron microscopy[3] and sediments at 20 S. It is composed of at least nine subunits[3,4] (Figure 1). The ATPase activity is associated with the two 410 kDa heavy chains, which, like those of flagellar dynein, are susceptible to vanadate mediated photocleavage[4,5]. Cytoplasmic dynein also contains what appear to be three subunits of

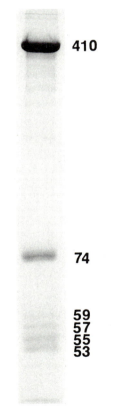

Figure 1. Electrophoretic gel of purified calf brain cytoplasmic dynein. Molecular masses are in kDa. The two heavy chains (410 kDa) are seen as one band at the top of the gel; note the complexity in the 74 kDa region[4].

Figure 2. Punctate organelle-like staining of rat saphenous nerve by using antibodies to cytoplasmic dynein[19].

74 kDa, and close to one each of 59, 57, 55, and 53 kDa[3,4]. Other apparent subunits of 173 kDa[6] or 150 kDa[7] have been observed to copurify with cytoplasmic dynein in variable amounts, as have additional polypeptides in the 40 to 50 kDa range[7]. While the heavy chains are similar in size to those of axonemal dyneins, the lower molecular weight subunits of the two forms of dynein are very different.

Cytoplasmic dynein is widely, if not universally, distributed, having been isolated from mammalian brain[4], testis[6,7], liver[7], and cultured cells[8], chicken brain[9], *C. elegans*[5], squid axoplasm[10,11], the giant amoeba *Reticulomyxa*[12], *Dictyostelium*[13] and *Paramecium*[14]. A cytoplasmic form of dynein found in sea urchin eggs has been postulated to be a ciliary precursor[15]. Cytoplasmic dynein was discovered as the retrograde counterpart to **kinesin**[16] and a growing body of evidence supports a role for cytoplasmic dynein in retrograde organelle transport. For example, retrograde organelle movements are blocked by the dynein inhibitors EHNA and vanadate[17,18] and by UV irradiation in the presence of vanadate[9,10].

Antibodies to cytoplasmic dynein produce a punctate, organelle-like staining pattern in mouse peripheral neurons[19] (Figure 2) and stain lysosomes in cultured fibroblasts[20] and endosomes. In dividing cells, prometaphase kinetochores are also stained[8,21] (Figure 3), suggesting a

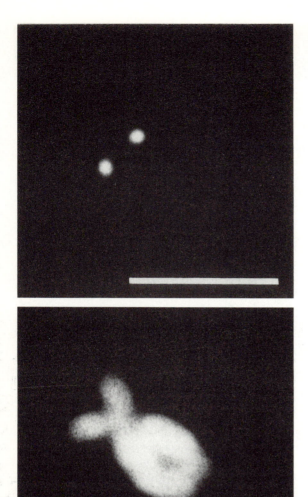

Figure 3. Kinetochore staining of isolated HeLa cell chromosome pair by using antibodies to cytoplasmic dynein. *Bottom panel*: antibody staining; *top panel*: DAPI staining[8].

role for cytoplasmic dynein in poleward chromosome movement in the early stages of mitosis. This is supported by rapid minus-end directed movements of microtubules along isolated prometaphase chromosomes[22].

■ PURIFICATION

Cytoplasmic dynein is purified by ATP extraction of taxol stabilized microtubules[4,5,23], following GTP extraction to remove kinesin and **dynamin**[4,23]. In some systems nucleotide depletion is first required to induce microtubule binding[5]. Subsequent purification of ATP-extracted cytoplasmic dynein is accomplished by sucrose density gradient centrifugation[4,5], though purification can also be achieved by chromatography with Bio-gel A1.5m, DEAE-Sepharose, and hydroxylapatite[4].

■ ACTIVITIES

Cytoplasmic dynein is a microtubule activated ATPase (basal ATPase ~50 nmol/min/mg; activated ATPase ~180-360 nmol/min/mg)[4,24]. Force production can be measured by a microtubule gliding assay. Rates of force production vary with species, ranging from 1.25[4]-2 μm/sec[5]. The direction of force production is retrograde[16] (toward the minus end of microtubules). Cytoplasmic dynein differs from axonemal dynein in the much greater sensitivity of the cytoplasmic enzyme to activation by microtubules (K_d for microtubules is 0.25 mg/ml) and in its ability to hydrolyse CTP, GTP and TTP at rates greater than ATP[24]. However, only ATP supports force production[16]. Cytoplasmic dynein is somewhat less sensitive to vanadate than axonemal dyneins (50% inhibition of cytoplasmic dynein ATPase occurs at 5-10 μM vanadate), and is more sensitive to NEM than kinesin[16,24]. Cytoplasmic dynein is also inhibited by EHNA[16,24].

■ ANTIBODIES

Polyclonal and monoclonal antibodies have been produced against cytoplasmic dynein of cow[19] and chicken[21] brain, HeLa cells[8], *D. discoideum*[13], and *D. melanogaster*[8].

■ GENES

Molecular cloning of the 150 kDa cytoplasmic dynein associated polypeptide has been accomplished[25] (Accession number X62160) and has revealed extensive homology with the product of the *Glued* gene in *Drosophila*. Partial cloning of the heavy chains of squid[26], *C. elegans*[27], *Dictyostelium*[28], mouse[29] and rat[29] cytoplasmic dyneins has also been reported. The 74kDa subunit has been cloned and sequenced[30].

■ REFERENCES

1. Vallee, R.B. and Shpetner, H.S. (1990) Ann. Rev. Bochem. 59, 909-932.
2. Vallee, R.B. and Bloom, G.S. (1991) Ann. Rev. Neurosci. 14, 59-92.
3. Vallee, R.B., Wall, J.S., Paschal, B.M. and Shpetner, H.S. (1988) Nature 332, 561-563.
4. Paschal, B.M., Shpetner, H.S. and Vallee, R.B. (1987) J. Cell Biol. 105, 1273-1282.
5. Lye, R.J., Porter, M.E., Scholey, J.M. and McIntosh, J.R. (1987) Cell 51, 309-318.
6. Neely, M.D. and Boekelheide, K. (1988) J. Cell Biol. 107, 1767-1776.
7. Collins, C.A. and Vallee, R.B. (1989) Cell Motil. Cytoskel. 14, 491-500.

8. Pfarr, C.M., Coue, M., Grissom, P.M., Hays, T.S., Porter, M.E. and McIntosh, J.R. Nature (1990) 345, 263-265.
9. Schroer, T.A., Steuer, E.R. and Sheetz, M.P. (1989) Cell 56, 937-946.
10. Schnapp, B.J. and Reese, T.S. (1989) Proc. Nat. Acad. Sci. (USA) 86, 1548-1552.
11. Gilbert, S.P. and Sloboda, R.D. (1989) J. Cell Biol. 109, 2379-2394.
12. Euteneuer, U., Koonce, M.P., Pfister, K.K. and Schliwa, M. (1988) Nature 332, 176-178.
13. Koonce, M.P. and McIntosh, J.R. (1990) Cell Motil. Cytoskel. 15, 51-62.
14. Schroeder, C.C., Fok, A.K. and Allen, R.D. (1989) J. Cell Biol. 111, 2553-2562.
15. Asai, D.J. (1986) Devel. Biol. 118, 416-424.
16. Paschal, B.M. and Vallee, R.B. (1987) Nature 330, 181-183.
17. Forman, D.S., Brown, K.J. and Livengood, D.R. (1983) J. Neurosci. 3, 1279-1288.
18. Forman, D.S., Brown, K.J. and Promesberger, M.E. (1983) Brain Res. 272, 194-197.
19. Hirokawa, N., Sato-Yoshitake, R., Yoshida, Y. and Kawashima, T. (1990) J. Cell Biol. 111, 1027-1037.
20. Lin, X.H. and Collins, C.A. (1992) J. Cell Sci. 101, 125-137.
21. Steuer, E., Wordeman, L., Schroer, T.A. and Sheetz, M.P. (1990) Nature 345, 266-268.
22. Hymen, A..A. and Mitchison, T.J. (1991) Nature 351, 206-211.
23. Paschal, B.M., Shpetner, H.S. and Vallee, R.B. (1991) Meth. Enzymol. 196, 181-191.
24. Shpetner, H.S., Paschal, B.M. and Vallee, R.B. (1988) J. Cell Biol. 107, 1001-1009.
25. Holzbaur, E.L.F., Hammarback, J.A., Paschal, B.M., Kravit, N.G., Pfister, K.K. and Vallee, R.B. (1991) Nature 351, 579-583.
26. Kronidou, N.G. and Sloboda, R.D. (1990) J. Cell Biol. 111, 25a.
27. Lye, R.J. and Waterston, R.H. (1990) J. Cell Biol. 111, 24a.
28. Koonoe, M.P., Grissom, P. and McIntosh, J.R. (1990) J. Cell Biol. 111, 24a.
29. Mikami, A., Paschal, B.M. and Vallee, R.B. (1991) J. Cell Biol. 115, 168a.
30. Paschal, B.M., Mikami, A., Pfister, K.K. and Vallee, R.B. (1992) J. Cell. Biol. 118, 1133-1143.

■ *Richard B. Vallee:*
Cell Biology Group,
Worcester Foundation for Experimental Biology,
Shrewsbury, MA, USA

Cytoplasmic Myosin II

Cytoplasmic myosin II defines a family of force generating enzymes that hydrolyze ATP to ADP and P$_i$ and uses the derived chemical energy to induce directed movement along actin filaments. The biological roles of these myosins include generating contractile forces in the furrow during cytokinesis (Figure 1). They have also been implicated in developmental processes where cellular shape changes and directed cell migrations appear to drive morphogenesis.

The myosin isoform isolated from nonmuscle tissue has a hexameric structure which is similar to that of **sarcomeric myosin**. It consists of two heavy chains (up to 240 kDa each) and two pairs of light chains (15-20 kDa each). The heavy chain can be divided into two principal domains, the N-terminal 95 kDa domain which contains the binding sites for ATP, **actin** and the light chains and can function as a motor *in vitro*, and the C-terminal tail which forms an α-helical coiled-coil structure that enables myosin molecules to form filaments. In some species, one light chain can be phosphorylated by a specific light chain kinase, which regulates the myosin ATPase and motor activities. By electron microscopy, cytoplasmic myosin II appears to be an elongated molecule (up to 200 nm in length) with two globular domains at the end of an extended stalk (Figure 2). This isoform of myosin has been isolated from a wide range of cells including yeast and mammalian cells[1-3]. Cytoplasmic myosin II can be found both as filaments and as soluble monomers within the same cell. Filament assembly is a multistep and highly regulated process. In some cases, phosphorylation at sites in the tail region appears to regulate filament disassembly. It is likely that the tail region also contains sites which play a role in determining the intracellular spatial and temporal localization of the myosin.

Disruption of the yeast myosin II gene results in cells with aberrant nuclear migration and defective bud formation[4]. Disruption of the *Dictyostelium* myosin II gene[5], interference with its expression[6] or complete deletion of the gene[7] gives rise to cells that are unable to divide in suspension culture and fail to complete development. These mutants are defective in many cellular motile events such as the constriction of cleavage furrows, capping of surface receptors and establishment of cell polarity[8]. Cell motility, chemotaxis and the movement of intracellular particles are also affected[9,10].

■ PURIFICATION

Cytoplasmic myosin is purified by utilizing the insolubility of myosin at low salt concentrations (~50 mM) and its solubility at high salt concentrations (>300 mM) or in sucrose[11]. Generally extracts are prepared in sucrose at 100 mM KCl and centrifuged. Dialysis of supernatants in low salt buffers (50 mM KCl) allows an actin-myosin complex to form which is collected by centrifugation. After resuspension in high salt - MgATP buffer, often containing 0.6 M KI, gel filtration can be used to remove most of the other proteins, principally actin. Further purification can be achieved by utilizing actin affinity or other chromatographic procedures.

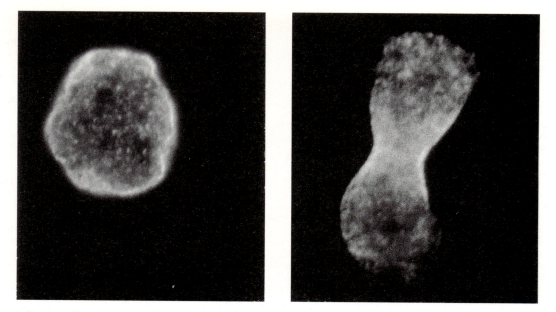

Figure 1. Immunofluorescence localization of *Dictyostelium* cytoplasmic myosin II in an interphase cell showing cortical and punctate staining (left) and in the furrow region of a dividing cell (right). (Pictures provided by T. Egelhoff).

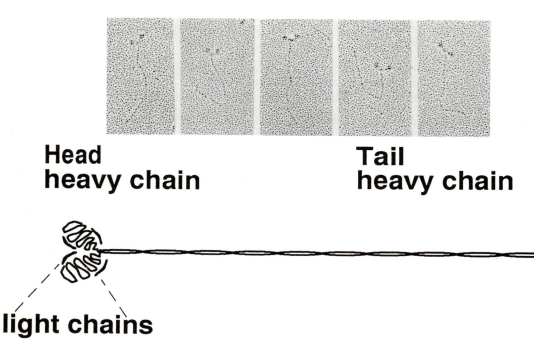

**Head
heavy chain**

**Tail
heavy chain**

light chains

Figure 2. Model of the structure of cytoplasmic myosin II. Light chains are clearly associated with the heavy chain head region, but their localization on the head is unknown. Inset shows rotary shadow electron micrographs of five examples of *Dictyostelium* cytoplasmic myosin II. (Pictures provided by P. Flicker).

ACTIVITIES

Cytoplasmic myosins bind to F-actin and the complex is dissociated by ATP. They also contain a MgATPase activity which can be stimulated up to 40-fold by F-actin. An apparently nonphysiologic Ca^{2+} and K^+-high salt ATPase can also be used as an assay. The motile activity of cytoplasmic myosin can be measured by attaching myosin to beads and examining their translocation along stationary actin bundles[12], or attaching myosin to a surface and observing the ATP dependent translocation of fluorescently labelled actin filaments along it[13]. Myosin moves toward the barbed end of actin filaments.

ANTIBODIES

Polyclonal sera against many cytoplasmic myosins are available[1]. There is some crossreaction with myosins from different species as well as with muscle myosins; the strongly antigenic epitopes are generally located in the tail region. Monoclonal antibodies to specific tail epitopes have been used to map functional sites in the tails of myosins[14-16]. Segments of the tail can be expressed at high levels in *E. coli*, and they have been used to characterize myosin assembly and function[17,18].

GENES

Cytoplasmic myosin II genes have been cloned and completely sequenced from *Dictyostelium* (GenBank M14628, M11938)[19], *Acanthamoeba* (GenBank Y00624, M12702, M12703, M19549)[20], yeast (GenBank X53947)[4], chicken (GenBank M26510)[21], human (GenBank M31013)[22] and *Drosophila* (GenBank M35012)[23].

REFERENCES

1. Korn, E.D. and Hammer, III, J.A. (1988) Annu. Rev. Biophys. Biophys. Chem. 17, 23-45.
2. Kiehart, D.P. (1990) Cell 60, 347-350.
3. Spudich, J.A. (1989) Cell Reg. 1, 1-11.
4. Watts, F.Z., Shiels, G. and Orr, E. (1985) EMBO J. 6, 3499-3505.
5. De Lozanne, A. and Spudich, J.A. (1987) Science 236, 1086-1091.
6. Knecht, D.A. and Loomis, W.F. (1987) Science 236, 1081-1086.
7. Manstein, D.J., Titus, M.A., De Lozanne, A. and Spudich, J.A. (1989) EMBO J. 8, 923-932.
8. Fukui, Y., De Lozanne, A. and Spudich, J.A. (1990) J. Cell Biol. 110, 367-378.
9. Wessels, D., Soll, D.R., Knecht, D., Loomis, W.F., De Lozanne, A. and Spudich, J.A. (1988) Dev. Biol. 128, 164-177.
10. Wessels, D. and Soll, D.R. (1990) J. Cell Biol. 111, 1137-1148.
11. Pollard, T.D. (1982) Methods Enzymol. 85, 331-356.
12. Scheetz, M.P. and Spudich, J.A. (1983) Nature 303, 31-35.
13. Kron, S.J. and Spudich, J.A. (1986) Proc. Natl. Acad. Sci. (USA) 83, 6272-6276.
14. Flicker, P.F., Peltz, G., Sheetz, M.P., Parham, P. and Spudich, J.A. (1985) J. Cell Biol. 100, 1024-1030.
15. Pagh, K. and Gerisch, G. (1986) J. Cell Biol. 103, 1527-1538.
16. Rimm, D.L., Kaiser, D.A., Bhandari, D., Maupin, P., Kiehart, D.P. and Pollard, T.D. (1990) J. Cell Biol. 111, 2405-2416.
17. O'Halloran, T.J., Ravid, S. and Spudich, J.A. (1990) J. Cell Biol. 110, 63-70.
18. Leinwand, L.A., Sohn, R., Frankel, S.A., Goodwin, E.B. and McNally, E.M. (1989) Cell Motil. Cytoskeleton 14, 3-11.
19. Warrick, H.M., De Lozanne, A., Leinwand, L.A. and Spudich, J.A. (1986) Proc. Natl. Acad. Sci. (USA) 83, 9433-9437.
20. Hammer III, J.A., Bowers, B., Paterson, B.M. and Korn, E.D. (1987) J. Cell Biol. 105, 913-925.
21. Shohet, R.V., Conti, M.A., Kawamoto, S., Preston, Y.A., Brill, D.A. and Adelstein, R.S. (1989) Proc. Natl. Acad. Sci. (USA) 85, 7726-7730.
22. Saez, C.G., Myers, J.C., Shows, T.B. and Leinwand, L.A. (1990) Proc. Natl. Acad. Sci. (USA) 87, 1164-1168.
23. Ketchum, A.S., Stewart, C.T., Stewart, M. and Kiehart, D.P. (1990) Proc. Natl. Acad. Sci. (USA) 87, 6316-6320.

■ *Hans M. Warrick and James A. Spudich:*
Departments of Cell Biology and Developmental Biology,
Stanford University School of Medicine
Stanford, CA, USA

Dynactin

Dynactin[1,2] is a cytosolic protein of approximately 150 kDa found in a variety of tissues and organisms. Dynactin is thought to facilitate dynein-mediated intracellular vesicle transport events such as retrograde axonal transport and movement of endosomes.

Dynactin (150-170 kDa) was first noted as a protein that copurified with 20 S **cytoplasmic dynein**[3-7]. Unlike true subunits of the enzyme, dynactin can be separated from cytoplasmic dynein by ion exchange chromatography. In embryonic chicken brain, dynactin is a component of the dynactin complex, a large (~1,000 kDa) heterooligomer containing five other polypeptides[1]. This complex is required for dynein-based vesicle motility *in vitro*; addition of the dynactin complex to highly purified dynein restores its ability to induce vesicle movement *in vitro*[7], while immunodepletion with anti-dynactin antibodies prevents movement[1]. The hypothesis that dynactin plays a role in vesicle movement in cells is supported by immunolocalization studies demonstrating that dynactin associates with internal membranes[1].

Chicken dynactins are encoded by a single gene which

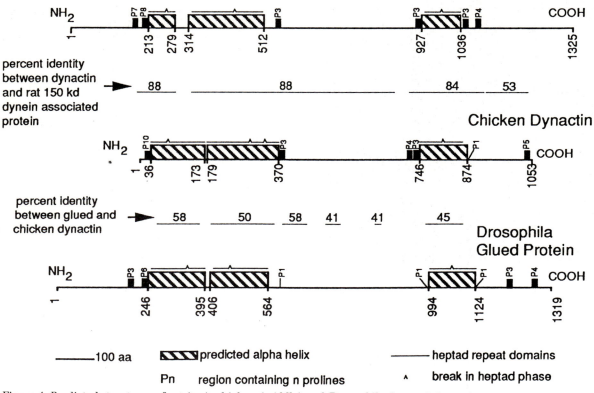

Figure 1. Predicted structures of rat (top), chicken (middle) and *Drosophila* (bottom) dynactins.

yields three transcripts and three protein isoforms, one of which is brain specific[1]. The amino acid sequences of chicken and rat dynactins[1,2] show significant homology to the *Drosophila* gene *Glued* (Figure 1). Mutations in *Glued* cause defects in neuronal function[8] and impair cell viability[9]. The primary sequences of all three dynactins predict long (100-200 AA) stretches of α-helix with the repeating heptad units characteristic of coiled-coils, suggesting that the protein is capable of forming homodimers or higher order structures. It is not known what region of the molecule is involved in binding membrane, dynein, or other polypeptides in the dynactin complex.

■ PURIFICATION

The dynactin complex copurifies with dynein by microtubule affinity and velocity sedimentation but can be isolated from dynein by anion exchange chromatography[7]. This method yields only approximately 2% of total dynactin, likely reflecting the poor interaction of dynactin with microtubules[1].

■ ACTIVITIES

The dynactin complex facilitates cytoplasmic dynein-mediated transport of membrane vesicles on microtubules although neither dynactin nor the dynactin complex is believed to have activity by itself[1,7]. Unlike dynein, the isolated dynactin complex does not demonstrate nucleotide-sensitive binding to microtubules.

■ ANTIBODIES

A monoclonal antibody (mAb 150.1) to chicken dynactin[10] (Figure 2) and mouse antiserum against bovine dynactin[2] are available.

■ GENES

The complete nucleotide sequence of rat brain dynactin (referred to as 150 kDa dynein-associated protein) has been published[2]. The nucleotide sequence of chicken dynactin[1] is also available (GenBank X62773).

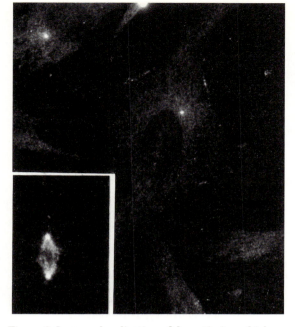

Figure 2. Immunolocalization of dynactin in a chick embryo fibroblast using the anti-dynactin antibody (mAb 150.1). The punctate and reticular staining pattern suggests that dynactin is associated with intracellular membranes. Inset: A mitotic cell stained with mAb 150.1. Photograph is from Gill et al. (1991)[1].

■ REFERENCES

1. Gill, S.R., Schroer, T.A., Szilak, I., Steuer, E.R., Sheetz, M.P. and Cleveland, D.W. (1991) J. Cell Biol. 115, 1639-1650.
2. Holzbaur, E.L.F., Hammarback, J.A., Paschal, B.M., Kravit, N.G., Pfister, K.K. and Vallee, R.B. (1991) Nature 351, 579-583.
3. Neely, M.D. and Boekelheide, K. (1988) J. Cell Biol. 107, 1767-1776.
4. Collins, C.A. and Vallee, R.B. (1989) Cell Motil. Cytoskel. 14, 491-500.
5. Gilbert, S.P. and Sloboda, R.D. (1989) J. Cell Biol. 109, 2379-2394.
6. Schnapp, B.J. and Reese, T.S. (1989) Proc. Natl. Acad. Sci. (USA) 86, 1548-1552.
7. Schroer, T.A. and Sheetz, M.P. (1991) J. Cell Biol. 115, 1309-1318.
8. Meyerowitz, E.M. and Kankel, D.R. (1978) Dev. Biol. 62, 112-142.
9. Harte, P.J. and Kankel, D.R. (1982) Genetics 101, 477-501.
10. Steuer, E.R., Schroer, T.A., Wordeman, L. and Sheetz, M.P. (1990) Nature 345, 266-268.

■ Trina A. Schroer:
Department of Biology,
The Johns Hopkins University,
Baltimore, MD 21218, USA

Dynamin

Dynamin is a nucleotide sensitive microtubule binding protein which hydrolyzes GTP. It is a member of a family of GTP-binding proteins with diverse functions.

Dynamin was identified as a microtubule associated protein in brain microtubules prepared by the taxol method without added nucleotide[1]. In purified form it crosslinks microtubules into regular, hexagonal arrays (Figure 1). The binding of dynamin to the microtubules appears to be highly cooperative, as revealed by long helical arrays of the protein on the microtubule surface. The projecting dynamin molecules in these arrays show a periodicity of 13 nm along the microtubule edge, appearing to follow the three-start left handed **tubulin** monomer helix[1].

Dynamin (Figure 2) is a 100 kDa polypeptide[1]. It contains three GTP-binding consensus elements located within an ~300 amino acid N-terminal domain[2]. This domain shows extensive homology (66% identity over 288 amino acids) with the product of the yeast gene VPS1 (also known as SP015) involved in vacuolar protein sorting[3] and meiosis[4]. Homology is also seen with the N-terminal domain of the interferon inducible Mx proteins, involved in inhibiting viral replication[5,6]. The C-terminal two thirds of the dynamin molecule is partially homologous with the the corresponding portion of the VPS1 gene product, but is unrelated to the Mx proteins. Dynamin also contains a unique 100 amino acid domain at its extreme C-terminus which is rich in proline (32%) and very basic (pI = 12.5) and is involved in microtubule binding[7].

Recent evidence has identified an ~100 kDa dynamin-like protein in *Drosophila* as the product of the *shibire* gene[8,9]. The rat and *Drosophila* proteins show a high degree of sequence conservation throughout their length (68% identical amino acids overall, 81% including conservative substitutions), and are, therefore, even more similar to each other than to VPS1 or Mx. Several alternative splicing products of *shi* were identified which differed in their tissue distribution.

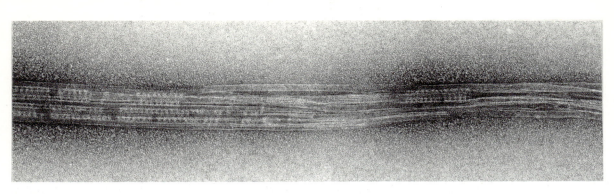

Figure 1. Bundle of microtubules crosslinked by purified dynamin. From Ref. **1**.

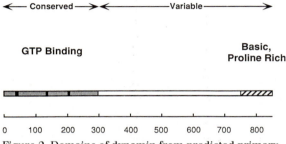

Figure 2. Domains of dynamin from predicted primary sequence[2]. Amino acid number is shown at bottom. N-terminal domain of ~300 kDa shared with VPS1p and Mx proteins is indicated.

shi[ts] mutants are well-known for their rapidly reversible paralytic phenotype, which results from a defect in synaptic vesicle recycling[10,11]. The mutants are now known to be blocked in endocytosis in both neuronal[11] and nonneuronal cells[12,13], specifically in the budding of coated and noncoated vesicles from the plasma membrane.

Taken together the existing data strongly suggest that dynamin is a member of a newly emerging family of membrane-sorting GTPases, though the details of how dynamin functions in the cell and the role of microtubules are uncertain. Dynamin is found in both soluble[1] and particulate[14] form. Whether this is indicative of a reversible or cyclical interaction with membranous organelles is unclear.

■ PURIFICATION

Dynamin is purified from brain tissue by GTP/AMP-PNP extraction of taxol-stabilized microtubules[1,15]. Subsequent purification is accomplished by rebinding to and reextraction from microtubules, followed by DEAE-Sepharose chromatography.

■ ACTIVITIES

Dynamin binds to microtubules in an ATP or GTP dependent manner[1]. As extracted from microtubules it exhibits microtubule activated ATPase activity. However, when completely purified it shows only a microtubule activated GTPase activity, and GTP is thought to be its physiological substrate[16]. An apparent cofactor found in partially purified dynamin preparations may be nucleoside diphosphokinase[16], which is known to be present in brain microtubule preparations. Dynamin crosslinked microtubule bundles come apart in the presence of nucleotide, with some bundles appearing to elongate[1].

■ ANTIBODIES

Existing antibodies include a rabbit autoimmune antibody[2] and rabbit immune antibodies raised against bovine[2] and rat[8,14,17] dynamin. One antibody shows very broad species crossreactivity, recognizing mammalian, *Drosophila*, and yeast forms of the protein[8,17].

■ GENES

Dynamin has been cloned from rat brain[2] (X54531) and *Drosophila* (X59448, X59449).

■ REFERENCES

1. Shpetner, H.S. and Vallee, R.B. (1989) Cell 59, 421-432.
2. Obar, R., Collins, C.A., Hammarback, J.A., Shpetner, H.S. and Vallee, R.B. (1990) Nature 347, 256-261.
3. Rothman, J.H., Raymond, C.K., Gilbert, T., O'Hara, P.J. and Stevens, T.H. (1990) Cell 61, 1063-1074.
4. Yeh, E., Driscoll, R., Coltrera, M., Olins, A. and Bloom, K. (1991) Nature 349, 713-715.
5. Staeheli, P., Haller, O., Boll, W., Lindermann, J. and Weissman, C. (1986) Cell 44, 147-158.
6. Staeheli, P. and Sutcliffe, J.G. (1988) Mol. Cell. Biol. 8, 4524-4528.
7. Herskovits, J.S., Schroeder, C.C. and Vallee, R.B. (1991) J. Cell Biol. 115, 34a.
8. Chen, M.S., Obar, R.A., Schroeder, C.C., Austin, T.W., Poodry, C.A., Wadsworth, S.C. and Vallee, R.B. (1991) Nature 351, 583-586.
9. Van der Bliek, A.M. and Meyerowitz, E.M. (1991) Nature 351, 411-413.
10. Poodry, C.A. and Edgar, L. (1979) J. Cell Biol. 81, 520-527.
11. Kosaka, T. and Ikeda, K. (1983) J. Neurobiol. 14, 207-225.
12. Kosaka, T. and Ikeda, K. (1983) J. Cell Biol. 97, 499-507.

In Figure 2:

Conserved ← → Variable

GTP Binding   Basic, Proline Rich

0   100   200   300   400   500   600   700   800

13. Kessell, I., Holst, B. and Roth, T.F. (1989) Proc. Natl. Acad. Sci. (USA) 86, 4968-4972.
14. Scaife, R. and Margolis, R.L. (1990) J. Cell Biol. 111, 3023-3033.
15. Shpetner, H.S. and Vallee, R.B. (1991) Meth. Enzymol. 196, 192-201.
16. Shpetner, H.S. and Vallee, R.B. (1992) Nature 355, 733-735.
17. Schroeder, C.C., Obar, R.A., Wadsworth, S.C. and Vallee, R.B. (1991) J. Cell Biol. 115, 34a.

■ Richard B. Vallee:
Cell Biology Group,
Worcester Foundation for Experimental Biology,
Shrewsbury, MA, USA

Kinesin

Kinesin[1,2] is a force generating enzyme that hydrolyzes ATP to ADP and P_i and uses the derived chemical energy to induce plus end directed movement along microtubules. This ubiquitous microtubule motor is thought to power anterograde organelle transport along microtubules in vivo.

Kinesin, a microtubule activated ATPase[3-5], was originally isolated from squid neural tissue as an activity that when adsorbed onto latex spheres caused them to move along microtubules and when adsorbed onto glass caused microtubules to translocate over the surface[6]. Subsequent studies have identified kinesin in a wide variety of cells, including *Dictyostelium*, *Drosophila*, sea urchin eggs and various mammalian tissues. Best characterized is the kinesin isoform isolated from bovine brain, which is a tetramer consisting of two heavy chains (120 kDa) and two light chains (62 kDa). By electron microscopy, kinesin appears as an elongated molecule (80 nm in length) with two globular heads, an extended stalk with a kink in the middle and a fan-shaped tail (Figure 2)[7,8]. The overall shape is reminiscent of **myosin**, although there is no significant amino acid homology between these two proteins. Secondary structure predictions of the heavy chain based upon the amino acid sequence reveals three principal domains: (1) an N-terminal 45 kDa globular domain that contains the microtubule and ATP binding site; (2) a central α-helical coiled-coil "stalk" that enables two heavy chains to dimerize, and (3) a 10 kDa C-terminal globular domain that, together with the light chains, comprises the fan-shaped tail. A variety of experiments indicate that the N-terminal globular region constitutes a functional motor. Most notably, the isolated N-terminal globular domain generated by proteolysis has microtubule activated ATPase activity[9,10], and a chimeric protein consisting of the *Drosophila* kinesin N-terminal region and a portion of α-**spectrin** translocates microtubules *in vitro*[11]. The C-terminal portion, in conjunction perhaps with the light chains, probably functions as an attachment domain that interacts with organelles or other cellular structures.

Kinesin has been widely implicated in the anterograde transport of membrane vesicles[6] (Figure 1) and in extending membrane tubules of the endoplasmic reticulum towards the cell periphery[12]. Consistent with this notion, antibodies to kinesin inhibit organelle transport in dissociated squid axoplasm[13] and disrupt pigment granule movements[14] and lysosome/endosome extension[15] when

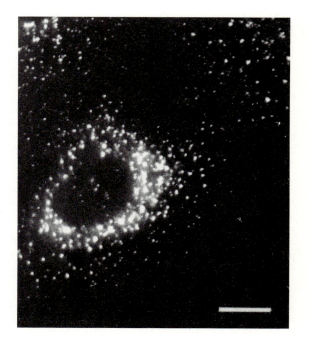

Figure 1. Immunolocalization of kinesin in PtK$_1$ cells using a monoclonal anti-kinesin antibody. The Triton X-100 soluble, punctate staining pattern suggests that kinesin is found associated with membranous organelles (photograph is from Pfister et al.[17]). Bar 10 μm.

microinjected into living cells. Furthermore, a disruption of the kinesin gene in *Drosophila* causes a neural defect that perhaps results from an impaired delivery of vesicles to nerve terminals[16]. Immunofluorescence experiments indicate that kinesin is associated with organelles[17], although a substantial amount of kinesin also exists in a soluble pool[18]. Highly purified kinesin stimulates transport of isolated chromaffin granules along microtubules[19];

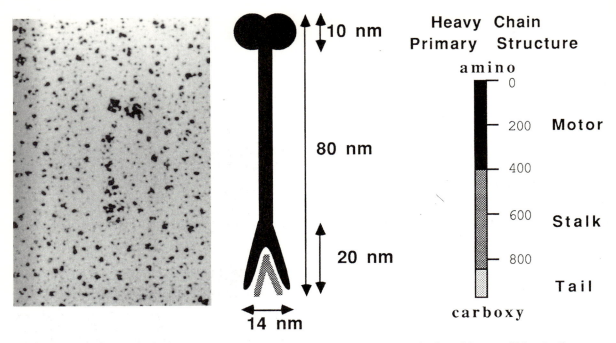

Figure 2. Kinesin structure. The electron micrograph on the left shows a rotary shadowed image of kinesin (from Hirokawa et al.[7]). The adjacent diagram shows the dimensions of the molecule and the positions of the heavy (dark) and light (shaded) chains. On the right is a linear map of the kinesin heavy chain (amino acid numbers are depicted) illustrating the approximate positions of the motor domain, the α-helical coiled-coil stalk, and the globular C-terminal domain. This assignment is based upon secondary structure predictions deduced from the predicted amino acid sequence and from functional studies.

however, in other systems, additional proteins may be required for this process[20]. With the exception of sea urchin eggs[21], kinesin is not present in high concentrations in the mitotic apparatus.

Proteins homologous to kinesin ("**kinesin related proteins**") have been identified through genetics and by fishing for related genes using PCR. These proteins contain an approximately 350 amino acid domain with 30-50% amino acid identity to the force generating domain of *Drosophila* kinesin but show no homology outside of this region. These findings suggest that there is a large family of kinesin motors that share similar force generating domains linked to different tails that may confer unique binding properties, which allow these different motors to move distinct objects within the cell.

■ PURIFICATION

Kinesin is purified routinely by microtubule affinity and conventional column chromatography[3,4,6,22] or by antibody affinity purification[23]. Approximately 1 mg of kinesin can be obtained from six bovine brains. Kinesin adheres to P11, DEAE-Sepharose, S-Sepharose and Affi-Gel Blue resins. The kinesin gene has also been expressed in *E. coli* and the expressed protein can be purified in an active form[11].

■ ACTIVITIES

Kinesin binds to microtubules in a nucleotide dependent manner (AMP-PNP induces rigor-like binding), and the ATPase activity is stimulated 1000-fold by microtubules (V_{max} turnover of 10-20 ATP/sec/heavy chain)[3,4]. The ATPase cycle is similar to that of myosin, with the exception that ADP release is the rate limiting step in the absence of microtubules. The motile activity of kinesin can be measured by attaching kinesin to glass and observing the ATP dependent translocation of microtubules along the surface[6] or by attaching kinesin to microspheres and examining their translocation along stationary microtubules[6,24]. Methods for studying motility induced by single kinesin molecules have been described[25,26]. Movement is always directed to the microtubule plus end, which distinguishes kinesin from the minus-end directed **dynein** motors. (Note: some members of the kinesin superfamily, notably *ncd* from *Drosophila* (see kinesin related proteins), induce minus-end directed movement.) Kinesin translocation and ATP activities are also inhibited at higher NEM (>2 mM) and vanadate (>100 mM) concentrations than the dyneins.

■ ANTIBODIES

Polyclonal sera against squid, bovine, adrenal medulla, *Dictyostelium* and sea urchin kinesins are available.

Polyclonal antibodies tend to react against kinesins of various species. Monoclonal antibodies to the heavy chain of sea urchin[9], squid[27] and bovine kinesin[17] are published, as are two monoclonals to the bovine kinesin light chain[17].

■ GENES

Kinesin heavy chain has been cloned and sequenced from *Drosophila*[28] (GenBank M24441), squid optic lobe[27] (GenBank J05258), sea urchin[29], and human[30] (GenBank X65873). The light chains have been cloned from rat (M75146, M75147, M75148)[31].

■ REFERENCES

1. Vale, R.D. (1990) Curr. Opin. Cell Biol. 2, 15-22.
2. Vale, R.D. (1987) Ann. Rev. Cell Biol. 3, 347-378.
3. Kuznetsov, S.A. and Gelfand, V.I. (1986) Proc. Natl. Acad. Sci. (USA) 83, 8530-8534.
4. Hackney, D.D. (1988) Proc. Natl. Acad. Sci. (USA) 85, 6314-6318.
5. Brady, S.T. (1985) Nature 317, 73-75.
6. Vale, R.D., Reese, T.S. and Sheetz, M.P. (1985) Cell 42, 39-50.
7. Hirokawa, N., Pfister, K.K., Yorifuji, H., Wagner, M.C., Brady, S.T. and Bloom, G.S. (1989) Cell 56, 867-878.
8. Scholey, J.M., Heuser, J., Yang, J.T. and Goldstein, L.S.B. (1989) Nature 338, 355-357.
9. Ingold, A.M., Cohn, S.A. and Scholey, J.M. (1989) J. Cell Biol. 105, 1453-1463.
10. Kuznetsov, S.A., Vaisberg, Y.A., Rothwell, S.W., Murphy, D.B. and Gelfand, V.I. (1989) J. Biol. Chem. 264, 589-595.
11. Yang, J.T., Saxon, W.M., Stewart, R.J., Raff, E.C. and Goldstein, L.S.B. (1990) Science 249, 42-47.
12. Debora, S.L. and Sheetz, M.P. (1988) Cell 54, 27-35.
13. Brady, S.T., Pfister, K.K. and Bloom, G.S. (1990) Proc. Natl. Acad. Sci. 87, 1061-1065.
14. Rodionov, V.I., Gyoeva, F.K. and Gelfand, V.I. (1991) Proc. Natl. Acad. Sci. 88, 4956-4960.
15. Hollenbeck, P.J. and Swanson, J.A. (1990) Nature 346, 864-866.
16. Saxton, W.M., Hicks, J., Goldstein, L.S.B. and Raff, E.C. (1991) Cell 64, 1093-1102.
17. Pfister, K.K., Wagner, M.C., Stenoien, D.L., Brady, S.T. and Bloom, G.S. (1989) J. Cell Biol. 107, 2657-2667.
18. Hollenbeck, P.J. (1989) J. Cell Biol. 108, 2335-2342.
19. Urritia, R., McNiven, M.A., Albanesi, J.P., Murphy, D.B. and Kachar, B. (1991) Proc. Natl. Acad. Sci. (USA) 88, 6701-6705.
20. Schroer, T.A., Schnapp, B.J., Reese, T.S. and Scheetz, M.P. (1988) J. Cell Biol. 107, 1785-1792.
21. Scholey, J.M., Porter, M.E., Grissom, P.M. and McIntosh, J.R. (1985) Nature 318, 483-486.
22. Wagner, M.C., Pfister, K.K., Bloom, G.S. and Brady, S.T. (1989) Cell Motil. Cytoskel. 12, 195-215.
23. Vale, R.D., Schnapp, B.J., Mitchison, T., Stever, E., Reese, T.S. and Sheetz, M.P. (1985) Cell 43, 623-632.
24. Gelles, J., Schnapp, B.J. and Sheetz, M.P. (1988) Nature 331, 450-453.
25. Howard, J.H., Hudspeth, A.J. and Vale, R.D. (1989) Nature 343, 154-158.
26. Block, S.M., Goldstein, L.S.B. and Schnapp, B.J. (1990) Nature 348, 348-352.
27. Kosik, K.S., Orecchio, L.D., Schnapp, B., Inouye, H. and Neve, R.L. (1990) J. Biol. Chem. 265, 3278-3283.
28. Yang, J.T., Laymon, R.S. and Goldstein, L.S.B. (1989) Cell 56, 879-889.
29. Wright, B.D., Hensen, J.H., Wedaman, K.P., Willy, P.J., Morand, J.N. and Scholey, J.M. (1991) J. Cell Biol. 113, 817-833.
30. Navone, F., Niclas, J., Hom-Bodner, N., Sparks, L., Bernstein, H.D., McCaffrey, G.M. and Vale, R.D. (1992) J. Cell. Biol. 117, 1263-1275.
31. Cyr, S.L., Pfister, K.K., Bloom, G.S., Slaughter, C.A. and Brady, S.T. (1991) Proc. Natl. Acad. Sci. (USA) 88, 10114-10118.

■ *Ronald D. Vale:*
Department of Pharmacology,
University of California,
San Francisco CA, USA

Kinesin Related Proteins

The kinesin related genes encode a family of proteins containing a domain with sequence similarity to the force generating microtubule motor domain of kinesin. Most of the genes were identified by mutations that specifically affect a variety of microtubule dependent processes (karyogamy, meiotic chromosome segregation and mitotic spindle formation) or that affect C. elegans behaviour.

Kinesin heavy chain contains an N-terminal domain that confers ATP dependent microtubule based movement[1]. Recently, several genes from a variety of organisms have been identified that, based on sequence data, contain kinesin related motor domains[2-9] (summarized in Table 1). Unlike the authentic kinesin genes from *Drosophila*[10] and squid[11], which are homologous throughout their sequence, the sequence similarity of the related genes is limited to approximately 340 residues that constitute the motor domain. The motor domain sequences may be at either the N-terminus or C-terminus of the predicted pro-teins. (The organization and predicted secondary structures of kinesin genes are shown in the Figure.) In one case, *KIP2*[8], a significant amount of flanking sequence is present on both sides of the motor domain. In general, the motor domains show from 30 to 40% amino acid identity to each other (Table 2); exceptions are *bimC*[6], *cut7*[7] and *KIP1*[8], whose motor domains are 60% identical, and *KAR3*[2] and *ncd*[3,4], which show 46% identity in their motor domains. One explanation for these groupings of motors with higher homology is that they merely reflect more recent divergence. A more interesting idea is that the sim-

Table 1. Properties of the Kinesin-Related Genes

Gene	Organism	Size	Presumed Functions	Orientation	Localization	Regulation	Subunits	Activity	Genbank#	Ref.
Kinesin	D. melanogaster	975	Movement of secretory granules, ER formation neural function	N-terminal	cytoplasmic	unknown	2 light chain 2 heavy chain	plus-end oriented vesicle movement	M24441	10
Kinesin	L. pealii	967	same	N-terminal			same	same	J05258	11
KAR3	S. cerevisiae	729	nuclear fusion mitosis, spindle elongation meiosis	C-terminal	cytoplasmic nuclear ND	induced constitutive induced	ND	ND	M31719	2
ncd	D. melanogaster	685, 700	meiosis, chromosome segregation	C-terminal	ND	female specific	ND	minus-end oriented	M33932	3,4
nod	D. melanogaster	666	meiosis, chromosome segregation	N-terminal	ND	female germarium	ND	ND	M36195	5
bimC	A. nidulans	1184	mitosis, SPB separation	N-terminal	ND	ND	ND	ND	M32075	6
cut7	S. pombe	1073	mitosis, SPB separation	N-terminal	ND	ND	ND	ND	X56022	7
KIP1	S. cerevisiae	1111	ND	N-terminal	ND	ND	ND	ND		8
KIP2	S. cerevisiae	706	ND	N-terminal	ND	ND	ND	ND		8
unc104	C. elegans	1584	neural function	N-terminal	ND	ND	ND	ND		9

ilarity within each group indicates a conservation of some underlying functional characteristic of the motor domain.

Excluding the authentic kinesins, there is a remarkable lack of similarity among the nonmotor domains, even among members of the same group. The significance of this extreme diversity is not yet clear. One appealing idea is that functional specificity is conferred by the nonmotor domains. In this sense, recruitment of kinesin related proteins in specific cellular processes arises from protein-protein interactions between the nonmotor domain and other proteins that couple the motors to macromolecular complexes. The prediction from this view is that each organism expresses a variety of motor proteins each containing distinct "specificity" domains which are themselves conserved amongst species.

At first blush, none of the kinesin related proteins would be considered to be authentic homologs based solely upon the criterion of similarity within the nonmotor domains. One possibility is that the "specificity" domains of the kinesin related proteins are comprised of relatively degenerate protein binding motifs and so have not been well conserved. Alternatively, it may be that the variety of kinesin related genes in each organism is quite large and only a small subset has been sampled. Indeed, although the total number of related genes in any given organism is not yet known, PCR amplification experiments suggest

Table 2. Pairwise similarity comparison between different kinesin-related genes.

		kinesin	KAR3	ncd	bimC	cut7	KIP1	KIP2	nod	unc104
kinesin	(4-339)	100	37	40	45	40	42	38	36	45
KAR3	(373-729)	37	100	46	35	39	36	36	31	35
ncd	(320-661)	40	46	100	39	36	38	33	29	36
bimC	(73-422)	45	35	39	100	59	65	36	35	41
cut7	(47-415)	40	39	36	59	100	54	38	32	36
KIP1	(41-417)	42	36	38	65	54	100	36	37	41
KIP2	(112-499)	38	36	33	36	38	36	100	33	38
nod	(1-326)	36	31	29	35	32	37	33	100	39
unc104	(1-353)	45	35	36	41	36	41	38	39	100

The indicated residues comprising each motor domain were compared using the GAP program of Devereux et al.[20]. Percent identities are indicated.

that there are at least five in S. cerevisiae[8] and possibly as many as 30 in Drosophila[12]. It will be interesting to determine whether any of these genes are actually functional or structural homologs. Ultimately, the resolution of this issue will require the isolation of complete sets of kinesin related genes from more than one organism. A description of the kinesin related genes and possible functions of the gene products follows below.

KAR3

Initially identified by a semi-dominant mutation that blocked karyogamy, KAR3[2] has an important function in mitosis and is essential in meiosis[13]. A significant fraction (40%) of mitotic cells terminally arrest in spindle elongation. In meiosis, cells arrest prior to formation of the first spindle. In karyogamy, the protein appears to be required for the interaction between sets of antiparallel cytoplasmic microtubules and subsequent movement of the two haploid nuclei. The behaviour of β-galactosidase hybrid proteins indeed suggests that KAR3 has two microtubule binding sites that could act to crossbridge microtubules. Dominant KAR3 alleles map to the conserved ATP binding/hydrolysis site in the motor domain. The distribution of the mutant protein in vivo suggests that it undergoes rigor binding to microtubules. The KAR3 gene is expressed

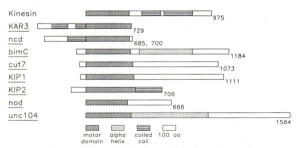

Figure. Comparison of kinesin related motor proteins. Proteins are aligned on the left side of the motor domains. Differences in the length of the motor domain are accounted for by small internal insertions as well as variation in the similarity extending into flanking sequences.

at a low level in mitotic cells (when the protein is nuclear), but is induced during both mating (when the protein is cytoplasmic) and meiosis[13].

ncd

Mutations in the ncd gene at the Drosophila claret locus[13,14] produce disturbances in female meiotic chromosome segregation and early zygotic mitosis. Some 80% of spindles show a diverse array of aberrant morphologies. Furthermore, the mitotic segregation defect shows a significant maternal effect, preferentially affecting the chromosomes derived from a mutant female parent but not those from the male parent. The ncd gene is expressed as a 2.2 kB mRNA in females[13], primarily in the ovary. A small amount of transcript may be expressed in males.

It is not yet clear where ncd acts during meiosis. A likely role[15] for a minus-end oriented motor is kinetochore to pole movement in anaphase A. The observation that maternally derived chromosomes are preferentially lost in early mitosis is consistent with a conservatively distributed kinetochore component. Alternatively, the disruption of spindle morphology in meiosis and the presence of a bundling activity led to the suggestion[16] that ncd may play a more direct role in spindle assembly, either for the half spindle, if it bundles parallel microtubules, or for the overlap zone, if it bundles antiparallel microtubules. The maternal effect on chromosome loss would then derive from the fact that karyogamy in Drosophila occurs only after the first mitotic division.

nod

Recessive mutations in the Drosophila nod locus specifically affect the segregation of chromosomes that have not engaged in meiotic recombinational exchange[5]. The usually nonrecombinant chromosome 4 is missegregated in 85% of meioses with loss events predominating. A dominant allele (nod^DTW) is a cold sensitive lethal causing a high rate of chromosome breakage and loss during mitosis. As for KAR3, the dominant mutation maps to the ATP binding/hydrolysis site[17]. Where nod acts in the distributive segregation system is uncertain. ncd and nod show genetic interactions[17].

The nod gene is expressed as a 2.4 kB mRNA in adult females[5]. The in situ hybridization signal reveals that the transcript is expressed weakly at the very earliest stages of oocyte development and more strongly at the time when nurse cells are providing the oocyte with maternal mRNAs. The transcript remains present at diminishing levels throughout embryonic development. For this reason and the effects of the nod^DTW mutation, it has been suggested that nod may also be involved in normal mitotic division[5].

bimC

A recessive temperature sensitive mutation in the Aspergillus nidulans bimC gene causes a block in mitosis[6]. Nuclei become arrested after spindle pole body duplication but prior to spindle formation. bimC, a gene closely related to cut7 and KIP1, may therefore be required for SPB separation.

cut7

Temperature sensitive mutations of the Schizosaccharomyces pombe cut7 gene block mitosis prior to spindle formation[7]. As in the bimC mutant phenotype, the spindle pole body appears to duplicate but fails to separate. Both recessive (cut7-332) and semidominant (cut7-744) alleles have been isolated.

KIP1

This gene was isolated from Saccharomyces cerevisiae by PCR amplification of primers encoding conserved residues from the motor domain[8]. Deletion of the gene has not yet revealed a function. However, the combination of a KIP1 mutation with a mutation in the CIN8 gene results in a lethal phenotype[8]. CIN8 may represent a second KIP1-like gene with overlapping function.

KIP2

This gene was identified by PCR amplification in the same manner as KIP1[8]. Deletion of the gene results in cold sensitive slow growth and increased sensitivity to benomyl, a microtubule depolymerizing drug. Double and triple mutants with deletions of KAR3 and KIP1 are viable.

unc104

Mutations in the C. elegans unc104 gene result in a spectrum of phenotypes[18] ranging from uncoordinated to totally paralyzed worms. Gene function is required during embryogenesis and severe mutants are paralyzed at birth. A variety of defects can be discerned at the cellular level including a decrease in the number of presynaptic vesicles, accumulation of vesicles in the neuronal cell body, decreased electron dense staining on the cytoplasmic side of the presynaptic membrane, loss of neuromuscular junctions, and abnormal growth of the muscle processes that target to the nerve cord. The nature of these defects suggests a general role for the protein in the movement of vesicular material to the presynaptic region. Secondary phenotypes such as the loss of muscle processes may result from the lack of normal synapse formation. Viable mutants arise from insertion of the Tc1 transposable element suggesting that the C-terminus is required for normal function. The protein is predicted to have three domains[9], an N-terminal motor domain, an α-helical stalk and a relatively basic C-terminal domain.

■ PURIFICATION

None of kinesin like proteins has been purified from their native source. Drosophila ncd[15,16] and S. cerevisiae KAR3[19] have been expressed and purified in E. coli.

■ ACTIVITIES

To date, ncd is the only kinesin related protein for which an in vitro activity has been demonstrated[15,16]. Although authentic kinesin moves exclusively towards the plus ends of microtubules, the *ncd* protein expressed in E. *coli* moves toward the minus ends at somewhat slower rates (4-15 μm/sec). Pharmacological data also indicate that *ncd* behaves less like kinesin and more like cytoplasmic dynein[15], and *ncd* generates torque[15] as it moves (3.3 rotations/μm of forward movement), an activity previously described only for 14 S **dynein**. The *ncd* protein expressed *in vitro* shows both ATP sensitive and ATP insensitive microtubule binding[3] and bundles microtubules in the absence of ATP, suggesting that *ncd* has two independent microtubule binding sites[16].

■ ANTIBODIES

Polyclonal antibodies to bacterially expressed *KAR3*[2] and *bimC*[6] fusion proteins have been prepared and the *KAR3* antibodies have been used successfully for immunofluorescence localization[2].

■ GENES

Sequences for *KAR3*[2] (GenBank M31719), *bimC*[6] (GenBank M32075), *ncd*[3,4] (GenBank M33932), *nod*[5] (GenBank M36195), *cut7*[7] (GenBank X56022), *unc104*[9] (GenBank M58582) are available.

■ UPDATE ON NEW KINESIN-RELATED MOTORS

Several new motors have been recently described that appear to be involved in mitosis. Eg5[21] is a plus-end directed kinesin-like motor from frog; antibody depletion studies indicate that it is involved in spindle formation in an in vitro system. CENP-E[22] is a large (312kDa), mammalian kinesin-like motor which accumulates in the G2 phase, is localized to the kinetochore in metaphase and is degraded at the end of mitosis. MKLP-1[23] is a mammalian plus-end directed kinesin-like motor that may be involved in anaphase B movements. SMY-1[24] is a divergent kinesin-like motor that was identified as a high-copy suppressor of Myo2 mutations in S. *cerevisiae*.

■ REFERENCES

1. Yang, J.T., Saxton, W.M., Stewart, R.J., Raff, E.C. and Goldstein, L.S.B. (1990) Science 249, 42-47.
2. Meluh, P.B. and Rose, M.D. (1990) Cell 60, 1029-1041.
3. McDonald, H.B. and Goldstein, L.S.B. (1990) Cell 61, 991-1000.
4. Endow, S.A., Henikoff, S. and Soler-Niedziela, L. (1990) Nature 345, 81-83.
5. Zhang, P., Knowles, B.A., Goldstein, L.S.B. and Hawley, R.S. (1990) Cell 62, 1053-1062.
6. Enos, A.P. and Morris, N.R. (1990) Cell 60, 1019-1027.
7. Hagen, I. and Yanagida, M. (1990) Nature 347, 563-566.
8. Roof, D.M., Meluh, P.B. and Rose, M.D. (1992) J. Cell Biol. 118, 95-108.
9. Otsuka, A.J., Jeyaprakash, A., García-Añoveros, J., Tang, L.Z., Fisk, G., Hartshorne, T., Franco, R. and Born, T. (1991) Neuron 6, 113-122.
10. Yang, J.T., Laymon, R.A. and Goldstein, L.S.B. (1989) Cell 56, 879-889.
11. Kosik, K.S., Orecchio, L.D., Schnapp, B., Inouye, H. and Neve, R.L. (1990) J.Biol. Chem. 265, 3278-3283.
12. Endow, S.A. and Hatsumi, M. (1991) Proc. Natl. Acad. Sci. 88, 4424-4427.
13. Yamamoto, A.H., Komma, D.J., Shaffer, C.D., Pirrotta, V. and Endow, S.A. (1989) EMBO J. 8, 3543-3552.
14. O'Tousa, J. and Szauter, P. (1980) Dros. Inf. Serv. 55, 119.
15. Walker, R.A., Salmon, E.D. and Endow, S.A. (1990) Nature 347, 780-782.
16. McDonald, H.B., Stewart, R.J. and Goldstein, L.S.B. (1990) Cell, 63, 1159-1165.
17. Hawley, S. personal communication.
18. Hall, D.H. and Hedgecock, E.M. (1991) Cell 65, 837-847.
19. Meluh, P.B. and Rose, M.D. unpublished observations
20. Devereux, J., Haeberli, P. and Smithies, O. (1984) Nucl. Acids Res. 12, 387-395.
21. Sawin, K.E., LeGuellec, K., Phillippe, M., and Mitchison, T.J. (1992) Nature 359, 540-543.
22. Yen, T.J., Li, G., Schaar, B.T., Szilak, I., and Cleveland, D.W. (1992) Nature 359, 536-539.
23. Nislow, C., Lombillo, V.A., Kuriyama, R., and McIntosh, J.R. (1992) Nature 359, 543-547.
24. Lillie, S.H. and Brown, S.S. (1992) Nature 355, 179-182.

■ *Mark D. Rose:*
Department of Molecular Biology,
Princeton University,
Princeton, New Jersey 08544-1014, USA

MYO2/Dilute Myosin

The MYO2 gene product from yeast[1] is an unconventional myosin[2] that functions in the polarized delivery of secretory vesicles to areas of active cell surface enlargement, such as the yeast bud. The MYO2 protein is similar in structure, and perhaps in function, to two recently identified unconventional myosins, the dilute protein of mouse[3] and the p190 protein from vertebrate brain[4]. The MYO2/dilute proteins define a novel class of unconventional myosin involved in polarized secretion.

Actin based transport systems are thought to utilize a variety of **myosin** related proteins to transport cellular cargo[5]. These myosin related molecules share closely related "head" domains capable of interacting with actin[6], but display a wide variety of apparently unrelated C-terminal domains that have been postulated to specify interactions with a transported component[5]. In particular, these unconventional myosin molecules have been suggested to serve as molecular "motors" to carry membrane vesicles to the cell surface for localized surface growth and secretion[2,5]. Characterization of a conditional mutant of the budding yeast *Saccharomyces cerevisiae* has provided experimental evidence for the proposed role for unconventional myosin molecules in polarized surface growth[1,7].

Morphogenesis by cells of the yeast *Saccharomyces cerevisiae* requires polarized cell surface growth[8]; an asymmetric pattern of growth is necessary in the budding mode of reproduction. The formation of a bud (the incipient progeny cell) begins with a localized evagination of the cell surface. Enlargement of the bud is also a spatially localized process, marked by the continued and directed delivery of virtually all new surface material to the developing bud[9,10]. Proliferation by this yeast therefore requires selection of a localized site for initiation of bud formation and subsequent polarized allocation of new surface components to this site. A conditional mutation in the yeast *MYO2* gene abolishes both of these polarized activities[1]. As a result, *myo2-66* mutant cells, with impaired MYO2 protein function, produce abnormally small progeny cells, because of a defect in the selective delivery of new surface components to the site of bud growth. Mutant cells then cease proliferation as cells without buds, reflecting an inability to initiate a new round of bud formation. The asymmetric growth during bud development results from the polarized transport of secretory vesicles to the bud[10-13]. It is this polarized nature of vesicle transport during bud formation that is impaired by the *myo2-66* mutation[1]. Cells with impaired MYO2 function continue to secrete and to accumulate cell surface components, but under these conditions surface growth is isotropic rather than localized to the bud. This unpolarized secretion by mutant cells is accompanied by an unusual accumulation of secretory vesicles within the cytoplasm, suggesting that the continued but undirected movement of vesicles to the cell surface in mutant cells is slow. A conditional defect in the MYO2 protein also deranges the actin cytoskeleton. These features of the *myo2-66* mutant phenotype suggest that the MYO2 myosin protein interacts with actin cables to direct secretory vesicles into the developing bud.

The *MYO2* gene product of yeast is a member of a novel class of unconventional myosin proteins. Like other unconventional myosins, the yeast MYO2 protein contains at its N-terminus a typical myosin head domain capable of force generating activity, while the C-terminal domain is dissimilar to that of conventional **sarcomeric myosin**[2]. The MYO2 C-terminal region is, however, strikingly similar to protein domains found in two myosin like molecules from vertebrates: the murine dilute protein[3] and the p190 protein from brain[4]. The C-terminal domain predicted for the p190 protein shares more than 90% amino acid identity with the murine dilute protein, and 28% identity (over a 200 amino acid stretch) with the yeast MYO2 protein[14]. Furthermore, in each of these three unconventional myosin proteins the N-terminal and C-terminal domains are joined by a short region of potential coiled-coil structure, suggesting that MYO2-like myosin proteins may undergo some degree of self assembly, perhaps to form a dimer.

The role of the *MYO2* gene product in polarized cell growth is not confined to the yeast system. Mutations at the murine dilute locus cause defects in dendritic development in melanocytes and neurological defects consistent with abnormal neuronal development[3]. Both of these dilute phenotypes may result from the inability of mutant cells to accomplish the polarized growth necessary to elaborate cellular processes. Moreover, immunofluorescence staining using antibody directed against p190 protein visualizes perinuclear regions and growing processes in cultured glial and neuronal cells[4]. The analyses of the structure and function of these three myosin like proteins reveal an unexpected degree of similarity, suggesting that the MYO2/dilute myosin proteins define a novel class of unconventional myosins for polarized secretion.

■ **ACTIVITIES**

The p190 protein displays an actin dependent ATPase activity[4]. Moreover, p190 protein binds calmodulin and can be phosphorylated by a type II calmodulin dependent protein kinase[15,16]. The ATPase activity of p190 protein is potentiated by calmodulin. The p190 protein also crosslinks actin filaments in an ATP dependent manner. The MYO2 protein from yeast and the dilute protein have yet to be characterized *in vitro*.

■ **ANTIBODIES**

None available

■ **GENES**

The genomic MYO2/dilute myosin gene has been cloned from yeast (EMBL accession number M35532), and cDNA clones have been obtained from mouse (EMBL accession number X57377) and vertebrate brain.

■ **REFERENCES**

1. Johnston, G.C., Prendergast, J.A. and Singer, R.A. (1991) J. Cell Biol. 113, 539-551.
2. Pollard, T.D., Doberstein, S.K. and Zot, H.G. (1991) Annu. Rev. Physiol. 53, 653-681.
3. Mercer, J.A., Seperack, P.K., Strobel, M.C., Copeland, N.G. and Jenkins, N.A. (1991) Nature 349, 709-713.
4. Espindola, F.S., Espreafico, E.M., Coelho, M.V., Martins, A.R., Costa, F.R., Mooseker, M.S. and Larson, R.E. (1992) J. Cell Biol. 118, 359-368.
5. Kiehart, D.P. (1990) Cell 60, 347-350.
6. Korn, E.D. and Hammer III, J.A. (1988) Annu. Rev. Biophys. Biophys. Chem. 17, 23-45.
7. Prendergast, J.P., Murray, L.E., Rowley, A., Carruthers, D.R., Singer, R.A. and Johnston, G.C. (1990) Genetics 124, 81-90.
8. Sloat, B.F., Adams, A. and Pringle, J.R. (1981) J. Cell Biol. 89, 395-405.
9. Johnston, G.C., Pringle, J.R. and Hartwell, L.H. (1977) Exp. Cell Res. 105, 79-98.
10. Thacz, J. and Lampen, J. (1972) J. Gen. Microbiol. 72, 243-247.
11. Thacz, J. and Lampen, J. (1973) J. Bacteriol. 113, 1073-1075.

12. Novick, P. and Schekman, R. (1979) Proc. Natl. Acad. Sci. (USA) 76, 1858-1862.
13. Field, C. and Schekman, R. (1980) J. Cell Biol. 86, 123-128.
14. Novick, P., Field, C. and Schekman, R. (1980) Cell 21, 205-215.
15. Cheney, R. and Espreafico, E. (personal communication).
16. Larson, R.E., Espindola, F.S. and Espreafico, E.M. (1990) J. Neurochem. 54, 1288-1294.

■ *Richard A. Singer and Gerald C. Johnston:*
Department of Biochemistry and Department of Microbiology,
Dalhousie University, Halifax,
N.S. Canada

Protozoan Myosin I

The myosins I of protozoa are low molecular weight, roughly globular, nonfilamentous, actin based mechanoenzymes. Recent studies suggest that these proteins may play key roles in driving amoeboid locomotion, pseudopod formation, phagocytosis, and intracellular vesicle movement.

The type I myosins of protozoa differ from **sarcomeric** and conventional **cytoplasmic myosins** in their small heavy chain (~111-128 kDa), monomeric nature (one heavy chain and one head/molecule), roughly globular shape, and inability to selfassemble into filaments[1,4]. Myosins I have been purified from both *Acanthamoeba castellanii* and *Dictyostelium discoideum*, although the *Acanthamoeba* enzymes are by far the best characterized[5]. Both organisms contain multiple isoforms (at least three in *Acanthamoeba* and five in *Dictyostelium*), and the heavy chain genes for four of these isoforms have been cloned and sequenced[6-9]. The deduced heavy chain amino acid sequences reveal an N-terminal ~680 residue domain, which is highly similar to the globular head (S1) domain of sarcomeric myosin, fused to a ~350-480 residue C-terminal domain, which shows no homology to the rod-like tail domain of conventional myosins. For three of the four sequenced myosin I isoforms this unique tail domain contains three distinct regions of sequence: an N-terminal ~220 residue region that is characterized by clusters of basic residues (tail-homology region1; TH-1), a middle ~185 residue region that is highly enriched in glycine, proline and alanine residues and is also basic (TH-2), and a C-terminal ~50 residue region (TH-3) that is also found in a diverse family of cytoskeleton associated proteins, including nonreceptor tyrosine kinases, phospholipase cγ, and **fodrin** (Figure). The positions of these conserved tail

sequences correlate very well with the locations of two apparent functional domains within the tail: a membrane binding site, which has been mapped to the N-terminal ~half of the tail[4,10,11] (i.e. within the TH-1 region) and a second **actin** binding site (the other being in the S1-domain), which has been mapped to the C-terminal ~half of the tail[5] (i.e. within the sequences corresponding to TH-2 plus TH-3) (Figure). These two potential anchoring sites for the S1-motor domain should allow myosin I to move one actin filament relative to another or membranes relative to actin. Both types of movement have been demonstrated *in vitro*.

Light microscopic immunofluorescence studies show that myosins I are concentrated just under the plasma membrane in *Acanthamoeba*[5], within pseudopods at the leading edge of migrating *Dictyostelium* cells, and at sites of particle ingestion (phagocytic cups) in *Dictyostelium*[12]. Immunogold electron microscopy of *Acanthamoeba* shows myosin I to be concentrated near the plasma membrane and on the inner surface of the contractile vacuole membrane[13].

■ PURIFICATION

The purification procedures for protozoan myosins I, as well as for the myosin I heavy chain kinase, have been summarized recently[14].

■ ACTIVITIES

Myosins I, like all myosins, are actin activated Mg^{2+}-ATPases. Expression of this activity requires phosphorylation of a single site on the heavy chain between the actin and ATP binding sites in the head by a specific myosin I heavy chain kinase. This kinase has been purified to homogeneity[14,15] and shown to be regulated by autophosphorylation and phospholipids[16]. Phosphorylated myosin I supports a number of motile activities *in vitro*. Myosin I when bound to latex beads moves these beads on *Nitella* actin cables[17]. By virtue of its two actin binding sites, monomeric myosin I crosslinks actin filaments into a gel,

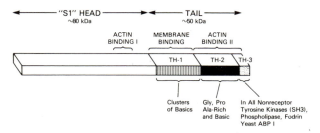

Figure. Schematic of protozoan myosin I heavy chain structure. One sequenced isoform, *Dictyostelium* myosin IA, is truncated, ending just after tail homology region I[9].

and, in the presence of ATP, causes the constriction of this gel (superprecipitation)[18]. Myosin I binds to both purified plasma membranes[11] and synthetic anionic phospholipid vesicles[10], and antibodies to myosin I block the movement of endogenous vesicles on actin filaments in *Acanthamoeba* cell extracts[19]. Preliminary reconstitution studies indicate that synthetic phospholipid vesicles coated with purified myosin I move on actin filaments[20]. *Dictyostelium* cells that are devoid of type II myosin but continue to express myosin I retain the ability to chemotax and to phagocytose[21]. The phenotype of *Dictyostelium* cells that lack one isoform of myosin I is consistent with the protein playing a role in chemotaxis and phagocytosis[22].

■ ANTIBODIES

Both polyclonal[5,12,13,22] and monoclonal[19] antibodies against various protozoan myosin I isoforms have been prepared. The polyclonal antisera often crossreact amongst protozoan myosin I isoforms, but the extent to which they can recognize putative myosins I in higher eukaryotes has not been reported.

■ GENES

The heavy chain sequences of *Acanthamoeba* myosins IB[7] and IC[6], and *Dictyostelium* myosins IA[9] and IB[8] have been published. A novel, high molecular weight form of *Acanthamoeba* myosin I has recently been described[23].

■ REFERENCES

1. Korn, E.D. and Hammer, J.A. III (1988) Annu. Rev. Biophys. Biophys. Chem. 17, 23-45.
2. Korn, E.D. and Hammer, J.A. III (1990) Curr. Opin. Cell Biol. 2, 57-62.
3. Pollard, T.D., Doberstein, S.K. and Zot, H.G. (1990) Ann. Rev. Physiol. 53, 653-681.
4. Adams, R.J. and Pollard, T.D. (1989) Cell Motil. Cytoskel. 14, 178-182.
5. Korn, E.D., Atkinson, M.A, Brzeska, H., Hammer, J.A. III, Jung, G. and Lynch, T.J. (1988) J. Cell Biochem. 36, 37-50.
6. Jung, G., Korn, E.D. and Hammer, J.A. III (1987) Proc. Natl. Acad. Sci. (USA) 84, 6720-6724.
7. Jung, G., Schmidt, C.J. and Hammer, J.A. III (1989) Gene 82, 269-280.
8. Jung, G., Saxe, C.L., Kimmel, A.R. and Hammer, J.A. III (1989) Proc. Natl. Acad. Sci. (USA) 86, 6186-6190.
9. Titus, M.A., Warrick, H.M. and Spudich, J.A. (1989) Cell Reg. 1, 53-63.
10. Adams, R.J. and Pollard, T.D. (1989) Nature 341, 328-331.
11. Miyata, H., Bowers, B. and Korn, E.D. (1989) J. Cell Biol. 109, 1519-1528.
12. Fukui, Y., Lynch, T.J, Brzeska, H. and Korn, E.D. (1989) Nature 341, 328-331.
13. Baines, I.C. and Korn, E.D. (1990) J. Cell Biol. 111, 1895-1904.
14. Lynch, T.J., Brzeska, H., Baines, I.C. and Korn, E.D. (1990) Meth. Enzym. 196, 12-23.
15. Hammer, J.A. III, Albanesi, J.P. and Korn, E.D. (1983) J. Biol. Chem. 258, 168-175.
16. Brzeska, H., Lynch, T.J. and Korn, E.D. (1990) J. Biol. Chem. 265, 3591-3594.
17. Albanesi, J.P., Fujisaki, H., Hammer, J.A. III, Korn, E.D., Jones, R. and Sheetz, M.P. (1985) J. Biol. Chem. 260, 8649-8652.
18. Fujisaki, H., Albanesi, J.P. and Korn, E.D. (1985) J. Biol. Chem. 260, 1183-1189.
19. Adams, R.J. and Pollard, T.D. (1986) Nature 322, 754-756.
20. Zot, H.G., Doberstein, S.K. and Pollard, T.D. (1992) J. Cell Biol. 116, 367-376.
21. Spudich, J.A. (1989) Cell Reg. 1, 1-11.
22. Jung, G. and Hammer, J.A. III (1990) J. Cell Biol. 110, 1955-1964.
23. Horowitz, J.A. and Hammer, J.A. III (1990) J. Biol. Chem. 265, 20646-20652.

■ *John A. Hammer III:*
Laboratory of Cell Biology,
National Heart, Lung, and Blood Institute,
National Institutes of Health,
Bethesda, MD, USA

Sarcomeric Myosins

Sarcomeric myosin is a large hexomeric protein with two globular heads and a long coiled-coil α-helical rod. A portion of the myosin rod aggregates to form the core of the thick filament found at the centre of the muscle sarcomere. The globular heads contain sites for interaction with both actin filaments and ATP, and it is this interaction which generates the force of muscle cells.

A myosin molecule consists of two heavy chains of ~200 kDa, and four light chains of ~20 kDa[1] (Figure 1). The C-terminal portions of the two heavy chains coil together to form a 150 nm long rod. The C-terminal end of this rod is known as light meromyosin (LMM). LMM is about 100 nm long, is insoluble at physiological ionic strength and aggregates to form the thick filament. The remaining portion known as S-2 is a 50 nm soluble region which connects the two globular heads to the core of the thick filament. Each of the globular heads (S-1) binds two different types of light chains. One type, the regulatory light chain, has a molecular weight of 20 kDa and is involved in the regulation or modulation of muscle contraction. In some muscles this function is effected via phosphorylation of a serine while in others it occurs via the binding of Ca^{2+} to the light chain. The second type of light chain is known as the alkali light chain and in skeletal muscle it consists of isomers of either 16 or 24 kDa. Neither light chain is

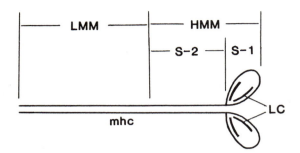

Figure 1. A schematic of the structure of the myosin molecule. Two head regions (S-1s) are attached to a rod (LMM plus S-2). Proteolytic cleavage can produce either S-1 or a two headed fragment heavy mero myosin (HMM), along with their respective fragments of the rod.

absolutely required for the ATPase activity of the myosin head, and both belong to a class of small Ca^{2+}-binding proteins which includes calmodulin and parvalbumin[2].

The analysis of crystals of the myosin head has shown it to be an elongated molecule with a rather large mass at its distal end tapering towards the attachment to the rod[3] (Figure 2). The myosin head is approximately 15-20 nm in length and 4-5 nm in width. The myosin molecule contains two protease sensitive regions, one close to the junction between the head and the rod, and a second one close to the hinge which separates the soluble and insoluble portions of the myosin rod. Experiments in which isolated myosin heads immobilized on a substrate interact with **actin** filaments have shown that the motile activity of the myosin molecule is contained entirely within the globular head region[4,5].

A number of isoforms of skeletal muscle myosin exist[6,7]. Within a single skeletal muscle cell there are at least two isoforms of myosin heavy chain which have been shown in some cells to be differentially localized along the thick filament with one isoform found predominantly in the centre of the filament, while the other is found at the ends[8]. In addition different heavy chain isoforms occur in fast and slow skeletal muscle, and in developing muscle[6,7,9-11]. Isoforms of the light chains are also found in various muscle cell types. The speed with which a muscle cell contracts appears to be determined largely by the isoform of the myosin heavy chain. The role played by different isoforms of the light chains is less well understood. The isoforms of cardiac and skeletal muscle are more closely related to each other than to the smooth muscle and nonmuscle isoforms[12]. Functionally these myosins resemble each other in their ability to participate in the formulation of sarcomeres and in the generation of rapid filament velocities. They also display a high degree of sequence homology and immunological crossreactivity. Different isoforms of cardiac myosin heavy chains are found in atria and ventricles.

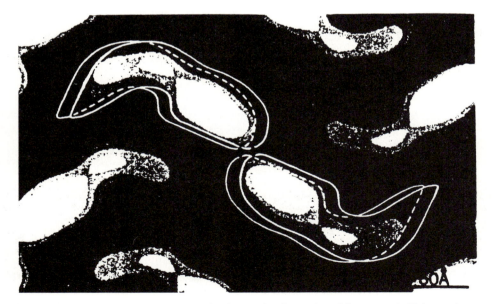

Figure 2. A filtered image of an electron micrograph of crystals of myosin subfragment-1. Thin sections were cut parallel to the [100] plane from tannic acid-embedded crystals. The stain-excluding light regions represent the protein. Two superimposed dimers are outlined by solid and dashed lines. (From Winkelmann, D.A., McKeel, H. and Rayment, I., ref. 3, with permission.)

■ PURIFICATION

Myosin is purified in high yield by differential solubility at high and low ionic strengths[13].

■ ACTIVITIES

Myosin hydrolyzes MgATP at a rate of approximately 0.03 s^{-1}. The ability of myosin to hydrolyze ATP in the absence of divalent cations is unique among ATP hydrolyzing enzymes. The addition of actin to myosin greatly accelerates the MgATPase activity to approximately 1 s^{-1}[14,15]. Much higher hydrolysis rates can be attained when the proteolytically soluble myosin head fragments interact with actin with activities reaching 20 s^{-1} per head. Isolated myosin heads can be bound to a substrate and retain the ability to translate actin filaments with velocities approximately that attained in muscle contraction[4]. Such immobilized heads can also generate forces which are again comparable to those generated in muscle[5]. Mechanical aspects of the actomyosin contractile interaction are measured most easily and accurately in muscle fibres.

■ ANTIBODIES

Polyclonal antisera have been raised against both myosin heavy chains and light chains. In addition a large number of monoclonal antibodies have also been described and some have been used to follow changes in myosin expression between different muscle types and during development[9,10,16].

■ GENES

A full length clone for the myosin heavy chains was first characterized in myosin from *C. elegans*[17]. Sarcomeric myosins are now known to form a large family with at least seven different genes in the mouse[18]. A number of these genes have been sequenced[6,7,12]. Multiple genes for the two classes of light chains from both cardiac and skeletal muscle fibres have also been identified and sequenced[7,16,19-22].

■ REFERENCES

1. Cooke, R. (1986) CRC Critical Reviews in Biochemistry 21, 53-118.
2. Collins, J.H. (1976) Nature (London) 259, 699.
3. Winkelmann, D.A., Meekel, H. and Rayment, I. (1985) J. Mol. Biol. 181, 487-501, 701-713.
4. Hynes, T.R., Block, S.M., White, B.T. and Spudich, J.A. (1987) Cell 48, 953-963.
5. Kishino, A. and Yanagida, T. (1988) Nature 334, 74-76.
6. Emerson, C.P. and Bernstein, S.I. (1987) Ann. Rev. Biochem. 56, 695-726.
7. Wade, R. and Kedes, L. (1989) Ann. Rev. Physiol. 51, 179-188.
8. Miller III, D.M., Ortiz, I., Berliner, G.C. and Epstein, H.F. (1983) Cell 34, 477-490.
9. Van Horn, R. and Crow, M.T. (1989) Dev. Biol. 134, 279-288.
10. Sweeney, L., Kennedy, J.M., Zak, R., Kokjohn, K., Kelley, S.W. (1989) Dev. Biol. 133, 361-374.
11. Stedman, H.H., Kelly, A.M. and Rudinstein, N.A. (1990) Ann. N.Y. Acad. Sci. 599, 119-126.
12. Strehler, E.E., Strehler-Page, M.A., Perriard, J.C., Periasamy, M. and Nadal-Ginard, B. (1986) J. Mol. Biol. 190, 291-317.
13. Margossian, S.S. and Lowey, S. (1982) Methods in Enzymol. 85, 55-77.
14. Pollard, T.D. (1982) Meth. in Enzym. 85, 123-129.
15. Greene, L.E. and Eisenberg, E. (1982) Meth. in Enzymol. 85, 709.
16. Winkelmann, D.A., Lowey, S. and Press, J.L. (1983) Cell 34, 295-306.
17. McLachlan, A.D. (1984) Annu. Rev. Biophys. Bioeng. 13, 167.
18. Wieczorek, D.F., Periasamy, M., Butler-Browne, G.S., Whalen, R.C., Nadal-Ginard, B. (1985) J. Cell Biol. 101, 618-629.
19. Wade, R., Feldman, D., Gunning, P. and Kedes, L. (1989) Mol. C. Bioch. 87, 119-136.
20. Muller, B., Maeda, K. and Wittinghofer, A. (1990) Nucleic Acids Res. 18, 6688.
21. Nudel, V., Calvo, J.M., Shani, M. and Levy, Z. (1984) Nucleic Acids Res. 12, 7175-7186.
22. McNally, E.M., Kraft, R., Bravo-Zehnden, M., Taylor, D.A. and Leinwand, L.A. (1989) J. Mol. Biol. 210, 665-671.
23. Tokunaga, M., Satoh, K., Toyoshima, C. and Wakabuyashi, T. (1987) Nature (London) 329, 635-638.

■ *Roger Cooke:*
Department of Biochemistry and Biophysics and the CVRI,
University of California,
San Francisco, CA, USA

Scallop Myosin

Scallop myosin is a Ca^{2+}-regulated contractile muscle protein that initiates contraction by the direct binding of calcium. Its regulatory subunits are the essential and regulatory light chains present in one to one stoichiometry with the myosin heavy chains.

Scallop myosin isolated from the striated adductor muscles consists of two heavy chains (225,553 Da), two regulatory light chains (R-LC; 17,683 Da) and two essential light chains (E-LC, 17,748 Da)[1,2]. Like other myosins it has a globular N-terminal head region and a tail piece made up of a two stranded α-helical coiled-coil. The head region contains the binding sites for nucleotide, **actin** and light chain, the ATPase sites and the triggering Ca^{2+}-binding

Figure 1. Dissociation of Regulatory Light Chains.

sites. The light chains function by inhibiting activity in the absence of Ca^{2+}.

Scallop myosin's R-LCs can be fully removed by EDTA treatment at room temperature. This is accompanied by a loss of regulation[3] (Figure 1). EDTA treated myosin readily recombine with foreign R-LCs, the hybrids regain Ca^{2+}-binding and regulation if the source of the foreign light chains is a regulated myosin[4]. All R-LCs contain a divalent cation binding site in domain-I[5], although this is not the specific Ca^{2+}-binding site[6]. The triggering Ca^{2+}-binding site is most likely located on domain-III of the E-LC[7], however the site needs to be stabilized by the interaction with the R-LC and the heavy chain. Polyclonal antibody against the E-LC interferes with regulation[8]. A regulatory domain consisting of the two light chains and a 10 kDa heavy chain fragment can be isolated after sequential digestion with papain and clostripain[9] (Figure 2). The regulatory domain retains the triggering Ca^{2+}-binding site and can be hybridized with foreign E-LCs. Therefore, properties of the light chains of various myosins can be tested. In activity or in rigor the N-terminal third of the R-LC moves relative to the E-LC; crosslinking the two light chains locks the system in the "on" position[10]. The light chains occupy the neck region of myosin[11]; the N-terminus of the R-LC lies at the junction of the myosin heads, while ATPase regulatory activity spans a considerable distance[12]. An intact regulatory domain is required for regulation, subfragment-1 is unregulated with high activity in the absence of Ca^{2+}, while heavy meromyosin and single-headed myosin are regulated molecules[13]. Molluscan myosins[14], including scallop striated myosin[15], can assume a folded 10 S configuration that further enhances trapping of nucleotides. Binding of Ca^{2+} by myosin reverses trapping and is associ-

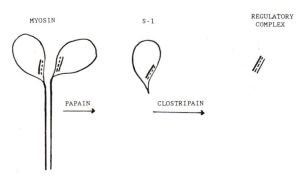

Figure 2. Formation of Regulatory Complex.

ated with a 100-fold activation of the MgATPase that is further stimulated 10-fold by actin[16]. Striated scallop myosin R-LCs cannot be phosphorylated in contrast to the R-LC of the **smooth (catch-)muscle myosin**[17].

Scallop striated muscle myosin heavy chain shows a greater sequence homology with **sarcomeric myosin** than with nonsarcomeric myosins, except the light chain binding region of the regulatory domain which resembles more the corresponding regions of other regulated myosins[18].

◼ PURIFICATION

Scallop myosin is purified by ammonium sulphate fractionation, followed by washing the filamentous myosin with low ionic strength salt solution[13]. Light chains can be obtained free of heavy chains after guanidine-HCl denaturation of myosin. E-LC and R-LC are separated by chromatography on anion exchange resin[19]. R-LCs can be selectively removed from myosin[20], myofibrils[3] or skinned fibre bundles[21].

◼ ACTIVITIES

Scallop myosin elicits muscle contraction by interacting with actin in a Ca^{2+}-dependent manner. Kinetic constants of the ATPase cycle[16], of the Ca^{2+}-binding[22] and light chain binding[23] have been determined. The role of the R-LC in tension generation has been measured[21].

◼ ANTIBODIES

Polyclonal IgG against heavy chain recognizes invertebrate myosins but not vertebrate myosins. Polyclonal IgG against scallop R-LC does not crossreact with other myosin R-LCs. Polyclonal IgG against scallop E-LC crossreacts weakly with *Mercenaria mercenaria* E-LC[24].

◼ GENES

Full length cDNAs for *Aequipecten* striated myosin heavy chain (EMBL X55714)[25] E-LC and R-LC are published[26]. Direct amino acid sequencing of the R-LC and E-LC has also been obtained[27].

◼ REFERENCES

1. Kendrick-Jones, J. and Scholey, J.M. (1981) J. Muscle Res. Cell Motil. 2, 347-372.
2. Szent-Györgyi, A.G. and Chantler, P.D. (1986) Myology ed. by Engel, A.G. and Banquer, B.Q. MacGraw Hill, 589-612.
3. Chantler, P.D. and Szent-Györgyi, A.G. (1980) J. Mol. Biol. 138, 473-492.
4. Sellers, J.R., Chantler, P.D. and Szent-Györgyi, A.G. (1980) J. Mol. Biol. 144, 223-245.
5. Bagshaw, C.R. and Kendrick-Jones, J. (1980) J. Mol. Biol. 140, 411-433.
6. Goodwin, E.B., Leinwand, L.A. and Szent-Györgyi, A.G. (1990) J. Mol. Biol. 216, 85-93.
7. Collins, J.H., Jakes, R., Kendrick-Jones, J., Leszyk, J., Barouch, W., Theibert, J.L., Spiegel, J. and Szent-Györgyi, A.G. (1986) Biochem. 25, 7651-7656.

8. Wallimann, T. and Szent-Györgyi, A.G. (1981) Biochem. 20, 1188-1197.

9. Kwon, H., Goodwin, E.B, Nyitray, L., Berliner, E., O'Neall-Hennessey, E., Melandri, F.D. and Szent-Györgyi, A.G.. (1990) Proc. Natl. Acad. Sci. (USA) 87, 4771-4775.

10. Hardwicke, P.M.D., Wallimann, T. and Szent-Györgyi, A.G. (1983) Nature 301, 478-482.

11. Flicker, P.F., Wallimann, T. and Vibert, P. (1983) J. Mol. Biol. 169, 723-741.

12. Wells, C., Warriner, K.E. and Bagshaw, C.R. (1985) Biochem. J. 231, 31-38.

13. Stafford, W.F. III., Szentkiralyi, E.M. and Szent-Györgyi, A.G. (1979) Biochem. 24, 5273-5280.

14. Castellani, L. and Cohen, C. (1987) Proc. Natl. Acad. Sci. (USA) 84, 4058-4062.

15. Ankrett, R.F., Rowe, A.J., Cross, R.A., Kendrick-Jones, J. and Bagshaw, C.R. (1991) J. Mol. Biol. 217, 323-335.

16. Jackson, A.P. and Bagshaw, C.R. (1988) Biochem. J. 251, 527-540.

17. Sohma, H., Yazawa, M. and Morita, F. (1985) J. Biochem. 98, 569-572.

18. Nyitray, L., Goodwin, E.B. and Szent-Györgyi, a.G. (1991) J. Biol. Chem. 266, 18469-18476.

19. Kendrick-Jones, J., Szentkiralyi, E.M. and Szent-Györgyi, A.G. (1976) J. Mol. Biol. 104, 747-775.

20. Ashiba, G. and Szent-Györgyi, A.G. (1985) Biochem. 24, 6618-6623.

21. Simmons, R.M. and Szent-Györgyi, A.G. (1985) J. Physiol. 358, 47-64.

22. Bennett, J. and Bagshaw, C.R. (1986) Biochem. J. 233, 173-177.

23. Bennett, J. and Bagshaw, C.R. (1986) Biochem. J. 233, 179-186.

24. Wallimann, T. and Szent-Györgyi, A.G. (1981) Biochem. 20, 1176-1187.

25. Nyitray, L., Goodwin, E.B. and Szent-Györgyi, A.G. (1990) Nucl. Ac. Res. 18, 7158.

26. Goodwin, E.B., Szent-Györgyi, A.G. and Leinwand, L.A. (1987) J. Biol. Chem. 262, 11051-11056.

27. Collins, J.H. (1991) J. Muscle Res. Cell Motil. 12, 3-25.

■ Andrew G. Szent-Györgyi:
Dept. of Biology,
Brandeis University,
Waltham, Massachusetts, USA

Smooth Muscle Myosin

Physiologic processes such as digestion, respiration and blood flow can be regulated by contraction and relaxation of smooth muscle cells, which contain the mechanochemical protein myosin. The signal for smooth muscles to contract is mediated by a cascade of enzymes that results in Ca^{2+}-dependent phosphorylation of the myosin light chain. The activated myosin interacts with actin leading to cell shortening and force generation.

Smooth muscle myosin is composed of two 200 kDa heavy chains, two 20 kDa regulatory phosphorylatable light chains, and two 17 kDa light chains. As seen by electron microscopy of metal shadowed images, the molecule has two globular heads approximately 20 nm long attached to a 155 nm long α-helical coiled-coil tail (Figure). The head or subfragment-1, which consists of the N-terminal 95 kDa region of the heavy chain and one of each class of light chains, binds **actin** and hydrolyzes MgATP. The insoluble tail is responsible for myosin's ability to form filaments at physiological ionic strength. The rod ends in a short (<50 amino acids), proline containing, nonhelical tailpiece. Two different tailpieces, whose function is unknown, have been identified in rabbit uterus, giving rise to two heavy chain isoforms[1].

Phosphorylation of Ser-19 on the 20 kDa regulatory light chain by Ca^{2+}-calmodulin-myosin light chain kinase causes a several hundred-fold increase in the actin activated ATPase activity of myosin or soluble heavy meromyosin[2,3], and is the key event required to trigger contraction in smooth muscle cells[4]. *In vitro*, phosphorylation also controls the conformational state of myosin. Myosin containing dephosphorylated regulatory light chains adopts a folded monomeric conformation in which the myosin tail is folded into approximately equal thirds (Figure)[5]. This form is enzymatically inactive, with a

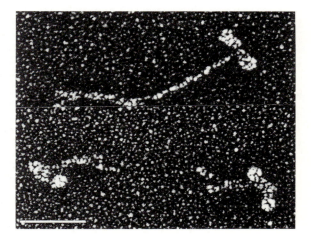

Figure. Metal-shadowed image of the extended and folded monomeric conformations adopted by smooth muscle myosin. Bar 50 nm.

turnover rate for MgATP of 0.0002-0.0005 sec[-1][6]. Under the same solvent conditions (0.1 M KCl, pH 7-7.5, 1 mM MgATP), phosphorylated myosin is enzymatically active and filamentous. It is not known to what extent this phosphorylation dependent myosin structural transition occurs *in vivo*, because dephosphorylated filaments are present in relaxed smooth muscle cells[7].

The three dimensional arrangement of smooth muscle myosin filaments within the cell is not well established. A sarcomere like arrangement of myosin and actin, analogous to skeletal muscle but considerably less well ordered, may constitute the basic contractile unit. In addition, there appears to be a cytoskeletal domain within the cell that contains actin, intermediate filaments and actin binding proteins, but which lacks myosin[8,9].

■ PURIFICATION

Smooth muscle myosin purified by standard methods (ammonium sulphate precipitation, cycles of low salt precipitation) contains trace amounts of Ca^{2+}-sensitive endogenous kinase and phosphatase. These activities can be removed by gel filtration chromatography on Sepharose-4B[10].

■ ACTIVITIES

Light chain phosphorylation has little effect on binding of actin to myosin[11], but increases the cycling rate for MgATP in the presence of actin[2,3]. Movement of single actin filaments in an *in vitro* motility assay is also regulated by phosphorylation: movement of actin by dephosphorylated myosin is not detectable[12,13], while phosphorylated myosin moves actin at ~0.3 µm/sec at 22°C.

■ ANTIBODIES

A number of monoclonal antibodies that react with the heavy chain (head or rod) of gizzard smooth muscle myosin have been characterized[14-16]. Some of the antibodies can detect, alter, or stabilize myosin conformational states, while others have effects on enzymatic activity.

■ GENES

The complete sequence of embryonic chicken gizzard myosin heavy chain (EMBL GenBank X06546)[17], and the partial sequence of rabbit uterus myosin heavy chain[18]

have been determined. The regulatory light chain (EMBL GenBank Y00983)[19] and the 17 kDa light chain (EMBL GenBank J02786)[20] have also been cloned and sequenced.

■ REFERENCES

1. Nagai, R., Kuro-o, M., Babij, P. and Periasamy, M. (1989) J. Biol. Chem. 264, 9734-9737.
2. Sellers, J.R. (1985) J. Biol. Chem. 260, 15815-15819.
3. Trybus, K.M. (1989) J. Cell Biol. 109, 2887-2894.
4. Kamm, K.E. and Stull, J.T. (1985) Ann. Rev. Pharmacol. Toxicol. 25, 593-620.
5. Trybus, K.M. and Lowey, S. (1984) J. Biol. Chem. 259, 8564-8571.
6. Cross, R.A., Cross, K.E. and Sobieszek, A. (1986) EMBO J. 5, 2637-2641.
7. Somlyo, A.V., Butler, T.M., Bond, M. and Somlyo, A.P. (1981) Nature 294, 567-569.
8. Draeger, A., Amos, W.B., Ikebe, M. and Small, J.V. (1990) J. Cell Biol. 111, 2463-2473.
9. Kargacin, G.J., Cooke, P.H., Abramson, S.B. and Fay, F.S. (1989) J. Cell Biol. 108, 1465-1475.
10. Sellers, J.R., Pato, M.D. and Adelstein, R.S. (1981) J. Biol. Chem. 256, 13137-13142.
11. Greene, L.E. and Sellers, J.R. (1987) J. Biol. Chem. 262, 4177-4181.
12. Umemoto, S. and Sellers, J.R. (1990) J. Biol. Chem. 265, 14864-14869.
13. Warshaw, D.M., Desrosiers, J.M., Work, S.S. and Trybus, K.M. (1990) J. Cell Biol. 111, 453-463.
14. Schneider, M.D., Sellers, J.R., Vahey, M., Preston, Y.A. and Adelstein, R.S. (1985) J. Cell Biol. 101, 66-72.
15. Trybus, K.M. and Henry, L. (1989) J. Cell Biol. 109, 2879-2886.
16. Ito, M., Pierce, P.R., Allen, R.E. and Hartshorne, D.J. (1989) Biochemistry 28, 5567-5572.
17. Yanagisawa, M., Hamada, Y., Katsuragawa, Y., Imamura, M., Mikawa, T. and Masaki, T. (1987) J. Mol. Biol. 198, 143-157.
18. Nagai, R., Larson, D.M. and Periasamy, M. (1988) Proc. Natl. Acad. Sci. (USA) 85, 1047-1051.
19. Messer, N.G. and Kendrick-Jones, J. (1988) FEBS Lett. 234, 49-52.
20. Nabeshima, Y., Nonomura, Y. and Fujii-Kuriyama, Y. (1987) J. Biol. Chem. 262, 10608-10612.

■ *Kathleen M. Trybus and Susan Lowey:*
Rosenstiel Research Center,
Brandeis University,
Waltham, MA, USA

5

Cytoskeletal Anchor Proteins

Vinculin in chicken lens cells. Immunofluorescence labeling of cultured chicken lens cells with antibodies agains vinculin displays both cell–cell and cell–extracellular matrix adhesion sites .

(courtesy of Dr Benjamin Geiger, Rehovot). For further details see 'Vinculin'.

Cytoskeletal Anchor Proteins

Cytoskeletal anchor proteins are a molecularly diversified class of cytoplasmic molecules whose primary function is to link, directly or indirectly, cytoskeletal filaments to the plasma membrane. In this capacity, anchor proteins play a central role both in the regulation of membrane structure and dynamics and in the assembly of the various cytoskeletal networks, features which play a pivotal role in cell adhesion and morphogenesis, polarization, motility, as well as in the regulation of growth and differentiation.

■ CYTOSKELETAL ANCHOR PROTEINS AS PERIPHERAL MEMBRANE COMPONENTS

In their 1972 paper, describing the fluid mosaic model of the structure of cell membranes, Singer and Nicolson[1] distinguished between two classes of membrane proteins; intrinsic and peripheral. They noted that proteins of the former type are indispensable for the basic structure of the membrane while the peripheral proteins may be associated with the integral components of the membrane either inside or outside of the cell. It is via these interactions that the peripheral proteins exert an effect on the properties and dynamics of membranes; they wrote: "The mosaic structure can be readily diversified in several ways. Although this diversification is a matter of speculation ... . Protein-protein interactions may be important in determining the properties of the membrane. Such interactions may result in the specific binding of a peripheral protein to the exterior exposed surface of a particular integral protein".

This insightful prediction proved right over the two decades since the formulation of the fluid mosaic model and interactions of extrinsic proteins with the integral components of the membrane are not a matter of "speculation" any more but rather a widely accepted mechanism for the generation of membranes domains.

The definition of peripheral membrane molecules is based on several, mostly indirect criteria including immunohistological localization at the light- and electron-microscope levels, biochemical analyses of isolated membranes and selective extractions. Thus, proteins which can be solubilized only after disruption of the lipid bilayer are considered integral components while those extractable at high or low salt or under extreme pH conditions are believed to reside at the periphery of the membrane, attached to its inner or outer surface. In addition, the recent advances in biochemical and molecular genetic analyses revealed, in integral (but not peripheral) proteins, one or several membrane-spanning domain(s) with characteristic hydrophobic amino acid sequences.

■ MEMBRANE-CYTOSKELETON INTERACTION

In this chapter, one particular family of peripheral membrane molecules will be discussed, namely those proteins which are involved in the linkage of cytoskeletal filaments to membranes[2]. The precise definition of these anchor proteins is not as straightforward as it may seem. Ideally, one might expect them to exhibit specific binding affinity to both the exposed cytoplasmic moieties of integral membrane constituents and to specific cytoskeletal components. However, the currently available information on the properties and binding specificity of these molecules is rather limited. Instead, an operational definition is usually adopted according to which cytoskeletal anchor proteins are identified by their apparent association with membrane-cytoskeleton interfaces, irrespective of whether they interact with the membrane or the cytoskeleton directly or indirectly. Moreover, cytoskeletal anchor proteins do not have to be strictly associated with the plasma membrane and may partly be diffusely distributed in the cytoplasm or associated with other regions along the cytoskeletal network. Therefore, one may find in this book *bona fide* anchor proteins classified under different titles: For example, some of the actin associated proteins including α-**actinin**, some of the actin binding proteins, **ezrin**, **dystrophin** and many others are highly enriched in the vicinity of the plasma membrane and could, in principle, fit into the category of anchor proteins. The same holds true for some of the associated components of the other cytoskeletal networks and of proteins classified here according to their functions. For example, motor proteins such as **brush border myosin I** may also play important role in linking the core bundle of microfilament to the microvillar membrane in intestinal epithelial cells, some of the organelle-associated proteins may be involved in linkage to the cytoskeleton, etc.

Notably, the cellular functions of cytoskeletal anchor proteins are highly diversified, ranging from the control of assembly of cytoskeletal networks and the transmembrane or transcellular transduction of mechanical forces, to the local modulation of membrane dynamics and signal transduction. Such activities are involved in major cellular features including motility, adhesion and morphogenesis as well as regulation of growth and differentiation.

■ INVOLVEMENT OF CYTOSKELETAL ANCHOR PROTEINS IN CELL ADHESION

One of the most prominent functions of cytoskeletal anchor proteins is the participation in adhesive interactions in cells and many of the proteins reviewed in this chapter are associated with specialized cell-matrix or cell-cell junctions. Such cellular interactions are characteristic of metazoan organisms and are responsible for the assembly of individual cells into functional tissues and organs. Attempts to elucidate the molecular mechanisms underlying cell adhesion usually address two distinct aspects, namely the specificity of interaction and the short- and long-range morphogenetic response triggered by it. The former depends primarily on the nature of the transmembrane receptors which directly mediate the adhesion (reviewed in the accompanying book under **"Cell adhesion and cell-cell contact proteins"**) and the latter, involves the assembly of the relevant force-generating cytoskeletal structures and their attachment to the cytoplasmic faces of the newly formed adhesions. The proper assembly of the entire transmembrane "adhesion complex" is of critical physiological importance for a wide variety of cellular processes including cell spreading and locomotion, embryonic cell sorting, mesenchymal-to-epithelial transformation or vice versa, tissue assembly, formation of transcellular barriers, establishment of transcellular forces, etc.

It is noteworthy that not only the post-binding cytoplasmic events, but also the formation of avid cellular adhesions depends on the transmembrane interactions with the cytoskeleton, mediated through the cytoplasmic "anchor proteins". It was shown that **integrin** or **cadherin** chains, missing parts of their cytoplasmic moieties (expressed in cells following transfection with truncated cDNAs) fail to mediate adhesion to the external surface[3,6]. While one cannot completely exclude the possibility that these truncations directly affected the intrinsic binding affinity, a more likely explanation is that cell adhesion requires multivalent interactions, for which the cytoplasmic association with the cytoskeleton is of a cardinal importance. Thus, assembly of adhesion complexes may be a two-way, multi-stage process, whereby the immobilization or clustering of the transmembrane adhesion molecules trigger the formation of a submembrane plaque and induce local cytoskeletal organization which, in turn, mechanically stabilize the contact area, increase the overall avidity by supporting multivalent interactions and even recruit new receptors to the nascent contact site. This model also predicts that the cytoskeletal anchor proteins are initially present in the cytoplasm in a diffusible form and bind to the membrane only after the establishment of cell contacts and the local assembly of the cytoskeleton[5,7]. It had been shown for some of these molecules that even after establishment of mature contact sites, an exchange of monomers occurs between the diffusible and membrane-bound pools of junctional molecules[8]. The mechanism responsible for the fine tuning of this two-way traffic is of great importance for cell adhesion and adhesion-dependent events and its elucidation is a great challenge for future research.

■ MOLECULAR AND FUNCTIONAL DIVERSITY OF CELL ADHESIONS.

Studies on cell adhesion, carried out over the last several decades, have indicated that this process is not only molecularly complex, involving many extrinsic and intrinsic components, but also highly diversified at the cellular and functional levels. Thus, cell adhesions cannot be regarded as one uniform entity but rather as a heterogenous family of related structures. There are many distinct "extracellular ligands" which provide the adhesive surface to which cells may attach. These include the various components of the extracellular matrix and basement membranes as well as specialized integral membrane-bound cell adhesion molecules. Interactions with these ligands trigger the formation of the two major classes of cell adhesions, namely, cell-matrix and cell-cell junctions, respectively. Furthermore, within each family there are several structurally- and functionally-distinct types of cell adhesions, identified, mainly, by electron microscopy and immunocytochemistry. Specialized matrix adhesions include focal contact-like structures[9] and hemidesmosomal adhesions. Focal contacts are associated with contractile elements of the cytoskeleton while hemidesmosomes are attached to the presumably non-contractile network of intermediate filaments. It is highly likely that the fine balance between interactions affect, in a major way, processes such as locomotion and adhesion.

Intercellular adhesion in vertebrate organisms is mediated through four major types of junctions, namely: tight junctions, adherens junctions, desmosomes, and gap junctions. Recent molecular studies have indicated that each of these contact sites contains a unique set of proteins which interact with each other, conferring on the particular adhesion region its unique structure and properties. Tight junctions, which usually occupy the most apical position in the junctional complex of epithelia, contain the proteins **cingulin** and **ZO-1** at their cytoplasmic aspects. It is, however, not entirely clear whether these molecules attach to specific components of any of the major cytoskeletal networks. Tight junctions are usually believed to be responsible for the transepithelial resistance as well as for membrane polarity (apical vs basolateral) in simple epithelial cells[10].

Adherens junctions, characterized by their specific association with the contractile microfilament system, contain a large variety of anchor proteins, some of which are common to both cell-cell and cell-matrix adhesions while others are exclusively associated with only one of the two. Thus, for example, proteins such as **vinculin**, α-actinin, **tenuin**, and **zyxin** are found in both types of adhesions. Other junctional proteins including **talin** and **paxillin** are differentially associated with matrix adhesions while **plakoglobin** and **catenins** are found in intercellular adherens junctions only. It is interesting to note that the cytoplasmic surfaces of adherens junctions also contain certain enzymes known to participate in signal transduction pathways. Among these are different tyrosine kinases (e.g. **pp60^{src}**), protein kinase C, phospholipase C-γ[11] and others[12]. Moreover, these junctions were shown to be major sites for tyrosine phosphorylation in normal and transformed cells[13].

The third class of cell contacts to be discussed here are desmosomes and basement-membrane associated hemidesmosomes. The studies on these junctions revealed families of anchor proteins including **desmoplakins**, plakoglobin (also associated with adherens junctions), **desmocalmin**, etc. which are most likely attached to the periphery of the intermediate filament system and to the cytoplasmic domains of the transmembrane proteins of desmosomes, namely **desmogleins** and **desmo-collins**.

The three junctions mentioned above (as well as gap junctions, which will not be discussed here) are distinct, both structurally, topologically and molecularly, yet they retain some characteristic spatial relationships and functional interdependence: In epithelial junctional complexes tight junctions occupy the most apical position, followed by adherens junctions and, then, desmosomes. Their interdependence is demonstrated by the fact that modulation of adherens junctions by antibodies to the uvomorulin (E-cadherin) leads also to breakdown of the neighbouring tight junctions[10,14].

In conclusion, cell adhesions appear to be major sites for cytoskeletal anchor protein organization in cells. It appears likely that in these membrane domains anchor proteins play major roles in the mechanical stabilization of the junction structure and in the assembly of different adhesions into coherent junctional complexes.

■ CELL ADHESIONS AS SIGNAL TRANSDUCING UNITS: THE ROLE OF CYTOSKELETAL ANCHOR PROTEINS.

The interrelationships between adhesion and the regulation of cell motility, growth and differentiation are among the most exciting and challenging questions in this field of research. It is now becoming apparent that the modulation of adherens junctions has dramatic effects on cell behavior and that changes in the signal transduction machinery (for example, modulation of protein phosphorylation) may dramatically affect cellular interactions. The exact mechanisms underlying these processes (namely, the identity of the relevant modified molecules and characterization of the molecular interactions affected) are not clear but the overall indications for their functional significance are most compelling. These include the facts that: (a) increase in tyrosine phosphorylation, induced by growth factors of phosphatase inhibitors, leads to disassembly of adherens junctions[15] (Volberg, Zick, Dror, Sabanay, Gilon, Levitzki and Geiger, unpublished data); (b) Nerve growth factor-independent neural differentiation of PC12 cells may be triggered by adhesive interactions, mediated by surface-associated NCAM or N-cadherin. This stimulation may be blocked by pertussis toxin, Ca^{2+}-channel blockers and certain kinase inhibitors[16]. (c) Protein kinase C inhibitors prevent chelator-mediated disassembly of tight- and adherens junctions[17]. (d) Transfection of cells, leading to increased levels of adhesion molecules (cadherins or integrins) can confer restrained growth behaviour on transformed cells[18,19]. (e) Conversely, blocking of adhesion molecules by specific antibodies may lead to the acquisition of an invasive phenotype[20]. In essentially all these systems the detailed molecular mechanisms responsible for the physiological effect is still not clear, yet one may predict that the vast knowledge which has accumulated on the mode of action of "classical" signal inducers, like hormones and growth factors, will be applied in the future to elucidate possible related pathways triggered by adhesive interactions.

■ ADHESION-INDEPENDENT CYTOSKELETAL-ANCHOR PROTEINS.

The high abundance of junction-associated proteins listed here as "cytoskeletal anchor proteins" may create the biased impression that cell adhesions are the only prominent site of action of such proteins. The main reason for this bias is the fact that junctions are easily recognizable membrane domains and thus more amenable to immunolocalization. However, long- or short-term anchorage of cytoskeletal structures occurs widely also in extrajunctional sites and is mediated, most likely, by specific anchor proteins. Among these are members of the **spectrin** family, **ankyrin**, **ABP120** of *Dictyostelium*, brush border myosin I and many additional proteins classified in this volume under different categories. Close and firm association of cytoskeletal filaments to membranes is a widely occurring phenomenon and one may expect the list of proteins involved in this process to grow considerably in the future.

■ ARE THERE DISTINCTIVE STRUCTURAL FEATURES COMMON TO CYTOSKELETAL ANCHOR PROTEINS?

The limited, yet significant information on the structure of different cytoskeletal anchor proteins points to considerable molecular diversity. Sequence homologies are limited and the size of the proteins varies significantly. Some exceptions to this rule are the similarities between vinculin and catenin[21,22] and the presence of *src*-homology domains on several junctional molecules[23]. Beside this, one generalization may be added, namely that these molecules are expected to contain multiple distinct functional domains. This notion is suggested by the mere fact that, by definition, anchor proteins should display binding sites to one or more components of the cytoskeleton, to the membrane and, some times, to additional anchor proteins. As will be specified below for specific anchor proteins, this prediction was indeed confirmed experimentally: Vinculin, for example, binds to talin (which also binds integrin[24]), to α-actinin and to itself[25-27]; α-actinin was shown to contain binding sites to vinculin, **actin** and integrin[24,28] etc. These multivalent interactions are apparently important for the formation of 3-dimensional networks through which membrane-cytoskeleton attachments occur. With the harnessing of molecular genetic approaches and the development of numerous immunochemical reagents it appears likely that within the next several years a much more complete molecular information on structure-function relationships in these proteins will become available. This includes a broader knowledge of the diverse types of anchor proteins and a deeper understanding of their mode of action at the molecular level.

■ REFERENCES

1. Singer, M. and Nicolson, G.L. (1972) Science 175, 720-731.
2. Geiger, B. (1983) Biochim. Biophys. Acta 737, 305-341.
3. Nagafuchi, A. and Takeichi, M. (1988) EMBO J. 7, 3679-3684.
4. Hayashi, Y., Haimovich, B., Reszka, A., Boettiger, D. and Horwitz, A. (1990) J. Cell Biol. 110, 175-184.
5. Geiger, B., Ginsberg, D., Salomon, D. and Volberg, T. (1990) Cell Differen. Develop. 32, 343-354.
6. Marcantonio, E.E., Guan, J.-L., Tevithick, J.E. and Hynes, R.O. (1990) Cell Regul. 1, 597-604.
7. Geiger, B., Volk, T., Volberg, T. and Bendori, R. (1987) J. Cell Sci. Suppl. 8, 251-272.
8. Kreis, T.E., Avnur, Z., Schlessinger, J. and Geiger, B. (1985) In: Cold Spring Harbor Symposium on "Molecular Biology of the Cytoskeleton". G. Borisy, D. Cleveland and D. Murphy, eds., pp. 45-57.
9. Burridge, K., Fath, K., Kelly, T., Nuckolls, G. and Turner, C. (1988) Ann. Rev. Cell Biol. 4, 487-525.
10. Gumbiner, B. (1987) Am. J. Physiol. 253, C749-C758.
11. McBride, K., Rhee, S.G. and Jaken, S. (1991) Proc. Natl. Acad. Sci. (USA) 88, 7111-7115.
12. Geiger, B. and Ginsberg, D. (1991) Cell Motil. Cytoskeleton 20, 1-6.
13. Volberg, T., Geiger, B., Dror, R. and Zick, Y. (1991) Cell Regul. 2, 105-120.
14. Gumbiner, B. and Simons, K. (1986) J. Cell Biol. 102, 457-468.
15. Boyer, B., Valles, A.M., Tucker, G.C., Jouanneau, J. and Thiery, J.P. (1990).J. Cell Biol. 111, 420a.
16. Doherty, P., Ashton, S.V., Moore, S.E. and Walsh, F.S. (1991) Cell 67, 21-33.
17. Citi, S. (1992) J. Cell Biol. 117, 169-178.
18. Giancotti, F.G. and Ruoslahti, E. (1991) Cell 60, 849-859.
19. Navarro, P., Gomez, M., Pizarro, A., Gamallo, C., Quintanilla, M. and Cano, A. (1991) J. Cell Biol. 115, 517-533.
20. Behrens, J., Mareel, M.M., VanRoy, F.M. and Birchmeier, W. (1989) J. Cell Biol. 108, 2435-2447.
21. Herrenknecht, K., Ozawa, M., Eckerskorn, C., Lottspeich, F. and Lenter, M. (1991) Proc. Natl. Acad. Sci. (USA) 88, 9156-9160.
22. Nagafuchi, A., Takeichi, M. and Tsukita, S. (1991) Cell 65, 849-857.
23. Koch, C.A., Anderson, D., Moran, M.F., Ellis, C. and Pawson, T. (1991) Science 252, 668-674.
24. Horwitz, A., Duggan, K., Buck, C., Beckerle, M.C. and Burrdige, K. (1986) Nature 320, 531-533.
25. Belkin, A.M. and Koteliansky, V.E. (1987) FEBS Lett. 220, 291-294.
26. Burridge, K. and Mangeat, P. (1984) Nature 308, 744-746.
27. Bendori, R., Salomon, D. and Geiger, B. (1989) J. Cell Biol. 108, 2383-2393.
28. Otey, C.A., Pavalko, F.M. and Burridge, K. (1990) J. Cell Biol. 111, 721-729.

■ *Benjamin Geiger:*
Department of Chemical Immunology,
Weizmann Institute of Science,
Rehovot 76100, Israel

Adducin

Adducin[1] (adducere: to bring together) was originally identified as a calmodulin binding protein associated with the erythrocyte membrane skeleton and has been purified from erythrocytes as a heteromeric molecule composed of equal amounts of an α-subunit (M_r 103 kDa) and a β-subunit (M_r 97 kDa, β-1). Adducin promotes spectrin/actin binding[2], is a substrate for protein kinases A and C[3,4], and is found in a wide variety of cells and tissues[5-7]. Alternatively spliced mRNAs encode distinct isoforms of the β-subunit[8,9]; erythrocyte adducin may represent one member of a family of proteins with diverse, tissue-specific functions.

cDNAs encoding human erythrocyte adducin[9] demonstrate that the amino acid sequences of the two subunits are remarkably similar with 49% identity and 66% homology, suggesting gene duplication in origin. The calculated molecular weights (81 kDa for α and 80 kDa for β-1) are lower than the observed mobilities on SDS-PAGE and this information, along with the size of purified adducin from negative stain and rotary shadowing electron microscopy suggests that the native molecule exists as a tetramer.

Studies from proteolytic digestion of purified erythrocyte adducin[10], amino acid sequence and native adducin's frictional coefficient of 1.5, suggest that both α and β-1 are composed of an N-terminal 40 kDa globular, hydrophobic, protease resistant core with a flexible, hydrophilic, protease sensitive C-terminal tail (Figure 1). Evidence suggests that sites for A and C kinase phosphorylation and calmodulin binding are located in the tail domains. The isolated protease resistant core is unable to crosslink **spectrin** and **actin**; these complex interactions may require contributions from core and tail domains simultaneously.

Adducin has a small region of similarity with the N-terminal actin binding domains of the actin crosslinking proteins (the spectrin superfamily and certain actin gelation proteins). α-**actinin** and β adducin are highly conserved for 22 amino acids and both subunits of adducin have a region of similarity with the 27 amino acid actin binding domain identified within the larger N-terminal actin binding region of the spectrin superfamily. The C-terminal ends of both α and β-1 adducin are identical for 22 amino acids and 11 of these are lysine residues. This highly basic segment is similar to a domain in the **MARCKS** protein[11] which contains protein kinase C phosphorylation sites and exhibits calmodulin binding.

cDNAs encoding human α adducin hybridize to a promi-

nent 4 kB mRNA in rat brain, spleen, kidney, liver, human reticulocytes and K562 cells. In contrast, cDNAs encoding human β adducin hybridize to several different size mRNAs including an 8.1 kB mRNA in rat brain, 3.7, 3.5, and 3.0 kB mRNAs in rat spleen, and 4.0 and 3.8 kB mRNAs in human reticulocytes and K562 cells. Three alternatively spliced cDNAs have been identified which encode isoforms of β adducin. They are designated in order of size of the predicted protein product as β-1 (80 kDa), β-2 (63 kDa) and β-3 (26 kDa). These three isoforms have identical N-terminal sequence, but different C-terminal regions. The highly basic C-terminus of β-1 is deleted in both β-2 and β-3. The tissue-specific expression and function of these isoforms is currently being studied.

Adducin has also been purified from bovine brain[5] and has similar, but slightly different properties compared to erythrocyte adducin. Using antibodies raised against purified adducin, immunoreactivity has been identified in a variety of cells and tissues by immunoblot and immunofluorescence microscopy. Adducin is found at lateral cell

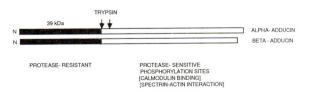
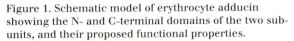

Figure 1. Schematic model of erythrocyte adducin showing the N- and C-terminal domains of the two subunits, and their proposed functional properties.

Figure 2. Immunofluorescence localization of adducin at sites of cell-cell contact in human epidermal keratinocytes grown in the presence of calcium.

borders of epithelial cells[6] including MDCK cells, keratinocytes (Figure 2), and human colon carcinoma cells. Adducin is one of the first proteins recruited to sites of Ca^{2+}-dependent cell-cell contact in human keratinocytes which suggests that adducin may play an important role in the molecular organization of specialized cell junctions. Treatment of keratinocytes with TPA results in phosphorylation of adducin and redistribution of adducin away from sites of cell-cell contact. Adducin is also abundant in all blood cells (which do not form junctions) and therefore is likely to have other roles such as local reorganization of the membrane skeleton at sites of phagocytosis. In some cells, α adducin may function independently[12], while in other cells it may function in association with one of the three (or more) isoforms of the β-subunit. The complexity of adducin subunit expression and function is just beginning to be understood.

PURIFICATION

Adducin has been purified from human erythrocytes[1] and bovine brain[5] using ion exchange, hydroxylapatite, and gel filtration chromatography.

ACTIVITIES

In vitro assays[2] suggest that adducin plays a role in the assembly and maintenance of a spectrin actin network by promoting the binding of spectrin and actin. Adducin binds to spectrin actin complexes with higher affinity than it binds either spectrin or actin alone and the formation of this ternary complex will promote binding of a second spectrin molecule to the complex. Recruitment of additional spectrin to spectrin-actin-adducin complexes is inhibited by calmodulin, suggesting a role for adducin in mediating Ca^{2+}-calmodulin dependent reorganization of the cytoskeleton. At higher concentrations, adducin has also been observed to bundle actin filaments[13].

ANTIBODIES

Polyclonal antibodies against human erythrocyte[1] and bovine brain[5] adducin have been raised; they crossreact with rat tissues and mammalian cell lines.

GENES

GenBank accession numbers for human adducin cDNAs are: alpha: X58141, beta-1: X58199. The α adducin gene has been localized to chromosome 4 by Southern blot analysis of hybrid cell lines.

REFERENCES

1. Gardner, K. and Bennett, V. (1986) J. Biol. Chem. 261, 1339-1348.
2. Gardner, K. and Bennett, V. (1987) Nature 328, 359-362.
3. Ling, E., Gardner, K. and Bennett, V. (1986) J. Biol. Chem. 261, 13875-13878.
4. Waseem, A. and Palfrey, H.C. (1988) Eur. J. Biochem. 178, 563-573.
5. Bennett, V., Gardner, K. and Steiner, J. (1988) J. Biol. Chem. 263, 5860-5869.
6. Kaiser, H.W., O'Keefe, E. and Bennett, V. (1989) J. Cell Biol. 109, 557-569.
7. Pinto-Correia, C., Goldstein, E.G., Bennett, V. and Sobel, J.S. (1991) Dev. Biol. 146, 301-311.
8. Tripodi, G., Piscone, A., Borsani, G., Tisminetzky, S., Salardi, S., Sidoli, A., James, P., Pongor, S., Blanchi, G. and Baralle, F.E. (1991) Biochem. Biophys. Res. Comm. 177, 939-947.
9. Joshi, R., Gilligan, D.M., Otto, E., McLaughlin, T. and Bennett, V. (1991) J. Cell Biol. 115, 665-675.
10. Joshi, R. and Bennett, V. (1990) J. Biol. Chem. 265, 13130-13136.
11. Graff, J.M., Stumpo, D.J. and Blackshear, P.J. (1989) J. Biol. Chem. 264, 11912-11919.
12. Waseem, A. and Palfrey, H.C. (1990) J. Cell Sci. 96, 93-98.
13. Mische, S.M., Mooseker, M.S. and Morrow, J.S. (1987) J. Cell Biol. 105, 2837-2845.

■ Diana M. Gilligan, Rashmi Joshi, and Vann Bennett:
Departments of Medicine and Biochemistry,
Howard Hughes Medical Institute,
Duke University Medical Center,
Durham, NC 27710, USA

ANKYRINS

Ankyrins are a family of proteins that are candidates to couple a variety of membrane-spanning cell surface proteins to the spectrin–actin skeleton on the cytoplasmic surface of the plasma membrane[1]. Integral proteins that are currently known to associate with ankyrins in in vitro assays and are co-localized with ankyrin in tissues include ion channels (anion exchanger, Na+K+–ATPase, amiloride-sensitive Na+–channel, voltage-dependent Na+–channel) and cell adhesion molecules (members of the neurofascin family)[1,2]. A physiological role for ankyrins has been proposed to be assembly and/or maintenance of cell surface proteins in specialized membrane domains.

Ankyrins isolated from erythrocytes[3] and brain[4] are the best characterized members of the family. These proteins have a molecular mass of 202 and 206 kDa, and are comprised of three domains: an N-terminal 89–95 kDa membrane-binding domain, a 62 kDa spectrin-binding domain, and a C-terminal domain that plays a regulatory role in erythrocyte ankyrin (Figure 1). Ankyrins in solution are monomers with moderate asymmetry reflected in a frictional ratio of about 1.5. A major contribution to the asymmetry is due to a 20 kDa subdomain located at the

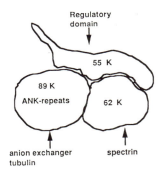

Figure 1. Schematic model of domain organization of erythrocyte ankyrin. Shapes of domains are based on hydrodynamic properties of the isolated 89 kDa domain, spectrin-binding domain, and ankyrin following calpain cleavage of a portion of the regulatory domain. ANK-repeats refer to the 33-residue repeats.

C-terminus, which is cleaved by **calpain**[5]. A striking feature shared by N-terminal domains of both brain and erythrocyte ankyrins (ankyrin$_B$ and ankyrin$_R$ respectively) is a repeated 33-amino acid motif present in 22 contiguous copies with the consensus sequence:-G-TPLH-AA--GH----V/A--LL--GA--N/D----[6-8] . Closely related repeats were originally noted in cell cycle control proteins, proteins regulating tissue differentiation, and also are present in certain transcription factors[9]. Physical properties of the N-terminal domain of ankyrin$_R$ suggest a nearly spherical shape with a CD spectrum consistent with a 30% α-helix content[10]. Since this domain is almost entirely comprised of 33-residue repeats, the repeats are not folded as a series of independent units configured as an extended chain.

■ PURIFICATION

Two alternatively spliced variants of erythrocyte ankyrin can be isolated and resolved from each other by DE53 and Mono Q anion exchange chromatography following selective extraction from human red blood cell membranes[5,11]. Typically 10–20 mg of the major spliced variant (protein 2.1) can be obtained from four units of whole blood in a procedure that requires a week. Brain ankyrin has been isolated from bovine and porcine brains obtained from local slaughterhouses by selective extraction, ammonium sulphate precipitation, gel filtration, affinity-chromatography with immobilized erythrocyte spectrin, and, finally hydroxylapatite chromatography[4]. 0.5–1 mg of brain ankyrin can be obtained from 400 g of tissue in about 10 days.

■ ACTIVITIES

Assays have been developed to measure association of radiolabeled ankyrins with membranes, the anion exchanger cytoplasmic domain, and spectrin[3-5, 11].

■ ANTIBODIES

Polyclonal antibodies have been raised in rabbits directed against erythrocyte ankyrin[3], bovine brain ankyrin[4], and bacterially expressed portions of the C-terminal domains of erythrocyte and brain ankyrins[7,16]. In general, the C-terminal domains of these proteins are the most immunogenic and best suited for distinguishing among family members.

■ GENES AND ALTERNATIVELY SPLICED VARIANTS

Two human genes have been defined at the level of cDNA sequences (Figure 2): ANK1 on chromosome 8p11 (Genebank X16609) which encodes ankyrin$_R$ (first isolated in erythrocytes, but expressed in a subset of neurons in the brain), and ANK2 on chromosome 4 q25-q27 (Genebank X56958, and X56957 for a portion of an alternative spliced variant which encodes ankyrin$_B$ (the major ankyrin in brain). Additional genes are likely to encode ankyrins at the nodes of Ranvier, and basolateral domains of epithelial tissues[1].

Ankyrin$_R$ is subject to modification by alternative splic-

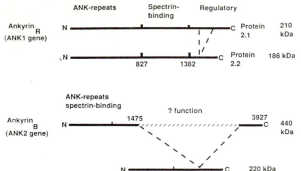

Figure 2. Alternatively spliced products of the ANK1 and ANK2 genes. Proteins 2.1 and 2.2 are terms derived from nomenclature based on mobility of human erythrocyte membrane proteins on SDS-polyacrylamide gels. Numbers refer to amino acid residues, while molecular masses are based on mobility on SDS-polyacrylamide gels. The dashed lines designate portions of the sequences which result from alternative splicing of pre-mRNAs. Note the difference in scale between ankyrin$_R$ and ankyrin$_B$.

ing of mRNA at sites located in the regulatory domain as well as the region between the N-terminal and spectrin-binding domains [6,7,12]. Alternative versions of the regulatory domain result from (A) deletion of a 163 amino acid region (missing in protein 2.2; Figure 2); (B) three alternate C-terminal sequences: either a highly basic stretch of 32 residues (pI greater than 10), or 33 acidic residues, or deletion of 24 amino acids near the C-terminus (1,849–1,873)[12]. An unresolved question in the case of human erythrocyte protein 2.2 is whether deletions of the 163 residue segment are associated with all three types of C-terminal sequence or if the 163 deletion is preferentially linked to one C-terminus.

A mutant mouse (nb/nb) is deficient in ankyrin$_R$ in erythrocytes as well as brain, and has provided a clear demonstration that the same gene is expressed in these different tissues[13,14]. Nb/nb mice exhibit a 90% reduction in expression of ankyrin$_R$ in erythrocytes and brain with a phenotype of severe hemolytic anemia and progressive cerebellar ataxia.

The ANK2 gene also is subject to alternative splicing of mRNA resulting in mRNA transcripts of 13, 7, and 4 kb in addition to the 9 kb major mRNA species[8]. The 13 kb mRNA encodes a 440 kDa alternatively spliced variant of ankyrin$_B$ which contains a sequence of a predicted size of 220 kDa inserted between the regulatory domain and spectrin/membrane-binding domains[15] (Figure 2). The inserted sequence encodes an uninterrupted stretch of polar residues, suggesting the possibility that 440 kDa anykrin$_B$ will have an extended rod-shaped domain in addition to the globular binding domains[16]. Expression of 440 kDa ankyrin$_B$ is maximal in the neonatal period of rat development, with a peak at day 10[15].

■ REFERENCES

1. Bennett, V. (1992) J. Biol. Chem. 267, 8703-8706.
2. Davis, J., T. Mclaughlin, and V. Bennett (1992). Mol. Biol. of the Cell 3, 265a.
3. Bennett, V. and P. Stenbuck. (1980) J. Biol. Chem. 255, 2540-2548.
4. Davis, J. and V. Bennett (1984) J. Biol. Chem. 259, 13550-13559.
5. Hall. T. and V. Bennett (1987) J. Biol Chem. 262, 10537-10545.
6. Lambert, S., H. Yu, J. Prchal et al. (1990) Proc. Nat. Acad. Sci. USA. 87, 1730-1734.
7. Lux, S., K. John, and V. Bennett. (1990) Nature 344, 36-42.
8. Otto, E., M. Kunimoto, T. Mclaughlin, and V. Bennett. (1991) J. Cell Biol. 114, 241-253.
9. Michaely. P. and V. Bennett. (1992) Trends in Cell Biol. 2, 127-129.
10. Davis L. and V. Bennett. (1990) J. Biol. Chem. 265, 10589-10596.
11. Bennet, V. (1983) Methods in Enzymology 96, 313-324 (Eds. Fleischer, S. and Fleischer, B.).
12. Lambert, S. and V. Bennett; unpublished data.
13. Kordeli, E. and V. Bennett (1991) J. Cell Biol. 114,1243-1259.
14. Peters, L., C. Birkenmeier, R. Bronson et al. (1991)J. Cell Biol. 114, 1233-1241.
15. Chan, W. and V. Bennett; unpublished data.
16. Kunimoto, M., E. Otto, and V. Bennett. (1991) J. Cell Biol. 115, 1319-1331.

■ *V. Bennett,*
Howard Hughes Medical Institute and Department of Biochemistry,
Duke University Medical Center,
Durham,

Band 6 Polypeptide

The band 6 polypeptide of ~75 kDa is a protein associated with the desmosomal plaque structures of suprabasal cell layers of several stratified and complex glandular epithelia.

Band 6 polypeptide (B6P) is a nonglycosylated protein cytoplasmically associated with the desmosomal plaque of suprabasal cell layers of diverse stratified and complex glandular epithelia[1-3](Figure). The protein has not been detected in simple epithelia, myocardium, meninges and lymph nodes and in cell culture lines derived from simple epithelial cells[3,4]. Biochemically, B6P is characterized by its positive charge under denaturing conditions (pI ~8) and its ability to bind type I **cytokeratins** *in vitro*[3,4].

By immunofluorescence microscopy B6P appears in the form of fluorescent dots, mostly restricted to the cell periphery[4].

■ PURIFICATION

B6P can be purified by elution from electrophoretically separated polypeptides of desmosomal fractions (for source see, e.g. **desmoplakin** and **desmoglein**) or by column chromatography[2,3].

■ ACTIVITIES

No specific activities for B6P have been characterized so far.

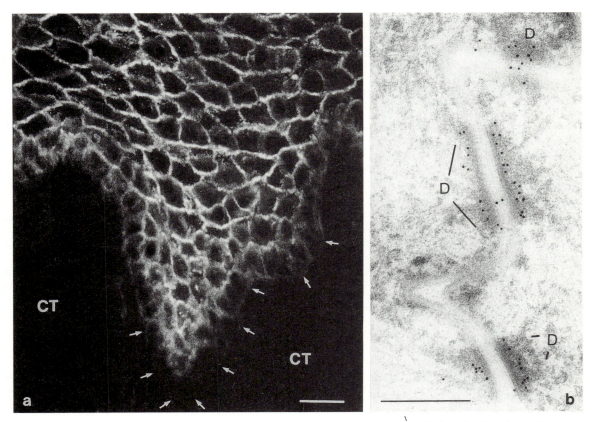

Figure. (a) Immunofluorescence microscopy of bovine muzzle epidermis with anti-B6P antibodies shows intense reaction of desmosomes in the suprabasal cell layers, whereas the desmosomes of the basal layer (demarcated by arrows) and the connective tissue (CT) are not stained. (b) Immunoelectron micrograph showing reaction of anti-B6P antibodies on ultrathin sections of Lowicryl-embedded bovine tongue mucosa. Colloidal gold particles conjugated to the secondary antibodies demarcate the desmosomal plaque structure of the desmosomes (D). (a) Bar 25 μm, (b) bar 0.5 μm.

■ ANTIBODIES

Only guinea pig antisera have been described so far[3].

■ GENES

In vitro translation experiments have indicated that B6P is a genuine mRNA product[2]. Partial amino acid sequence data and peptide maps also show its distinct character[5,6].

■ REFERENCES

1. Franke, W.W., Schmid, E., Grund, C., Müller, H., Engelbrecht, I., Moll, R., Stadler, J. and Jarasch, E.-D. (1981) Differentiation 20, 217-241.
2. Franke, W.W., Müller, H., Mittnacht, S., Kapprell, H.-P. and Jorcano, J.L., (1983) EMBO J. 2, 2211-2215.
3. Kapprell, H.-P., Owaribe, K. and Franke, W.W. (1988) J. Cell Biol. 106, 1679-1691.
4. Kapprell, H.-P., Duden, R., Owaribe, K., Schmelz, M. and Franke, W.W. (1990) In Morphoregulatory Molecules (G.M. Edelman et al. eds.) Wiley, New York 285-314.
5. Müller, H. and Franke, W.W. (1983) J. Mol. Biol. 163, 647-671.
6. Theis, D. and Franke, W.W. unpublished data.

■ *J. Kartenbeck and W.W. Franke:*
Institute of Cell and Tumour Biology,
German Cancer Research Centre,
D-6900 Heidelberg, Germany

Catenins

Catenins[1] are peripheral cytoplasmic proteins which have been identified by their ability to interact with the cytoplasmic region of the cell adhesion molecule uvomorulin/E-cadherin[2,3]. Catenins mediate the cytoplasmic anchorage of E-cadherin to the microfilament network. These proteins are found in a variety of different cell types from several species including human, mouse and chicken[2].

α-, β-, and γ-catenins are structurally distinct cytoplasmic proteins with apparent molecular masses of 102, 88 and 80 kDa, respectively. They have been detected by immunoprecipitation analysis with antibodies directed against the Ca^{2+}-dependent cell adhesion molecule uvomorulin/**E-cadherin**[2-4]. Transfection of uvomorulin full length cDNA into uvomorulin negative mouse NIH3T3 and L-cells and subsequent analysis revealed that catenins of host origin can complex with the introduced uvomorulin[1]. When uvomorulin cDNA was introduced into cell lines from other species, such as HeLa cells and avian fibroblasts, the expressed protein was also associated with endogenous 102, 88 and 80 kDa proteins and each of these proteins showed structural similarities to the respective mouse catenins[1]. Using cDNA constructs coding for uvomorulin with cytoplasmic or extracellular deletions it was shown that the 102, 88 and 80 kDa proteins complex with the cytoplasmic domain of uvomorulin. Analysis of various mutant uvomorulin polypeptides expressed in mouse L cells suggested that this association is mediated by a specific 72 amino acid domain in the cytoplasmic region[1,5]. Chimeric proteins between H-2K^d and this 72 amino acid domain of uvomorulin were shown, by immunoprecipitation with anti-H-2K^d antibodies, to complex with catenin α, β, and γ[5].

Analysis of the molecular organization of the uvomorulin/catenin complex revealed that β-catenin binds directly to uvomorulin and that α-catenin mediates the association with **actin**, whereas γ-catenin seems to be located in the periphery of the complex.

■ PURIFICATION

Catenins can be purified by immunoprecipitation and affinity column chromatography using anti-uvomorulin antibodies and subsequent SDS-PAGE analysis. Protein sequences were obtained from large scale preparations and microsequencing analysis.

■ ACTIVITIES

The full biological functions of catenins are not completely known. Catenins connect uvomorulin/E-cadherin/, and most likely also other **cadherins** with actin bundles[5]. This association regulates the strength of the uvomorulin mediated cell adhesion[6]. More recently it was shown that only uvomorulin associated with catenins induces a redistribution of other integral membrane proteins, suggesting that catenins are involved in a submembraneous network formation[7].

■ ANTIBODIES

Catenins have low immunogenicity and conventional immunizations have not been successful. Anti-peptide antibodies were raised against α-catenin[8]. These antibodies recognize α-catenin in immunoprecipitation, immunoblot, and immunofluorescence experiments in different cell types from various species[8,9]. Also, anti-α-catenin antibodies immunoprecipitate complexes which contain human N-, mouse P-cadherin, chicken A-CAM, or *Xenopus* U-cadherin demonstrating that α-catenin is complexed with other cadherins.

GENES

Complete coding sequences for α-catenin were isolated and the primary structure was determined. Sequence comparison revealed homology to vinculin[8,9]. GenBank EMBL accession No. X59990.

REFERENCES

1. Ozawa, M., Baribault, H. and Kemler, R. (1989) EMBO J. 8, 1711-1717.
2. Vestweber, D. and Kemler, R. (1984) Cell Differ. 15, 269-273.
3. Peyrieras, N., Louvard, D. and Jacob, F. (1985) Proc. Natl. Acad. Sci. USA 82, 8067-8071.
4. Vestweber, D., Gossler, A., Boller, K. and Kemler, R. (1987) Devel. Biol. 124, 451-456.
5. Ozawa, M., Ringwald, M. and Kemler, R. (1990) Proc. Natl. Acad. Sci. USA 87, 4246-4250.
6. Kemler, R., Ozawa, M. and Ringwald, M. (1989) Curr. Opin. Cell Biol. 1, 892-897.
7. McNeill, H., Ozawa, M., Kemler, R. and Nelson, W.J. (1990) Cell 62, 309-316.
8. Herrenknecht, K., Ozawa, M., Eckerskorn, C., Lottspeich, F., Lentner, M. and Kemler, R. (1991) Proc. Natl. Acad. Sci. USA 88, 9156-9160.
9. Nagafuchi, A., Takeichi, M. and Tsukita, S. (1991) Cell 65, 849-857.

■ Rolf Kemler:
Max-Planck-Institut für Immunbiologie,
7800 Freiburg,
Germany

Cingulin

Cingulin[1,2] is a rod shaped dimeric phosphoprotein localized in the cytoplasmic domain of vertebrate tight junctions. Its biophysical properties and sequence indicate a protein with a coiled-coil structure. Cingulin has been detected in native and cultured single- and multi-layered epithelia of avian and mammalian species, whereas it is absent from tissues of mesenchymal origin.

Cingulin isolated from chicken intestinal epithelial cells is a disulphide linked dimer of two subunits, each with an apparent molecular mass of 108 kDa[1,2]. In the electron microscope, purified cingulin appears as an elongated, flexible rod of about 130 nm in length and 2 nm in width[1] (Figure 1). The molecule is resistant to denaturation by heat, ethanol and acid, has a pI of about 3.5[2] and is phosphorylated on serine *in vivo* (unpublished). It is immunologically distinct from other cytoskeletal proteins, and shows unique amino acid composition[1]. Analysis of

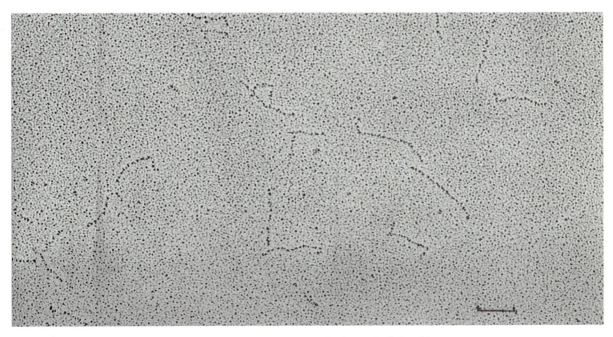

Figure 1. Electron micrograph showing rotary shadowed purified cingulin[1]. Bar 50 nm.

Figure 2. Immunofluorescent localization of cingulin in primary cultures of chicken kidney epithelial cells. Bar 10 µm.

microscopy. In addition, it is differentially expressed in the layers of stratified epithelia, e.g. cornea[2]. It is absent from fibroblasts, muscle cells, and other tissues of mesenchymal origin. Immunofluorescent labelling of cultured kidney epithelial cells reveals the typical subcellular localization of cingulin (Figure 2). Cingulin is expressed in human normal colon tissue and, in increased levels, in colon adenocarcinomas[4]. By immunological crossreactivity, cingulin has also been detected in tissues from several other mammalian species[2], whereas it has not been detected in invertebrate species, including *Drosophila melanogaster*.

■ PURIFICATION

Cingulin can be purified from chicken intestinal epithelial cells by ammonium sulphate fractionation of cellular extracts followed by gel filtration and heat treatment[2]. Ion exchange chromatography has also been used as a purification step exploiting the acidic isoelectric point of cingulin[1].

■ ACTIVITIES

The function of cingulin has not yet been determined.

■ ANTIBODIES

Mouse monoclonal antibodies and a rabbit polyclonal antiserum against chicken cingulin have been described, which have been used for immunoblotting, immunofluorescence and immunoelectron microscopy[1,2]. The polyclonal antiserum crossreacts well with mammalian cingulin[4].

■ GENES

cDNA clones encoding 2.7 kB of the 3' end of the cingulin gene have been isolated from chicken cDNA libraries in λgt11 and λgt10[3].

■ REFERENCES

1. Citi, S., Sabanay, H., Jakes, R., Geiger, B. and Kendrick-Jones, J. (1988) Nature 333, 272-276.
2. Citi, S., Sabanay, H. Kendrick-Jones, J. and Geiger, B. (1989) J. Cell Science 93, 107-122.
3. Citi, S., Kendrick-Jones, J. and Shore, D. (1990) J. Cell Biol. 111, 409a.
4. Citi, S., Amorosi, A., Franconi, F., Giotti, A. and Zampi, G.C. (1991) Am. J. Pathol. 138, 781-789.

■ *Sandra Citi:*
Department of Cell Biology and Anatomy,
Cornell University Medical College,
New York, NY, USA

chicken cingulin cDNAs reveals a high degree of α-helical structure, and the characteristic heptad repeat pattern of charged and hydrophobic residues typical of coiled-coil proteins[3].

Monoclonal and polyclonal antibodies against cingulin recognize by immunoblotting two major polypeptides of apparent molecular mass of 140 kDa and 108 kDa in chicken intestinal cells, the latter may be a proteolytic product of the former[1], and one major polypeptide of 140 kDa in human colonic cells[1,4]. Cingulin is extracted from intestinal cells at approximately physiological ionic strength and without detergents, and is therefore defined as a peripheral component of the tight junction membrane[1]. Immunoelectron microscopic localization shows that cingulin is located at a distance of about 40 nm from the midline of the plasma membrane[1].

Cingulin is localized in the junctional region of essentially all epithelial tissues where tight junctions have been described by classical freeze-fracture electron

Desmocalmin

Desmocalmin is a plaque constitutive protein of desmosomes which can bind to calmodulin in a Ca²⁺-dependent manner[1]. It can also bind to keratin filaments in vitro. Desmocalmin probably functions as a key protein responsible for the formation of desmosomes in a calmodulin dependent manner.

Desmocalmin reveals an apparent molecular mass of 240 kDa by SDS-PAGE[1]. In the low angle, rotary shadowing electron microscope, the desmocalmin molecules look like flexible rods ~100 nm long consisting of two polypeptide chains lying side by side, indicating that the desmocalmin molecule is a homodimer (Figure 1). Desmocalmin molecules can bind to the reconstituted **keratin** filaments *in vitro* at their ends in an end-on fashion and crosslink these filaments. In 2D gel electrophoresis of isolated desmosomes of bovine muzzle epidermis, desmocalmin is isoelectrically focused at an acidic pH of ~5.5, while **desmoplakin** I and II[2] are isoelectric at approximately neutral pH. Desmocalmin is trapped in a calmodulin affinity column in the presence of calcium and eluted with EGTA.

The antibody raised against the purified desmocalmin can specifically recognize desmocalmin, not desmoplakin I and II[2]. With this antibody, desmocalmin was shown by both immunoelectron and immunofluorescence microscopy to be localized at the desmosomal plaque just beneath the plasma membrane (Figure 2). The rodlike structures can be clearly identified just beneath the plasma membrane of desmosomes by deep-etch replica technique. Taken together, desmocalmin may play a pivotal role in forming and maintaining desmosomes in a calmodulin dependent manner.

■ PURIFICATION

Desmocalmin can be purified from isolated desmosomes of bovine muzzle epidermis without denaturing agents[1,3]. Desmocalmin is effectively extracted from isolated desmosomes with a low salt solution at pH 9.5-10.5. After ammonium sulphate fractionation and gel filtration, desmocalmin is finally purified by calmodulin affinity column chromatography. Desmocalmin can bind to calmodulin in the presence of calcium, so that desmocalmin is eluted from the Affigel-calmodulin column with a solution containing EGTA.

■ ACTIVITIES

Desmocalmin can bind to calmodulin in a Ca²⁺-dependent manner[1]. This protein can also bind to the reconstituted keratin filaments *in vitro* in the presence of Mg²⁺, but not to **actin** filaments.

■ ANTIBODIES

Polyclonal antibody against bovine keratinocyte desmocalmin is published[1]. This polyclonal antibody crossreacts with human desmocalmin.

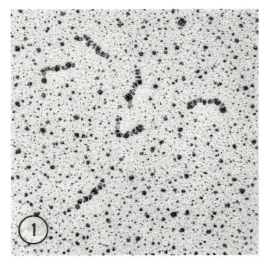
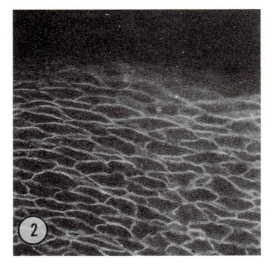

Figure 1. Rotary shadow electron micrograph of desmocalmin (x 120,000).
Figure 2. Immunofluorescence localization of desmocalminin in bovine muzzle epidermis (x 230).

■ GENES

No sequence data of desmocalmin has yet been published.

■ Sachiko Tsukita and Shoichiro Tsukita:
Department of Information Physiology,
National Institute for Physiological Sciences,
Okazaki, Japan

■ REFERENCES

1. Tsukita, S. and Tsukita, S. (1985) J. Cell Biol. 101, 2070-2080.
2. Mueller, H. and Franke, W.W. (1983) J. Mol. Biol. 163, 647-671.
3. Gorbsky, G. and Steinberg, M.S. (1981) J. Cell Biol. 90, 243-248.

Desmoplakins

Desmoplakins are major proteins of desmosomes (maculae adhaerents) where they are found throughout the entire cytoplasmic plaque structure. While desmoplakin I (~250 kDa) is an abundant constitutive component in the desmosomal plaque of all desmosomes, desmoplakin II (~215 kDa) is a major plaque protein only in certain cell types and tissues but relatively minor or absent in others.

Desmoplakin I (DP I) is a major, non-N-glycosylated, non-membranous constituent of the cytoplasmic plaque of all desmosomes, including those of epithelia, myocardial and Purkinje fibre cells of the heart, arachnoidal cells of meninges, dendritic reticulum cells of germinal centres in lymph nodes, and certain glial cells of lower vertebrates[1-10]. Most of the amino acid sequence of human and a part of bovine desmoplakin I have been determined[11,12] and a structural model for a dimeric, partly coiled-coil, rod-like arrangement has been derived therefrom[12], taking into consideration the biochemical characterization of isolated desmoplakin[13]. The protein may be highly phosphorylated[9] and also occurs in a soluble cytoplasmic form of ~9 S. The concentration of the cytoplasmic form is very low in normally grown cells and tissues but very high in cells grown in media with reduced Ca^{2+} concentrations[14-16].

Desmoplakin II (DP II) also appears as a major desmosomal protein in several stratified epithelia and cultured cell lines but it does not occur in a stable stoichiometry with respect to desmoplakin I and is produced at very low concentrations in some epithelial cell lines (e.g. human A-431 cells) and not at all in other desmosome forming tissues such as myocardium[2,3,17].

In one portion of the molecule desmoplakins I and II show significant sequence homology with the **pemphigoid antigens** characteristic of the plaques of hemidesmosomes[12,18,19].

By immunofluorescence microscopy desmoplakins can be identified in typical punctate arrays, mostly restricted along cell-cell boundaries (Figure). In some cultured cell lines, fluorescent dots are also often seen in the cytoplasm, probably representing internalized desmosomal halves[6,20].

■ PURIFICATION

DP I and II can be isolated by elution from electrophoretically separated polypeptides of desmosome fractions, or in urea by the method of O'Keefe et al.[13]. The most commonly used tissue is the stratified epithelium of the multilayered *stratum spinosum* of the perinostril epithelium of bovine muzzle or porcine snout. Various procedures to isolate desmosomal structures have been published (see under **desmoglein**).

■ ANTIBODIES

Monoclonal and polyclonal antibodies, both with a wide range of crossreactivity, have been described[2,3] and are commercially available from several companies.

■ GENES

A partial bovine cDNA clone[19] and a complete human cDNA clone[12,21] have been published. By hybridization reaction either a single mRNA (~9.2 kB) desmoplakin I or two mRNAs (~9.2 kB and 7.5 kB) have been identified in the different cells and tissues. DP II mRNA is discussed to be derived from a larger DP I-encoding transcript of a common gene by alternative splicing[12,21] (GenBank M77830).

■ REFERENCES

1. Cowin, P. and Garrod, D.R. (1983) Nature 302, 148-150.
2. Cowin, P., Kapprell, H.-P. and Franke, W.W. (1985) J. Cell Biol. 101, 1442-1454.
3. Cowin, P., Franke, W.W., Grund, C. and Kapprell, H.-P. (1985) In The Cell in Contact (G. Edelman, J.P. Thierry, eds.) Wiley, New York, 427-460.
4. Franke, W.W., Moll, R., Schiller, D.L., Schmid, E., Kartenbeck, J. and Müller, H. (1982) Differentiation 23, 115-127.
5. Franke, W.W., Schiller, D.L., Hatzfeld, M. and Winter, S. (1983) Proc. Natl. Acad. Sci. (USA) 80, 7113-7117.
6. Kartenbeck, J., Franke, W.W., Moser, J.G. and Stoffels, U. (1983) EMBO J. 2, 735-742.
7. Kartenbeck, J., Schwechheimer, K., Moll, R. and Franke, W.W. (1984) J. Cell Biol. 98, 1072-1081.
8. Moll, R., Cowin, P., Kapprell, H.-P. and Franke, W.W. (1986) Lab. Invest. 1, 4-25.

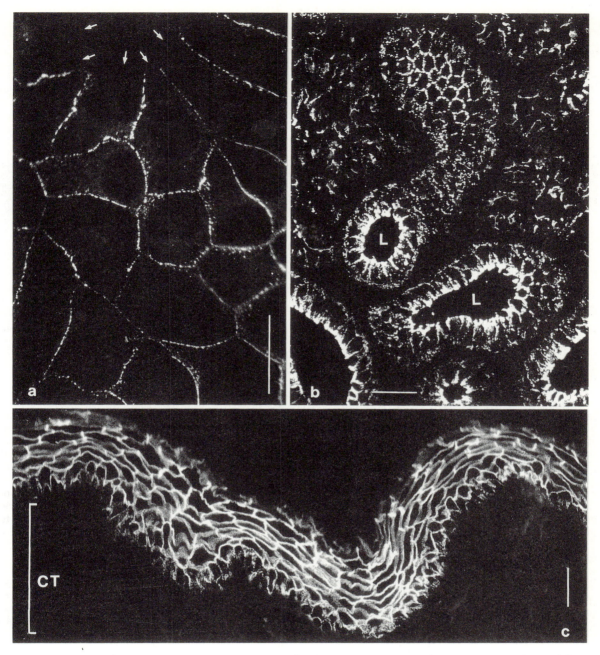

Figure. Appearance of desmosomal structures in immunofluorescence microscopy after staining with desmoplakin antibodies in various cells and tissues. (a) cultured canine kidney epithelial cells of line MDCK (arrows demarcate free cell boundaries); (b) cross and grazing section of ducts and acinar cells of a bovine salivary gland (L, lumen); (c) stratified epithelia cells of rat esophagus (bracket demarcates connective tissue, CT). Bar 25 µm.

9. Müller, H. and Franke, W.W. (1983) J. Mol. Biol. 163, 647-671.
10. Rungger-Brändle, E., Achtstätter, T. and Franke, W.W. (1989) J. Cell Biol. 109, 705-716.
11. Franke, W.W., Goldschmidt, M.D., Zimbelmann, R., Müller, H.M. and Schiller, D.L. (1989) Proc. Natl. Acad. Sci. (USA) 86, 4027-4031.
12. Green, K.J., Parry, D.A.D., Steinert, P.M., Virata, M.L.A. and Wagner, R.M. (1990) J. Biol. Chem. 265, 2603-2612.
13. O'Keefe, E.J., Erickson, H.P. and Bennett, V. (1989) J. Biol. Chem. 264, 8310-8318.
14. Duden, R. and Franke, W.W. (1988) J. Cell Biol. 107, 1049-1063.
15. Pasdar, M. and Nelson, W.J. (1988) J. Cell Biol. 106, 677-685.

16. Pasdar, M. and Nelson, W.J. (1988) J. Cell Biol. 106, 687-695.
17. Angst, B.D., Nilles, L.A. and Green, K.J. (1990) J. Cell Sci. 97, 247-257.
18. Amagai, M., Hashimoto, T., Tajima, S., Inokuchi, Y., Shimizi, N., Saito, M., Miki, K. and Nishikawa, T., (1990) J. Invest. Dermatol. 95, 252-259.
19. Schwarz, M.A., Owaribe, K., Kartenbeck, J. and Franke, W.W. (1990) Annu. Rev. Cell Biol. 6, 461-491.
20. Kartenbeck, J., Schmid, E., Franke, W.W. and Geiger, B. (1982) EMBO J. 1, 725-732.
21. Virata, M.L.A., Wagner, R.M., Parry, D.A.D. and Green, K.J. (1992) Proc. Natl. Acad. Sci. (USA) 89, 544-548.

■ *J. Kartenbeck and W.W. Franke:*
Institute of Cell and Tumour Biology,
German Cancer Research Center,
6900 Heidelberg, Germany

Micro-Calpain and Milli-Calpain (EC 3.4.22.17)

There are two isoforms of the ubiquitous, nonlysosomal, Ca^{2+}-dependent cysteine proteinases known as calpains. The two calpains are distinguished from one another by unique catalytic subunits and by their different affinities for calcium ions. Each calpain is present in all mammalian cells examined thus far, except enucleated erythrocytes that contain only micro-calpain. The functions of the enzymes and the factors controlling their activities in vivo are not yet known.

Micro-calpain and milli-calpain are found in variable amounts (absolute and relative) in different mammalian cells[1-4]. The enzymes have been extensively characterized in mammals and chicken. Several reports describe homologous activities, but not necessarily homologous proteins, from invertebrates (including *Drosophila*), amphibians, and plants. The proteinases are predominantly cytoplasmic, however they may associate transiently with intracellular membrane surfaces, the membrane skeleton and/or the cytoskeleton. Colocalization of milli-calpain with **talin** in adhesion plaques of several cell types is suggestive of the possible significance of proteolytic modification of talin in regulating adhesion plaque function[5]. Each calpain has a catalytic subunit of approximately 80 kDa and a regulatory subunit of approximately 30 kDa. The complete cDNA sequences of the catalytic subunit from human[6,7] and the 30 kDa subunit of several species (human[1], rabbit[1], porcine[8], bovine[9]) are known.

Each subunit contains a domain homologous to the calmodulin family of Ca^{2+}-binding proteins (IV and IV'), with four potential EF-hands per subunit. Domain II is homologous to cysteine proteinases, although not closely related to either papain or cathepsin B. Calpain activity can be modulated *in vitro* by a variety of mechanisms. The proteinase is inactive in the absence of calcium. Upon Ca^{2+}-binding, autoproteolytic modification of both subunits occurs and results in proteolytic activity characterized by an increased sensitivity to calcium. *In vitro*, phospholipids decrease the calcium ion requirement for autoproteolysis of milli-calpain. Calpastatin, an endogenous proteinaceous inhibitor specific for calpains, is also present in all cells that contain calpain[1-4]. Its concentration, relative to the proteinases, varies considerably. The cDNA sequence of calpastatin reveals four repeats (corresponding to 140 amino acids), each of which has functional activity against the calpains[10-12]. Calpastatin, binds, reversibly, to each calpain only when the enzymes are in their Ca^{2+}-bound conformations. The regulation of calpain-calpastatin interactions *in vivo* are not known. A stimulatory activity for calpain has been described in both soluble and cytoskeletal fractions from brain[1,2].

■ PURIFICATION

Calpains can be purified from a variety of tissues or cells. Kidney and muscle (cardiac, smooth or skeletal) are the common sources of milli-calpain. Conventional anion exchange and gel filtration chromatography, in combination with affinity chromatography using immobilized reactive red, will purify milli-calpain[1,2]. Micro-calpain can be purified from erythrocytes, kidney, brain, and cardiac muscle. The purification from tissues, using multiple, conventional, chromatography procedures and hydrophobic interaction chromatography, is considerably more laborious than purification of milli-calpain[2].

■ ACTIVITIES

Proteolytic activity *in vitro* is maximal between pH 7-7.6, and is dependent on a reducing environment and calcium ion concentration. Milli-calpain activity is measured at 30°C or less, micro-calpain can be assayed up to 37°C. The production of acid soluble fragments of α-casein is the most common assay for proteolytic activity[1,2]. This assay is sensitive, but not specific, for calpain. Synthetic peptides, in general are poorly hydrolyzed. Talin, **spectrins**, **ankyrin**, **filamin**, many protein kinases, including protein kinase C, and numerous other proteins are cleaved to characteristic fragments by both calpains. The cleavage site specificity of calpain is poorly defined, with no well-defined differences between the two isoforms[1,2]. *In vivo*, calpains may participate in a variety of cellular processes as either regulatory enzymes or as degradative enzymes[1-4]. Proteolytic modification of talin and filamin during the aggregation of platelets by micro-calpain is the best documented function of either calpain[1,13]. Current areas of investigation include degradation of the *c-mos* protooncogene product, proteolysis of protein kinase C, modification and/or degradation of spectrin/fodrin, talin, neurofilaments, microtubule-associated proteins and other cytoskeletal proteins.

■ INHIBITORS

The only inhibitor known to be specific for calpains is calpastatin. Effective nonpeptide inhibitors of thiol-proteinases are iodoacetate, N-ethyl-maleimide, and mersalyl (a mercurial). Numerous peptide based inhibitors are effective against, but not specific for, calpains including epoxysuccinyl peptides, peptide aldehydes, such as leupeptin and antipain, peptide chloromethanes and peptide diazomethanes[1,2].

■ ANTIBODIES

There are numerous antisera against mammalian and chicken calpains. Polyclonal sera have been raised against both the native holoenzymes and SDS-PAGE purified subunits[1-3]. Antisera raised against a specific mammalian calpain usually crossreact with the antigen from any other mammal[5,14] and may crossreact weakly with nonmammalian species. Antisera raised against chicken calpain crossreact with *Xenopus*. Antisera raised against native milli-calpain frequently crossreact with both subunits of micro-calpain. Antisera elicited against SDS-PAGE purified subunits may be less crossreactive. Isoform specific antisera have been generated by affinity-purification of antibodies. Using synthetic peptides as antigens, highly specific antisera have been raised[15]. Monoclonal antibodies against mammalian calpains are also available and may be isoform specific[1-4,13].

■ GENES

The genes for the catalytic subunits for micro-calpain and milli-calpain map to human chromosomes 11 and 1 respectively[16]. The gene for the 30 kDa subunit maps to chromosome 19. The gene for human milli-calpain is published (GenBank J04700). A novel member of the calpain family, specific to skeletal muscle, has been identified by cDNA hybridization. This calpain mapped to human chromosome 15[16].

■ REFERENCES

1. Mellgren, R.L. and Murachi, T. eds. CRC Press, Boca Raton FL (1990) 1-288.
2. Croall, D.E. and DeMartino, G.N. (1991) Physiol. Rev. 71, 813-847.
3. Murachi, T. (1989) Biochem. Internat. 2, 651-656.
4. Suzuki, K., Imajoh, S., Emori, Y., Kawasaki, H., Minami, Y. and Ohno, S. (1987) FEBS Lett. 220, 271-277.
5. Beckerle, M.C., Burridge, K., DeMartino, G.N. and Croall, D.E. (1987) Cell 51, 569-577.
6. Aoki, K., Imajoh, S., Ohno, S., Emori, Y., Koike, K., Kosaki, G. and Suzuki, K. (1986) FEBS Lett. 205, 313-317.
7. Imajoh, S., Aoki, K., Ohno, S., Emori, Y., Kawasaki, H., Sugihara, H. and Suzuki, K. (1988) Biochemistry 27, 8122-8128.
8. Sakihama, T., Kakidani, H., Zenita, K., Yumota, N., Kikuchi, T., Sasaki, T., Kannagi, R., Nakanishi, S., Ohmori, K., Takio, K., Titani, K. and Murachi, T. (1985) Proc. Natl. Acad. Sci. 82, 6075-6079.
9. McClelland, P., Lash, J.A. and Hathaway, D.R. (1989) J. Biol. Chem. 264, 17428-17431.
10. Emori, Y., Kawasaki, H., Imajoh, S., Imahori, K. and Suzuki, K. (1987) Proc. Natl. Acad. Sci. 84, 3590-3594.
11. Maki, M., Takano, E., Osawa, T., Ooi, T. and Murachi, T. (1988) J. Biol. Chem. 263, 10254-10261.
12. Maki, M., Bagci, H., Hamaguchi, K., Ueda, M., Murachi, T. and Hatanaka, M. (1989) J. Biol. Chem. 264, 17428-17431.
13. Samis, J.A., Zboril, G. and Elce, J.S. (1987) Bochem. J. 246, 481-488.
14. Schollmeyer, J.E. (1986) Exp. Cell Res. 162, 411-422.
15. Croall, D.E., Slaughter, C.A., Wortham, H.S., Skelly, C.M., DeOgny, L. and Moomaw, C.R. (1992) Biochim. Biophys. Acta, 1121, 47-53.
16. Ohno, S., Minoshima, S., Kudoh, J., Fukuyama, R., Shimizu, Y., Ohmi-Imajoh, S., Shimizu, N. and Suzuki, K. (1990) Cytogenet. Cell Genet. 53, 225-229.

■ Dorothy E. Croall:
Department of Biochemistry,
Microbiology and Molecular Biology,
University of Maine, Orono,
Maine, USA

Paxillin

Paxillin is a focal adhesion protein that has been shown to interact with vinculin in vitro. The localization of paxillin and its interaction with vinculin suggests that it is involved in linking actin filaments to the plasma membrane.

Paxillin is a 68 kDa cytoskeletal protein that localizes to focal adhesions in cells grown in culture[1,2] (Figure). These are sites of cell-substratum adhesion where **actin** containing stressfibres terminate at the plasma membrane. In contrast, paxillin is absent from the cell-cell adherens junctions formed by epithelial cells[2]. Immunoblot analysis reveals paxillin to be most abundant in smooth muscle tissue and to a lesser extent in cardiac

A B

Figure. Panel A, immunolocalization of paxillin to focal adhesions in a cultured chick embryo fibroblast. Panel B, actin stressfibres in the same cell as (A) labelled with fluorescein-phalloidin.

and skeletal muscle. It is present in liver and brain at very low levels[2]. Immunolabelling of tissue sections localizes paxillin to the dense plaques of smooth muscle and the myotendinous junctions of skeletal muscle[3]. These are both regions where actin filaments associate, via *integrin* molecules, with the extracellular matrix[4-6] and are regions where force generated during muscle contraction is transmitted across the membrane[7]. Paxillin is also concentrated at the postsynaptic side of the neuromuscular junction in skeletal muscle[3]. The function of focal adhesion proteins at the neuromuscular junction is unclear. Paxillin is tyrosine phosphorylated in a developmentally regulated manner[12].

■ PURIFICATION

Paxillin is readily extracted in low ionic strength buffer from chicken gizzard smooth muscle and can be purified by a combination of conventional ion exchange and antibody affinity chromatography[3]. Paxillin purified from smooth muscle has a molecular mass of approximately 68 kDa. The protein runs as a diffuse band on SDS-PAGE and this may be due to the presence of multiple isoforms seen upon 2D gel analysis[3]. These are probably the results of a posttranslational modification (e.g. phosphorylation). Paxillin is not abundant and yields are 50-100-fold lower than for another focal adhesion protein, **talin**.

■ ACTIVITIES

Using a combination of *in vitro* blot overlay and solid phase microtiter well binding assays, paxillin has been shown to interact with **vinculin** with an apparent K_d of 6 X 10^{-8}M. The binding site for paxillin on vinculin has been located to the C-terminal rod domain of vinculin[3]. This is the opposite end of vinculin to which talin binds[8-10]. Paxillin is heavily phosphorylated on tyrosine following Rous *sarcoma* virus (RSV) transformation of chick embryo fibroblasts[1]. The significance of this event in the genera-

tion of the transformed phenotype is in question since a similar phosphorylation of paxillin occurs in a nontransforming mutant fibroblast line[11].

■ ANTIBODIES

Monoclonal antibodies to paxillin were originally described that react with a 76 kDa phosphoprotein from RSV transformed chick embryo fibroblasts[1].

■ GENES

The sequencing of paxillin cDNA is in progress.

■ REFERENCES

1. Glenney, J.R. and Zokas, L. (1989) J. Cell Biol. 108, 2401-2408.
2. Turner, C.E., Glenney, J.R. and Burridge, K. (1990) J. Cell Biol. 111, 1059-1068.
3. Turner, C.E., Kramarcy, N., Scalock, R. and Burridge, K. (1991) Exp. Cell Res. 192, 651-655.
4. Bozyczko, D., Decker, L., Muschler, J. and Horwitz, A.F. (1989) Exp. Cell Res. 183, 72-91.
5. Kelly, T., Molony, L. and Burridge, K. (1987) J. Biol. Chem. 262, 17189-17199.
6. Swasdison, S. and Mayne, R. (1989) Cell Tiss. Res. 257, 537-543.
7. Tidball, J., O'Halloran, T. and Burridge, K. (1986) J. Cell Biol. 103, 1465-1472.
8. Price, D., Jones, P., Davison, M.D., Patel, B., Bendori, R., Geiger, B. and Critchley, D.R. (1989) Biochem. J. 259, 453-461.
9. Turner, C.E. and Burridge, K. (1989) Eur. J. Cell Biol. 49, 202-206.
10. Groesch, M.E. and Otto, J.J. (1990) Cell Motil. Cytoskel. 15, 41-50.
11. Glenney, J.R. (1989) J. Biol. Chem. 264, 20163-20166.
12. Turner, C.E. (1991). J. Cell Biol. 115, 201-207.

■ *Christopher, E. Turner:*
Dept. of Anatomy and Cell Biology,
Suny Health Science Center at Syracuse,
Syracuse, NY
13210, USA

Pemphigoid Antigens

The pemphigoid antigens are polypeptides that react with the sera of patients with an autoimmune skin-blistering disease, bullous pemphigoid. It is now known that they are major components of hemidesmosomes in the basal cells of stratified and complex epithelia.

The pemphigoid antigens were originally defined as target molecule(s) that were recognized by autoimmune sera of patients suffering from diseases of the pemphigoid group, the most common of which is *bullous pemphigoid* (BP), characterized by subepidermal blistering. The antigens were found to be present in the epidermal basement membrane zone (BMZ) of these patients. It is now known that these antigens are normal components of certain epithelial BMZ, and present also in cultured keratinocytes. They have been found in the hemidesmosome of stratified epithelial cells (epidermal, esophageal, corneal, etc.) and complex epithelial cells (tracheal, bladder, glandular myoepithelial, etc.), but not in simple epithelial cells that lack typical hemidesmosomes, including vascular endothelial cells and cardiac cells[1-3].

The BP autoimmune sera have been found to immunoprecipitate only a 230 kDa polypeptide (BP230)[4,5]. By immunoblotting, however, sera from different patients also reacted with a 180 kDa protein (BP180), as well as some other polypeptides[6]. Antibodies specific to BP230 and BP180 can be affinity purified separately and they did not crossreact by immunoblot analysis. This indicates that the two BP antigens are different proteins[7-9]. BP230 and BP180 are so far the only proteins recognized as BP or pemphigoid antigens.

BP230 is a basic protein with a pI of about 8[5], localized in hemidesmosomal plaques of stratified and complex epithelial basal cells[2,3,7]. The cDNA of BP230 (8.9 kB) encodes a polypeptide of 2,649 amino acids[10]. Analysis of the deduced amino acid sequence predicts the presence of a putative signal peptide of 43 amino acids, a putative transmembrane domain of 17 amino acids, and several potential sites for N-glycosylation and phosphorylation. It is predicted that BP230 is an integral membrane protein. It probably forms an α-helical coiled-coil dimeric structure in the central rod portion, flanked by terminal globular domains. The predicted amino acid sequence shows high homology in the carboxyterminal domain with desmoplakin[11,12].

BP180 is a glycoprotein[13], epitopes of which are found in the intracellular, extracellular, or both regions of hemidesmosomes, depending on the antisera used[7,8,14]. A partial cDNA clone has been isolated from a human keratinocyte library[15]. The clone recognizes a 6 kB RNA transcript, distinct from the 9 kB mRNA of BP230. Sequence analysis of this partial cDNA has revealed the presence of two collagen domains[16].

■ PURIFICATION

The pemphigoid antigens (BP230 and BP180) have not been purified yet. Hemidesmosome-enriched fractions

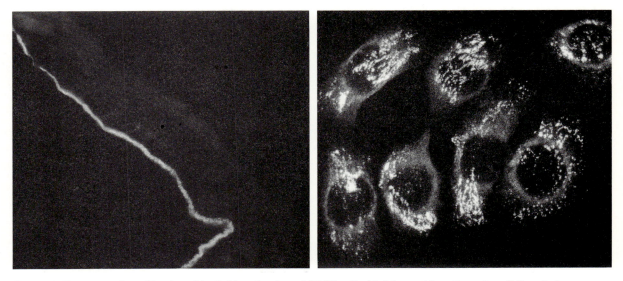

Figure 1. Frozen section of bovine skin (left) and cultured FRSK cells (right), a rat keratinocyte cell line, in immunofluorescence microscopy using BP autoantibodies.

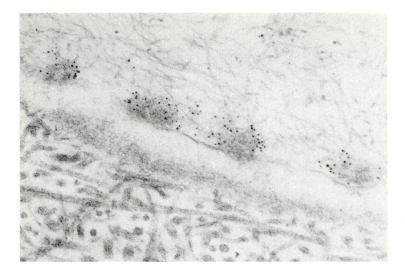

Figure 2. Immunoelectron microscopy of bovine cornea stained with BP230 autoantibodies showing specific decoration of hemidesmosomal plaques.

can be prepared from bovine corneal epithelial cells, in which both BP230 and BP180 are major components on SDS-PAGE[13].

■ ACTIVITIES

No specific activities for pemphigoid antigens have been described so far.

■ ANTIBODIES

Because *bullous pemphigoid* is rather common in the field of dermatology, BP autoimmune sera can be obtained from dermatologists. The sera usually recognize either BP230, BP230 and BP180, and rarely only BP180. They usually show wide species crossreactivity. Some polyclonal and monoclonal antibodies to each protein have been reported[13,15,17,18].

■ GENES

The full length cDNA for BP230 has been published (M69225)[10]. Only partial cDNAs for BP180 have been isolated so far[15,16].

■ PATHOLOGY

Other pemphigoids, *cicatricial pemphigoid* (CP), *vesicular pemphigoid* and *pemphigoid* (or herpes) *gestationis*, are clinically classified as diseases different from BP, but some of them may well recognize the same antigens, BP230 and/or BP180[19,20]. Approximately a third of BP patients have significant mucosal lesions, some of which are similar to CP, suggesting some overlap in the two diseases.

■ REFERENCES

1. Mutasim, D.F., Takahashi, Y., Labib, R.S., Anhalt, G.J., Patel, H.P. and Diaz, L.A. (1985) J. Invest. Dermatol. 84, 47-53.
2. Westgate, G.E., Weaver, A.C. and Couchman, J.R. (1985) J. Invest. Dermatol. 84, 218-224.
3. Owaribe, K., Kartenbeck, J., Stumpp, S., Magin, T.M., Krieg, T., Diaz, L.A. and Franke, W.W. (1990) Differentiation 45, 207-220.
4. Stanley, J.R., Hawley-Nelson, P., Yuspa, S.H., Shevach, E.M. and Katz, S.I. (1981) Cell 24, 897-903.
5. Mueller, S., Klaus-Kovtun, V. and Stanley, J.R. (1989) J. Invest. Dermatol. 92, 33-38.
6. Labib, R.S., Anhalt, G.J., Patel, H.P., Mutasim, D.F. and Diaz, L.A. (1986) J. Immunol. 136, 1231-1235.
7. Klatte, D.H., Kurpakus, M.A., Grelling, K.A. and Jones, J.C.R. (1989) J. Cell Biol. 109, 3377-3390.
8. Cook, A.L, Hanahoe, T.H.P., Mallett, R.B. and Pye, R.J. (1990) Br. J. Dermatol. 122, 435-444.
9. Robledo, M.A., Kim, S.C., Korman, N.J., Stanley, J.R., Labib, R.S., Futamura, S. and Anhalt, G.J. (1990) J. Invest. Dermatol. 94, 793-797.
10. Sawamura, D., Li, K., Chu, M.-L. and Uitto, J. (1991) J. Biol. Chem. 266, 17784-17790.
11. Schwarz, M.A., Owaribe, K., Kartenbeck, J. and Franke, W.W. (1990) Annu. Rev. Cell Biol. 6, 461-491.
12. Jones, J.C.R. and Green, K.J. (1991) Curr. Opin. Cell Biol. 3, 127-132.
13. Owaribe, K., Nishizawa, Y. and Franke, W.W. (1991) Exp. Cell Res. 192, 622-630.
14. Shimizu, H., McDonald, J.N., Kennedy, A.R. and Eady, R.A.J. (1989) Arch. Dermatol. Res. 281, 443-448.
15. Diaz, L.A., Ratrie III, H., Saunders, W.S., Futamura, S., Squiquera, H.L., Anhalt, G.J. and Giudice, G.J. (1990) J. Clin. Invest. 86, 1088-1094.
16. Giudice, G.J., Squiquera, H.L., Elias, P.M. and Diaz, L.A. (1991) J. Clin. Invest. 87, 734-738.
17. Sugi, T., Hashimoto, T., Hibi, T. and Nishikawa, T. (1989) J. Clin. Invest. 84, 1050-1055.
18. Tanaka, T., Korman, N.J., Shimzu, H., Eady, R.A.J., Klaus-Kovtun, V., Cehrs, K. and Stanley, J.R. (1990) J. Invest. Dermatol. 94, 617-623.

19. Bernard, P., Prost, C., Lecerf, V., Intrator, L., Combemale, P., Bedane, C., Roujeau, J.-C., Revuz, J., Bonnetblanc, J.-M. and Dubertret, L. (1990) J. Invest. Dermatol. 94, 630-635.
20. Kelly, S.E., Bhogal, B.S., Wojnarowska, F., Whitehead, P., Leigh, I.M. and Black, M.M. (1990) Br. J. Dermatol. 122, 445-449.

■ Katsushi Owaribe:
Department of Molecular Biology,
School of Science, Nagoya University,
Nagoya 464-1, Japan

Plakoglobin

Plakoglobin is found underneath the membrane of all symmetrical intercellular adhesive junctions and in a cytosolic form in most cell types. Its localization suggests a structural, membrane reinforcement role. However its codistribution with members of the cadherin family of morphoregulatory cell adhesion molecules and its resemblance to the product of the Drosophila segment polarity gene Armadillo suggest that plakoglobin is an essential component in the intracellular cascade of response to positional information.

Plakoglobin (82 kDa; pI 5.3) is an abundant, cytoplasmic protein that exists as both a soluble, globular dimer and in a membrane associated form[1]. It is enriched in the electron dense submembranous plaques of symmetrical adhesive junctions e.g. desmosomes[2,3], adherens, lens and endothelial junctions[1]. In contrast, plakoglobin is absent from junctions between cells and matrix[1] e.g. hemidesmosomes and focal contacts. It therefore colocalizes with the various members of the **cadherin** family that include desmosomal adhesion molecules[4,5].

Plakoglobin is highly conserved, with 63% sequence homology between human and *Drosophila* forms[6,7]. In both species it is encoded by two mRNA's that are produced from a single gene by alternative splicing of the noncoding regions[6,8]. *Xenopus* β-**catenin**, a protein that associates directly with the cadherin cytoplasmic tail, has been sequenced and forms a third member of the plakoglobin family. This protein has 63% identity to plakoglobin and 70% identity to the *Armadillo* protein[9].

Structurally, plakoglobin has a central highly conserved hydrophobic region containing 12.5 42 amino acid repeats flanked by hydrophilic N- and C-domains that are less conserved. The internal repeats of the central domain have two subdomains that contain alternating hydrophobic regions and hydrophilic zones[6,8]. The *Drosophila* form has additional glycine rich sequences in its C-tail that are not present in the human or bovine homologues[7,8].

Plakoglobin is also described in the literature as desmosomal band 5 and desmosomal protein (DPIII)[10].

The precise role of plakoglobin is presently open to speculation. The presence of plakoglobin at a number of structurally similar yet biochemically diverse junctions suggests that it plays a central role in plaque structure and function. However this role is clearly limited to plaques associated with cadherin molecules. Plakoglobin has been shown to coprecipitate as a complex with **desmoglein**, a desmosomal cadherin molecule[11]. This raises the possibility that plakoglobin may interact with cadherin-type proteins and in so doing may regulate junction assembly, stability and hence modulate cell interactions from within the cell.

Mutants of the *Drosophila* homologue known as *Armadillo* show a segment polarity phenotype. Expression of the product of another segment polarity gene *wingless*, a secreted morphogen, causes accumulation of *Armadillo* protein prior to exerting its effect on pattern formation[12]. This genetic evidence suggests that plakoglobin may be involved in intracellular reception or implementation of the *wingless/Int-1* signal although its precise place in this hypothetical signaling pathway is not clear at present[7]. The presence of *Armadillo*/plakoglobin in most cells, however, argues for a capacity to respond to a wider range of factors and signals. A dynamic equilibrium between the soluble and membrane bound forms of plakoglobin is compatible with either a structural, stabilizing role or one involving some form of signal transduction.

■ PURIFICATION

Soluble plakoglobin (20-30% of total plakoglobin and 0.05% of total soluble proteins) can be purified from a variety of cell types by collecting the 100,000X g supernatant from tissues homogenized in physiological buffers. Enriched fractions are then obtained by gel filtration, ion exchange chromatography and immunoabsorption techniques. Membrane bound plakoglobin is readily dissociated from nondesmosomal junctions by extracting cells with either low or high salt buffers[13]. Desmosomal plakoglobin, in contrast, is remarkably resistant to extraction and must be purified by elution from SDS-polyacrylamide gels following electrophoretic separation of enriched desmosomal fractions[13].

■ ACTIVITIES

Defined activities of plakoglobin have so far not been characterized.

■ ANTIBODIES

Polyclonal and monoclonal antibodies which crossreact with plakoglobin of most vertebrate species have been described[1,2].

■ GENES

cDNA's encoding bovine, human (EMBL/GenBank M23410) plakoglobin, *Drosophila Armadillo* (GenBank X54468) and *Xenopus* β-catenin (GenBank M77013) have been reported.

■ REFERENCES

1. Cowin, P., Kapprel, H.P., Franke, W.W., Tamkun, J. and Hynes, R.O. (1986) Cell 46, 1063-1073.
2. Cowin, P. and Garrod, D.R. (1983) Nature 302, 148-150.
3. Cowin, P., Kapprel, H.P. and Franke, W.W. (1984) J. Cell Biol. 101, 1442-1454.
4. Goodwin, L., Raynor, K., Hill, J., Raszi, L., Manabe, M. and Cowin, P. (1990) Biochem. Biophys. Res. Comm. 173, 1224-1230.
5. Mechanic, S., Raynor, K., Hill, J. and Cowin, P. (1991) Proc. Natl. Acad. Sci. 88, 4476-4480.
6. Franke, W.W., Goldschmidt, M.D., Zimbelman, R., Mueller, H.M., Schiller, D.L. and Cowin, P. (1989) Proc. Natl. Acad. Sci. (USA) 86, 4027-4031.
7. Pfeifer, M. and Wieschaus, E. (1990) Cell 63, 1167-1178.
8. Riggleman, R. and Wieschaus, E. (1989) Genes Dev. 3, 96-113.
9. McCrea, P., Turck, C.W. and Gumbiner, B. (1991) Science 254, 1359-1361.
10. Gorbsky, G., Cohen, S. and Steinberg, M.S. (1985) Proc. Natl. Acad. Sci. (USA) 82, 810-814.
11. Korman, N.J., Eyre, R.W., Klaus-Kovtun, V. and Stanley, J.R. (1989) N. Engl. J. Med. 321, 631-635.
12. Riggleman, R., Schedl, P. and Wieschaus, E. (1990) Cell 63, 549-560.
13. Kapprell, H.P., Cowin, P. and Franke, W.W. (1987) Eur. J. Biochem. 166, 505-517.

■ *Pamela Cowin:*
Departments of Cell Biology and Dermatology,
NYU Medical Center,
New York, NY, USA

pp60[c-src]

pp60[c-src], the 60 kDa product of the c-src gene, is a protein tyrosine kinase (PTK), which is highly conserved throughout the vertebrate kingdom. pp60[c-src] is widely expressed with the highest levels being in neurons and platelets. pp60[c-src] is a member of a family of closely related PTKs, which are all anchored to the inner face of cytoplasmic membranes, and which may function as signal transducing subunits of cell surface receptors that lack their own catalytic domain. These PTKs are composed of three main functional domains - an N-terminal membrane association domain linked to a hypervariable region, a regulatory domain, and a protein kinase catalytic domain.

The c-*src* gene is a member of a family of closely related PTK genes, which currently has eight members (the *src, yes, fgr, lck, hck, fyn, lyn,* and *blk* genes)[1,2]. There are related genes in simpler organisms such as *Drosophila*. All the *src* family genes encode PTKs of about 525 amino acids in length with very similar structural organizations. These proteins are ~80% identical over their C-terminal 450 residues, which includes the catalytic domain, but diverge almost completely in their N-terminal 80 residues (exon 1) (Figure). The conserved N-terminal Gly is myristoylated, and this modification is essential for membrane association. The N-terminal 14 residues of pp60[c-src] provide a necessary and sufficient signal for myristoylation. The myristoylated N-terminus binds to a 32 kDa membrane "receptor" that may be required for membrane localization of pp60[c-src] following its synthesis on soluble ribosomes[3]. To the C-terminal side of the unique region lie two regulatory domains, SH3 and SH2 (SH = src homology) (Figure)[4]. The SH3 domain (residues ~88-140), plays a negative regulatory role in pp60[c-src] function, and is related in sequence to similar domains in the *abl* and *fps* PTKs, as well as in a series of other proteins, including regulatory enzymes and cytoskeletal proteins. The SH2 domain (residues ~141-250) is also regulatory, and plays a role in

substrate selection, probably as a result of its ability to bind to phosphotyrosine containing sequences in other proteins. The *src* family SH2 domain is related to similar domains in the *abl* and *fps* PTKs, and in several regulatory enzymes such as (GAP(p120[ras-GAP])) phospholipase C-γ and PI-3' kinase. There are two additional forms of pp60[c-src] generated by alternate splicing in neuronal cells in the CNS. These contain six (exon NI) or 17 (exon NI + exon NII)

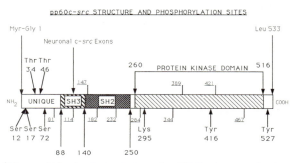

pp60c-*src* STRUCTURE AND PHOSPHORYLATION SITES

Figure. Schematic depiction of pp60[c-src] using numbering for the chicken protein. The residue numbers of the exon boundaries are underlined.

additional residues inserted at position 114 (chicken pp60[c-src] numbering). Sequences within the N-terminal half have been identified which are responsible for localizing pp60[c-src] to different regions of the cytoplasm. Mild proteolytic treatment of pp60[c-src] releases an active ~30 kDa catalytic domain.

Three oncogenically activated forms of the c-src gene have been found in acutely-transforming avian retroviruses (RSV, S1 and S2), which all cause sarcomas[5]. These v-src genes have been activated by different mutations, but in every case the C-terminal tail including the negative regulatory Tyr527 site has been deleted and replaced by a different sequence. Additional point mutations have been incurred in each case, some of which cause oncogenic activation in their own right (Arg95 to Trp; Thr338 to Ile; Glu378 to Gly; Ile441 to Phe). pp60[v-src] associates with the cytoskeleton, and complexes with hsp90 and an unidentified 50 kDa protein; pp60[c-src] has neither of these two properties.

In vertebrates, pp60[c-src] is expressed in most cell types from a single ~4 kB mRNA. pp60[c-src] is localized to cytoplasmic membranes as deduced by cell fractionation and immunofluorescence staining. Typically, pp60[c-src] is less than 0.005% of total cell protein. pp60[c-src] is generally distributed on the cytosolic face of cytoplasmic membranes, but is enriched at focal contacts and in the perinuclear region, with a specific population being associated with the MTOC and centrosomes[6]. pp60[v-src] is distributed in a fashion similar to pp60[c-src] as well as at sites of cell-cell contact, although by virtue of its deregulated PTK activity many subcellular structures are disorganized in cells expressing pp60[v-src]. In the animal, pp60[c-src] is present at the highest level in platelets and neurons, where pp60[c-src] is enriched in axon terminals and growth cones. In situ hybridization for c-src mRNA and immunolocalization[7] of the neuronal form of pp60[c-src] in the brain show high concentrations are present in the hippocampus, mesencephalon, pons, medulla, olfactory bulb and cerebellum.

pp60[c-src] is phosphorylated at multiple sites[1,2]. Some but not all of these sites have been identified, and many lie in the unique N-terminal domain. The known sites are (chicken pp60[c-src] numbering): Ser12 - protein kinase C - effect unknown; Ser17 - cAMP-dependent protein kinase - effect unknown; Thr34, Thr46 (chicken only), and Ser72 - cyclin B/p34[cdc2] - effect unknown, but pp60[c-src] from mitotic cells shows increased activity, possibly due to decreased phosphorylation at Tyr527; Ser72 - phosphorylated only in neuronal cells by an unidentified neuronal cell protein kinase - effect unknown; Tyr416 - autophosphorylation site - increased activity; Tyr527 - unknown protein-tyrosine kinase (possibly CSK) - phosphorylation negatively regulates pp60[c-src] PTK activity. Mutations have been made in all these sites; the only mutation that has a significant effect is the mutation of Tyr527 to Phe, which prevents negative regulation, thus increasing PTK activity about ten-fold and converting pp60[c-src] into a transforming protein[5].

■ PURIFICATION

pp60[c-src] has been partially purified from chick brains[8] and human platelets[9,10] by conventional means, involving several sequential column chromatographies. pp60[v-src] has also been purified conventionally from RSV-induced tumours and RSV-transformed cells. pp60[c-src] has been expressed in E. coli (largely insoluble)[11], in yeast (toxic at high levels)[12], in insect cells using a baculovirus vector (~10% is membrane bound and the rest is soluble)[13], and in chicken and mammalian cells[14] using a variety of vectors. It has been purified from these sources both by mAb 327 affinity chromatography[13], affinity chromatography on poly (Glu/Tyr at 4:1) and conventional column chromatography. The fraction of pp60[c-src] that is soluble and active varies greatly between different sources. In general the baculovirus-expressed protein is most active, having a low level of phosphorylation at Tyr527.

■ ACTIVITIES

Although pp60[c-src] clearly acts as a PTK, its true function is unknown. The gene has been disrupted in the mouse, and a complete absence of pp60[c-src] results in a relatively mild phenotype, namely osteopetrosis (it is possible that the fyn and yes proteins, which are coexpressed with pp60[c-src] in most cells, may substitute)[15]. pp60[c-src] is localized to the inner face of plasma membranes, and it is assumed that like pp56[lck], which interacts with the cytoplasmic tail of **CD4** and **CD8** in T cells, pp60[c-src] may act as the catalytic subunit of cell surface receptor lacking its own catalytic domain. This could involve the unique N-terminal region of pp60[c-src] as is the case for the CD4/CD8 and p56[lck] interaction. The 32 kDa membrane receptor identified for pp60[c-src] could also serve as a signalling molecule[3]. Although src-related genes have been identified in Drosophila and C. elegans, genetic analysis has not yet revealed their function.

pp60[c-src] PTK activity can be measured either in immunoprecipitates or in solution using a variety of peptide and protein substrates. pp60[c-src] will use ATP (K_m ~5 mM), dATP or GTP. PTK activity is optimal at 2-3 mM Mn^{2+} (or 10 mM Mg^{2+}), and at pH 7.0-7.5 [Val[5]]-angiotensin II, acid denatured rabbit muscle enolase and poly (Glu:Tyr) (4:1) are the most commonly used substrates. pp60[c-src] appears to be entirely specific for tyrosine residues in its substrates. The physiological substrates of pp60[c-src] are not known, but several potentially relevant pp60[v-src] substrates have been identified, including PI-3' kinase and GAP, and the cytoskeletal proteins **talin**, **vinculin**, **ezrin**, **connexin** 43, **clathrin** heavy chain and **integrin** β_5.

■ ANTIBODIES

Numerous antibodies have been raised against pp60src. Anti-pp60src antibodies were first detected in the serum of RSV tumour bearing rabbits[16]. Most of these anti-tumour sera recognize both pp60[v-src] and chicken pp60[c-src]; some of these sera crossreact with mammalian pp60[c-src]. A number of polyclonal sera against intact recombinant pp60[c-src], and against separate N- and C-terminal fragments have been raised[11,17]. Likewise many mAbs against pp60[c-src] have been obtained[18,19]. The immunodominant epitopes in pp60[c-src] lie in the N-terminal 120 residues, and most of the mAb-binding sites map to this region. Two main classes of mAb have been identified binding to residues 28-38 (e.g. EB7, EC10, 19A6), and to residues 92-128 (e.g. GD11, EB8, 327, 16E6). One unusual mAb recognizes both N- and C-terminal sequences in a combined epitope (R2D2). Anti-C-terminal peptide sera specific for either pp60[c-src] or pp60[v-src] have been made. Anti-peptide antibodies against other regions have also been made including the N-terminal residues 2-17 (mAb LA022), the neuronal NI exon (anti-peptide and anti-idiotype), the Tyr416 autophosphorylation site region, and residues 498-512 (inhibit PTK activity).

Most of the antibodies can be used for immunoprecipitation, and several different anti-tumour sera, anti-recombinant pp60[c-src] sera, and mAbs have been used for immunofluorescence staining. Only some of the antibodies are suitable for immunoblotting (e.g. mAbs 327, 19A6, 16E6).

■ GENES

Genomic and cDNA clones and sequences for c-src have been obtained from humans[20] (GenBank M16237, M16243-5, K03212-9), mice (GenBank M17031), chickens[21] (GenBank J00344), fish and Xenopus (GenBank J04822). Related genes have been cloned from Drosophila[22] (GenBank M11917), C. elegans, and Hydra, but these genes are equally closely related to all of the src family genes. The c-src gene has 11 coding exons, with two alternately spliced exons being utilized in certain neuronal cells in the CNS, and at least two 5' noncoding exons. The promoter region of c-src has not been fully characterized, and there may be more than one promoter. The human c-src gene maps to chromosome 20q13.3.

■ REFERENCES

1. Cooper, J.A. (1990) In "Peptides and Protein Phosphorylation" (ed. B.E. Kemp, CRC Press) pp 86-113.
2. Parsons, J.T. and Weber, M.J. (1989) Curr. Top. Microbiol. Immunol. 147, 79-127.
3. Resh, M.D. and Ling, H.P. (1990) Nature 346, 84-86.
4. Koch, C.A., Anderson, D., Moran, M.F., Ellis, C. and Pawson, T. (1991) Science 252, 668-674.
5. Hunter, T. (1987) Cell 49, 1-4.
6. David-Pfeuty, T. and Nouvian-Dooghe, Y. (1990) J. Cell Biol. 111, 3097-3116.
7. Sugrue, M.M., Brugge, J.S., Marshak, D.R., Greengard, P. and Gustafson, E.L. (1990) J. Neurosci. 10, 2513-2527.
8. Purchio, A.F., Erikson, E., Collet, M.S. and Erikson, R.L. (1981) Cold Spring Harbor Conferences on Cell Proliferation 8, 1203-1215.
9. Presek, P., Reuter, C., Findik, D. and Bette, P. (1988) Biochim. Biophys. Acta 969, 271-280.
10. Feder, D. and Bishop, J.M. (1990) J. Biol. Chem. 265, 8205-8211.
11. Gilmer, T.M. and Erikson, R.L. (1983) J. Virol. 45, 462-465.
12. Kornbluth, S., Jove, R. and Hanafusa, H. (1987) Proc. Natl. Acad. Sci. (USA) 84, 4455-4459.
13. Morgan, D.O., Kaplan, J.M., Bishop, J.M. and Varmus, H.E. (1991) Meth. Enzymol. 200, 645-660.
14. Johnson, P.J., Coussens, P.M., Danko, A.V. and Shalloway, D. (1985) Mol. Cell. Biol. 5, 1073-1083.
15. Soriano, P., Montgomery, C., Geske, R. and Bradley, A. (1991) Cell 62, 693-702.
16. Brugge, J. and Erikson, R. (1977) Nature 269, 346-348.
17. Resh, M.D. and Erikson, R.L. (1985) J. Virol. 55, 242-245.
18. Lipsich, L., Lewis, A.J. and Brugge, J.S. (1983) J. Virol. 48, 352-360.
19. Parsons, S.J., McCarley, D.J., Ely, C.M, Benjamin, D.C. and Parsons, J.T. (1984) J. Virol. 51, 272-282.
20. Tanaka, A., Gibbs, C.P., Arthur, R.R., Anderson, S.K., Kung, H. and Fujita, D. (1987) Mol. Cell. Biol. 7, 1978-1983.
21. Takeya, T. and Hanafusa, H. (1983) Cell 32, 881-890.
22. Simon, M.A., Drees, B., Kornberg, T. and Bishop, J.M. (1985) Cell 42, 831-840.

■ Tony Hunter and Suzanne Simon:
The Salk Institute, P.O. Box 85800,
San Diego, CA, USA

Talin

Talin[1,2] is a high molecular weight protein concentrated at regions where bundles of actin filaments attach to and transmit tension across the plasma membrane to the extracellular matrix. Talin binds in vitro to the cytoplasmic domains of the integrin family of extracellular matrix receptors, to vinculin and to actin. It has been proposed as one of several proteins linking bundles of actin filaments to the plasma membrane.

Talin has been isolated and studied from chicken gizzard smooth muscle[3] and human blood platelets[4]. The apparent molecular mass on SDS-PAGE is 225-235 kDa with the avian form of the protein having a slightly smaller mass than the mammalian form. The gene for mouse fibroblast talin has been cloned and sequenced[5] and indicates a true molecular mass of 270 kDa. Electron microscopy of rotary shadowed talin molecules reveals an elongated, flexible protein, about 60 nm long, which is more coiled up in low ionic strength conditions[3]. Below 0.7 mg/ml talin exists as a monomer. Above this concentration it begins to self-associate to form dimers[3]. The protein is readily cleaved into two domains by many proteases. On SDS-PAGE these fragments have apparent molecular masses of 47 and 190-200 kDa. Sequence analysis has revealed that the 47 kDa N-terminal domain[5] is highly homologous to domains in **ezrin** and the erythrocyte **protein 4.1**. Like talin, both of these proteins are associated with the plasma membrane and the cytoskeleton. The large C-terminal domain of talin is unusually rich in alanine and is predicted to contain significant α-helix content[5]. It also contains the **vinculin** binding site[6,7].

In many types of cultured cells, talin is concentrated in focal adhesions[8] (focal contacts, adhesion plaques), regions where bundles of **actin** filaments attach to the cytoplasmic face of the plasma membrane and where the external face of the membrane adheres most tightly to the underlying substratum. Talin is also found in ruffling membranes and subjacent to bundles of extracellular matrix on the cell surface[8]. Talin is concentrated *in vivo* at the cytoplasmic face of the plasma membrane where cells interact with and transmit tension to the extracellular matrix; for example, it is enriched at the myotendinous junctions of skeletal muscle[9] and the dense plaques of smooth muscle[10]. Together with other focal adhesion proteins, it is found at the postsynaptic face of some but not all neuromuscular junctions[11,12]. Talin is present but less concentrated at the basal surface of epithelial cells where they adhere to the basement membrane[10]. It is very abundant in platelets (greater than 3% of total platelet protein)[4]. In resting platelets talin is distributed diffusely throughout the cytoplasm, but in response to activation it is redistributed to the cortex underlying the plasma membrane[13]. Although talin is present in cell-extracellular matrix junctions, it is absent from many cell-cell junctions, such as the zonula adherens junctions of epithelial cells[14]. However, microinjection of the C-terminal domain of talin into epithelial cells in culture results in an accumulation of this talin fragment in the zonula adherens junctions of these cells[15]. Talin is concentrated in the cell-cell adhesions made by lymphocytes, for example in cytotoxic lymphocytes where these are adhering to target cells[16]. Notably, these lymphocyte cell-cell adhesions are also sites where **integrins** are concentrated. The amount of talin appears to be low in neuronal cells, although it has been reported in nerve growth cones[17].

Transformation of cells in culture by tumour viruses frequently results in a loss of focal adhesions and an altered distribution of talin[8]. In cells transformed by Rous sarcoma virus (RSV), talin accumulates in podosomes (rosette adhesions) and shows elevated phosphotyrosine content[18,19]. However, this increase in phosphotyrosine also occurs in cells infected with nontransforming mutants of RSV, suggesting that this modification of talin may not contribute to the transformed phenotype[19]. Talin is also a substrate for protein kinase C *in vitro*[20,21], and the level of talin phosphorylation is elevated in cells treated with tumour promoters which stimulate protein kinase C[21,22]. Talin is readily cleaved into its two domains by Ca^{2+}-dependent proteases. The cleavage of talin has been observed during platelet aggregation[23], but its significance has not been resolved.

■ PURIFICATION

Talin has been purified from low ionic strength extracts of chicken gizzard smooth muscle[3] and from Triton X-100 lysates of human platelets[4,7]. The purification involves conventional chromatography in ion exchange resins and gel filtration. The extreme sensitivity of talin to cleavage by Ca^{2+}-dependent proteases can be a problem and the removal of calcium by chelation is recommended, as well as the inclusion of protease inhibitors such as leupeptin.

■ ACTIVITIES

Talin has been shown to bind to integrin *in vitro* with a low affinity ($K_d \sim 10^{-6}M$)[24], to vinculin with a higher affinity ($K_d \sim 10^{-8}M$)[25] and to actin (the affinity has not been determined)[26]. Taken together these *in vitro* binding results suggest a role for talin in linking actin filaments to integrins in the plasma membrane. Introduction of anti-talin antibodies into chicken embryo fibroblasts, spreading on a **fibronectin** substratum, inhibits cell spreading and causes the cells to round up (Nuckolls and Burridge, unpublished observations). This suggests that talin functions in the adhesion of cells to fibronectin.

■ ANTIBODIES

Polyclonal antibodies against chicken gizzard talin[8] cross-react with amphibian talin, but bind only poorly to mammalian talin. Polyclonal antibodies against human platelet talin[22] have been generated, but the human platelet protein is a very poor immunogen. Monoclonal antibodies are available against chicken gizzard talin, one of which (8d4) crossreacts with mammalian talin[27].

■ GENES

Mouse fibroblast talin has been cloned and sequenced[5] (GenBank/EMBL/DDBJ X56123).

■ REFERENCES

1. Burridge, K. and Molony, L. (1990) Adv. in Cell Biol. 3, 95-109.
2. Beckerle, M.C. and Yeh, R.K. (1990) Cell Motil. Cytoskel. 16, 7-13.
3. Molony, L., McCaslin, D., Abernethy, J., Paschal, B. and Burridge, K. (1987) J. Biol. Chem. 262, 7790-7795.
4. Collier, N.C. and Wang, K. (1982) J. Biol. Chem. 257, 6937-6943.
5. Rees, D.J.G., Ades, S.E., Singer, S.J. and Hynes, R.O. (1990) Nature 347, 685-689.
6. O'Halloran, T. and Burridge, K. (1986) Biochim. Biophys. Acta 869, 337-349.
7. Turner, C.E. and Burridge, K. (1989) Eur. J. Cell Biol. 49, 202-206.
8. Burridge, K. and Connell, L. (1983) J. Cell Biol. 97, 359-367.
9. Tidball, J.G., O'Halloran, T. and Burridge, K. (1986) J. Cell Biol. 103, 1465-1472.
10. Drenckhahn, D., Beckerle, M., Burridge, K. and Otto, J. (1988) Eur. J. Cell Biol. 46, 513-522.
11. Sealock, R., Paschal, B., Beckerle, M. and Burridge, K. (1986) Exp. Cell Res. 163, 143-150.
12. Rochlin, M.W., Chen, Q., Tobler, M., Turner, C.E., Burridge, K. and Peng, H.B. (1989) J. Cell Sci. 92, 461-472.
13. Beckerle, M.C., Miller, D.E., Bertagnolli, M.E. and Locke, S.J. (1989) J. Cell Biol. 109, 3333-3346.
14. Geiger, B., Volk, T. and Volberg, T. (1985) J. Cell Biol. 101, 1523-1531.
15. Nuckolls, G.H., Turner, C.E. and Burridge, K. (1990) J. Cell Biol. 110, 1635-1644.
16. Kupfer, A., Singer, S.J. and Dennert, G. (1986) J. Exp. Med. 163, 489-498.
17. Letourneau, P.C. and Shattuck, T.A. (1989) Development 105, 505-519.
18. Pasquale, E.B., Maher, P.A. and Singer, S.J. (1986) Proc. Natl. Acad. Sci. (USA) 83, 5507-5511.
19. DeClue, J.E. and Martin, G.S. (1987) Mol. Cell Biol. 7, 371-378.
20. Litchfield, D.W. and Ball, E.H. (1986) Biochem. Biophys. Res. Comm. 134, 1276-1283.
21. Beckerle, M.C. (1990) Cell Reg. 1, 227-236.
22. Turner, C.E., Pavalko, F.M. and Burridge, K. (1989) J. Biol. Chem. 264, 11938-11944.
23. Fox, J.E.B. and Goll, D.E. (1985) J. Biol. Chem. 260, 1060-1066.
24. Horwitz, A., Duggan, K., Buck, C., Beckerle, M.C. and Burridge, K. (1986) Nature 320, 531-533.
25. Burridge, K. and Mangeat, P. (1984) Nature 308, 744-746.
26. Muguruma, M., Matsumura, S. and Fukazawa, T. (1990) Biochem. Biophys. Res. Comm. 171, 1217-1223.
27. Otey, C., Griffiths, W. and Burridge, K. (1990) Hybridoma 9, 57-62.

■ Keith Burridge:
Department of Cell Biology and Anatomy, University of North Carolina at Chapel Hill, Chapel Hill, NC 27599, USA

Tensin

Tensin, a 170 kDa polypeptide purified from chicken gizzard, has been found to interact in vitro with the barbed ends of actin filaments. Immunofluorescence experiments have shown that material crossreactive with tensin is located at such places as adhesion plaques of fibroblasts, Z-lines of skeletal and cardiac muscle, and dense plaques of smooth muscle. These observations suggest that tensin may play a specific role in transmission of contractile forces in different cells and tissues by linking the ends of actin filaments to other cellular structures.

The discovery of tensin stems from studies on the interaction of F-**actin** with **vinculin**, a protein found at adhesion plaques of fibroblasts. In early experiments, vinculin preparations from chicken gizzard inhibited actin polymerization in a manner suggesting a direct interaction of the protein with the barbed ends of actin filaments[1]. Subsequent studies indicated that this phenomenon is attributable to the activity of a heterogenous group of peptides (designated HA1; 20-45 kDa) contaminating the vinculin preparations[2]. Antibodies raised against HA1 were shown to crossreact with high molecular weight bands (150-200 kDa) in immunoblots of proteins from different tissues, suggesting that HA1 peptides are proteolytic fragments of a larger protein[3]. A 170 kDa polypeptide which crossreacts with anti-HA1 antibodies has since been purified from chicken gizzard[4]. This polypeptide was named "tensin" on the basis of its putative role in maintaining tension in actin filaments by linking them to other structures[4].

Substoichiometric concentrations of purified tensin have been found to inhibit actin polymerization at the barbed ends[5]. This effect can be explained by the specific

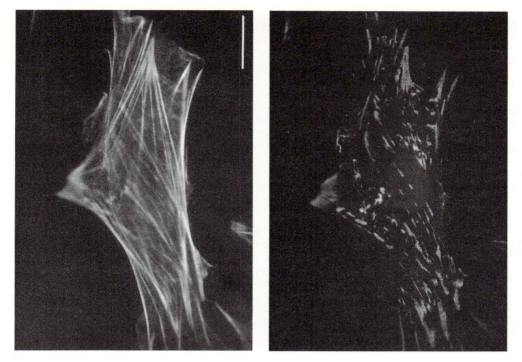

Figure 1. (Left) Fluorescence staining of F-actin in chicken embryo fibroblast using fluorescein-labelled phalloidin. Bar 20 μm. (Right) Indirect immunofluorescence staining of the same cell using a monoclonal antibody to tensin followed by rhodamine-labelled goat anti-mouse antibody.

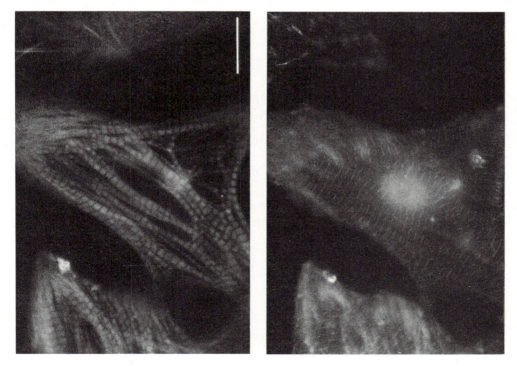

Figure 2. (Left) Fluorescence staining of F-actin in cultured chicken cardiac myocytes (bottom and centre) using fluorescein-labelled phalloidin. Bar 20 μm. (Right) Indirect immunofluorescence staining of the same cells with a monoclonal antibody against tensin followed with rhodamine-labelled goat anti-mouse antibody. (Micrographs provided by Dr. D.C. Lin, Johns Hopkins University.)

binding of the protein to the barbed ends of the filament, a proposition substantiated by the cytochalasin-sensitive cosedimentation of tensin with F-actin[6]. Tensin has also been shown to bind vinculin on nitrocellulose blots[4]; the specificity of this interaction requires further investigation.

Immunofluorescence studies with anti-HA1 antibodies and anti-tensin antibodies have demonstrated the presence of crossreactive material at different locations where the ends of actin filaments associate with other structures[3,7,9]: adhesion plaques and extracellular matrix contacts of fibroblasts (Figure 1), cell-cell contacts of epithelial cells, Z-lines of skeletal and cardiac muscle (Figure 2), intercalated discs of cardiac muscle, dense plaques of smooth muscle, and costameres of skeletal and cardiac muscle. In double immunofluorescence experiments on well spread chicken embryo fibroblasts, tensin is often found at the same locations as vinculin, **talin**, and *integrin*[7-9]. A model based on interactions of these proteins for the transmembrane linkage of actin filaments to extracellular matrix proteins has been proposed[9,10].

Another 170 kDa polypeptide with a more acidic pI (7.3) and a larger Stokes radius (79Å) than those of tensin (pI = 8.0; 68Å) has also been purified from chicken gizzard[6]. This polypeptide has a peptide map with a significant amount of similarity to that of tensin. However, the polypeptide does not appear to interact with F-actin in polymerization and cosedimentation assays[6]. Gizzard extracts also contain a 210 kDa polypeptide which cross-reacts with anti-tensin antibodies and has a peptide map very similar to that of tensin[11]. The degree to which tensin and these two polypeptides are related genetically, structurally and functionally remains to be defined.

cDNA clones have been isolated from a chicken embryo fibroblast library[12] and an adult chicken heart library[13]. Overlapping cDNA clones covering enough nucleotides to encode a polypeptide of about 200 kDa have been obtained from the chicken heart library and some of the nucleotide sequences are very similar to that of a clone from the chicken embryo fibroblast library. In each case, the deduced amino acid sequence of tensin contains a region highly homologous with the "src homology 2" (SH2) domain found in many nonreceptor protein tyrosine kinases (see **pp60**[c-src]), the *crk* oncogene product, phospholipase C-γ, and the *ras* GTPase activating protein[12,13]. Moreover, tensin itself was found to contain phosphotyrosine, the level of which increases with transformation by pp60[v-src12]. Whether these properties of tensin reflect involvement of the protein in cytoskeletal rearrangements in oncogenic transformation or a role in signal transduction is an intriguing question which warrants further investigation.

■ PURIFICATION

Tensin can be purified from a high salt extract of chicken gizzard with the use of a series of different chromatographic columns[4,11].

■ ACTIVITIES

In the pyrene-actin polymerization assay, substoichiometric concentrations of tensin inhibit elongation of actin filaments at the barbed ends[5]. The protein cosediments with F-actin; this interaction is inhibited by cytochalasin D, presumably by competition for binding sites at the barbed ends of the filaments[6].

■ ANTIBODIES

A preparation of rabbit polyclonal antibodies against HA1 was used in the earlier studies[3]. More recently, preparations of rabbit polyclonal antibodies[7-9] and monoclonal antibodies (Figures 1 and 2) against purified tensin have been made. These preparations have been tested in immunofluorescence staining of fibroblasts and muscle cells and were found to give essentially the same results as anti-HA1 antibodies.

■ GENES

Overlapping cDNA clones encoding essentially the entire sequence of tensin have been isolated from an adult chicken heart library[13], and a 3.5 kB clone has been isolated from a chicken embryo fibroblast library[12].

■ REFERENCES

1. Wilkins, J.A. and Lin, S. (1982) Cell 28, 83-90.
2. Wilkins, J.A. and Lin, S. (1986) J. Cell Biol. 102, 1085-1092.
3. Wilkins, J.A., Risinger, M.A. and Lin, S. (1986) J. Cell Biol. 103, 1483-1494.
4. Wilkins, J.A., Risinger, M.A. and Lin, S. (1987) J. Cell Biol. 105, 130a.
5. Butler, J.A. and Lin, S. (1989) J. Cell Biol. 109, 172a.
6. Butler, J.A. and Lin, S. (1990) Biophys. J. 57, 214a.
7. Risinger, M.A., Wilkins, J.A. and Lin, S. (1987) J. Cell Biol. 105, 130a.
8. Risinger, M.A. and Lin, S. (1988) J. Cell Biol. 107, 256a.
9. Lin, S., Risinger, M.A. and Butler, J.A. (1989) In Cytoskeletal and Extracellular Matrix Proteins, Springer Series in Biophysics, Vol. 3, U. Aebi and J. Engel, eds., Springer-Verlag, London, pp. 341-344.
10. Butler, J.A. and Lin, S. (1990) J. Cell. Biochem. Suppl. 14A, 234.
11. Butler, J.A. and Lin, S. manuscript in preparation.
12. Davis, S., Lu, M.L., Lo, S.H., Lin, S., Butler, J.A., Druker, B.J., Roberts, T.M., An, Q. and Chen, L.B. (1991) Science 252, 712-715.
13. Chuang, J.Z., Lin, D.C., Carter, W.L., Cunningham, M.E., Butler, J.A. and Lin, S. (1991) J. Cell Biol. 115, 166a.

■ *Shin Lin:*
Department of Biophysics,
Johns Hopkins University,
Baltimore, MD 21218, USA

Vinculin

Vinculin is a 117 kDa microfilament associated protein located at the cytoplasmic aspects of adherens type junctions (AJ). These adhesion sites are specialized membrane domains through which cells form tight and stable adhesions to either the surrounding extracellular matrix networks or directly to the membrane of other cells. Based mainly on its ubiquitous localization in AJ, it was proposed that vinculin plays an important role in linking actin to the junctional membrane and thus promotes the establishment and stabilization of cell adhesions.

Vinculin was initially isolated from chicken gizzard extract as a by-product during the purification of α-**actinin**[1]. Immunolabelling of cultured fibroblasts and a variety of tissues indicated that the protein is ubiquitously associated with AJ, both at cell-cell and cell-extracellular matrix interaction sites[2,3] and may be used as a hallmark for these junctions (see Figure). It had further been established that vinculin is present in a very broad spectrum of cells including lymphoid cells and platelets which do not form stable adhesions.

Among the first studies on vinculin was the analysis of

Figure. Immunofluorescence labelling for vinculin of cultured chicken lens cells displays both sites of cell-extracellular matrix adhesions (mostly focal contacts) and cell-cell adherens. Magnification, x890.

its distribution in cells following microinjection with fluorescently tagged protein[4]. This, and later microinjection studies, established the fact that vinculin maintains two cellular pools including a diffusible cytoplasmic pool and a fraction which is associated with the junctional membrane[5]. These two pools exhibit a dynamic equilibrium between them which may affect the size and stability of AJ in cells.

Attempts to elucidate the specific role of vinculin in AJ formation involved binding and cosedimentation assays of purified vinculin with various AJ molecules. Among the vinculin binding proteins detected in these experiments are **talin**[6-8], α-actinin[9], **paxillin**[10] and possibly vinculin itself[7,8]. It is noteworthy however that the *in vivo* interactions of vinculin are more complex in as much as they display a strict spatial selectivity. In cell-extracellular matrix interaction sites, for example, both talin and vinculin are present, while the former is absent from cell-cell AJ. The presence of multiple binding sites on vinculin, suggesting that it might interact, in adhesion sites, with several junctional molecules simultaneously was demonstrated by cDNA transfection experiments in which different nonoverlapping segments (C- or N-terminal) of the molecule were shown to independently associate *in vivo* with adhesion sites[11].

Electrophoretic analysis indicated the presence of multiple isoelectrophoretic forms of vinculin. By two-dimensional electrophoretic analysis at least three isoforms of vinculin were detected (α, β and γ). The latter two are specifically expressed in muscle cells and the α-form appears to be phosphorylated[12]. In addition, a higher molecular weight variant of vinculin was detected in muscle and denoted meta-vinculin[13-15]. Recent sequence data indicated that this molecule contains an extra segment of 68 amino acids located near the proline-rich region, between the N-terminal globular head domain and the C-terminal tail[16]. The functional significance of this isoform heterogeneity is still unclear.

Sequencing of vinculin cDNA from various sources indicated that the protein has a molecular weight of ~117.000 with a large N-terminal head domain which contains the talin binding site and three ~110 amino acids repeats. This region is separated from the C-terminal domain by several stretches of prolines in which the major proteolytic cleavage sites are located[17-19].

Vinculin was shown to be post-translationally modified by phosphorylation on either serine (by protein kinase C[20]) or tyrosine (by **pp60**[src])[21] residues. Vinculin was also shown to interact with lipids[22,23], though the molecular basis for this interaction and its physiological significance are still unclear.

■ PURIFICATION

The primary source of vinculin was smooth muscle from which it was purified largely as described by Geiger[1] or by Burridge and Feramisco[4]. In addition, vinculin was purified, following similar procedures from other sources such as human smooth muscle and pig smooth muscle.

■ ACTIVITIES

Vinculin was shown to bind several AJ proteins including talin[6-8], α-actinin[9], paxillin[10], and itself[7,8]. The latter was also supported by electron microscopy which indicated that oligomers can be frequently detected in concentrated vinculin solutions[24,25]. In addition to its binding to adhesion-related molecules vinculin was shown to directly interact with membrane lipids.

■ ANTIBODIES

Antibodies originally used for vinculin localization were raised in rabbits and guinea pigs[1,2]. In recent years many monoclonal antibodies were produced which react with vinculins from a large variety of vertebrates from fish to man. Several of these antibodies are now commercially available.

■ GENES

Vinculins from different sources have been cloned and sequenced including chicken[17-19], human[26] and *Caenorhabditis elegans*[27]. Partial sequences of mouse vinculin were presented by Ben-Ze'ev et al.[28]. The specific extra-sequences found in meta-vinculin were derived from protein and DNA sequence analysis[16,29].

■ REFERENCES

1. Geiger, B. (1979) Cell 18, 193-205.
2. Geiger, B., Tokuyasu, K.T., Dutton, A.H. and Singer S.J. (1980) Proc. Natl. Acad. Sci. (USA) 77, 4127-4131.
3. Otto, J.J. (1990) Cell Mot. Cytoskeleton 16, 1-6.
4. Burridge, K. and Feramisco, J.R. (1980) Cell 19, 587-595.
5. Kreis, T.E., Avnur, Z., Schlessinger, J. and Geiger, B. (1985) Cold Spring Harbor Symp. on "Molecular Biology of the Cytoskeleton" (ed. G. Borisy, D. Cleveland and D. Murphy) pp 45-57.
6. Burridge, K. and Mangeat, P. (1984) Nature 308, 744-746.
7. Belkin, A.M. and Koteliansky, V.E. (1987) FEBS Lett. 220, 291-294.
8. Otto, J.J. (1983) J. Cell Biol. 97, 1283-1287.
9. Wachsstock, D.H., Wilkins, J.A. and Lin, S. (1987) Biochem. Biophys. Res. Commun. 146, 554-560.
10. Turner, C.E., Glenney Jr., J.R. and Burridge, K. (1990) J. Cell Biol. 111, 1059-1068.
11. Bendori, R., Salomon, D. and Geiger. B. (1989) J. Cell Biol. 108, 2383-2394.
12. Geiger, B. (1982) J. Mol. Biol. 159, 685-701.
13. Siliciano, J.D. and Craig, S.W. (1987) J. Cell Biol. 104, 473-482.
14. Belkin, A.M., Ornatsky, O.I., Kabakov, A.E., Glukhova, M.A. and Koteliansky, V.E. (1988) J. Biol. Chem. 263, 6631-6635.
15. Gimona, M., Small, J.V., Moeremans, M., Van Damme, J., Puype, M. and Vandekerckhove, J. (1988b) Protoplasma 145, 133-140.
16. Gimona, M., Small, J.V., Moeremans, M., Van Damme, J., Puype, M. and Vandekerckhove, J. (1988a) EMBO J. 7, 2329-2334.
17. Price, G.J., Jones, P., Davison, M.D., Patel, B., Eperon, I.C. and Critchley, D.R. (1987) Biochem. J. 245, 595-603.
18. Price, G.J., Jones, P., Davison, M.D., Patel, B., Bendori, R., Geiger, B. and Critchley, D.R. (1989) Biochem. J. 259, 453-461.
19. Coutu, M.D. and Craig, S.W. (1988) Proc. Natl. Acad. Sci. (USA) 85, 8535-8539.
20. Werth, D.K., Niedel, J.E. and Pastan, I. (1983) J. Biol. Chem. 258, 11423-11426.
21. Sefton, B., Hunter, T., Ball, E. and Singer, S.J. (1981) Cell 24, 165-174.
22. Niggli, V., Dimitrov, D.P., Brunner, J. and Burger, M.M. (1986) J. Biol. Chem. 261, 6912-6918.
23. Burn, P. and Burger, M.M. (1987) Science 235, 476-479.
24. Miliam, L.M. (1985) J. Mol. Biol. 184, 543-545.
25. Molony, L. and Burridge, K. (1985) J. Cell Biochem. 29, 31-36.
26. Weller, P.A., Ogryzko, E.P., Corben, E.B., Zhidkova, N.I., Patel, B., Price, G.J., Spurr, N.K., Koteliansky, V.E. and Critchley, D.R. (1990) Proc. Natl. Acad. Sci. (USA) 87, 5667-5671.
27. Barstead, R.J. and Waterston, R.H. (1989) J. Biol. Chem. 264, 10177-10185.
28. Ben-Ze'ev, A., Reiss, R., Bendori, R. and Gorodecki, B. (1990) Cell Regul. 1, 621-636.
29. Koteliansky, V.E., Ogryzko, E.P., Zhidkova, N.I., Weller, P.A., Critchley, D.R., Vancompernolle, K., Vanderkerckhove, J., Strasser, P., Way, M., Gimona, M. and Small, J.V. (1992) Eur. J. Biochem. 204, 676-772.

■ *Benjamin Geiger:*
Department of Chemical Immunology,
The Weizmann Institute of Science,
Rehovot, Israel

ZO-1

ZO-1[1] is a high molecular weight, peripheral membrane protein, intimately morphologically associated with the zonula occludens, or tight junction, in a wide variety of vertebrate epithelia. The biological functions of ZO-1 are unknown, but the protein shows a requirement of cell-cell interaction[2,3] involving E-cadherin (L-CAM, uvomorulin[4]) in order to localize at the cell surface concomitant with junction formation.

ZO-1 is identified by three monoclonal antibodies[1,5] (R26.4C, R40.40D3, and R40.76) raised against a junctional complex enriched subfraction[6] from murine liver. Immunolocalization studies show that the protein is localized on the cytoplasmic surfaces[1] corresponding to the points of membrane fusion[7,8] of this intercellular junction (Figures 1, 2). **Cingulin**, a polypeptide distinct from ZO-1, is also localized close to the tight junction[9]. In the glomerulus of the kidney, ZO-1 has been shown to be localized to the cytoplasmic surfaces of the membranes beneath the slit diaphragms joining the pedicels of the

podocytes comprising the visceral layer of Bowman's capsule, suggesting an interesting modification of the *zonula occludens* in the embryogenesis of these filtration structures[10].

The ZO-1 polypeptide has a molecular mass of 225 kDa in mouse tissues and 210 kDa in cultured Madin-Darby canine kidney (MDCK) epithelial cells[5]. ZO-1 is a peripheral membrane protein and can be extracted from plasma membranes with either 6 M urea or high pH. The molecule is partially solubilized by 0.3 M KCl, and relatively insoluble in the nonionic detergents Triton X-100 and

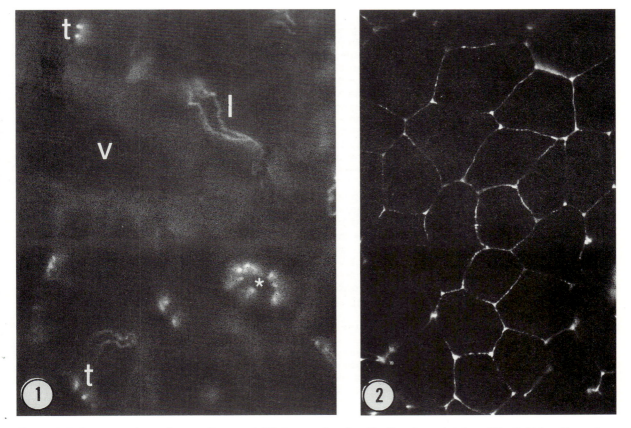

Figure 1. In frozen sections of mouse liver, anti-ZO-1 monoclonal antibodies show staining of the tight junctions at the bile canaliculi as two dots in transverse section (t) and linear threads in longitudinal section (I), consistent with the zonular nature of the structure. At a triad, a branch of the portal vein (v) shows no detectable staining, while the junctional complexes between the bile duct cells (*)stain intensely. Figure 2. In monolayer culture, the anti-ZO-1 antibodies stain the sites of cell-cell interaction between adjacent Madin Darby canine kidney cells.

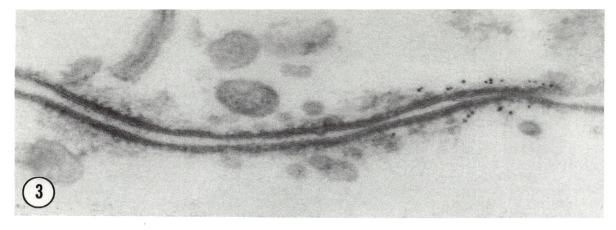

Figure 3. Anti-ZO-1 immunocytochemistry of subfractionated mouse liver membranes reveals staining with colloidal gold-labelled secondary antisera exclusively on the cytoplasmic surfaces of the tight junction.

octyl-β-D-glucopyranoside. Electrophoretically purified, [^{35}S]methionine-labelled MDCK ZO-1 has an $S_{20,w}$ of 5.3 and a Stokes radius of 8.6 nm, suggesting an asymmetric monomeric molecule. The mouse liver corresponding values are 9 nm and 6 $S_{20,w}$. Metabolic labelling of MDCK cells with [^{32}P]orthophosphate reveals that ZO-1 is a phosphoserine containing phosphoprotein; comparisons of MDCK cell subclones with different transepithelial resistances and identical tight junction structure and ZO-1 content[11] show a possible correlation of the state of ZO-1 phosphorylation with transepithelial resistance[12]. Scatchard analysis of competitive and saturable binding of two of the monoclonal antibodies indicates a copy number of 3 x 10^4 molecules of ZO-1/cell, which agrees within a factor of two with estimates of the number of freeze fracture particles/μm of junctional fibre length[5].

The pooled monoclonal antibodies were used to screen a rat kidney cDNA library in the λ_{gt11} expression vector[3], resulting in the isolation of a partial cDNA. Northern analysis reveals a 7.5 kB message in both rat tissues and in the Caco-2 human intestinal epithelial cell line. Preliminary studies[13] reveal alternate splicing of ZO-1 mRNA generates two isoforms differing by 240 bp, both of which are coexpressed in the T84 human enterocyte cell line.

■ PURIFICATION

ZO-1 is purified by electroelution from SDS-polyacrylamide gels or by immunoprecipitation following extraction of MDCK cells with 6 M urea and 0.1% Triton X-100 using monoclonal antibodies covalently coupled to sepharose beads[5].

■ ACTIVITIES

No known enzymatic activities are associated with ZO-1.

■ ANTIBODIES

Monoclonal antibodies are available to different ZO-1 epitopes[1,5]. An anti-fusion protein polyclonal antibody has been developed which shows identical staining properties to the monoclonal antisera[3].

■ GENES

Only a partial cDNA for ZO-1 has been reported[3]. A full nucleotide sequence is not yet available.

■ REFERENCES

1. Stevenson, B.R., Siliciano, J.D., Mooseker, M.S. and Goodenough, D.A. (1986) J. Cell Biol. 103, 755-766.
2. Siliciano, J.D. and Goodenough, D.A. (1988) J. Cell Biol. 107, 2389-2399.
3. Anderson, J.M., Van Itallie, C.M., Peterson, M.D., Stevenson, B.R., Carew, E.A. and Mooseker, M.S. (1989) J. Cell Biol. 109, 1047-1056.
4. Gumbiner, B. and Simons, K. (1986) J. Cell Biol. 102, 457-468.
5. Anderson, J.M., Stevenson, B.R., Jesaitis, L.A., Goodenough, D.A. and Mooseker, M.S. (1988) J. Cell Biol. 106, 1141-1149.
6. Stevenson, B.R. and Goodenough, D.A. (1984) J. Cell Biol. 98, 1209-1221.
7. Farquhar, M.G. and Palade, G.E. (1963) J. Cell Biol. 17, 375-412.
8. Goodenough, D.A. and Revel, J.-P. (1970) J. Cell Biol. 45, 272-290.
9. Stevenson, B.R., Heintzelman, M.B., Anderson, J.M., Citi, S. and Mooseker, M.S. (1989) Am. J. Physiol. 257, C621-C628.
10. Schnabel, E., Anderson, J.M. and Farquhar, M.G. (1990) J. Cell Biol. 111, 1255-1263.
11. Stevenson, B.R., Anderson, J.M., Goodenough, D.A. and Mooseker, M.S. (1988) J. Cell Biol. 107, 2401-2408.
12. Stevenson, B.R., Anderson, J.M., Braun, I.D. and Mooseker, M.S. (1989) Biochem. J. 263, 597-599.

13. Willott, E., Van Itallie, C.M., Kasraian, P., Jameson, B., Heintzelman, M. and Anderson, J.M. (1990) J. Cell Biol. 111, 409a.

■ Daniel A. Goodenough:
*Department of Anatomy and Cellular Biology,
Harvard Medical School,
220 Longwood Avenue,
Boston, MA 02115, USA*

Zyxin

Zyxin[1] is an 82 kDa adherens junction component that was originally identified[2] by characterization of a nonimmune rabbit serum that stained focal contacts by indirect immunofluorescence. It has been identified in a wide variety of avian tissues and is most prominent in smooth muscle and fibroblasts. Zyxin is present in low abundance relative to the other adhesion plaque components vinculin, talin and α-actinin. Its subcellular localization at a number of different sites of actin-membrane interaction suggests that zyxin may play some role in the regulation or organization of these membrane-cytoskeletal attachments.

Zyxin is a cytoplasmic protein that is found in a number of distinct types of adherens junctions including the adhesion plaques or focal contacts of cultured cells, the dense plaques of smooth muscle cells, and the apical junctional

A

B

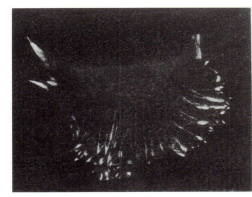

Figure 1. (A) A chicken embryo fibroblast viewed by interference reflection microscopy to visualize the adhesion plaques which appear black by this approach. (B) Indirect immunofluorescence localization of zyxin.

complex of pigmented retinal epithelial cells. Because a comprehensive immunocytochemical analysis of the subcellular distribution of zyxin in different cell types has not yet been completed, it is not clear whether zyxin is a constitutive component of both cell-substratum and cell-cell adherens junctions. The distribution of zyxin in cultured cells (Figure 1) is particularly interesting since the protein is found at the adhesion plaques, sites of very close cell-substratum contact defined by interference reflection microscopy, as well as along the **actin** filament bundles (stress-fibres) near where they terminate at the adhesion plaques. Zyxin colocalizes with a subset of the actin filament bundles, but is notably absent from arcs, microspikes, and even the terminal portions of some stress-fibres; the molecular basis for this heterogeneous distribution of zyxin along actin filament arrays remains to be determined.

Zyxin has been isolated from avian smooth muscle and many of its biochemical and biophysical properties have been determined. On SDS-PAGE, zyxin migrates with an apparent molecular mass of 82 kDa (Figure 2). The protein has a Stokes radius of 5.6 nm and a sedimentation coefficient of 3.0 S. The hydrodynamic properties of zyxin suggest that it is a monomer with a calculated molecular weight of 69 kDa. The derived frictional ratio (f/f_0) of 2.1 indicates that the protein is asymmetric in shape. It has an average pI of 6.9, but displays a number of isoelectric variants in the range of 6.4-7.2. The protein is phosphorylated on multiple sites *in vivo*; this posttranslational modification may contribute to its heterogeneity in isoelectric focusing gels. The protein fractionates with the aqueous phase in Triton X-114 phase partition experiments and does not appear to be glycosylated. The properties of zyxin as well as its primary amino acid sequence (see below) distinguish it from all other previously described adherens junction components. The term "zyxin" is derived from a word root meaning "a joining" and refers to the fact that this 82 kDa protein is localized extensively at areas where actin filaments are joined to the plasma membrane at sites of cell adhesion.

Zyxin was originally identified and characterized in avian cells because the original non-immune serum (from rabbit F396) was, for some unknown reason, chicken specific. Because of the low titer and lack of crossreactivity of the F396 serum, a comprehensive survey of the presence of zyxin in other organisms has not yet been completed. However, preliminary indirect immunofluorescence studies using a recently developed zyxin specific polyclonal antiserum reveal the presence of an immunoreactive adhesion plaque component in both bovine and simian cells.

■ PURIFICATION

Zyxin can be purified from either fresh or frozen chicken gizzard, however yields are significantly increased (up to five-fold) when fresh tissue is used. The protein is extracted from smooth muscle homogenates under low

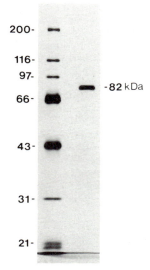

Figure 2. Silver stained SDS-PAGE gel showing purified zyxin from avian smooth muscle.

ionic strength, alkaline pH conditions. Subsequent ammonium sulphate fractionation, followed by chromatography on DEAE-Cellulose, Phenyl-Sepharose CL-4B and Hydroxylapatite results in purified protein as illustrated in the silver-stained gel shown in Figure 2. This procedure[1] yields approximately 20 µg of purified zyxin from 300 g of fresh smooth muscle.

■ ACTIVITIES

Zyxin has been shown to interact with α-**actinin** in vitro[3]. The interaction between zyxin and α-actinin has been demonstrated by solution and solid phase binding assays. The association between zyxin and α-actinin is direct, specific, saturable and of moderate affinity ($K_d=1.1\mu M$). Zyxin and α-actinin are extensively colocalized in vivo raising the possibility that zyxin is targeted to particular regions in the cell at least to some extent by virtue of its association with α-actinin.

■ ANTIBODIES

The original rabbit antiserum is not an efficient reagent for most purposes; it appears to recognize only an avian antigen and has a fairly low titer. One additional anti-zyxin antibody having broader crossreactivity has been described[1], but because it is a murine polyclonal antiserum, it is in limited supply.

■ GENES

cDNAs encoding zyxin have recently been isolated and characterized (Sadler and Beckerle, manuscript in preparation). Sequence analysis of zyxin cDNA clones reveals that zyxin is a previously uncharacterized protein with interesting structural features. Zyxin exhibits an unusually high proline content, with one 200 amino acid region of the protein consisting of >30% proline. In addition, three contiguous LIM-repeats[4,5] are found at the C-terminus of the protein; LIM repeats are found in certain transcription factors as well as in proteins thought to be important in cell differentiation. LIM repeats have been postulated to be Zinc-finger domains[6] and we have found that purified zyxin is indeed a Zinc-binding metalloprotein[7].

■ REFERENCES

1. Crawford, A.W. and Beckerle, M.C. (1991) J. Biol. Chem. 266, 5847-5853.
2. Beckerle, M.C. (1986) J. Cell Biol. 103, 1679-1687.
3. Crawford, A.W., Michelsen, J.W. and Beckerle, M.C. (1992) J. Cell Biol. 116, 1381-1393.
4. Freyd, G., Kim, S.W. and Horvitz, H.R. (1990) Nature 344, 876-879.
5. Karlsson, O., Thor, S., Norberg, T., Ohlsson, H. and Edlund, T. (1990) Nature 344, 879-882.
6. Liebhaber, S.A., Emery, J.G., Urbanek, M., Wang, X. and Cooke, N.E. (1990) Nucleic Acids Res. 18, 3871-3879.
7. Sadler-Riggleman, I. and Beckerle, M.C. (1991) J. Cell Biol. 115, 394a.

■ Aaron W. Crawford and Mary C. Beckerle:
Department of Biology,
University of Utah,
Salt Lake City, UT 84112, USA

Organelle Membrane Associated Structural Proteins

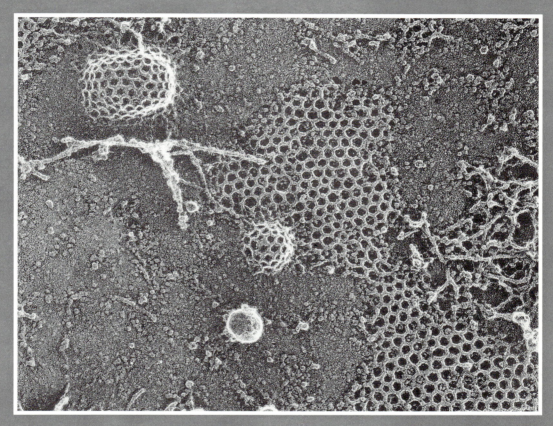

Polygonal clathrin lattices are seen on the inner surface of the plasma membrane of a cultured cell, exposed by attaching the cell to glass and then breaking it open by sonication to wash-out its cyoplasmic contents. In preparation for platinum replication and electron microscopy, the cell membrane was aldehyde-fixed and was then freeze-dried while still attached to glass. Captured thusly, the clathrin lattices of the cell appear in many differing forms and degrees of curvature, indicating different stages in the formation of clathrin-coated vesicles, the process by which membrane surface receptors are sequestered and internalized.

(Courtesy of Dr John Heuser, Washington University, St Louis).

Organelle Membrane Associated Structural Proteins

Intracellular membrane traffic involves rounds of sorting of cargo into specific domains of an organelle, budding of carrier vesicles and directed movement of these transport intermediates to their corresponding acceptor membrane where fusion occurs. Cytoskeletal and organelle membrane associated structural components regulate these processes. These include membrane bound coat proteins, proteins linking organelles to cytoskeletal filaments and organelle associated scaffold proteins. Although most of these processes and structures are yet poorly understood, a growing number of proteins involved have been identified and their functions are being characterized.

Interactions of cytoskeletal structures with membranes are manyfold and serve various functions. The best characterized examples include the anchoring of cytoskeletal filaments to specific domains at the plasma membrane, and an increasing number of "anchor proteins" mediating such interactions are known (see section on **Cytoskeletal Anchor Proteins**). These proteins play important roles in a variety of cell-cell interactions including junction formation, cell adhesion and signal transduction (for further details see the accompanying book ***"GUIDEBOOK TO THE EXTRACELLULAR MATRIX AND CELL ADHESION PROTEINS"***). In addition to these "classical" membrane-cytoskeleton interactions, however, other membrane-associated structural components have been characterized or postulated. These proteins are usually (supposedly) associated with membranes of cytoplasmic organelles and mediate interactions with cytoskeletal filaments (e.g. microtubules), play structural roles, or regulate processes of intracellular membrane traffic. Rather few of these proteins have been characterized so far, and relatively little is known about their functions.

The most prominent cytoplasmic organelles include those of the exocytic pathway, endocytic organelles, as well as several others like the nucleus and mitochondria. They have distinct morphological features (e.g. the stacked cisternae of the Golgi complex) and unique protein compositions which are maintained despite extensive membrane traffic. Organelle membrane associated scaffolds may be involved in these functions. Furthermore, the location of some of these organelles in the cytoplasm is well defined and maintained during various forms of cellular motility. Increasing evidence suggests that the cytoskeleton, particularly microtubules, are responsible for the specific positioning of these organelles[1]; in fact, specific linker proteins may associate cytoplasmic organelles with cytoskeletal structures and anchor them in defined regions of the cytoplasm.

Another group of proteins regulates membrane traffic via vesicular carriers in between the different membrane-bounded cytoplasmic compartments. These proteins are usually components of coats associated with the vesicular carriers and the donor [and (?) acceptor] compartments. In principle, these various components can thus be grouped into the following three classes of organelle membrane associated structural proteins: (1) the relatively well characterized coat proteins, (2) linker proteins, which are now being identified and described and (3) the putative organelle associated scaffold proteins.

■ COAT PROTEINS

Vesicular carriers are the transport intermediates in the net unidirectional transport of cargo from one membrane compartment to another. The best characterized of these transport vesicles are coated on their cytoplasmic surfaces with proteins which are recruited from a cytosolic pool. The major functions of these coat proteins appear to be the sorting of the specific vesicular cargo and the regulation of vesicle budding (see **2** for a recent review). Three types of vesicles with different coats have so far been identified, two types of **clathrin**-coated vesicles and the nonclathrin- or **COP**-coated vesicles.

Two types of clathrin-coated vesicles

The coats of the clathrin-coated vesicles consist of complexes of clathrin and **clathrin adaptor proteins**. Of the two types of clathrin-coated vesicles one mediates internalization of material from the plasma membrane and the other transports newly synthesized protein from the trans-Golgi network (TGN) to endosomes. The Golgi complex and plasma membrane associated clathrin-coated vesicles contain different adaptor complexes, HA1 and HA2, respectively (for reviews see **3,4**). Clathrin probably plays a mechanical role in vesicle budding, while the HA1 or HA2 adaptor proteins are thought to attach the clathrin to the cytoplasmic domains of a selected group of TGN or plasma membrane proteins such as the mannose-6-phosphate (MPR) or LDL receptor (for a review see **3**).

Clathrin is the structural subunit which builds up the characteristic polyhedral cages on the cytoplasmic face of the clathrin-coated vesicles[5]. Three clathrin heavy chains and three light chains form the typical triskelion structure, the unit building block of the fibrous coat of these vesicles. Additional proteins, **auxilin**[6] and **AP180**[7], have been identified in brain which modulate/regulate the assembly of clathrin into polyhedral coat structures.

The clathrin adaptor protein complexes (HA1 and HA2) attach the clathrin coat to the (vesicle) membrane via interaction with both clathrin and specific transmembrane receptors recruited into the clathrin coated pits. HA1 and HA2 are heterotetramers; the former consists of γ-, β'-adaptin, a 47 and a 20 kDa protein, the latter of α-, β-adaptin, a 50 and a 17 kDa protein. Substantial homology has been found between the comparable subunits[8-10]. The structure of the adaptors consists of a central "head", flanked by two ear-like appendages[11]. The ears which can be removed by proteolysis are the C-terminal domains of the adaptins[11,12]. Binding of the adaptor complex to clathrin has been suggested to occur via the N-terminal domains of the adaptins, located in the head domain[13,14].

So far, very little is known about the molecular mechanism(s) which target the adaptor complexes to their corresponding membrane compartments. It appears unlikely, however, that this targetting is accomplished simply by the binding of the adaptors to the cytoplasmic tails of transmembrane receptor (cargo) proteins (e.g. LDL or MPR receptor), since these proteins are also present on membranes of other compartments. The fungal metabolite brefeldin A (BFA) rapidly induces redistribution of γ-adaptin (a subunit of the TGN HA1 adaptor) away from the Golgi complex, but has no effect on HA2-associated α-adaptin[15]. Thus, binding of the two adaptors to their different target membranes is regulated individually. Using a permeabilized cell system it has been shown that G proteins participate in the association of HA1 (and β-COP) with the Golgi membrane[15]. This semi in vitro system should prove very helpful for the further characterization of the molecular machinery involved in targetting of the coat proteins to their corresponding cytoplasmic membrane.

COP-coated vesicles

The coat proteins (COPs) of the Golgi complex derived nonclathrin-coated vesicles are less well characterized than those of the clathrin-coated vesicles. COP-coated vesicles which were first isolated only about three years ago[16] have been implicated in signal-independent "bulk-flow" membrane transport through the Golgi complex (for a review see **17**). COP-coated vesicles are also structurally different from the clathrin-coated vesicles, although at least one of the COPs (β-COP) is homologous to β-adaptin, an HA1 adaptor subunit[10].

The COPs form a complex of 13-14S, with a Stokes radius of ~10 nm and an estimated molecular weight of ~550,000[10]. The native complex can be immunoisolated (Duden and Kreis, unpublished) and it consists of α- (160 kDa), β- (110 kDa), γ- (105 kDa) and δ- (60 kDa) COP, and a few smaller proteins of 30-40 kDa and ~20 kDa. A protein complex (coatomer) with virtually the same protein composition has been purified[18]. These proteins are similar in molecular weight to the clathrin adaptor proteins, and the homology of β-COP with β-adaptin suggests that COPs and adaptors are proteins with homologous structure and function.

Nothing is known about the mutual interactions of the COPs in the complex. It has been shown, however, that proteins belonging to the family of trimeric G proteins play important roles in binding of the COPs to membranes of the Golgi complex[19] (for reviews see **20, 21**).

Furthermore, members of the family of ADP-ribosylation factors (ARFs) may themselves be components of the COP-coat and may also be involved in the regulation of coat association with membranes of the Golgi complex[22]. COP-coated vesicles have been implicated in nonselective "bulk-flow" membrane traffic[17]. It has, however, not so far been ruled out that proteins (e.g. vesicular stomatitis virus glycoprotein, VSV-G), which appear to be transported to the cell surface via nonreceptor-mediated bulk-flow, may be selectively incorporated into COP-coated vesicles. If COPs and adaptors were homologous systems, one could assume that both protein complexes also have similar functions with respect to sorting of (different) cargo into the respective carriers.

COPs may be involved in the budding of vesicles, in analogy to the presumed function of the clathrin-coated vesicle associated proteins. In an alternative and perhaps less likely function which is based on the effects of BFA, β-COP is part of a membrane associated "scaffold" which regulates membrane traffic in a restricitive way. BFA rapidly (<60 sec) induces redistribution of this protein (and also γ-adaptin[15]) away from the Golgi complex, so that it appears diffuse throughout the cytoplasm[23]. BFA most likely interferes with the rebinding of cytosolic COPs to the target membranes (for a recent review see[20]). COP-coated structures associated with membranes of the Golgi complex disappear[24], TGN and endosomal membranes fuse[25-27], and upon more prolonged incubations the Golgi complex disintegrates and Golgi resident proteins return to the endoplasmic reticulum[28,29]. Forward membrane traffic may thus be halted in the absence of the budding machinery. In the alternative model, the membrane associated coat needs to be locally dissociated such that budding (mediated by other factors) or fusion can occur. Trimeric G proteins might play an essential role in these processes[20,21]. BFA completely removes this coat, "fusogenic" sites on organelle membranes are made accessible and extensive uncontrolled fusion (e.g. TGN/endosomes, ER/Golgi) occurs[20,30]. Furthermore, the extent and specificity of membrane microtubule-based motor protein interactions may be altered[20,30].

It has been postulated that populations of vesicles other than the three types described above may also be coated with specific (related?) proteins[20,25-27,30]. But so far, such vesicles have neither been substantially characterized, nor have other additional coat proteins been identified.

Other vesicle associated proteins

A number of other vesicles have been identified; no morphologically distinct coated structures, however, have been detected on their membranes. The so far best characterized of these vesicles are the synaptic vesicles. The most prominent proteins associated with these vesicles are those of the **synapsin** family (synapsin Ia, b, IIa and b)[31]. Synapsins are phosphoproteins associated with the cytoplasmic face of these vesicles, and synapsin I interacts with cytoskeletal filaments, particularly F-**actin**, in a phosphorylation dependent manner[31-33]. They have been implicated in the regulation of neurotransmitter release from nerve endings. It has been suggested that influx of Ca^{2+} activates a Ca^{2+}-calmodulin dependent protein kinase which phosphorylates synapsin I (amongst other proteins). Phosphorylated synapsin I is released from the membrane allowing docking of the vesicles to the presynaptic membrane[34].

■ CYTOPLASMIC LINKER PROTEINS

Microtubules play an essential role in the movement of cytoplasmic organelles, and microtubule-based motor proteins have been characterized (see section on **Motor Proteins**). The modes of interaction of these organelles with microtubules is so far only poorly understood, but it is probably diverse. On the one hand, they include interactions involving the families of microtubule-dependent motors, and on the other hand, anchoring of organelles, for example to microtubules in the region of the microtubule organizing center, may be accomplished by motor independent mechanisms. The binding of organelles to microtubules may be mediated by cytoplasmic linker proteins (CLIPs).

In vitro microtubule - organelle binding assays have been established and used to identify such linker proteins. Two proteins, **CLIP-170** and a protein of 50 kDa, have been identified mediating interactions with microtubules of endocytic carrier vesicles and lysosomes, respectively. Whereas the 50 kDa protein has so far not been well characterized[35], the structure of CLIP-170 has been deduced from its amino acid sequence[36]. CLIP-170 is a very elongated (2.5 by 110 nm), homodimeric molecule, its structure resembling the molecular motor proteins **myosin** and **kinesin**. In contrast to these mechanoenzymes, however, none of the known consensus nucleotide binding sites could be identified on CLIP-170, and it is thus unlikely that it has motor activity. A conserved, tandemly repeated motif has been identified in its N-terminal domain which is involved in microtubule binding; one of these repeats is also present with high homology in other proteins (DP-150[37]; **dynactin**; Glued[38]; BIK1[39]), which all have been implicated in interactions of microtubules with membranes.

A third putative organelle - microtubule linker protein - has been identified in a microtubule binding protein fraction obtained from rat liver. This 58 kDa protein (**58K**)[40] binds to polymerized **tubulin** *in vitro* and is associated with the Golgi complex in tissue culture cells. It has been suggested that it is involved in stabilizing the position of the Golgi complex in the juxtanuclear region[40].

The interaction of each class of cytoplasmic organelles with microtubules may be mediated by specific CLIPs. These CLIPs may form an initial specific contact between vesicles or organelles and microtubules, so that the motors can act subsequently and translocate these membrane bounded structures. As a corollary to this model, the binding force of the CLIPs to microtubules has to be weakened (by specific, motor protein dependent (?), modifications of the CLIPs), so that the motors can then perform their function. In fact, binding of CLIP-170 to microtubules is sensitive to phosphorylation by a kinase present in the microtubule binding protein fraction[41]. Clearly, further work will be required to test this hypothetical model of microtubule - organelle interactions.

■ ORGANELLE SCAFFOLD PROTEINS

The maintenance of the specific morphological structure and membrane protein composition of cytoplasmic organelles, like the endoplasmic reticulum, tubular endosomes or the Golgi complex, probably depends on specific organelle associated skeletal network or "scaffold" proteins. The tubular morphology of some endosomes, for example, appears to be independent of microfilaments and microtubules[42], and the cisternae of the Golgi complex seem to be kept in stacks by intercisternal "glue" material visible by electron-microscopy[43,44].

Virtually nothing is known about these putative organelle scaffold proteins. These proteins may interact with resident organelle membrane proteins (e.g. CLIP receptors) and ensure their specific retention. They may also anchor organelles like the Golgi complex in a microtubule independent manner in the perinuclear region of the microtubule organizing centre[45] (Ho and Kreis, unpublished). The Golgi complex fragments at the onset of mitosis[46-48] when intracellular membrane transport is essentially shut down[49]. A likely scenario implicates mitotic kinases (e.g. $p34^{cdc2}$ protein kinase) which phosphorylate these organelle scaffold proteins and thus induce disintegration of their membrane associated skeleton. An equal portion of the fragmented Golgi complex will be received by the two daughter cells upon division, protein modification is reversed and its interphase structure is regained. This process may be in several aspects analogous to the well characterized disassembly of the nuclear lamina at the onset of mitosis[50-52] when **lamins** are phosphorylated by the $p34^{cdc2}$ protein kinase[53-56]. Subsequent to this phosphorylation the nuclear lamina disassembles, lamins A and C are solubilized and the nuclear envelope breaks down. Further work will lead to the identification and characterization of the, so far enigmatic scaffold proteins, associated with other cytoplasmic organelles.

■ REFERENCES

1. Kreis, T.E. (1990) Cell Motil. Cytoskel. 15, 67-70.
2. Kreis, T.E. (1992) Curr. Opinion Cell Biol. 4, 609-615.
3. Pearse, B.M.F. and Robinson, M.S. (1990) Ann. Rev. Cell Biol. 6, 151-171.
4. Morris, S.A., Ahle, S. and Ungewickell, E. (1989) Curr. Opinion Cell Biol. 1, 684-690.
5. Crowther, R.A. and Pearse, B.M.F. (1987) J. Cell Biol. 91, 790-797.
6. Ahle, S. and Ungewickell, E. (1990) J. Cell Biol. 111, 19-29.
7. Ahle, S. and Ungewickell, E. (1986) EMBO J. 5, 3143-3149.
8. Ahle, S., Mann, A., Eichelsbacher, U. and Ungewickell, E. (1988) EMBO J. 7, 919-929.
9. Robinson, M.S. (1990) J. Cell Biol. 111, 2319-2326.
10. Duden, R., Griffiths, G., Frank, R., Argos, P. and Kreis, T.E. (1991) Cell 46, 649-665.
11. Heuser, J.E. and Keen, J. (1988) J. Cell Biol. 107, 877-886.

12. Kirchhausen, T., Nathanson, K.L., Matsui, W, Vaisberg, A., Chow, E.P., Burne, C., Keen, J.H. and Davis, A.E. (1989) Proc. Natl. Acad. Sci. (USA) 86, 2612-2616.
13. Ahle, S. and Ungewickell, E. (1989) J. Biol. Chem. 264, 20089-20093.
14. Keen, J.H. and Beck, K.A. (1989) Biochem. Biophys. Res. Commun. 158, 17-23.
15. Robinson, M.S and Kreis, T.E. (1992) Cell 69, 129-138.
16. Malhotra, V., Serafini, T., Orci, L., Shepherd, J.C. and Rothman, J.E. (1989) Cell 58, 329-336.
17. Rothman, J.E. and Orci, L. (1992) Nature 355, 409-415.
18. Waters, M.G., Serafini, T. and Rothman, J.E. (1991) Nature 349, 248-251.
19. Donaldson, J.G., Kahn, R.A., Lippincott-Schwartz, J. and Klausner, R.D. (1991) Science 254, 1197-1199.
20. Klausner, R.D., Donaldson, J.G. and Lippincott-Schwartz (1992) J. Cell Biol. 1071-1080.
21. Barr, F.A., Leyte, A. and Huttner, W.B. (1992) Trends Cell Biol. 2, 91-94.
22. Serafini, T., Orci, L., Amherdt, M., Brunner, M., Kahn, R.A. and Rothman, J.E. (1991) Cell 67, 239-253.
23. Donaldson, J.D., Lippincott-Schwartz, J., Bloom, G.S., Kreis, T.E. and Klausner, R.D. (1990) J. Cell Biol. 111, 2295-2306.
24. Orci, L., Tagaya, M., Amherdt, M., Perrelet, A., Donaldson, J.G., Lippincott-Schwartz, J., Klausner, R.D. and Rothman, J.R. (1991) Cell 64, 1183-1195.
25. Lippincott-Schwartz, J., Yuan, L., Tipper, C., Amherdt, M., Orci, L. and Klausner, R.D. (1991) Cell 67, 601-616.
26. Wood, S.A., Park, J.E. and Brown, W.J. (1991) Cell 67, 591-600.
27. Hunziker, W., Whitney, J.A. and Mellman, I. (1991) Cell 67, 617-627.
28. Lippincott-Schwartz, J., Yuan, L.C., Bonifacino, J.S. and Klausner, R.D. (1989). Cell 56, 801-813.
29. Doms, R.W., Russ, G. and Yewdell, J.W. (1989). J. Cell Biol. 109, 61-72.
30. Pelham, H.R.B. (1991) Cell 67, 449-451.
31. DeCamilli, P., Benfenati, F., Valtorta, F. and Greengard, P. (1990) Ann. Rev. Cell Biol. 6, 433-460.
32. Bähler, M. and Greengard, P. (1987) Nature 326, 704-707.
33. Petrucci, T.C. and Morrow, J.S. (1987) J. Cell Biol. 105, 1355-1363.
34. Llinas, R., McGuiness, T., Leonard, C.S., Sugimori, M. and Greengard, P. (1985) Proc. Natl. Acad. Sci. (USA) 82, 3035-3039.
35. Mithieux, G. and Rousset, B. (1989) J. Biol. Chem. 264, 4664-4668.
36. Pierre, P., Scheel, J., Rickard, J.E. and Kreis, T.E. (1992) Cell 70, 887-900.
37. Holzbaur, E.L.F., Hammarback, J.A., Paschal, B.M., Kravit, N.G., Pfister, K.K. and Vallee, R.B. (1991) Nature 351, 579-583.
38. Swaroop, A., Swaroop, M. and Garen, A. (1987) Proc. Nat. Acad. Sci. (USA) 84, 6501-6505.
39. Berlin, V., Styles, C.A. and Fink, G.R. (1990) J. Cell Biol. 111, 2573-2586.
40. Bloom, G.S. and Brashear, T.A. (1989) J. Biol. Chem. 264, 16083-16092.
41. Rickard, J.E. and Kreis, T.E. (1990) J. Biol. Chem. 110, 1623-1633.
42. Tooze, J. and Hollinshead, M. (1991) J. Cell Biol. 115, 635-653.
43. Franke, W.W, Kartenbeck, J., Krien, S., van der Woude, W.J., Scheer, U. and Morre, D.J. (1972) Z. Zellforsch. 132, 365-380.
44. Mollenhauer, H.H. and Morre, D.J. (1978) Subcell. Biochem. 5, 327-359.
45. Turner, J.R. and Tartakoff, A.M. (1989) J. Cell Biol. 109, 2081-2088.
46. Robbins, E. and Gonatas, N.K. (1964) J. Cell Biol. 21, 429-463.
47. Melmed, R.N., Benitez, C.J. and Holt, S.J. (1973) J. Cell Sci. 12, 163-173.
48. Lucocq, J.M., Berger, E.G. and Warren, G. (1989) J. Cell Biol. 109, 463-474.
49. Warren, G. (1985) Trends Biochem. Sci. 10, 439-443.
50. Gerace, L. and Blobel, G. (1980) Cell 19, 277-287.
51. Gerace, L. and Burke, B. (1998) Ann. Rev. Cell Biol. 4, 335-374.
52. Nigg, E.A. (1989) Int. Rev. Cytol. 110, 27-92.
53. Peter, M., Nakagawa, J., Doree, M., Labbe, J.C. and Nigg, E.A. (1990) Cell 61, 591-602.
54. Ward, G.E. and Kirschner, M.W. (1990) Cell 61, 561-577.
55. Heald, R. and McKeon, F. (1990) Cell 61, 579-589.
56. Enoch, T., Peter, M., Nurse, P. and Nigg, E.A. (1992) J. Cell Biol. 112, 797-807.

■ *Thomas E. Kreis:*
Department of Cell Biology,
Sciences III, University
CH-1211, Geneva, Switzerland

AP180 (AP3, NP185)

AP180 is a clathrin associated protein that has been purified from bovine brain coated vesicles. In vitro it binds to clathrin and drives its assembly into polyhedral coat structures. Other functions may include an association with the clathrin coated vesicle adaptor HA2 and with tubulin. The expression of AP180 is induced by NGF in PC12 cells.

AP180 is a monomeric protein with a molecular weight of 115.000-120.000[1,2]. The protein was first purified from bovine brain **clathrin** coated vesicles[1]. AP180 migrates as a 170-190 kDa polypeptide in SDS-PAGE and is therefore difficult to resolve from the clathrin heavy chain. In polyacrylamide gels (T=11,1%) with a low proportion of crosslinker (C=0.9%) such as described by Neville[3], AP180 separates well from the clathrin heavy chain, which behaves in this electrophoretic system like an ~150 kDa polypeptide[4]. AP180 binds to clathrin triskelia with a stoichiometry of one per clathrin triskelion (trimer of heavy chains)[2] and thereby induces assembly of clathrin into polyhedral cages[1] (Figure). By immunological criteria and 2-D peptide mapping AP180 bears no close structural relationship to any of the other known coated vesicle proteins[1]. Two neuronal proteins AP-3[5,6] and NP185[7] were recently shown to be identical with AP180[8]. An interaction between NP185 and the clathrin coated vesicle adaptor complex HA2[7] (**clathrin adaptor proteins**) has been reported as having an interaction with **tubulin**[9]. AP180 is

phosphorylated *in vitro* at serine residue(s) by coated vesicle associated casein kinase II[10]. The protein is also phosphorylated *in situ*[5]. So far, AP180 has only been described in cells of neuronal origin[1,5,7]. In PC12 cells the expression of AP180 is induced by NGF[7]. Its cellular concentration in differentiated PC12 cells was shown to exceed that of clathrin[7]. ~70% of the AP180 in brain is not associated with membranes[4].

■ PURIFICATION

AP180 is extracted from coated vesicles with 0.5 M Tris together with clathrin, and other coat proteins. After gel filtration, ion exchange chromatography on MonoQ resin and hydroxylapatite chromatography the protein **auxilin** constitutes the major contaminating species[1,11]. This becomes partially removed upon gel filtration on Superose[6].

■ ACTIVITIES

AP180 binds to clathrin triskelions and promotes their assembly into regular coat structures. This activity can be quantitated in a sedimentation assay by either pelleting or fractionation in 5-30% sucrose gradients. Assembly of clathrin is also readily monitored by electron microscopy of negatively stained specimens.

■ ANTIBODIES

Murine monoclonal antibodies to AP180 which can be used for immunoblotting and immunoprecipitation have been described[1,7].

■ GENES

There is no published sequence information on AP180.

■ REFERENCES

1. Ahle, S. and Ungewickell, E. (1986) EMBO J. 5, 3143-3149.
2. Prasad, K. and Lippoldt, R.E. (1988) Biochemistry 27, 6098-6104.
3. Neville, D. (1971) J. Biol. Chem. 246, 6328-6334.
4. Ungewickell, E. and Oestergaard, L. (1989) Anal. Biochem. 179, 352-356.
5. Keen, J.H. and Black, M.M. (1986) J. Cell Biol. 102, 1325-1333.
6. Keen, J.H. (1987) J. Cell Biol. 105, 1989-1998.
7. Kohtz, S.D. and Puszkin, S. (1988) J. Biol. Chem. 263, 7418-7425.

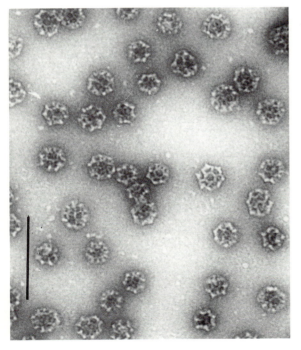

Figure. Electron micrograph of negatively stained clathrin cages assembled in the presence of AP180. Bar 250 nm.

8. Murphy, J., Pleasure, I.T., Puszkin, S., Prasad, K. and Keen, J.H. (1991) J. Biol. Chem. 266, 4401-4408.
9. Kohtz, S.D. and Puszkin, S. (1989) J. Neurochem. 52, 285-295.
10. Morris, S.A., Mann, A. and Ungewickell, E. (1990) J. Biol. Chem. 265, 3354-3357.
11. Ahle, S. and Ungewickell, E. (1990) J. Cell Biol. 111, 19-29.

■ Ernst Ungewickell:
Department of Pathology,
Washington University,
School of Medicine,
660 South Euclid Avenue,
St. Louis, MO, USA

Auxilin

Auxilin is a single chain clathrin associated protein that has been identified in the coat of clathrin coated vesicles from rat and bovine brain. In vitro it drives the assembly of clathrin triskelia into polyhedral coat structures.

Auxilin is an ~86 kDa protein which has been purified from bovine brain **clathrin** coated vesicles[1]. In SDS-PAGE it behaves anomalously like a 110 kDa polypeptide and is therefore difficult to resolve from the 100-115 kDa sub-units of the **clathrin adaptor protein** complexes, also present in clathrin coated vesicles. Upon inclusion of 6-8 M urea into the separation gel, auxilin's electrophoretic mobility shifts to that of a 126 kDa polypeptide and resolves well from other coat proteins. Auxilin binds to clathrin triskelia with a stoichiometry of one per clathrin heavy chain and thereby induces the assembly of clathrin into polyhedral cages. The protein is phosphorylated *in vitro* at serine residue(s) by coated vesicle associated casein kinase II[2]. So far, auxilin has only been described in clathrin coated vesicles from neuronal tissue. By immunological criteria and 2-D peptide mapping auxilin was shown to bear no close structural relationship to any of the other known coated vesicle proteins[1].

■ PURIFICATION

Auxilin is extracted from coated vesicles with 0.5 M Tris together with clathrin, adaptor proteins and **AP180**. After gel filtration and hydroxylapatite chromatography AP180 constitutes the major contaminating species. This is partially removed upon gel filtration on Superose[6]. Alternatively, auxilin is readily purified by immunoaffinity chromatography using an immobilized monoclonal antibody to the protein (mAb 100/4)[1].

■ ACTIVITIES

Auxilin binds to clathrin triskelions and promotes their assembly into regular coat structures. This activity can be quantitated in a sedimentation assay by either pelleting or fractionation in 5-30% sucrose gradients. Assembly of clathrin is also readily monitored by electron microscopy of negatively stained specimens.

■ ANTIBODIES

A murine monoclonal antibody (mAb 100/4) to auxilin which can be used for immunoblotting and immunoprecipitation has been described[1].

■ GENES

There is no sequence information on auxilin.

■ REFERENCES

1. Ahle, S. and Ungewickell, E. (1990) J. Cell Biol. 111, 19-29.
2. Morris, S.A., Mann, A. and Ungewickell, E. (1990) J. Biol. Chem. 265, 3354-3357.

■ Ernst Ungewickell:
Department of Pathology,
Washington University,
School of Medicine,
660 South Euclid Avenue,
St. Louis, MO, USA

Clathrin

Clathrin[1-3] is the structural unit which forms the characteristic outer polyhedral cage on the cytoplasmic surface of coated vesicles. It has the extraordinary shape of a three-legged triskelion[4]: its flexibility allows it to coat economically a variety of membranes from planar to small spherical vesicles[5]; its mechanical strength provides the structural framework in combination with the other coat structural units, the adaptors, necessary to bud a vesicle into the cytoplasm. Coated vesicles participate in selective recycling of membrane proteins at the plasma membrane, during endocytosis and at the trans-Golgi network on an intracellular transport pathway.

Clathrin isolated from mammalian brain (e.g. pig, bullock) or other tissues (e.g. human placenta) has a trimeric structure consisting of three copies of the heavy chain (180 kDa) and three copies of a light chain (Figure 1). In yeast and mammalian sources there is a single clathrin heavy chain gene[6]. The clathrin light chains in mammalian tissues are more polymorphic. There are two related light chain genes in bovine tissues. In bovine brain, alternative splicing of both light chain transcripts occurs, which leads to slightly larger products than in other tissues[7,8].

Clathrin occurs in most eukaryotic cells. Immuno-fluorescence staining of fibroblasts shows the presence of thousands of coated pits and vesicles per cell (Figure 2). Many of these are randomly distributed on the plasma membrane but a discrete population of coated pits is found in the region of the Golgi apparatus, particularly on the trans-Golgi network. The subsets of coated pits are distinguished by the presence of different adaptors[9-11].

The function of the coated vesicle coat (consisting of clathrin and **clathrin adaptor proteins**) is to provide a structure to concentrate certain receptors in the membrane of the vesicle which is budded into the cytoplasm, thus separating these receptors from the parent membrane and its resident proteins[12]. The coat is then released and its components, including clathrin, are recycled to form another coated pit. The uncoated vesicle fuses with a target membrane compartment, e.g. the endosome, thus delivering its contents into the lumen of that compartment.

In yeast (*Saccharomyces cerevisiae*), drastic consequences result from elimination of the clathrin heavy chain gene[13,14]. A variable portion of the mutant cells are inviable; those that do grow, do so slowly and accumulate autophagic vacuoles. At least one sorting process has been shown to be disrupted; retention of the endoprotease Kex2 in the Golgi apparatus is defective with the consequence that precursor α-factor remains unprocessed and is secreted as inactive molecules[15]. Disruption of a yeast clathrin light chain gene also leads to a slow growth phenotype which is however unstable in that faster growing variants often arise[16].

■ PURIFICATION

Functional clathrin triskelions are extracted from coated vesicles and purified by conventional column chromatography. Typically, clathrin is separated from clathrin adaptor proteins (the other major structural units of the

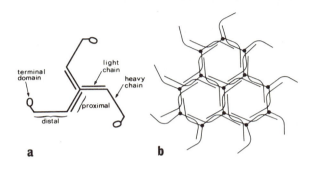

a   b

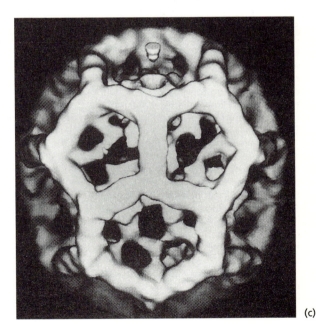

(c)

Figure 1. (a) Schematic drawing showing the modular structure of the triskelion. (b) Packing diagram showing how triskelions form a hexagonal lattice. For simplicity the terminal domains, which pack under the vertices, have been omitted. (c) 3-D map of a clathrin cage, computed from electron micrographs of unstained specimens embedded in vitreous ice. The terminal domains are visible in the interior of the cage.

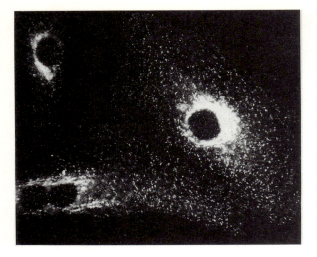

Figure 2. Immunofluorescence localization of clathrin in bovine fibroblasts.

coat) by gel filtration through Sepharose CL-4B[17]. After concentration (e.g. by ammonium sulphate precipitation) the clathrin can be repolymerized to form polyhedral cages[4,5].

ACTIVITIES

Clathrin triskelions extracted in 10 mM TrisCl at pH 7.5 assemble into cages within seconds when exposed by rapid mixing to polymerization conditions (2 mM $MgCl_2$, pH 6.2). The cages produced are a heterogeneous set of polyhedra ranging in diameter from 50 nm to 120 nm, the mean size depending largely on the clathrin concentration[5].

In certain conditions, the clathrin adaptor proteins promote the assembly of clathrin and have a dramatic effect on the size distribution of particles formed[18,19]. The coats containing adaptors are much more homogeneous in size (50-80 nm in diameter) than the cages with clathrin alone. Electron microscopy of these reconstituted coats embedded in vitreous ice has shown that the adaptors, in stoichiometric ratio to clathrin, form an inner shell of material, underneath the clathrin cage, surrounding the volume normally occupied by a vesicle[20,21].

Little is known about the control of coat assembly and disassembly *in vivo*. A member of the heat shock family of proteins, hsp70, has been observed to slowly promote clathrin depolymerization from cages *in vitro*, dependent on ATP hydrolysis and the presence of clathrin light chains[3,22]. A conformationally labile domain of clathrin light chain LCa, like many other peptides, binds to hsp70, stimulating ATP hydrolysis[23]. In the absence of ATP, this protein has been seen trapped at the vertices of triskelions[24]. In this position, hsp70 would presumably interfere with interactions necessary for maintenance of the clathrin lattice and at least prevent formation of empty clathrin cages in the cytoplasm.

ANTIBODIES

A number of monoclonal antibodies against determinants either on the clathrin heavy chain or the clathrin light chain have been characterized[25]. In general polyclonal sera and monoclonal antibodies raised against bovine brain clathrin crossreact with clathrin from a number of other mammalian species reflecting the high degree of conservation of protein sequence. Monoclonal antibodies have also been reported against yeast clathrin heavy chain which do not crossreact with mammalian clathrin. However, two mABs raised against bovine clathrin apparently do bind to yeast clathrin heavy chains and one crossreacts with sea urchin clathrin[26].

GENES

Clathrin heavy chain has been cloned and sequenced from rat[6] (EMBL/GenBank J03583) and cloned and partly sequenced from yeast[13,14]. Clathrin light chains have also been cloned and sequenced from rat[7,8] (EMBL/GenBank Y00265) and yeast[16] (EMBL/GenBank DDB X 52272).

REFERENCES

1. Pearse, B.M.F. and Crowther, R.A. (1987) Annu. Rev. Biophys. Biophys. Chem. 16, 49-68.
2. Pearse, B.M.F. and Robinson, M.S. (1990) Ann. Rev. Cell Biol. 6, 151-171.
3. Keen, J.H. (1990) Ann. Rev. Biochem. 59, 415-438.
4. Ungewickell, E. and Branton, D. (1981) Nature 289, 420-422.
5. Crowther, R.A. and Pearse, B.M.F. (1981) J. Cell Biol. 91, 790-797.
6. Kirchhausen, T., Harrison, S.C., Ping Chow, E., Mattaliano, R.J., Ramachandran, K.L., Smart, J. and Brosius, J. (1987) Proc. Natl. Acad. Sci. (USA) 84, 8805-8809.
7. Jackson, A.P., Seow, H.-F., Holmes, N., Drickamer, K. and Parham, P. (1987) Nature 326, 154-159.
8. Kirchhausen, T., Scarmato, P., Harrison, S.C., Monroe, J.J., Chow, E.P., Mattaliano, R.J., Ramachandran, K.L., Smart, J.E., Ahn, A.H. and Brosius, J. (1987) Science 236, 320-324.
9. Robinson, M.S. and Pearse, B.M.F. (1986) J. Cell Biol. 102, 48-54.
10. Robinson, M.S. (1987) J. Cell Biol. 104, 887-895.
11. Ahle, S., Mann, A., Eichelsbacher, U. and Ungewickell, E. (1988) EMBO J. 7, 919-929.
12. Bretscher, M.S., Thomson, J.N. and Pearse, B.M.F. (1980) Proc. Natl. Acad. Sci. (USA) 77, 4156-4159.
13. Payne, G.S. and Schekman, R. (1985) Science 230, 1009-1014.
14. Lemmon, S. and Jones, E.W. (1987) Science 238, 504-509.
15. Payne, G.S. and Schekman, R. (1989) Science 245, 1358-1365.
16. Silveira, L.A., Wong, D.H., Masiarz, F.R. and Schekman, R. (1990) J. Cell Biol. 111, 1437-1449.
17. Keen, J.H., Willingham, M.C. and Pastan, I.H. (1979) Cell 16, 303-312.
18. Zaremba, S. and Keen, J.H. (1983) J. Cell Biol. 97, 1339-1347.
19. Pearse, B.M.F. and Robinson, M.S. (1984) EMBO J. 3, 1951-1957.
20. Vigers, G.P.A., Crowther, R.A. and Pearse, B.M.F. (1986) EMBO J. 5, 529-534.
21. Vigers, G.P.A., Crowther, R.A. and Pearse, B.M.F. (1986) EMBO J. 5, 2079-2085.
22. Schmid, S.L., Braell, W.A., Schlossman, D.M. and Rothman, J.E. (1984) Nature (London) 311, 228-231.

23. DeLuca-Flaherty, C., McKay, D.B., Parham, P. and Hill, B.L. (1990) Cell 62, 875-887.
24. Heuser, J. and Steer, C.J. (1989) J. Cell Biol. 109, 1457-1466.
25. Brodsky, F.M. (1985) J. Cell Biol. 101, 2047-2054.
26. Lemmon, S.K., Lemmon, V.P. and Jones, E.W. (1988) J. Cell. Biochem. 36, 329-340.

■ Barbara M.F. Pearse:
Medical Research Council,
Laboratory of Molecular Biology,
Hills Road, Cambridge, CB2 2QH,
UK

Clathrin Adaptor Proteins

Adaptor proteins constitute a family of ubiquitous multi subunit proteins which bind to clathrin, promote its assembly into polyhedral structures and attach it to cellular membranes. Two major adaptors have been described, one associated with endocytic clathrin coated membranes and the other associated with Golgi-derived clathrin coated membranes. Direct interactions between adaptor proteins and cytoplasmic portions of certain receptors have been reported. The proposed biological function of adaptor proteins involves the specific recruitment of proteins into clathrin coated membrane domains.

Two major types of **clathrin** adaptor proteins have been described, which are complementary in intracellular distribution[1-5]. The first is referred to as the HA2 adaptor[1] or assembly protein 2 (AP2)[5]. It is predominantly present in clathrin coated structures on or near the plasma membrane[2,3] (Figure). HA2 is a heterotetramer with the subunit composition $\alpha/\beta/50$ kDa/17 kDa[3,4]. α- and β-subunits, also termed α- and β-adaptins[6], run on SDS-PAGE as 100-112

kDa polypeptides. Four α-isoforms (α_{a1}, α_{a2}, α_{c1} and α_{c2}) which are highly homologous in sequence have been described[3]. α_a (107,605 Da) and α_c (104,017 Da) are encoded by different genes[7]. α_a-type isoforms are predominantly expressed in neuronal cells, while α_c isoforms appear to be the more widely distributed subunit[3,7]. The molecular basis for the electrophoretic differences seen between α_{a1} and α_{a2} and α_{c1} and α_{c2}, respectively, are not

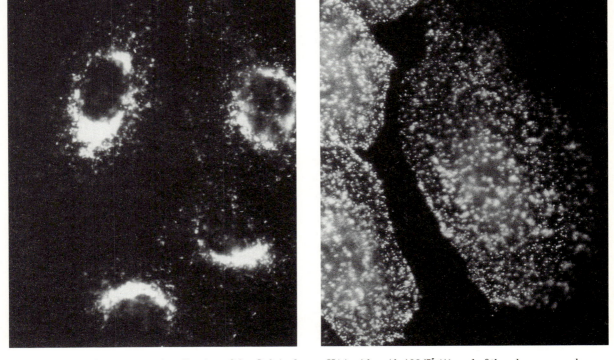

A   B

Figure. Immunofluorescence localization of the Golgi adaptor HA1 with mAb 100/3[5] (A) and of the plasma membrane adaptor HA2 with mAb AP.6[20] (B) in MDBK cells.

known. The β-subunit (104,700 Da) of mammalian brain contains a 14 residue insert which results from alternative splicing[8]. Serine residues of α_a and β are phosphorylated by a coated vesicle associated kinase activity[9] which bears the characteristics of casein kinase II[10]. The 50 kDa subunit is phosphorylated at threonine[11,12] by a hitherto uncharacterized kinase. The biological role of adaptor protein phosphorylation is not known. Although α- and β-subunits share no significant homologies in sequence, their overall organization is strikingly similar. Both are structured into 60-70 kDa N-terminal domains and 30-40 kDa C-terminal head domains[13] which appear when the adaptor is viewed by electron microscopy as appendages emanating from a brick-like structure[14]. The appendages are joined via proline and glycine rich stalks to the rest of the molecule[7,8,13]. Intersubunit interactions between α/β, 50 kDa and 17 kDa subunits involve only the 60-70 kDa domains[15]. The β-subunit contains a binding site for clathrin[16].

The second adaptor protein, referred to as HA1 adaptor[1] or AP1[5], is largely concentrated to the region of the Golgi apparatus[3] (Figure). It is a heterotetramer with the subunit composition β'/γ/47 kDa/19 kDa[3]. β'- and γ-subunits (β'- and γ-adaptins[6]) run on SDS-PAGE as 115 and 104 kDa polypeptides, respectively[3]. β' is highly homologous to the β-subunit[3] of HA2 while γ (91,352 Da) shares only about 25% overall identity with the α-subunit of HA2[17]. The sequence analysis of the γ-subunit suggests also a domain organization similar to that of the α- and β-subunits of the HA2 adaptor[17]. The β'- and γ-subunits can be phosphorylated *in vitro* at serine residues[9].

PURIFICATION

Both clathrin adaptors are obtained from bovine brain coated vesicles by extraction with 0.5 M Tris and are purified to homogeneity by gel filtration, hydroxylapatite[1,18] and anion exchange chromatography on MonoQ[3,4]. The most efficient step for the separation of HA2 from HA1 is the hydroxylapatite column to which HA2 adsorbs strongly. Affinity purification of the HA2 adaptor using clathrin immobilized to sepharose has been described[5]. It is claimed that this procedure fully preserves the clathrin assembly promoting activity of this adaptor (see below).

ACTIVITIES

Adaptors bind to clathrin triskelia and thereby promote their assembly into regular coat structures. This activity can be quantitated in a sedimentation assay. *In vitro* interactions between adaptors and membranes[4] and cytoplasmic receptor tails have been described[6,19].

ANTIBODIES

Several reports describe monoclonal antibodies to the α-[2,3,20], β/β'-[3] and γ-[3] adaptor subunits. mAb 100/3 (anti-γ)[3] and AP.6 (anti-α)[20] can be used for immunoprecipitation, immunofluorescence and immunoblotting. mAb 100/3 does not recognize the γ-subunit from mouse and rat.

GENES

Complete cDNA sequences are available for mouse HA2 α_a (α_a-adaptin) and α_c adaptor subunits (α_c-adaptin) (GenBank/EMBL X14971 and X14972)[7], a full length rat brain cDNA for a β-type subunit (J04527)[13], partial rat brain (J04528)[13] and full length clones for the β-subunit from human fibroblasts (M34175)[8] and rat lymphocytes (M34176)[8], complete rat brain cDNA clone for the 50 kDa subunit of HA2 (AP50) (M23674)[11], a complete rat brain cDNA clone for HA2γ (EMBL X54424)[17].

REFERENCES

1. Pearse, B.M.F. and Robinson, M.S. (1984) EMBO J. 3, 1951-1957.
2. Robinson, M.S. (1987) J. Cell Biol. 104, 887-895.
3. Ahle, S., Mann, A., Eichelsbacher, U. and Ungewickell, E. (1988) EMBO J. 7, 919-929.
4. Virshup, D.M. and Bennett, V. (1988) J. Cell Biol. 106, 39-50.
5. Keen, J.H. (1987) J. Cell Biol. 105, 1989-1998.
6. Pearse, B.M.F. (1988) EMBO J. 7, 3331-3336.
7. Robinson, M.S. (1989) J. Cell Biol. 108, 833-842.
8. Ponnambalam, S., Robinson, M., Jackson, A.P., Peiperl, L. and Parham, P. (1990) J. Biol. Chem. 265, 4814-4820.
9. Morris, S.A., Mann, A. and Ungewickell, E. (1990) J. Biol. Chem. 265, 3354-3357.
10. Bar-Zvi, D., Mosley, S.T. and Branton, D. (1988) J. Biol. Chem. 263, 4408-4415.
11. Thurieau, C., Brosius, J., Burne, C., Jolles, P., Keen, J.H., Mattaliano, R.J., Cow, E.P., Ramachandran, K.L. and Kirchhausen, T. (1988) DNA 7, 663-669.
12. Keen, J.H., Chestnut, M.H. and Beck, K.A. (1987) J. Biol. Chem. 262, 3864-3871.
13. Kirchhausen, T., Nathanson, K.L., Matsui, W., Vaisberg, A., Chow, E.P., Burne, C., Keen, J.H. and Davis, A.E. (1989) Proc. Natl. Acad. Sci (USA) 86, 2612-2616.
14. Heuser, J.E. and Keen, J. (1988) J. Cell Biol. 107, 877-886.
15. Zaremba, S. and Keen, J.H. (1985) J. Cell Ciochem. 28, 47-58.
16. Ahle, S. and Ungewickell, E. (1989) J. Biol. Chem. 264, 20089-20093.
17. Robinson, M.S. (1990) J. Cell Biol. 111, 2319-2326.
18. Manfredi, J.J. and Bazari, W.L. (1987) J. Biol. Chem. 262, 12182-12188.
19. Glickman, J.N., Conibear, E. and Pearse, B.M.F. (1989) EMBO J. 8, 1041-1047.
20. Chin, D.J., Straubinger, R.M., Acton, S., Näthke, I. and Brodsky, F. (1989) Proc. Natl. Acad. Sci. (USA) 86, 9289-9293.

Ernst Ungewickell:
Department of Pathology,
Washington University,
School of Medicine,
660 South Euclid Avenue,
St. Louis, MO, USA

β-COP

β-COP, a 110 kDa protein, is a major component associated with the coat of nonclathrin coated vesicles derived from the Golgi complex. It shares significant homology to the clathrin adaptor protein β-adaptin[1]. β-COP may be involved in vesicle budding or be part of an exoskeleton associated with the membranes of the Golgi complex.

β-COP was originally identified with a monoclonal antibody, M3A5, as a 110 kDa peripheral membrane protein associated with the cytoplasmic face of the Golgi complex[2]. By immunoelectronmicroscopy, β-COP is localized throughout the Golgi complex predominantly at the rims of Golgi *cisternae* and on nonclathrin coated vesicles[1] (Figure). The nonclathrin coated vesicles are thought to mediate bulk flow transport through the Golgi complex[3] and have recently been purified utilizing a cell free assay that reconstitutes intracisternal transport, where they accumulate in the presence of the nonhydrolyzable nucleotide analogue GTPγS[4,5]. The coat of these nonclathrin coated vesicles consists of four major coat proteins (COPs), α-, β-, γ- and δ-COP (160, 110, 98, and 61 kDa, respectively), as well as proteins of smaller molecular weight[5] (Table). The cytosolic form of β-COP is in a nonglobular complex of ~550 kDa[1]. This complex, containing the COPs, can be immunoprecipitated in a native form with an antibody against a peptide of β-COP (Duden and Kreis, unpublished).

The N-terminal half of β-COP shares significant homology with the **clathrin adaptor protein** β-adaptin[2]. This region on β-adaptin appears to bind **clathrin** with high affinity *in vitro*[6]. This homology and the comparable molecular weights of the COPs and the clathrin coated vesicle associated proteins (Table) suggest that these proteins may be related structurally and may have similar functions[2,5,7,8]. Yet, cargo and selectivity of the two classes classes of vesicles are different[7,9].

Treatment of cells with the drug brefeldin A[10], which interferes with intracellular membrane traffic and leads to the morphological disappearance of the Golgi complex[11,12], induces a rapid (<1 min) redistribution of β-COP[13] and γ-adaptin[14], and probably dissociates the nonclathrin coat from Golgi and Golgi complex derived membranes. The association of β-COP with membranes is thus probably dynamic[1] and it appears to be regulated by trimeric GTP binding proteins[15].

Various models for β-COP function have been proposed[16], β-COP may be directly involved in the budding of

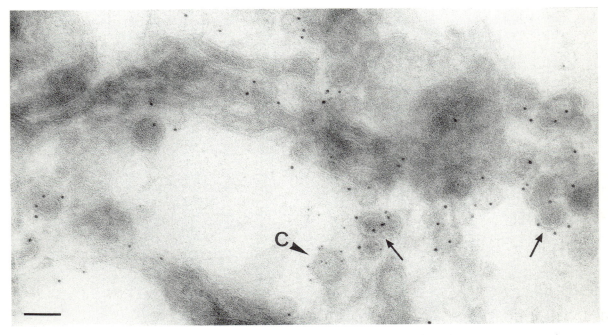

Figure. Immunolocalization of β-COP (and clathrin) on a GTPγS-treated rat liver Golgi fraction.
Immunogold labelling of β-COP with anti-EAGE (9 nm gold) and clathrin with an antiserum against clathrin light chains (5 nm gold) on ultrathin frozen sections of GTPγS-treated rat liver Golgi fraction is shown. The coats of the vesicular profiles contain either β-COP (arrows) or clathrin ("C" arrow head). Bar 100 nm.

TABLE
Comparison of the proteins associated with clathrin and nonclathrin coated vesicles.

Clathrin coated vesicles	Nonclathrin coated vesicles
Clathrin heavy chain (180kd)	α-COP (160kd)
	Clathrin light chains (30-40kd)
β-adaptin (105kd)	β-COP (110kd)
α- or γ-adaptin (92-108kd)	γ-COP (98kd)
~50kd protein (47-50kd)	δ-COP (61kd)
~20kd protein (17-20kd)	
	36kd*
	35kd*
	20kd*

The coat of clathrin coated vesicles consists of clathrin "triskelions" and clathrin adaptor proteins. The clathrin triskelion is made up from three heavy and three light chains, whereas the adaptor complexes are heterotetramers, containing one β-adaptin, one α- or γ-adaptin, and one copy each of α ~50 kDa and a 20 kDa protein. The major components of nonclathrin coated vesicles are α-, β-, γ-, and δ-COPs. The three smaller proteins (indicated by asterisks) have been identified in a cytosolic complex containing the COPs (Waters et al., 1991). It is unclear whether they are also present on the nonclathrin coated vesicles.

Golgi complex derived nonclathrin coated vesicles, or alternatively, it may be part of an exoskeleton, peripherally associated with the membranes of the Golgi complex, which defines domains that regulate membrane traffic through the Golgi complex.

■ PURIFICATION

β-COP can be isolated by affinity purification with specific antibodies[1]. It has also been purified as a component of a cytosolic complex using ammonium sulphate precipitation, followed by DEAE, hydroxylapatite and Mono Q chromatography, and isoelectric precipitation[17]. β-COP constitutes about 0.2% of soluble cytosolic protein[17].

■ ACTIVITIES

None have been determined so far.

■ ANTIBODIES

A monoclonal antibody, M3A5[2], and several polyclonal anti-peptide antibodies[1] against β-COP have been described. These antibodies generally react with β-COP from avian and mammalian species (some also react with amphibians) by immunoblotting and immunofluorescence.

■ GENES

The full length cDNA clone of rat liver β-COP has been sequenced[1] (GenBank X57228).

■ REFERENCES

1. Duden, R., Griffiths, G., Frank, R., Argos, P. and Kreis, T.E. (1991) Cell 64, 649-665.
2. Allan, V.J. and Kreis, T.E. (1986) J. Cell Biol. 103, 2229-2239.
3. Orci, L., Glick, B.S. and Rothman, J.E. (1986) Cell 46, 171-184.
4. Malhotra, V., Serafini, T., Orci, L., Shepherd, J.C. and Rothman, J.E. (1989) Cell 58, 329-336.
5. Serafini, T., Stenbeck, G., Brecht, A., Lottspeich, F., Orci, L., Rothman, J.E. and Wieland, F.T. (1991) Nature 349, 215-220.
6. Keen, J.H. and Beck, K.A. (1989) Biochem. Biophys. Res. Comm. 158, 17-23.
7. Duden, R., Allan, V. and Kreis, T.E. (1991) Trend in Cell Biol. 1, 14-19.
8. Robinson, M.S. (1991) Nature 349, 743-744.
9. Pearse, B.M.F. and Robinson, M.S. (1990) Annu. Rev. Cell Biol. 6, 151-172.
10. Harri, E., Loeffler, W., Sigg, H.P., Stahelin, H. and Tamm, H. (1963) Helv. Chem. Acta 46, 1235-1243.
11. Lippincott-Schwartz, J., Yuan, L.C., Bonifacino, J.S. and Klausner, R.D. (1989) Cell 56, 801-813.
12. Doms, R.W., Russ, G. and Yewdell, J.W. (1989) J. Cell Biol. 109, 61-72.
13. Donaldson, J.G., Lippincott-Schwartz, J., Bloom, G.S., Kreis, T.E. and Klausner, R.D. (1990) J. Cell Biol. 111, 2295-2306.
14. Robinson, M.S. and Kreis, T.E. (1992) Cell 69, 129-138.
15. Klausner, R.D., Donaldson, J.G. and Lippincott-Schwartz, J. (1992) J. Cell Biol. 116, 1071-1080.
16. Kreis, T.E. (1992) Curr. Op. Cell Biol. 4, 609-615.
17. Waters, M.G., Serafini, T. and Rothman, J.E. (1991) Nature 349, 248-251.

■ Rainer Duden and Thomas E. Kreis:
Department of Cell Biology
Sciences III, University
CH-1211 Geneva,
Switzerland

CLIP-170

CLIP-170 (cytoplasmic linker protein with 170kDa) is a microtubule-binding protein regulated by phosphorylation and characterized by its specific accumulation towards the plus ends of a subset of microtubules in interphase cells. The cellular role of CLIP-170 may be the linkage of endosomes to microtubules.

In HeLa cells, CLIP-170 colocalizes by immunofluorescence with short segments at the peripheral (plus) ends of microtubules (Figure) and is also found in the mitotic spindle[1]. A similar distribution has also been found in several other tissue culture cells, including Vero cells and human skin fibroblasts. However, association of CLIP-170 with nonmicrotubular cytoplasmic structures is suggested by its nondiffuse localization in cells with depolymerized microtubules and in mitotic cytoplasm[1]. CLIP-170 is essential for the cytosol-dependent binding of endocytic vesicles to microtubules *in vitro*, and the protein colocalizes with transferrin receptor-positive endosomes *in vivo*[2]. CLIP-170 also accumulates at desmosomal plaques in polarizing epithelial cells in culture[3]. The protein may, therefore, be involved in linking microtubules to other cellular structures and in determining their cellular location. It could also be involved in regulating microtubule plus end dynamics by mediating an interaction with peripherally located organelles.

The microtubule association of CLIP-170 *in vitro* is regulated by phosphorylation, with increased phosphate incorporation inhibiting microtubule binding[4]. The protein is phosphorylated on serine residues *in vivo*, and turnover of the phosphate group is much faster than for total cellular protein[4]. Phosphorylation of CLIP-170 *in vivo* is regulated by microtubules; depolymerization of microtubules by treatment of cells with

nocodazole leads to rapid dephosphorylation of the protein[4].

Sequence analysis of the CLIP-170 cDNA suggests that dimerization occurs via a central α-helical coiled-coil domain[2]. The N-terminal sequence of CLIP-170 contains two copies of a motif of 57 amino acids responsible for the microtubule-binding activity of the protein[2]. This motif is also found in one copy in the *Drosophila* glued protein[5] (homologous to DP-150[6] and **dynactin**[7]) and the BIKI protein from *S. cerevisiae*[8], indicating that it represents a conserved microtubule-binding motif[2]. CLIP-170 is composed only of the 170 kDa polypeptide as determined by lack of coprecipitating polypeptides during immunoisolation of native protein[4]. The sedimentation coefficient of CLIP-170 on sucrose gradients is ~5.7 S[1] and calculations from hydrodynamic data indicate that it exists as a homodimer in solution[2].

■ PURIFICATION

CLIP-170 has been purified from HeLa cells almost to homogeneity by antibody affinity purification[4].

■ ACTIVITIES

The only functional *in vitro* assay for CLIP-170 activity is binding to microtubules, analyzed by cosedimentation.

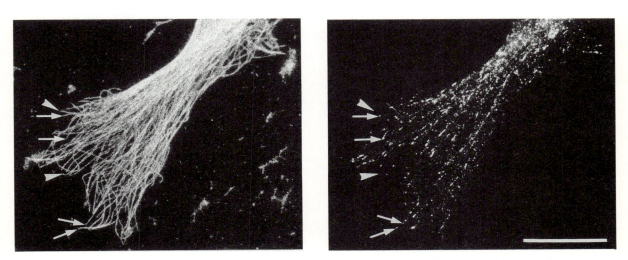

Figure. HeLa cells labelled by double immunofluorescence for tubulin (left) and CLIP-170 (right). Arrows and arrowheads indicate microtubule ends labelled or unlabelled, respectively, by the antibodies to CLIP-170. Bar 20 µm.

■ ANTIBODIES

Rabbit polyclonal and mouse monoclonal antibodies have been raised against CLIP-170 from HeLa cells. They have been used for immunofluorescence, immunoblotting, immunoprecipitation and affinity purification of the protein in HeLa cells[1,4]. Some of the antibodies crossreact with other mammalian species. The protein has not so far been identified immunologically in nonmammalian species.

■ GENES

CLIP-170 has been cloned and sequenced from HeLa cells[2]. (GenBank M97501). The sequence data indicate that it is a novel microtubule binding protein.

■ REFERENCES

1. Rickard, J.E. and Kreis, T.E. (1990) J. Cell Biol. 110, 1623-1633.
2. Pierre, P., Scheel, J., Rickard, J.E. and Kreis, T.E. (1992) Cell 70, 887-900.
3. Wacker, I.U., Rickard, J.E., De Mey, J.R. and Kreis, T.E. (1992) J. Cell Biol. 117, 813-824.
4. Rickard, J.E. and Kreis, T.E. (1991) J. Biol. Chem. 266, 17597-17605.
5. Swaroop, A., Swaroop, M. and Garen, A. (1987) Proc. Natl. Acad. Sci. (USA) 84, 6501-6505.
6. Holzbauer, E.L.F., Hammarback, J.A., Paschal, B.M., Kravit, N.G., Pfister, K.K. and Vallee, R.B. (1991) Nature 351, 579-583.
7. Gill, S.R., Schroer, T., Szilak, I., Steuer, E.R., Sheetz, M.P. and Cleveland, D.W. (1991) J. Cell Biol. 115, 1639-1650.
8. Trueheart, J., Boeke, J.D. and Fink, G. (1987) Mol. Cell. Biol. 7, 2316-2328.

■ Janet E. Rickard and Thomas E. Kreis:
Department of Cell Biology
Sciences III, University
CH-1211 Geneva,
Switzerland

58K

58K[1] is a cytoplasmically oriented, peripheral membrane protein of the Golgi apparatus. The protein was originally identified in a tissue extract based upon its microtubule binding activity. In light of these properties, the function of 58K may be to anchor the Golgi to microtubules.

The structural integrity of the Golgi apparatus and its location within the cell are determined by interactions of the Golgi apparatus with microtubules. Drugs or other treatments which cause microtubules to depolymerize or assume abnormal distributions invariably cause the Golgi apparatus to fragment and become scattered through the cytoplasm[2-4]. Moreover, many cells contain one or a small number of discrete, perinuclear microtubule organizing centers (MTOCs), which maintain a close association with the Golgi apparatus, even as their exact location in the cell is experimentally manipulated[5-8].

To account for these phenomena, it is likely that specific proteins are involved in linking the Golgi apparatus to microtubules and maintaining this organelle near the MTOC. So far, two proteins, 58K[1] and β-**COP** (110K)[9], have been proposed to perform such a function. SDS-PAGE of purified 58K indicated that the protein is composed of a single type of ~58 kDa subunit, although the number of subunits per molecule remains to be determined[1]. The initial enrichment step for 58K is based on its affinity for microtubules in a crude tissue extract. Purified 58K also binds to microtubules, in a saturable manner of one 58K monomer per α/β-**tubulin** heterodimer[1]. This level of saturation binding is atypically high for microtubule associated proteins (MAPs)[10-12], is indicative of the modest affinity of 58K for microtubules relative to conventional MAPs, and signifies that 58K is unlikely to regulate microtubule assembly or stability in vivo.

Instead, the presumptive function of 58K was revealed

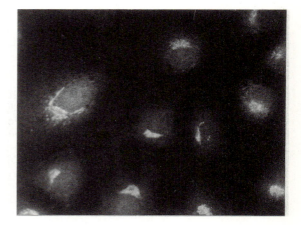

Figure. The intracellular distribution of 58K. Shown here is a field of cultured PtK$_1$ cells stained with a monoclonal anti-58K antibody followed by TRITC-labelled goat anti-mouse IgG. 58K is localized to the Golgi apparatus.

initially by immunofluorescence microscopy of cultured hepatoma cells, which demonstrated that 58K is localized on the Golgi apparatus[1] (Figure). Staining of the Golgi apparatus by anti-58K antibodies has since been observed in numerous other cultured cell types[13,14], including all mammalian cell lines that have been examined so far; those which also have been surveyed by immunoblotting have been shown to contain immunoreactive 58K (Bloom, unpublished). Further evidence of a Golgi apparatus localization for 58K came from studies of Golgi membranes isolated from rat liver. These preparations were found to contain stably bound 58K, and subfractionation of the Golgi membranes indicated that 58K is a peripheral membrane protein which faces the cytoplasm. In this orientation, 58K is well placed to interact with microtubules *in vivo*, and thereby perform its suspected function.

That function may also include helping to maintain the Golgi apparatus near the MTOC[1]. A potential mechanism for keeping the Golgi apparatus at this location would be to have the Golgi coated with a relatively weak ligand for microtubules, like 58K. Associations between the Golgi apparatus and microtubules might then be favored where the number of weak (58K-mediated) interactions between them would be maximal. This should occur where microtubules are most concentrated, in the immediate vicinity of the MTOC.

■ PURIFICATION

A method has been developed for purifying 58K from liver[1], the richest known tissue source of the protein (Bloom, unpublished). Seven rat livers weighing ~100 g typically yield 1.5-2.0 mg of purified 58K. A high speed extract of the tissue is supplemented with microtubules polymerized from purified brain tubulin with the aid of taxol. Centrifugation is then used to collect the brain microtubules, which are decorated with liver MAPs and MAP-like factors, such as 58K. A salt wash of the microtubules releases the nontubulin proteins, including 58K, into the supernatant. Gel filtration chromatography of this fraction yields pure 58K.

■ ACTIVITIES

Purified 58K binds to MAP-free microtubules in a saturable manner, and cofractionates with cytoplasmically oriented peripheral membrane proteins of the Golgi apparatus[1]. In addition, 58K is a weak stimulator of tubulin assembly[1]. Like many other Golgi proteins, 58K appears to redistribute to the endoplasmic reticulum within 10-30 minutes in cells that are treated with brefeldin A. In this respect, 58K is very distinct from β-COP which is removed from the Golgi apparatus within 1-2 minutes by brefeldin A[13,14].

■ ANTIBODIES

Five monoclonal mouse IgG1 antibodies to 58K are available[1]. One of these recognizes 58K in all mammalian cells that have been examined so far. The others react with 58K only in hepatocytes or in a limited number of other cell types.

■ GENES

Several peptides derived from purified 58K have been sequenced, but the complete sequence of the protein has not yet been obtained.

■ REFERENCES

1. Bloom, G.S. and Brashear, T.A. (1989) J. Biol. Chem. 264, 16083-16092.
2. Wehland, J., Henkart, M., Klausner, R. and Sandoval, I. (1983) Proc. Natl. Acad. Sci. (USA) 80, 4286-4290.
3. Rogalski, A.A. and Singer, S.J. (1984) J. Cell Biol. 99, 1092-1100.
4. Ho, W.C., Allan, V.J., van Meer, G., Berger, E.G. and Kreis, T.E. (1989) Eur. J. Cell Biol. 48, 250-263.
5. Kupfer, A., Louvard, D. and Singer, S.J. (1982) Proc. Natl. Acad. Sci. (USA) 79, 2603-2607.
6. Kupfer, A., Dennert, G. and Singer, S.J. (1983) Proc. Natl. Acad. Sci. (USA) 80, 7224-7228.
7. Tassin, A.M., Paintrand, M., Berger, E.G. and Bornens, M. (1985) J. Cell Biol. 101, 630-638.
8. Nemere, I., Kupfer, A. and Singer, S.J. (1985) Cell Motil. 5, 17-29.
9. Allan, V.J. and Kreis, T.E. (1986) J. Cell Biol. 103, 2229-2239.
10. Kim, H., Jensen, C.G. and Rebhun, L.I. (1986) Ann. N.Y. Acad. Sci. 466, 218-239.
11. Murofushi, H., Kotani, S., Aizawa, H., Hisanaga, S., Hirokawa, N. and Sakai, H. (1986) J. Cell Biol. 103, 1911-1919.
12. Hirokawa, N. and Hisanaga, S. (1987) J. Cell Biol. 104, 1553-1561.
13. Donaldson, J.G., Lippincott-Schwartz, J., Bloom, G.S., Kreis, T.E. and Klausner, R.D. (1990) J. Cell Biol. 111, 2295-2306.
14. Ktistakis, N.T., Roth, M.G. and Bloom, G.S. (1991) J. Cell Biol. 113, 1009-1023.

■ *George S. Bloom:*
Department of Cell Biology and Neuroscience,
University of Texas
Southwestern Medical Center,
5323 Harry Hines Blvd.,
Dallas, TX 75235, USA

Synapsins

The synapsins are a family of closely related synaptic vesicle associated phosphoproteins (termed synapsin Ia, Ib, IIa and IIb) that have been implicated in the regulation of neurotransmitter release from nerve endings[1,2].

The synapsins were initially discovered as prominent cellular target molecules for endogenous protein phosphorylation in the brain. Specifically, the synapsins were demonstrated to be major substrate proteins for cAMP dependent protein kinase[3]. Later, they were also shown to be phosphorylated by Ca^{2+} calmodulin dependent protein kinases[4-6]. The synapsin family includes synapsin I and synapsin II (previously called Protein I and Protein III, respectively). They are encoded by different genes, and each of them exists in two alternatively spliced variants, termed synapsins Ia (84 kDa) and Ib (80 kDa) and synapsins IIa (74 kDa) and IIb (55 kDa), respectively[7]. Common to all four proteins is the so-called head region, a globular, collagenase resistant region. Synapsin I contains in addition an elongated, very basic, collagenase sensitive tail region (Figure 1). In mature neurons, the synapsins are localized on the cytoplasmic side of synaptic vesicles and appear to be present in virtually all nerve terminals (Figure 2) irrespective of the neurotransmitter released[8-10]. Recent studies using monospecific antibodies provide evidence for a differential distribution of the four forms in nerve terminals[7]. A number of physiological and pharmacological stimuli which enhance synaptic efficacy have been shown to increase the phosphorylation state of the synapsins[11]. Experiments performed with synapsin I at the squid giant synapse argue that dephosphosynapsin I acts as an inhibitory determinant for neurotransmitter release. Thus, injection of the dephospho form of synapsin I inhibited release of neurotransmitter[12]. This effect was abolished by phosphorylation of synapsin I in the tail region prior to its injection.

■ PURIFICATION

Synapsin I is purified from brain tissue by acid or salt/detergent extraction, carboxymethylcellulose (step with highest purification factor), hydroxylapatite and gel filtration chromatography[13,14]. Synapsin II is more difficult to purify. Conventional chromatographic techniques alone or in combination with antibody affinity columns are used[15].

■ ACTIVITIES

Synapsin I binds with high affinity and saturability to purified synaptic vesicles (from which endogenous synapsin I has been removed)[13] and phospholipid vesicles containing negatively charged phospholipids[16]. It bundles F-actin in a phosphorylation dependent manner[14,17]. Furthermore, it has been reported to interact with **spectrin,** microtubules, neurofilaments and calmodulin[2].

■ ANTIBODIES

Polyclonal antibodies raised against purified synapsin I, synapsin II, various synthetic peptides (including phosphorylation sites) and monoclonal antibodies have been described[2,7].

■ GENES

Full length cDNAs for rat synapsin Ia (GenBank M27812), Ib (M27924), bovine synapsin Ia (M27810), Ib (M27811), rat synapsin IIa (M27925), IIb (M27926) and a genomic clone for human synapsin I (J05431) have been published[7,18]. The genes for human and murine synapsin I have been mapped to the X-chromosome[19].

■ REFERENCES

1. Bähler, M., Benfenati, F., Valtorta, F. and Greengard, P. (1990) BioEssays 12, 259- 263.
2. DeCamilli, P., Benfenati, F., Valtorta, F. and Greengard, P. (1990) Ann. Rev. Cell Biol. 6, 433-460.
3. Ueda, T., Maeno, H. and Greengard, P. (1973) J. Biol. Chem. 248, 8295-8305.
4. Huttner, W.B. and Greengard, P. (1979) Proc. Natl. Acad. Sci. (USA) 76, 5402-5406.
5. McGuinness, T.L., Lai, Y. and Greengard, P. (1985) J. Biol. Chem. 260, 1696-1704.
6. Nairn, A.C. and Greengard, P. (1987) J. Biol. Chem. 262, 7273-7281.
7. Südhof, T.C., Czernik, A.J., Kao, H.-T., Takei, K., Johnston, P.A., Horiuchi, A., Kanazir, S.D., Wagner, M.A., Perin, M.S., DeCamilli, P. and Greengard, P. (1989) Science 245, 1474-1480.

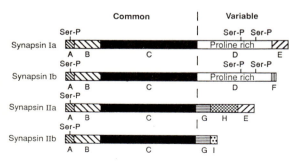

Figure 1. Diagram of shared and individual domains in the four synapsins. Each domain is given a letter symbol and represented in scale. Numbers to the left indicate the synapsin subspecies. The large central homologous domain (domain C) is highly conserved, and domain E is found at the C-terminus of both synapsins Ia and IIa (Taken from ref. 7).

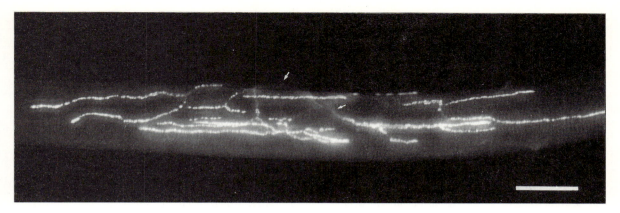

Figure 2. Immunofluorescent staining of synapsin I in a neuromuscular junction on a single fibre teased from frog cutaneous pectoris muscle. Immunoreactivity is highly concentrated in the nerve terminal region and virtually no fluorescence is associated with the muscle fibre or with the unmyelinated preterminal axon (arrows). (Picture provided by Dr. F. Valtorta). Bar 50 μm.

8. De Camilli, P., Cameron, R. and Greengard, P. (1983) J. Cell Biol. 96, 1337-1354.

9. De Camilli, P., Harris, S.M., Huttner, W.B. and Greengard, P. (1983) J. Cell Biol. 96, 1355-1373.

10. Huttner, W.B., Schiebler, W., Greengard, P. and DeCamili, P. (1983) J. Cell Biol. 96, 1374-1388.

11. Nestler, E.J. and Greengard, P. (1984) John Wiley & Sons Inc., New York.

12. Llinas, R., McGuinness, T., Leonard, C.S., Sugimori, M. and Greengard, P. (1985) Proc. Natl. Acad. Sci. (USA) 82, 3035-3039.

13. Schiebler, W., Jahn, R., Doucet, J.-P., Rothlein, J. and Greengard, P. (1986) J. Biol. Chem. 261, 8383-8390.

14. Bähler, M. and Greengard, P. (1987) Nature 326, 704-707.

15. Huang, C.-K., Browning, M.D. and Greengard, P. (1982) J. Biol. Chem. 257, 6524-6528.

16. Benfenati, F., Greengard, P., Brunner, J. and Bähler, M. (1989) J. Cell Biol. 108, 1851-1862.

17. Petrucci, T.C. and Morrow, J.S. (1987) J. Cell Biol. 105, 1355-1363.

18. Südhof, T.C. (1990) J. Biol. Chem. 265, 7849-7852.

19. Yang-Feng, T.L., DeGennaro, L.J. and Francke, U. (1986) Proc. Natl. Acad. Sci. (USA) 83, 8679-8683.

■ *Martin Bähler and Paul Greengard:*
Laboratory of Molecular and Cellular Neuroscience,
The Rockefeller University,
New York, NY, USA

7

Other Proteins

Major Sperm Proteins

Major sperm protein (MSP) defines a family of small, basic polypeptides found exclusively in the sperm of nematodes. MSP forms the array of fine filaments that pack the pseudopod of these actomyosin deficient amoeboid cells.

Major sperm proteins comprise ~15% of the total protein[1,2] and form the cytoskeleton in nematode sperm, unique cells that lack **actin** filaments, microtubules, and intermediate filaments[3,4]. The protein includes a family of highly conserved, 14.2-14.3 kDa isoforms; except for a minor pI 7.1 isoform in *Caenorhabditis elegans*, the MSPs are basic proteins with pIs in the range of 8.3-8.9[5]. Even the most dissimilar MSPs are over 80% homologous and most of the amino acid differences are conservative substitutions[6]. *C. elegans*, for example, contains over 30 MSP genes expressed in roughly equal amounts that produce only three separable isoforms of the protein[5]. Sperm from *Ascaris suum* contain two isoforms, each consisting of 126 amino acids that differ at only four residues (King, K.L., Stewart, M., Roberts, T.M. and Seavy, M., unpublished). Apart from an acetylated N-terminal alanine in both isoforms from *Ascaris*, no other posttranslational modifications have been detected[5]. None of the MSPs exhibit even limited sequence homology to other structural proteins.

MSP is synthesized in late spermatocytes where it assembles into parallel bundles of filaments called fibrous bodies[7]. These structures segregate to the developing spermatids and disassemble[8]. During spermiogenesis, MSP reassembles into a filamentous network in the spermatozoan pseudopod (Figure 1). This filament array is particularly well ordered in *Ascaris* sperm where the filaments are arranged into 15-20 branched fibre complexes that span the length of the pseudopod (Figure 2). Individual filaments splay out laterally from each complex and interdigitate with similar filaments from neighbouring complexes. In live sperm, the filament system treadmills rearward, at the same rate as the cell crawls forward due to assembly and disassembly at opposite ends of the fibre complexes[9]. At present, we do not know if centripetal flow of the MSP cytoskeleton is assisted by a molecular motor.

In ethanol, 2-methyl-2,4-pentanediol (MPD), and other water-miscible alcohols both isoforms of *Ascaris* MSP assemble into filaments 10 nm wide with characteristic substructure repeated axially at 9 nm. These filaments are identical to native fibres isolated from detergent-lysed

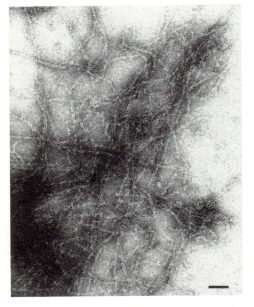

Figure 1. Negative-stained MSP filaments isolated from the pseudopod of an Ascaris spermatozoon lysed with Triton X-100. The area shown includes the remnant of the distal end of a fibre complex. Bar 50 nm.

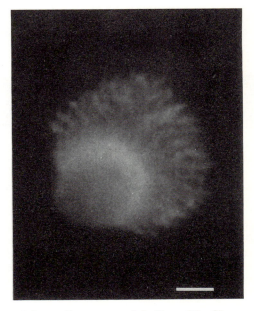

Figure 2. Imunofluorescence labelling of the fibre complexes in an *Ascaris* spermatozoon with an anti-MSP monoclonal antibody AZ10. Bar 10 μm.

sperm and exhibited a critical concentration for assembly of 0.2 mM in 30% ethanol. In polyethylene glycol (Mol.wt.~18,500) both *Ascaris* MSPs form needle shaped crystals composed of helical fibres with the same 9 nm axial repeat observed in isolated filaments (King, K.L., Stewart, M., Roberts, T.M. and Seavy, M., unpublished).

■ PURIFICATION

The MSP cytoskeleton is highly labile and released as ~40% of the soluble protein fraction following cellular homogenization. The small size of the protein allows substantial enrichment by gel permeation chromatography. Phosphocellulose chromatography and chromatofocusing columns have been used to complete purification[10]. The two isoforms from *Ascaris* separate readily by either cation exchange or reversed phase HPLC.

■ ACTIVITIES

The only known activity of MSP is its capacity to assemble into filaments[8,9].

■ ANTIBODIES

Polyclonal and monoclonal antibodies have been raised against both *C. elegans* and *Ascaris* MSP[11,8-10]. None of these antibodies crossreact with other nematode tissues or various other amoeboid cells. Crossreaction with sperm from other species of nematodes has not been tested.

■ GENES

MSP genes have been identified in 16 species of nematodes[12]. Complete cDNA sequences of several *C. elegans*

MSP genes[11,13,14] (GenBank K02617, K02618), one *Ascaris* MSP (GenBank M15680)[11], and two *Onchocerca volvulus* MSPs (GenBank J04662, J04663)[15] have been published.

■ REFERENCES

1. Klass, M. and Hirsh, D. (1981) Devel. Biol. 84, 299-312.
2. Nelson, G.A. and Ward, S. (1981) Exp. Cell Res. 131,149-160.
3. Nelson, G.A., Roberts, T.M. and Ward, S. (1982) J. Cell Biol. 92, 121-131.
4. Roberts, T.M. (1987) Cell Motil. Cytoskel. 8, 130-142.
5. Burke, D.J. and Ward, S. (1983) J. Bol. Biol. 171, 1-29.
6. Bennett, K.L. and Ward, S. (1986) Devel. Biol. 118, 141-147.
7. Ward, S. and Klass, M. (1982) Devel. Biol. 92, 203-208.
8. Ward, S., Roberts, T.M., Strome, S., Pavalko, F.M. and Hogan, E. (1986) J. Cell Biol. 102, 1787-1796.
9. Sepsenwol, S., Ris, H. and Roberts, T.M. (1989) J. Cell Biol. 108, 55-66.
10. Ward, S. and Klass, M. (1986) Methods Enzymol. 134, 414-420.
11. Klass, M., Ammons, D. and Ward, S. (1988) J. Mol. Biol. 199, 15-22.
12. Scott, A.L., Dinman, J., Susman, D.J. and Ward, S. (1989) Parasitol. 98, 471-478.
13. Klass, M.R., Kinsley, S. and Lopez, L.C. (1984) Mol. Cell. Biol. 4, 529-537.
14. Ward, S., Burke, D.J., Sulston, J.E., Coulson, A.R., Albertson, D.G., Ammons, D., Klass, M. and Hogan, E. (1988) J. Mol. Biol. 199, 1-13.
15. Scott, A.L., Dinman, J., Susman, D.J., Yenbutr, P. and Ward, S. (1989) Mol. Biochem. Parasitol. 36, 119-126.

■ *Thomas M. Roberts:*
Department of Biological Science,
Florida State University
Tallahasse, Florida, USA

Tektins

Tektins are a family of filamentous proteins that are associated with a specific set of tubulin protofilaments in ciliary and flagellar microtubules[1]. Antibodies to tektins crossreact with basal bodies, centrioles, centrosomes, mitotic spindles and midbodies from a variety of species including humans; certain anti-tektin antibodies also crossreact with intermediate filament (IF) proteins. Sequencing studies indicate that tektins share primary and secondary structural homology with IF proteins.

Tektins were first isolated from sea urchin sperm flagellar axonemes[2]. Extraction of axonemal microtubules with 0.5% sarkosyl yields stable ribbons of three protofilaments (pf); the pf-ribbons are composed of α/β-**tubulin**, tektins, and several other polypeptides[3]. Extraction with 0.5% sarkosyl plus 2 M urea yields a filamentous material, free of tubulin and composed of tektins A, B and C in equimolar amounts[4]; by negative stain EM this material appears as 2 nm diameter fibrils and bundles thereof[2]. Tektins A (55 kDa), B (51 kDa) and C (47 kDa) have iso-

electric points of ~6.9, 6.2 and 6.15 respectively[5]; by sequence determination tektin A has a molecular weight of 53,000. Tektins A, B and C are very similar in their amino acid compositions and tryptic peptide maps, but they are sufficiently different to have arisen from different genes[4]. Unfixed microtubules do not easily stain with anti-tektin antibodies, although antibodies do decorate thin fibrils extending from the pf-ribbons[6]. These results have been interpreted to indicate that the tektins are assembled as filaments, extending axially along the pf-ribbon but struc-

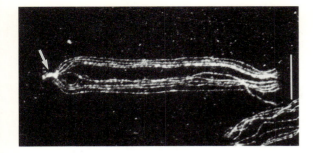

Figure. Spermatozoan of *L. pictus*, with the flagellar axoneme splayed out and stained with affinity purified anti-tektin B antibodies (from Steffen and Linck, ref.[8]). Such patterns demonstrate that all nine doublet microtubules contain each of the tektins; the intermittent staining along the filaments is probably due to random masking of tektins by tubulin in this methanol-fixed preparation. The central pair microtubules are not preserved in these preparations. Note the intense anti-tektin antibody staining of the basal body (arrow). Bar 10 um.

turally integrated into the microtubule wall, wherein they are largely unexposed to antibody. The tektin containing pf-ribbons form the part of the α-tubule wall that joins with the inner wall of the β-tubule of doublet microtubules (Figure); the nine-fold array of doublet microtubules is apparently maintained by nexin filaments linked to the tektin pf-ribbons[7]. Some evidence suggests that the central pair microtubules also contain tektins[7,8].

The presence of tektins in other systems besides sea urchin sperm flagella has been studied by immunofluorescence microscopy, immuno-EM and SDS-PAGE immunoblotting. Preliminary evidence suggests that tektin or tektin-like proteins are also associated with microtubules of cilia (in ctenophores[9], molluscs and echinoderms[5,7]), basal bodies, centrioles and centrosomes (in mammals including humans)[1,8], mitotic spindles (in mammals)[1], and midbodies (in mammals)[1]. It may be significant that in all cases so far examined, anti-tektin antibodies crossreact with components of cold stable microtubules. A number of polyclonal and monoclonal anti-tektin antibodies also crossreact with certain intermediate filament proteins, including human **keratins** and nuclear **lamins**[1,10,11].

The expression of tektin A has been studied in some detail in sea urchin embryogenesis. Tektin A is synthesized *de novo* and in a quantal amount at the onset of ciliogenesis; it is resynthesized during ciliary regeneration[12,13]. Analysis of tektin A mRNA levels, using cDNA probes, reveals that it is not present in unfertilized eggs, but present at very low levels after fertilization; it is maximally expressed immediately prior to the synthesis of tektin A during ciliogenesis and it is reexpressed during ciliary regeneration[14,15]. These results support the original hypothesis that the amount of tektin synthesized may limit the length of ciliary axoneme growth[12].

■ PURIFICATION

Tektin can be purified from sea urchin sperm flagellar axonemes by extracting axonemes with sarkosyl and urea, followed by sedimentation at 100,000 g for 90 min to obtain a pellet of filaments composed of nearly equimolar amounts of tektins[4]. Extraction conditions must be varied for other species where more than three tektins may occur with different solubilities[5]. Separation of individual tektins can be accomplished by SDS-PAGE or reverse phase chromatography[4].

■ ACTIVITIES

Given their nature, tektins would appear to act as structural proteins; however, their specific functions have not been determined. Since tektins are associated with stable microtubules, and indeed with a particularly stable region of the microtubule, they may function in stabilizing microtubules; however, tektins may have additional functions as reviewed elsewhere[1].

■ ANTIBODIES

Affinity purified rabbit polyclonal antibodies have been characterized against tektins A, B and C from the sea urchin species *Lytechinus pictus* and and *Strongylocentrotus purpuratus*[1,5,8]. Monoclonal antibodies have been prepared against tektins from *S. purpuratus*[11,16]. These antibodies crossreact with certain microtubule systems and intermediate filament systems in a variety of species ranging from marine invertebrates to humans.

■ GENES

Several cDNAs for tektins, including a full length 2750 base cDNA for tektin A1, have been cloned from an expression library constructed from the blastula stage of *S. purpuratus* embryos undergoing ciliogenesis. These clones have been used to study the expression of tektin A during embryogenesis in sea urchins[15]. The 2750 base cDNA has been sequenced and analyzed for possible tubulin binding domains and for its homology to intermediate filament proteins[17].

■ REFERENCES

1. Steffen, W. and Linck, R.W. (1989) In Cell Movement, Vol. 2, Eds. R.D. Warner & J.R. McIntosh, Alan R. Liss, New York, pp. 67-81.
2. Linck, R.W. and Langevin, G.L. (1982) J. Cell Sci. 58, 1-22.
3. Linck, R.W. (1976) J. Cell Sci. 20, 405-439.
4. Linck, R.W. and Stephens, R.E. (1987) J. Cell Biol. 104, 1069-1075.
5. Linck, R.W., Goggin, M.J., Norrander, J.M. and Steffen, W. (1987) J. Cell Sci. 88, 453-466.

6. Linck, R.W., Amos, L.A. and Amos, W.B. (1985) J. Cell Biol. 100, 126-135.
7. Stephens, R.E., Oleszko-Szuts, S. and Linck, R.W. (1989) J. Cell Sci. 92, 391-402.
8. Steffen, W. and Linck, R.W. (1987) Proc. Natl. Acad. Sci. (USA) 85, 2643-2647.
9. Linck, R.W., Stephens, R.E. and Tamm, S.L. (1991) in press.
10. Steffen, W. and Linck, R.W. (1989) Cell Motil. Cytoskel. 14, 359-371.
11. Chang, X. and Piperno, G. (1987) J. Cell Biol. 104, 1563-1568.
12. Stephens, R.E. (1977) Dev. Biol. 61, 311-329.
13. Stephens, R.E. (1989) J. Cell Sci. 92, 403-413.
14. Norrander, J.M., Stephens, R.E. and Linck, R.W. (1988) J. Cell Biol. 107, 20a.
15. Norrander, J.M. et al. (1992) submitted.
16. Steffen, W. and Linck, R.W. (1989) Electrophoresis 10, 714-718.
17. Norrander, J.M., Amos, L.A. and Linck, R.W. (1992) Proc. Natl. Acad. Sci. (USA) 89, 8567-8571.

■ *Richard, W. Linck:*
Department of Cell Biology & Neuroanatomy,
University of Minnesota,
Minneapolis, MN, USA

Index

ACT2 gene product see actin 13
actin binding protein-50 (ABP-50) 15
actin binding protein-120 (ABP-120) 16
actin binding protein-280 (ABP-280; nonmuscle filamin) 18
actin depolymerizing factor (ADF) 20
actin-RPV (actin-related protein vertebrate) see actins 13
actinogelin see α-actinin 22
actins 13
actobindin 23
actolinkin 25
actophorin see depactin 43
adducin 219
ADF see depactin 43
α-actinins 22
α/β-tubulin 127
α-internexin 157
anchorin CII see annexins 26
ankyrins 221
annexins 26
AP180 255
armadillo gene product (Drosophila) see plakoglobin 235
assembly protein 2 (AP2) see clathrin adaptor proteins 259
auxilin 256
axonemal dyneins 185

band 4.9 see dematin 42
band 6 polypeptide 223
β-COP 261
brush border myosin I 187
buttonin see sea urchin MAPs 120

c-proteins 39
calcimedin see annexin 26
caldesmons 29
calpactin see annexin 26
calphobindin see annexin 26
calponin 32
calspectrin see spectrin 76
caltractin 188
CamBP see adducin 219
capactin see capping protein 34
capping proteins 34
CapZ see capping protein 34
catenins 224
CDC31 gene product see caltractin 188
centractin see actin 13
centrin see caltractin 188
chartins 107
chromobindin see annexins 26
cingulin 225
clathrin 257
clathrin adaptor proteins 259
CLIP-170 263

cofilin 35
coronin 37
cytokeratins 145
cytoplasmic dynein (MAP1C) 191
cytoplasmic myosin II 193
cytovillin see ezrin 47

dematins 42
depactin 43
desmin 148
desmocalmin 227
desmoplakins 228
destrin see depactin 43
dynactin 195
dynamin 197
dystrophin 45

elongation factor 1a see ABP-50 15
endonexin see annexins 26
epinemin 151
ezrin 47

fascin 49
filaggrins 152
filamin see ABP-280 18
filensin 153
fimbrin 50
fodrin see spectrin 76
fragmin see severin 73
Fx see thymosin β4 79

gCap39 (macrophage capping protein MCP) 51
gelactins see small actin crosslinking proteins 74
gelation factor see ABP-120 16
gelsolins 52
GFAP 155
glued gene product (Drosophila) see dynactin 195

HA2 adaptor see clathrin adaptor proteins 259
hisactophilin 54

inner arm dynein see axonemal dynein 185
insertin 55

keratin see cytokeratin 145
kinesin 199
kinesin related proteins 201

lamins 158
lipocortin see annexin 26

macrophage capping protein see gCap39 51
major sperm proteins 271

MAP1A 108
MAP1B/MAP5 110
MAP1C see cytoplasmic dynein 191
MAP1X see MAP1B 110
MAP2 111
MAP3 113
MAP4 (MAP-U) 115
MAP5 see MAP1B 110
MARCKS 56
MARPs 117
micro-calpain and milli-calpain (EC 3.4.22.17) 230
mini-myosin see protozoan myosin 206
MYO2 myosin 204
myomesin and M-protein 58

nebulin 59
nestin 160
neurofilament triplet proteins 161
nuclear actin binding protein (NAB) 60

outer arm dynein see axonemal dynein 185

p81 see annexin 26
paramyosin 62
paranemin 164
paxillin 231
pemphigoid antigens 233
pericentrin 118
peripherin 165
PK1, PK2 see adducin 219
plakoglobin 235
plastin see fimbrin 50
plectin 166
ponticulin 64
pp60c-src 236
profilins 66
projectin see titin 81
proteins 4.1 68
protozoan myosin I 206

radial spoke proteins 119
radixin 70

sarcomeric M-creatine kinase 71
sarcomeric myosins 207
scallop myosin 209
sea urchin MAPs and microtubule motors 120
severin 73
shibere gene product (Drosophila) see dynamin 197
small actin crosslinking proteins 74
smooth muscle myosin 211
spasmin see caltractin 188

spectrins (fodrin) 76
STOPs 122
synapsin 266
syncolin 124
synemin 167
synexin *see* annexin 26

talin 239
tau 125
τ-tubulin 130
tektins 272
tensin 240
tenuin 78
terminal web (TW) 260/240 *see* spectrin 76
thymosin β4 (Tβ4) 79
titin 81

tropomodulin 83
tropomyosins 85
troponins 87
tubulin tyrosin ligase (TTL) and tubulin carboxypeptidase (TCP) 131
twitchin *see* titin 81

UMAP *see* MAP4 115

VAC-β *see* annexins 26
villin 89
vimentin 169
vinculin 243
vitamin D binding/Gc protein (DBP/Gc) 92
VPS1 (*S. Cerevisiae*) *see* dynamin 197

X-MAP 132

ZO-1 245
zyxin 247

25 kDa inhibitor of actin polymerization (25 kDa IAP) 94
43 kDa protein 95
58K 264
110K/CM *see* brush border myosin 187
115/110 *see* adducin 219
205K MAP (Drosophila) 133